Graduate Texts in Mathematics 104

B. A. Dubrovin
A. T. Fomenko
S. P. Novikov

Modern Geometry—Methods and Applications

Part II. The Geometry and Topology of Manifolds

Translated by Robert G. Burns

With 126 Illustrations

Springer

B. A. Dubrovin
Department of Mathematics and Mechanics
Moscow University
Leninskie Gory
Moscow 119899
Russia

A. T. Fomenko
Moscow State University
V-234 Moscow
Russia

S. P. Novikov
Institute of Physical Sciences and Technology
Maryland University
College Park, MD 20742-2431
USA

R. G. Burns (Translator)
Department of Mathematics
Faculty of Arts
York University
4700 Keele Street
North York, ON, M3J 1P3
Canada

Mathematics Subject Classification (1991): 53-01, 53B50, 57-01, 58Exx
(Revised for Vol. 2)
Dubrovin, B. A.
Modern geometry—methods and applications.
(Springer series in Soviet mathematics) (graduate texts in mathematics; 93-)
"Original Russian edition ... Moskva: Nauka, 1979"—T.p. verso.
Includes bibliographies and indexes.
1.Geometry. I. Fomenko, A. T. II. Novikov, Sergei Petrovich. III. Title. IV. Series.
V. Series: Graduate texts in mathematics; 93, etc.
QA445.D82 1984 516 83-16851

This book is part of the Springer Series in Soviet Mathematics.

Original Russian edition: Sovremennaja Geometria: Metody i Prilozenia. Moskva: Nauka, 1979.

Dubrovin et. al.: Modern Geometry—Methods and Applications
Part II: The Geometry and Topology of Manifolds

First Indian Reprint 2010

ISBN 978-81-8489-659-6

This edition is published by Springer (India) Private Limited, (a part of Springer Science+Business Media), Registered Office: 3rd Floor, Gandharva Mahavidyalaya, 212, Deen Dayal Upadhyaya Marg, New Delhi – 110 002, India

Printed in India by Rakmo Press (P) Ltd., New Delhi.

Preface

Up until recently, Riemannian geometry and basic topology were not included, even by departments or faculties of mathematics, as compulsory subjects in a university-level mathematical education. The standard courses in the classical differential geometry of curves and surfaces which were given instead (and still are given in some places) have come gradually to be viewed as anachronisms. However, there has been hitherto no unanimous agreement as to exactly how such courses should be brought up to date, that is to say, which parts of modern geometry should be regarded as absolutely essential to a modern mathematical education, and what might be the appropriate level of abstractness of their exposition.

The task of designing a modernized course in geometry was begun in 1971 in the mechanics division of the Faculty of Mechanics and Mathematics of Moscow State University. The subject-matter and level of abstractness of its exposition were dictated by the view that, in addition to the geometry of curves and surfaces, the following topics are certainly useful in the various areas of application of mathematics (especially in elasticity and relativity, to name but two), and are therefore essential: the theory of tensors (including covariant differentiation of them); Riemannian curvature; geodesics and the calculus of variations (including the conservation laws and Hamiltonian formalism); the particular case of skew-symmetric tensors (i.e. "forms") together with the operations on them; and the various formulae akin to Stokes' (including the all-embracing and invariant "general Stokes formula" in n dimensions). Many leading theoretical physicists shared the mathematicians' view that it would also be useful to include some facts about manifolds, transformation groups, and Lie algebras, as well as the basic concepts of visual topology. It was also agreed that the course should be given in as simple and concrete a language as possible, and that wherever practicable the terminology should be that used by physicists. Thus it was along these lines that the archetypal course was taught. It was given more permanent form as duplicated lecture notes published under the auspices of Moscow State University as:

Differential Geometry, Parts I and II, by S. P. Novikov, Division of Mechanics, Moscow State University, 1972.

Subsequently various parts of the course were altered, and new topics added. This supplementary material was published (also in duplicated form) as:

Differential Geometry, Part III, by S. P. Novikov and A. T. Fomenko, Division of Mechanics, Moscow State University, 1974.

The present book is the outcome of a reworking, re-ordering, and extensive elaboration of the above-mentioned lecture notes. It is the authors' view that it will serve as a basic text from which the essentials for a course in modern geometry may be easily extracted.

To S. P. Novikov are due the original conception and the overall plan of the book. The work of organizing the material contained in the duplicated lecture notes in accordance with this plan was carried out by B. A. Dubrovin. This accounts for more than half of Part I; the remainder of the book is essentially new. The efforts of the editor, D. B. Fuks, in bringing the book to completion, were invaluable.

The content of this book significantly exceeds the material that might be considered as essential to the mathematical education of second- and third-year university students. This was intentional: it was part of our plan that even in Part I there should be included several sections serving to acquaint (through further independent study) both undergraduate and graduate students with the more complex but essentially geometric concepts and methods of the theory of transformation groups and their Lie algebras, field theory, and the calculus of variations, and with, in particular, the basic ingredients of the mathematical formalism of physics. At the same time we strove to minimize the degree of abstraction of the exposition and terminology, often sacrificing thereby some of the so-called "generality" of statements and proofs: frequently an important result may be obtained in the context of crucial examples containing the whole essence of the matter, using only elementary classical analysis and geometry and without invoking any modern "hyperinvariant" concepts and notations, while the result's most general formulation and especially the concomitant proof will necessitate a dramatic increase in the complexity and abstractness of the exposition. Thus in such cases we have first expounded the result in question in the setting of the relevant significant examples, in the simplest possible language appropriate, and have postponed the proof of the general form of the result, or omitted it altogether. For our treatment of those geometrical questions more closely bound up with modern physics, we analysed the physics literature:

books on quantum field theory (see e.g. [35], [37]) devote considerable portions of their beginning sections to describing, in physicists' terms, useful facts about the most important concepts associated with the higher-dimensional calculus of variations and the simplest representations of Lie groups; the books [41], [43] are devoted to field theory in its geometric aspects; thus, for instance, the book [41] contains an extensive treatment of Riemannian geometry from the physical point of view, including much useful concrete material. It is interesting to look at books on the mechanics of continuous media and the theory of rigid bodies ([42], [44], [45]) for further examples of applications of tensors, group theory, etc.

In writing this book it was not our aim to produce a "self-contained" text: in a standard mathematical education, geometry is just one component of the curriculum; the questions of concern in analysis, differential equations, algebra, elementary general topology and measure theory, are examined in other courses. We have refrained from detailed discussion of questions drawn from other disciplines, restricting ourselves to their formulation only, since they receive sufficient attention in the standard programme.

In the treatment of its subject-matter, namely the geometry and topology of manifolds, Part II goes much further beyond the material appropriate to the aforementioned basic geometry course, than does Part I. Many books have been written on the topology and geometry of manifolds: however, most of them are concerned with narrowly defined portions of that subject, are written in a language (as a rule very abstract) specially contrived for the particular circumscribed area of interest, and include all rigorous foundational detail often resulting only in unnecessary complexity. In Part II also we have been faithful, as far as possible, to our guiding principle of minimal abstractness of exposition, giving preference as before to the significant examples over the general theorems, and we have also kept the interdependence of the chapters to a minimum, so that they can each be read in isolation insofar as the nature of the subject-matter allows. One must however bear in mind the fact that although several topological concepts (for instance, knots and links, the fundamental group, homotopy groups, fibre spaces) can be defined easily enough, on the other hand any attempt to make nontrivial use of them in even the simplest examples inevitably requires the development of certain tools having no forbears in classical mathematics. Consequently the reader not hitherto acquainted with elementary topology will find (especially if he is past his first youth) that the level of difficulty of Part II is essentially higher than that of Part I; and for this there is no possible remedy. Starting in the 1950s, the development of this apparatus and its incorporation into various branches of mathematics has proceeded with great rapidity. In recent years there has appeared a rash, as it were, of nontrivial applications of topological methods (sometimes in combination with complex algebraic geometry) to various problems of modern theoretical physics: to the quantum theory of specific fields of a geometrical nature (for example, Yang–Mills and chiral fields), the theory of fluid crystals and superfluidity, the general theory of relativity, to certain physically important nonlinear wave equations (for instance, the Korteweg–de Vries and sine–Gordon equations); and there have been attempts to apply the theory of knots and links in the statistical mechanics of certain substances possessing "long molecules". Unfortunately we were unable to include these applications in the framework of the present book, since in each case an adequate treatment would have required a lengthy preliminary excursion into physics, and so would have taken us too far afield. However, in our choice of material we have taken into account which topological concepts and methods are exploited in these applications, being aware of the need for a topology text which might be read (given strong enough motivation) by a young theoretical physicist of the modern school, perhaps with a particular object in view.

The development of topological and geometric ideas over the last 20 years has brought in its train an essential increase in the complexity of the algebraic apparatus used in combination with higher-dimensional geometrical intuition, as also in the utilization, at a profound level, of functional analysis, the theory of partial differential equations, and complex analysis; not all of this has gone into the present book, which pretends to being elementary (and in fact most of it is not yet contained in any single textbook, and has therefore to be gleaned from monographs and the professional journals).

Three-dimensional geometry in the large, in particular the theory of convex figures and its applications, is an intuitive and generally useful branch of the classical geometry of surfaces in 3-space; much interest attaches in particular to the global problems of the theory of surfaces of negative curvature. Not being specialists in this field we were unable to extract its essence in sufficiently simple and illustrative form for inclusion in an elementary text. The reader may acquaint himself with this branch of geometry from the books [1], [4] and [16].

Of all the books on the topology and geometry of manifolds, the classical works *A Textbook of Topology* and *The Calculus of Variations in the Large*, of Siefert and Threlfall, and also the excellent more modern books [10], [11] and [12], turned out to be closest to our conception in approach and choice of topics. In the process of creating the present text we actively mulled over and exploited the material covered in these books, and their methodology. In fact our overall aim in writing Part II was to produce something like a modern analogue of Seifert and Threlfall's *Textbook of Topology*, which would however be much wider-ranging, remodelled as far as possible using modern techniques of the theory of smooth manifolds (though with simplicity of language preserved), and enriched with new material as dictated by the contemporary view of the significance of topological methods, and of the kind of reader who, encountering topology for the first time, desires to learn a reasonable amount in the shortest possible time. It seemed to us sensible to try to benefit (more particularly in Part I, and as far as this is possible in a book on mathematics) from the accumulated methodological experience of the physicists, that is, to strive to make pieces of nontrivial mathematics more comprehensible through the use of the most elementary and generally familiar means available for their exposition (preserving, however, the format characteristic of the mathematical literature, wherein the statements of the main conclusions are separated out from the body of the text by designating them "theorems", "lemmas", etc.). We hold the opinion that, in general, understanding should precede formalization and rigorization. There are many facts the details of whose proofs have (aside from their validity) absolutely no role to play in their utilization in applications. On occasion, where it seemed justified (more often in the more difficult sections of Part II) we have omitted the proofs of needed facts. In any case, once thoroughly familiar with their applications, the reader may (if he so wishes), with the help of other sources, easily sort out the proofs of such facts for himself. (For this purpose we recommend the book [21].) We have, moreover, attempted to break down many of these omitted proofs into soluble pieces which we have placed among the exercises at the end of the relevant sections.

In the final two chapters of Part II we have brought together several items from the recent literature on dynamical systems and foliations, the general theory of relativity, and the theory of Yang–Mills and chiral fields. The ideas expounded there are due to various contemporary researchers; however in a book of a purely textbook character it may be accounted permissible not to give a long list of references. The reader who graduates to a deeper study of these questions using the research journals will find the relevant references there.

Homology theory forms the central theme of Part III.

In conclusion we should like to express our deep gratitude to our colleagues in the Faculty of Mechanics and Mathematics of M.S.U., whose valuable support made possible the design and operation of the new geometry courses; among the leading mathematicians in the faculty this applies most of all to the creator of the Soviet school of topology, P. S. Aleksandrov, and to the eminent geometers P. K. Raševskiĭ and N. V. Efimov.

We thank the editor D. B. Fuks for his great efforts in giving the manuscript its final shape, and A. D. Aleksandrov, A. V. Pogorelov, Ju. F. Borisov, V. A. Toponogov and V. I. Kuz'minov, who in the course of reviewing the book contributed many useful comments. We also thank Ja. B. Zel'dovič for several observations leading to improvements in the exposition at several points, in connexion with the preparation of the English and French editions of this book.

We give our special thanks also to the scholars who facilitated the task of incorporating the less standard material into the book. For instance the proof of Liouville's theorem on conformal transformations, which is not to be found in the standard literature, was communicated to us by V. A. Zorič. The editor D. B. Fuks simplified the proofs of several theorems. We are grateful also to O. T. Bogojavlenskiĭ, M. I. Monastyrskiĭ, S. G. Gindikin, D. V. Alekseevskiĭ, I. V. Gribkov, P. G. Grinevič, and E. B. Vinberg.

Translator's acknowledgments. Thanks are due to Abe Shenitzer for much kind advice and encouragement, to several others of my colleagues for putting their expertise at my disposal, and to Eadie Henry for her excellent typing and great patience.

Contents

CHAPTER 1
Examples of Manifolds

§1. The Concept of a Manifold

1.1. Definition of a Manifold

The concept of a manifold is in essence a generalization of the idea, first formulated in mathematical terms by Gauss, underlying the usual procedure used in cartography (i.e. the drawing of maps of the earth's surface, or portions of it).

The reader is no doubt familiar with the normal cartographical process: The region of the earth's surface of interest is subdivided into (possibly overlapping) subregions, and the group of people whose task it is to draw the map of the region is subdivided into as many smaller groups in such a way that:

(i) each subgroup of cartographers has assigned to it a particular subregion (both labelled i, say); and

(ii) if the subregions assigned to two different groups (labelled i and j say) intersect, then these groups must indicate accurately on their maps the rule for translating from one map to the other in the common region (i.e. region of intersection). (In practice this is usually achieved by giving beforehand specific names to sufficiently many particular points (i.e. land-marks) of the original region, so that it is immediately clear which points on different maps represent the same point of the actual region.)

Each of these separate maps of subregions is of course drawn on a flat sheet of paper with some sort of co-ordinate system on it (e.g. on "squared" paper). The totality of these flat "maps" forms what is called an "atlas" of the region of the earth's surface in question. (It is usually further indicated on each map how to calculate the actual length of any path in the subregion represented by that map, i.e. the "scale" of the map is given. However the basic concept of a manifold does *not* include the idea of length; i.e. as it is usually defined, a manifold does not *ab initio* come endowed with a metric; we shall return to this question subsequently.)

The above-described cartographical procedure serves as motivation for the following (rather lengthy) general definition.

1.1.1. Definition. A *differentiable n-dimensional manifold* is an arbitrary set M (whose elements we call "points") together with the following structure on it. The set M is the union of a finite or countably infinite collection of subsets U_q with the following properties.

(i) Each subset U_q has defined on it co-ordinates x_q^α, $\alpha = 1, \ldots, n$ (called *local co-ordinates*) by virtue of which U_q is identifiable with a region of Euclidean n-space with Euclidean co-ordinates x_q^α. (The U_q with their co-ordinate systems are called *charts* (rather than "maps") or *local co-ordinate neighbourhoods.*)

(ii) Each non-empty intersection $U_p \cap U_q$ of a pair of such subsets of M thus has defined on it (at least) two co-ordinate systems, namely the restrictions of (x_p^α) and (x_q^α); it is required that under each of these co-ordinatizations the intersection $U_p \cap U_q$ is identifiable with a region of Euclidean n-space, and further that each of these two co-ordinate systems be expressible in terms of the other in a one-to-one differentiable manner. (Thus if the *transition* or *translation functions* from the co-ordinates x_q^α to the co-ordinates x_p^α and back, are given by

$$\begin{aligned} x_p^\alpha &= x_p^\alpha(x_q^1, \ldots, x_q^n), \qquad \alpha = 1, \ldots, n; \\ x_q^\alpha &= x_q^\alpha(x_p^1, \ldots, x_p^n), \qquad \alpha = 1, \ldots, n, \end{aligned} \tag{1}$$

then in particular the Jacobian $\det(\partial x_p^\alpha / \partial x_q^\beta)$ is non-zero on the region of intersection.) The general smoothness class of the transition functions for all intersecting pairs U_p, U_q, is called the *smoothness class of the manifold M* (with its accompanying "atlas" of charts U_q).

Any Euclidean space or regions thereof provide the simplest examples of manifolds. A region of the complex space $\mathbb{C}^n$ can be regarded as a region of the Euclidean space of dimension $2n$, and from this point of view is therefore also a manifold.

Given two manifolds $M = \bigcup_q U_q$ and $N = \bigcup_p U_p$, we construct their *direct product* $M \times N$ as follows: The points of the manifold $M \times N$ are the ordered pairs (m, n), and the covering by local co-ordinate neighbourhoods is given by

$$M \times N = \bigcup_{p,q} U_q \times V_p,$$

where if x_q^α are the co-ordinates on the region U_q, and y_p^β the co-ordinates on V_p, then the co-ordinates on the region $U_q \times V_p$ are (x_q^α, y_p^β).

These are just a few (ways of obtaining) examples of manifolds; in the sequel we shall meet with many further examples.

It should be noted that the scope of the above general definition of a manifold is from a purely logical point of view unnecessarily wide; it needs to be restricted, and we shall indeed impose further conditions (see below). These conditions are most naturally couched in the language of general topology, with which we have not yet formally acquainted the reader. This could have been avoided by defining a manifold at the outset to be instead a smooth non-singular surface (of dimension n) situated in Euclidean space of some (perhaps large) dimension. However this approach reverses the logical order of things; it is better to begin with the abstract definition of manifold, and then show that (under certain conditions) every manifold can be realized as a surface in some Euclidean space.

We recall for the reader some of the basic concepts of general topology.

(1) A *topological space* is by definition a set X (of "points") of which certain subsets, called the *open sets* of the topological space, are distinguished; these open sets are required to satisfy the following three conditions: first, the intersection of any two (and hence of any finite collection) of them should again be an open set; second, the union of any collection of open sets must again be open; and thirdly, in particular the empty set and the whole set X must be open.

The complement of any open set is called a *closed set* of the topological space.

The reader doubtless knows from courses in mathematical analysis that, exceedingly general though it is, the concept of a topological space already suffices for continuous functions to be defined: A map $f: X \to Y$ of one topological space to another is *continuous* if the complete inverse image $f^{-1}(U)$ of every open set $U \subseteq Y$ is open in X. Two topological spaces are *topologically equivalent* or *homeomorphic* if there is a one-to-one and onto map between them such that both it and its inverse are continuous.

In Euclidean space $\mathbb{R}^n$, the "Euclidean topology" is the usual one, where the open sets are just the usual open regions (see Part I, §1.2). Given any subset $A \subset \mathbb{R}^n$, the *induced topology* on A is that with open sets the intersections $A \cap U$, where U ranges over all open sets of $\mathbb{R}^n$. (This definition extends quite generally to any subset of any topological space.)

1.1.2. Definition. The *topology* (or *Euclidean topology*) *on a manifold M* is given by the following specification of the open sets. In every local co-ordinate neighbourhood U_q, the open (Euclidean) regions (determined by the given identification of U_q with a region of a Euclidean space) are to be open in the topology on M; the totality of open sets of M is then obtained by admitting as open also arbitrary unions of countable collections of such regions, i.e. by closing under countable unions.

With this topology the continuous maps (in particular real-valued functions) of a manifold M turn out to be those which are continuous in the usual sense on each local co-ordinate neighbourhood U_q. Note also that any open subset V of a manifold M inherits, i.e. has induced on it, the structure of a manifold, namely $V = \bigcup_q V_q$, where the regions V_q are given by

$$V_q = V \cap U_q. \tag{2}$$

(2) "Metric spaces" form an important subclass of the class of all topological spaces. A *metric space* is a set which comes equipped with a "distance function", i.e. a real-valued function $\rho(x, y)$ defined on pairs x, y of its elements ("points"), and having the following properties:

(i) $\rho(x, y) = \rho(y, x)$;
(ii) $\rho(x, x) = 0$, $\rho(x, y) > 0$ if $x \neq y$;
(iii) $\rho(x, y) \leq \rho(x, z) + \rho(z, y)$ (the "triangle inequality").

For example n-dimensional Euclidean space is a metric space under the usual Euclidean distance between two points $x = (x^1, \ldots, x^n)$, $y = (y^1, \ldots, y^n)$:

$$\rho(x, y) = \sqrt{\sum_{\alpha=1}^{n} (x^\alpha - y^\alpha)^2}.$$

A metric space is topologized by taking as its open sets the unions of arbitrary collections of "open balls", where by *open ball* with centre x_0 and radius ε we mean the set of all points x of the metric space satisfying $\rho(x_0, x) < \varepsilon$. (For n-dimensional Euclidean space this topology coincides with the above-defined Euclidean topology.)

An example important for us is that of a manifold endowed with a Riemannian metric. (For the definition of the distance between two points of a manifold with a Riemannian metric on it, see §1.2 below.)

(3) A topological space is called *Hausdorff* if any two of its points are contained in disjoint open sets.

In particular any metric space X is Hausdorff; for if x, y are any two distinct points of X then, in view of the triangle inequality, the open balls of radius $\frac{1}{2}\rho(x, y)$ with centres at x, y, do not intersect.

We shall henceforth assume implicitly that all topological spaces we consider are Hausdorff. Thus in particular we now supplement our definition of a manifold by the further requirement that it be a Hausdorff space.

(4) A topological space X is said to be *compact* if every countable collection of open sets covering X (i.e. whose union is X) contains a finite subcollection already covering X. If X is a metric space then compactness is equivalent to the condition that from every sequence of points of X a convergent subsequence can be selected.

(5) A topological space is *(path-)connected* if any two of its points can be joined by a continuous path (i.e. map from $[0, 1]$ to the space).

(6) A further kind of topological space important for us is the "space of

mappings" $M \to N$ from a given manifold M to a given manifold N. The topology in question will be defined later on.

The concept of a manifold might at first glance seem excessively abstract. In fact, however, even in Euclidean spaces, or regions thereof, we often find ourselves compelled to introduce a change of co-ordinates, and consequently to discover and apply the transformation rule for the numerical components of one entity or another. Moreover it is often convenient in solving a (single) problem to carry out the solution in different regions of a space using different co-ordinate systems, and then to see how the solutions match on the region of intersection, where there exist different co-ordinate systems. Yet another justification for the definition of a manifold is provided by the fact that not all surfaces can be co-ordinatized by a single system of co-ordinates without singular points (e.g. the sphere has no such co-ordinate system).

An important subclass of the class of manifolds is that of "orientable manifolds".

1.1.3. Definition. A manifold M is said to be *oriented* if for every pair U_p, U_q of intersecting local co-ordinate neighbourhoods, the Jacobian $J_{pq} = \det(\partial x_p^\alpha / \partial x_q^\beta)$ of the transition function is positive.

For example Euclidean n-space $\mathbb{R}^n$ with co-ordinates $x^1, \ldots, x^n$ is by this definition oriented (there being only one local co-ordinate neighbourhood). If we assign different co-ordinates $y^1, \ldots, y^n$ to the points of the same space $\mathbb{R}^n$, we obtain another manifold structure on the same underlying set. If the co-ordinate transformation $x^\alpha = x^\alpha(y^1, \ldots, y^n)$, $\alpha = 1, \ldots, n$, is smooth and non-singular, then its Jacobian $J = \det(\partial x^\alpha / \partial y^\beta)$, being never zero, will have fixed sign.

1.1.4. Definition. We say that the co-ordinate systems x and y define the *same orientation* of $\mathbb{R}^n$ if $J > 0$, and *opposite orientations* if $J < 0$.

Thus Euclidean n-space possesses two possible orientations. In the sequel we shall show that more generally any connected orientable manifold has exactly two orientations.

1.2. Mappings of Manifolds; Tensors on Manifolds

Let $M = \bigcup_p U_p$, with co-ordinates x_p^α, and $N = \bigcup_q V_q$, with co-ordinates y_q^β, be two manifolds of dimensions n and m respectively.

1.2.1. Definition. A mapping $f: M \to N$ is said to be *smooth of smoothness class* k, if for all p, q for which f determines functions $y_q^\beta(x_p^1, \ldots, x_p^n) = f(x_p^1, \ldots, x_p^n)_q^\beta$, these functions are, where defined, smooth of smoothness class k (i.e. all their partial derivatives up to those of kth order exist and are continuous). (It follows that the smoothness class of f cannot exceed the maximum class of the manifolds.)

Note that in particular we may have $N = \mathbb{R}$, the real line, whence $m = 1$, and f is a real-valued function of the points of M. The situation may arise where a smooth mapping (in particular a real-valued function) is not defined on the whole manifold M, but only on a portion of it. For instance each local co-ordinate x_p^α (for fixed α, p) is such a real-value function of the points of M, since it is defined only on the region U_p.

1.2.2. Definition. Two manifolds M and N are said to be *smoothly equivalent* or *diffeomorphic* if there is a one-to-one, onto map f such that both $f: M \to N$ and $f^{-1}: N \to M$, are smooth of some class $k \geq 1$. (It follows that the Jacobian $J_{pq} = \det(\partial y_q^\beta / \partial x_p^\alpha)$ is non-zero wherever it is defined, i.e. wherever the functions $y_q^\beta = f(x_p^1, \ldots, x_p^n)_q^\beta$ are defined.)

We shall henceforth tacitly assume that the smoothness class of any manifolds, and mappings between them, which we happen to be considering, are sufficiently high for the particular aim we have in view. (The class will always be assumed at least 1; if second derivatives are needed, then assume class ≥ 2, etc.)

Suppose we are given a curve segment $x = x(\tau)$, $a \leq \tau \leq b$, on a manifold M, where x denotes a point of M (namely that point corresponding to the value τ of the parameter). That portion of the curve in a particular co-ordinate neighbourhood U_p with co-ordinates x_p^α is described by the parametric equations

$$x_p^\alpha = x_p^\alpha(\tau), \qquad \alpha = 1, \ldots, n,$$

and in U_p its *velocity* (or *tangent*) *vector* is given by

$$\dot{x} = (\dot{x}_p^1, \ldots, \dot{x}_p^n).$$

In regions $U_p \cap U_q$ where two co-ordinate systems apply we have the two representations $x_p^\alpha(\tau)$ and $x_q^\beta(\tau)$ of the curve, where of course

$$x_p^\alpha(x_q^1(\tau), \ldots, x_q^n(\tau)) \equiv x_p^\alpha(\tau).$$

Hence the relationship between the components of the velocity vector in the two systems is expressed by

$$\dot{x}_p^\alpha = \sum_\beta \frac{\partial x_p^\alpha}{\partial x_q^\beta} \dot{x}_q^\beta. \tag{3}$$

As for Euclidean space, so also for general manifolds this formula provides the basis for the definition of "tangent vector".

1.2.3. Definition. A *tangent vector to an n-manifold M* at an arbitrary point x is represented in terms of local co-ordinates x_p^α by an n-tuple (ξ^α) of "components", which are linked to the components in terms of any other system x_q^β of local co-ordinates (on a region containing the point) by the formula

$$\xi_p^\alpha = \sum_{\beta=1}^{n} \left(\frac{\partial x_p^\alpha}{\partial x_q^\beta} \right)_x \xi_q^\beta. \tag{4}$$

The set of all tangent vectors to an n-dimensional manifold M at a point x forms an n-dimensional linear space $T_x = T_x M$, the *tangent space* to M at the point x. We see from (3) that the velocity vector at x of any smooth curve on M through x is a tangent vector to M at x. From (4) it can be seen that for any choice of local co-ordinates x^α in a neighbourhood of x, the operators $\partial/\partial x^\alpha$ (operating on real-valued functions on M) may be thought of as forming a basis $e_\alpha = \partial/\partial x^\alpha$ for the tangent space T_x.

A smooth map f from a manifold M to a manifold N gives rise for each x, to an *induced linear map of tangent spaces*

$$f_*: T_x \to T_{f(x)},$$

defined as sending the velocity vector at x of any smooth curve $x = x(t)$ (through x) on M, to the velocity vector at $f(x)$ to the curve $f(x(t))$ on the manifold N. In terms of local co-ordinates x^α in a neighbourhood of $x \in M$, and local co-ordinates y^β in a neighbourhood of $f(x) \in N$, the map f may be written as

$$y^\beta = f^\beta(x^1, \ldots, x^n), \qquad \beta = 1, \ldots, m.$$

It then follows from the above definition of the induced linear map f_* that its matrix is the Jacobian matrix $(\partial y^\beta/\partial x^\alpha)_x$ evaluated at x, i.e. that it is given by

$$\xi^\alpha \to \eta^\beta = \frac{\partial f^\beta}{\partial x^\alpha} \xi^\alpha. \tag{5}$$

For a real-valued function $f: M \to \mathbb{R}$, the induced map f_* corresponding to each $x \in M$ is a real-valued linear function (i.e. linear functional) on the tangent space to M at x; from (5) (with $m = 1$) we see that it is represented by the gradient of f at x and is thus a covector. Interpreting the differential of a function at a point in the usual way as a linear map of the tangent space, we see that f_* at x is just df.

1.2.4. Definition. A *Riemannian metric on a manifold* M is a point-dependent, positive-definite quadratic form on the tangent vectors at each point, depending smoothly on the local co-ordinates of the points. Thus at each point $x = (x_p^1, \ldots, x_p^n)$ of each region U_p with local co-ordinates x_p^α, the metric is given by a symmetric matrix $(g_{\alpha\beta}^{(p)}(x_p^1, \ldots, x_p^n))$, and determines a (symmetric) scalar product of pairs of tangent vectors at the point x:

$$\langle \xi, \eta \rangle = g_{\alpha\beta}^{(p)} \xi_p^\alpha \eta_p^\beta = \langle \eta, \xi \rangle,$$

$$|\xi|^2 = \langle \xi, \xi \rangle,$$

where as usual summation is understood over indices recurring as superscript and subscript. Since this scalar product is to be co-ordinate-independent, i.e.

$$g^{(p)}_{\alpha\beta}\xi^\alpha_p\eta^\beta_p = g^{(q)}_{\gamma\delta}\xi^\gamma_q\eta^\delta_q,$$

it follows from the transformation rule for vectors that the coefficients $g^{(p)}_{\alpha\beta}$ of the quadratic form transform (under a change to co-ordinates x^ν_q) according to the rule

$$g^{(q)}_{\gamma\delta} = \frac{\partial x^\alpha_p}{\partial x^\gamma_q} g^{(p)}_{\alpha\beta} \frac{\partial x^\beta_p}{\partial x^\delta_q}. \tag{6}$$

The definition of a *pseudo-Riemannian metric on a manifold M* is obtained from the above by replacing the condition that the quadratic form be at each point positive definite, by the weaker requirement that it be non-degenerate. (It then follows from the smoothness assumption that, provided M is connected, the index of inertia of the quadratic form is constant (cf. §3.2 of Part I).)

1.2.5. Definition. A *tensor of type (k, l) on a manifold* is given in each local co-ordinate system x^α_p by a family of functions

$${}^{(p)}T^{i_1 \dots i_k}_{j_1 \dots j_l}(x)$$

of the points x. In other local co-ordinates x^β_q (embracing the point x) the components ${}^{(q)}T^{s_1 \dots s_k}_{t_1 \dots t_l}(x)$ of the (same) tensor are related to its components in the system x^α_p by the transformation rule

$${}^{(q)}T^{s_1 \dots s_k}_{t_1 \dots t_l} = \frac{\partial x^{s_1}_q}{\partial x^{i_1}_p} \cdots \frac{\partial x^{s_k}_q}{\partial x^{i_k}_p} \frac{\partial x^{j_1}_p}{\partial x^{t_1}_q} \cdots \frac{\partial x^{j_l}_p}{\partial x^{t_l}_q} {}^{(p)}T^{i_1 \dots i_k}_{j_1 \dots j_l}. \tag{7}$$

All of the definitions and results of Chapter 3 of Part I pertaining to tensors defined on regions of Cartesian n-space, now apply without change to tensors on manifolds.

A metric $g_{\alpha\beta}$ on a manifold provides an example of a tensor of type (0, 2) (compare (6) and (7)). On an oriented manifold such a metric gives rise to a *volume element*

$$T_{\alpha_1 \dots \alpha_n} = \sqrt{|g|}\,\varepsilon_{\alpha_1 \dots \alpha_n}, \qquad g = \det(g_{\alpha\beta}),$$

where $\varepsilon_{\alpha_1 \dots \alpha_n}$ is the skew-symmetric tensor of rank n such that $\varepsilon_{12 \dots n} = 1$ (see §18.2 of Part I). It follows (as in §18.2 of Part I) that the volume element is a tensor with respect to co-ordinate changes with positive Jacobian, and so is indeed a tensor on our manifold-with-orientation. As in Part I, so also in the present context of general manifolds, it is convenient to write the volume element in the notation of differential forms (in arbitrary co-ordinates defining the same orientation):

$$\Omega = \sqrt{|g|}\, dx^1 \wedge \cdots \wedge dx^n.$$

A Riemannian metric dl^2 on a (connected) manifold M gives rise to a metric space structure on M with distance function $\rho(P, Q)$ defined by

$$\rho(P, Q) = \min_{\gamma} \int_{\gamma} dl,$$

where the infimum is taken over all piecewise smooth arcs joining the points P and Q. We leave it to the reader to verify that the topology on M defined by this metric-space structure coincides with the Euclidean topology on M.

It follows from the results of §29.2 of Part I, that any two points of a manifold (with a Riemannian metric defined on it) sufficiently close to one another can be joined by a geodesic arc. For points far apart this may in general not be possible, though if the manifold is connected such points can be joined by a broken geodesic.

1.3. Embeddings and Immersions of Manifolds. Manifolds with Boundary

1.3.1. Definition. A manifold M of dimension m is said to be *immersed* in a manifold N of dimension $n \geq m$, if there is given a smooth map $f: M \to N$ such that the induced map f_* is at each point a one-to-one map of the tangent plane (or in other words if in terms of local co-ordinates the Jacobian matrix of the map f at each point has rank m). The map f is called an *immersion* of the manifold M into the manifold N. (In its image in N, self-intersections of M may occur.)

An immersion of M into N is called an *embedding* if it is one-to-one. Abusing language slightly, we shall then call M a *submanifold* of N.

We shall always assume that any submanifold M we consider is defined in each local co-ordinate neighbourhood U_p of the containing manifold N by a system of equations

$$\left.\begin{array}{l} f_p^1(x_p^1, \ldots, x_p^n) = 0, \\ \cdots\cdots\cdots\cdots\cdots\cdots \\ f_p^{n-m}(x_p^1, \ldots, x_p^n) = 0, \end{array}\right\} \quad \text{where rank} \left(\frac{\partial f_p^\alpha}{\partial x_p^\beta}\right) = n - m,$$

with the property that on each intersection $U_q \cap U_p$, the systems $(f_p^\alpha = 0)$ and $(f_q^\alpha = 0)$ have the same set of zeros. It follows that throughout each neighbourhood U_p of N we can introduce new local co-ordinates $y_p^1, \ldots, y_p^n$ satisfying

$$y_p^{m+1} = f_p^1(x_p^1, \ldots, x_p^n), \ldots, y_p^n = f_p^{n-m}(x_p^1, \ldots, x_p^n).$$

In terms of these co-ordinates the submanifold M is in each U_p given by the equations

$$y_p^{m+1} = 0, \ldots, y_p^n = 0,$$

while $y_p^1, \ldots, y_p^m$ will serve as local co-ordinates on the submanifold M.

1.3.2. Definition. A closed region A of a manifold M, defined by an inequality of the form $f(x) \leq 0$ (or $f(x) \geq 0$) where f is a smooth real-valued function on M, is called a *manifold-with-boundary*. (It is assumed here that the boundary ∂A, given by the equation $f(x) = 0$, is a non-singular submanifold of M, i.e. that the gradient of the function f does not vanish on that boundary.)

Let A and B be manifolds with boundary, both given, as in the preceding definition, as closed regions of manifolds M and N respectively. A map $\varphi: A \to B$ is said to be a *smooth map of manifolds-with-boundary* if it is the restriction to A of a smooth map

$$\tilde{\varphi}: U \to N, \qquad \tilde{\varphi}|_A = \varphi,$$

of an open region U of M, containing A. (If A is defined in M by the inequality $f(x) \leq 0$, then U is usually taken to be $U_\varepsilon = \{x | f(x) < \varepsilon\}$ where $\varepsilon > 0$.

We conclude this section by mentioning yet another widely used term: a compact manifold without boundary is called *closed*.

§2. The Simplest Examples of Manifolds

2.1. Surfaces in Euclidean Space. Transformation Groups as Manifolds

A non-singular surface of dimension k in n-dimensional Euclidean space is given by a set of $n - k$ equations

$$f_i(x^1, \ldots, x^n) = 0, \qquad i = 1, \ldots, n - k, \tag{1}$$

where for all x the matrix $(\partial f_i/\partial x^j)$ has rank $n - k$. If at a point $(x_0^1, \ldots, x_0^n)$ on this surface the minor $J_{j_1 \ldots j_{n-k}}$ made up of those columns of the matrix $(\partial f_i/\partial x^j)$ indexed by $j_1, \ldots, j_{n-k}$, is non-zero, then as local co-ordinates on a neighbourhood of the surface about the point we make take

$$(y^1, \ldots, y^k) = (x^1, \ldots, \hat{x}^{j_1}, \ldots, \hat{x}^{j_{n-k}}, \ldots, x^n), \tag{2}$$

where the hatted symbols are to be omitted (see §7.1 of Part I). Since the surface is presupposed non-singular, it follows that it is covered by the regions of the form $U_{j_1 \ldots j_{n-k}}$, where this symbol denotes the set of all points of the surface at which the minor $J_{j_1 \ldots j_{n-k}}$ does not vanish.

2.1.1. Theorem. *The covering of the surface* (1) *by the regions*

$$U_{j_1 \dots j_{n-k}}, \qquad 1 \le j_1 < \cdots < j_{n-k} \le n,$$

each furnished with local co-ordinates (2), *defines on the surface the structure of a smooth manifold.*

PROOF. Throughout the region $U_{j_1 \dots j_{n-k}}$ of the surface (1) equations of the following form hold:

$$x^{j_i} = \varphi^i(y^1, \dots, y^k), \qquad i = 1, \dots, n-k,$$

where the φ^i are (smooth) functions. Similarly, in the region $U_{s_1 \dots s_{n-k}}$ with coordinates

$$(z^1, \dots, z^k) = (x^1, \dots, \hat{x}^{s_1}, \dots, \hat{x}^{s_{n-k}}, \dots, x^n),$$

we have

$$x^{s_i} = \psi^i(z^1, \dots, z^k), \qquad i = 1, \dots, n-k,$$

where again the ψ^i are smooth functions. Throughout the region of intersection of $U_{j_1 \dots j_{n-k}}$ and $U_{s_1 \dots s_{n-k}}$, we have the following smooth transition functions $y \to z$ and $z \to y$ (where for ease of expression we are assuming $1 < j_1 < s_1 < j_2 < \cdots$; the general case is clear from this):

$$\begin{aligned}
y^1 &= z^1 && (= x^1), \\
&\dots\dots\dots\dots \\
y^{j_1 - 1} &= z^{j_1 - 1} && (= x^{j_1 - 1}), \\
\varphi^1(y^1, \dots, y^k) &= z^{j_1} && (= x^{j_1}), \\
y^{j_1} &= z^{j_1 + 1} && (= x^{j_1 + 1}), \\
&\dots\dots\dots\dots \\
y^{s_1 - 2} &= z^{s_1 - 1} && (= x^{s_1 - 1}), \\
y^{s_1 - 1} &= \psi^1(z^1, \dots, z^k) && (= x^{s_1}), \\
y^{s_1} &= z^{s_1} && (= x^{s_1 + 1}), \\
&\dots\dots\dots\dots \\
y^k &= z^k && (= x^n).
\end{aligned} \tag{3}$$

It is immediate that the two transition functions displayed here are mutual inverses, completing the proof of the theorem. □

Remark 1. It is not difficult to calculate the Jacobian of the transition function $y \to z$: it is given (up to sign) by

$$J_{(y) \to (z)} = \pm \frac{J_{s_1 \dots s_{n-k}}}{J_{j_1 \dots j_{n-k}}} \neq 0.$$

Remark 2. It is easy to see (much as in §7.2 of Part I) that the tangent space to the manifold (1) is identifiable with the linear subspace of $\mathbb{R}^n$ consisting of the solutions of the system of equations

$$
\begin{aligned}
&\frac{\partial f_1}{\partial x^\alpha}\xi^\alpha = 0,\\
&\dots\dots\dots\dots\\
&\frac{\partial f_{n-k}}{\partial x^\alpha}\xi^\alpha = 0.
\end{aligned}
\tag{4}
$$

The (co)vectors grad $f_i = (\partial f_i/\partial x^\alpha)$, $i = 1, \dots, n-k$, are orthogonal (in the sense of the standard Euclidean metric on $\mathbb{R}^n$) to the surface at each point.

Our next goal will be that of showing that a non-singular surface in Euclidean space can be oriented. For this purpose we need to introduce an alternative definition of an orientation of a manifold.

To begin with, consider at any point x of an n-manifold M the various frames (i.e. ordered bases) τ for the tangent space to M at x each consisting, of course, of n independent tangent vectors in some order. Any two such frames τ_1, τ_2 are linked to one another via a non-singular linear transformation A which sends the vectors in τ_2 to those in τ_1 in order. We shall say that the ordered bases τ_1, τ_2 *lie in the same orientation class* if $\det A > 0$, and *lie in opposite orientation classes* if $\det A < 0$. (Thus at each point x of the manifold M, there are exactly two orientation classes of ordered bases of the tangent space at x.) Since a frame τ for the tangent space at x can be moved continuously from x to take up the positions of frames for the tangent spaces at nearby points, it makes sense to speak of an orientation class as depending continuously on the points of the manifold. We are now ready for our alternative definition of orientation.

2.1.2. Definition. A manifold is said to be *orientable* if it is possible to choose at every point of it a single orientation class depending continuously on the points. A particular choice of such an orientation class for each point is called an *orientation* of the manifold, and a manifold equipped with a particular orientation is said to be *oriented*. If no orientation exists the manifold is *non-orientable*. (Imagine a frame moving continuously along a closed path in the manifold, and returning to the starting point with the opposite orientation.)

2.1.3. Proposition. *Definition* 1.1.3 *is equivalent to the above definition of an orientation on a manifold.*

PROOF. If the manifold M is oriented in the sense of Definition 1.1.3, then at each point x of M we may choose as our orienting frame the ordered n-tuple $(e_{1j}, \dots, e_{nj})$ consisting of the standard basis vectors tangent to the co-ordinate axes of the local co-ordinate system $x_j^1, \dots, x_j^n$ on the local co-ordinate neighbourhood U_j in which x lies. If x lies in two local co-ordinate neighbourhoods U_j and U_k then we shall have two orienting frames chosen at x; however, since M is oriented in the sense of Definition 1.1.3, the Jacobian of the transition function from the local co-ordinates on U_j to those on U_k is positive, so that (in view of the transformation rule for vectors) the two frames lie in the same orientation class.

Conversely, suppose that M is oriented in the sense of Definition 2.1.2 above, and that there is given at each point x a frame lying in the orientation class of the given orientation of M. Around each point x there is an open neighbourhood (in the Euclidean topology on M, and of size depending on x) sufficiently small for there to exist (new) co-ordinates $x^1, \ldots, x^n$ on the neighbourhood with the property that at each point of it the standard (ordered) basis $(e_1, \ldots, e_n)$ of vectors tangent to the axes of $x^1, \ldots, x^n$ in order, lies in the given orientation class; this is so in view of the continuity of the dependence of the given orientation class on the points of M. If we choose one such neighbourhood (with the new co-ordinates introduced on it) for each point of M, then their totality forms a covering of the manifold by local co-ordinate neighbourhoods; furthermore, the transition functions for the regions of overlap all have positive Jacobians, since at each point of such regions the standard frames lie in the same orientation class (namely the one given beforehand on M). This completes the proof. □

2.1.4. Theorem. *A smooth non-singular surface M^k in n-dimensional space $\mathbb{R}^n$, defined by a system of equations of the form* (1), *is orientable.*

PROOF. Let τ denote a point-dependent tangent frame to the surface M^k. Obviously the (ordered) n-tuple $\hat{\tau} = (\tau, \operatorname{grad} f_1, \ldots, \operatorname{grad} f_{n-k})$ of vectors is linearly independent at each point (since the (co)vectors grad f_i are linearly independent among themselves and orthogonal to the surface). Now choose τ at each point of the surface M^k in such a way that the frame $\hat{\tau}$ (for the tangent space of $\mathbb{R}^n$) lies in the same orientation class as the standard frame $(e_1, \ldots, e_n)$. Since this orientation class is certainly continuously dependent on the points of $\mathbb{R}^n$, so also will the orientation class of τ depend continuously on the points of M^k. This completes the proof. □

The simplest example of a non-singular surface in $\mathbb{R}^{n+1}$ is the *n-dimensional sphere* S^n, defined by the equation

$$x_1^2 + \cdots + x_{n+1}^2 = 1;$$

it is a compact n-manifold. Convenient local co-ordinates on the n-sphere are obtained by means of the stereographic projection (see §9 of Part I). Thus let U_N denote the set of all points of the sphere except for the north pole $N = (0, \ldots, 0, 1)$, and similarly let U_S be the whole sphere with the south pole $S = (0, \ldots, 0, -1)$ removed. Local co-ordinates $(u_N^1, \ldots, u_N^n)$ on the region U_N are obtained by stereographic projection, from the north pole, of the sphere onto the hyperplane $x^{n+1} = 0$; similarly, projecting stereographically from

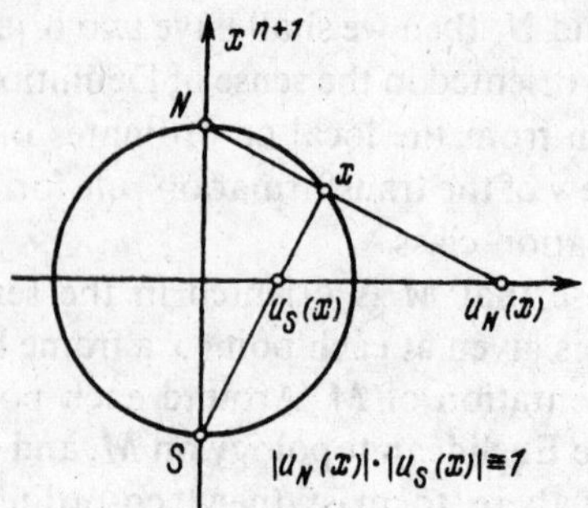

Figure 1. Local co-ordinates on the sphere via stereographic projections.

the south pole onto the same hyperplane yields co-ordinates $(u_S^1, \ldots, u_S^n)$ for the region U_S (see Figure 1). It is clear from Figure 1 that the origin and the two points $u_N(x)$ and $u_S(x)$ in the plane x^{n+1} are collinear, and that the product of the distances of $u_N(x)$ and $u_S(x)$ from the origin is unity. From this and a little more it follows easily that the transition function from the co-ordinates $(u_N^1, \ldots, u_N^n)$ to the co-ordinates $(u_S^1, \ldots, u_S^n)$ is given by (verify it!)

$$(u_S^1, \ldots, u_S^n) = \left(\frac{u_N^1}{\sum_{\alpha=1}^{n} (u_N^\alpha)^2}, \ldots, \frac{u_N^n}{\sum_{\alpha=1}^{n} (u_N^\alpha)^2} \right), \tag{5}$$

while the transition functions in the other direction are obtained by interchanging the letters N and S in this formula.

The n-sphere bounds a manifold with boundary, denoted by D^{n+1} and called the (closed) $(n+1)$-*dimensional disc* (or *ball*), defined by the inequality

$$f(x) = x_1^2 + \cdots + x_{n+1}^2 - 1 \leq 0.$$

Note finally that the sphere S^n separates the whole space $\mathbb{R}^{n+1}$ into two non-intersecting regions defined by $f(x) < 0$ and $f(x) > 0$.

Finally (before turning to the consideration of the classical transformation groups) we introduce the concept of "two-sidedness".

2.1.5. Definition. A connected $(n-1)$-dimensional submanifold of Euclidean space $\mathbb{R}^n$ is called *two-sided* if a (single-valued) continuous field of unit normals can be defined on it. We shall call such a submanifold a *two-sided hypersurface*. (See the remark below for the justification of this.)

2.1.6. Theorem. *A two-sided hypersurface in* $\mathbb{R}^n$ *is orientable.*

PROOF. Let v be a continuous field of unit normal vectors to a two-sided hypersurface M. At each point of M choose an ordered basis τ for the tangent space in such a way that the frame (τ, v) and the standard tangent frame $(e_1, \ldots, e_n)$ of $\mathbb{R}^n$ lie in the same orientation class of $\mathbb{R}^n$. It follows that the orientation class of τ must be continuously dependent on the points of M, yielding the desired conclusion. □

Remark. It will be shown in §7 that any two-sided hypersurface in $\mathbb{R}^n$ is defined by a single non-singular equation $f(x)=0$ (and hence is indeed a hypersurface), whence it follows that such a hypersurface always bounds a manifold-with-boundary. Somewhat later, in Chapter 3, it will also be proved that any closed hypersurface in $\mathbb{R}^n$ is two-sided.

The transformation groups introduced in §14 of Part I constitute important instances of manifolds defined by systems of equations in Euclidean space. Thus in particular:

(1) the general linear group $GL(n, \mathbb{R})$, consisting of all $n \times n$ real matrices with non-zero determinant, is clearly a region of $\mathbb{R}^{n^2}$;

(2) the special linear group $SL(n, \mathbb{R})$ of matrices with determinant $+1$ is the hypersurface in $\mathbb{R}^{n^2}$ defined by the single equation
$$\det A = 1;$$

(3) the orthogonal group $O(n, \mathbb{R})$ is the manifold defined by the system of equations
$$AA^{\mathrm{T}} = 1.$$

(4) the group $U(n)$ of unitary matrices is defined in the space of dimension $2n^2$ of all complex matrices by the equations
$$A\bar{A}^T = 1,$$
where the bar denotes complex conjugation.

In §14 of Part I it was shown that these groups (and others) are smooth non-singular surfaces in $\mathbb{R}^{n^2}$ (or $\mathbb{R}^{2n^2}$); we can now therefore safely call them smooth manifolds.

Note that all of these "group" manifolds G have the following property, linking their manifold and group structures: the maps φ: $G \to G$, defined by $\varphi(g) = g^{-1}$ (i.e. the taking of inverses), and ψ: $G \times G \to G$ defined by $\psi(g, h) = gh$ (i.e. the group multiplication), are smooth maps.

2.1.7. Definition. A manifold G is called a *Lie group* if it has given on it a group operation with the property that the maps φ, ψ defined as above in terms of the group structure, are smooth.

All of the transformation groups considered in Part I are in fact Lie groups.

2.2. Projective Spaces

We define an equivalence relation on the set of all non-zero vectors of $\mathbb{R}^{n+1}$ (regarded as a vector space) by taking two non-zero vectors to be equivalent if they are scalar multiples of one another. The equivalence classes under this relation are then taken to be the points of (*real*) *projective space of dimension n*, denoted by $\mathbb{R}P^n$. (Each projective space comes with a natural manifold structure, which will be precisely defined below.)

We now give an alternative (topologically equivalent) description of $\mathbb{R}P^n$. Consider the set of all straight lines in $\mathbb{R}^{n+1}$ passing through the origin. Since such a straight line is completely determined by any direction vector, and since any non-zero scalar multiple of any particular direction vector serves equally well, we may take these straight lines as the points of $\mathbb{R}P^n$. Now each of these straight lines intersects the sphere S^n (with equation $(y^0)^2 + \cdots + (y^n)^2 = 1$) at exactly two (diametrically opposite) points. Thus the points of $\mathbb{R}P^n$ are in one-to-one correspondence with the pairs of diametrically opposite points of the n-sphere. We may therefore think of projective space $\mathbb{R}P^n$ as obtained from S^n by "glueing", as they say, that is by identifying, diametrically opposite points. (We note in passing the consequence that functions on $\mathbb{R}P^n$ may be considered as even functions on the sphere S^n: $f(y) = f(-y)$.)

Examples. (a) The *projective line* $\mathbb{R}P^1$ has as its points pairs of diametrically opposite points of the circle S^1. Since every point of the upper semicircle (where $y > 0$) has its partner in the lower semicircle, we can obtain (a topologically equivalent space to) $\mathbb{R}P^n$ by taking only the bottom semicircle (together with the points where $x = \pm 1$) and identifying its end points $x = \pm 1$. Clearly the result is again a circle; we have thus constructed a one-to-one correspondence (which is in fact a topological equivalence) between $\mathbb{R}P^1$ and the circle S^1 (see Figure 2).

The analogous construction can be carried out in the general case, i.e. for $\mathbb{R}P^n$. One takes the disc D^n (obtained as the lower half of the sphere S^n) and identifies diametrically opposite points of its boundary. (The case $n = 2$ is illustrated in Figure 3.)

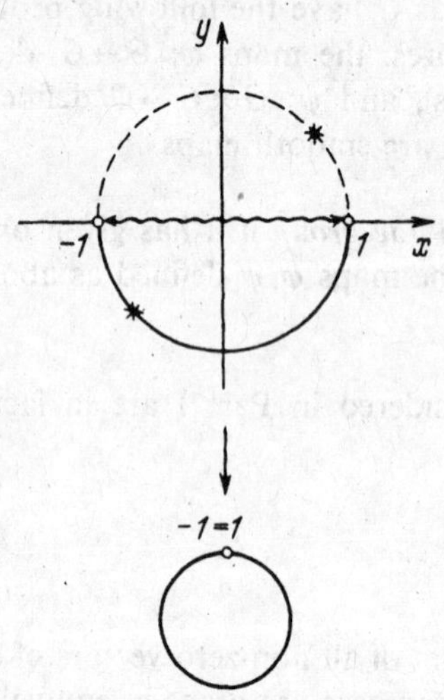

Figure 2

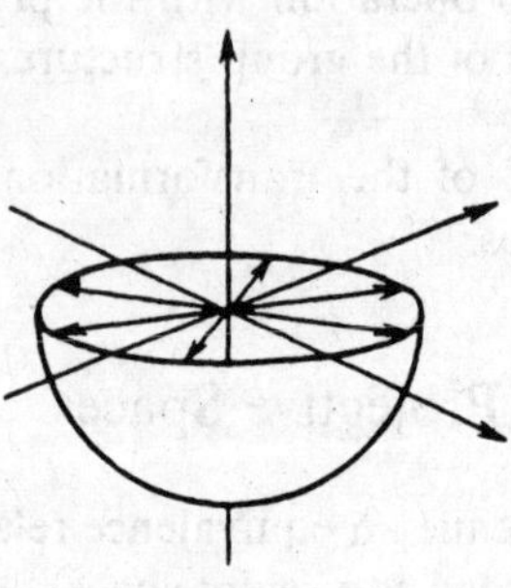

Figure 3

(b) In §14.3 of Part I a homomorphism from the group $SU(2)$ onto the group $SO(3)$ was constructed, under which each matrix A of $SU(2)$ has the same image as $-A$ (i.e. having kernel $\{\pm 1\}$), or, in other words, identifying the points A and $-A$ of the manifold $SU(2)$ in the image manifold $SO(3)$. We saw in §14.1 of Part I that there is a homeomorphism between the manifold $SU(2)$ and the 3-sphere S^3 under which matrices A and $-A$ are sent to diametrically opposite points of S^3. Hence we obtain an identification (in fact topological equivalence) of $SO(3)$ with projective 3-space $\mathbb{R}P^3$.

We now introduce explicitly a (natural) manifold structure on the projective spaces $\mathbb{R}P^n$.

For this purpose we return to our original characterization of $\mathbb{R}P^n$ as consisting of equivalence classes of non-zero vectors in the space $\mathbb{R}^{n+1}$ with co-ordinates $y^0, \ldots, y^n$. For each $q = 0, 1, \ldots, n$, let U_q denote the set of equivalence classes of vectors $(y^0, \ldots, y^n)$ with $y^q \neq 0$. On each such region U_q of $\mathbb{R}P^n$ we introduce the local co-ordinates $x_q^1, \ldots, x_q^n$ defined by

$$\begin{aligned} x_q^1 &= \frac{y^0}{y^q}, \ldots, x_q^q = \frac{y^{q-1}}{y^q}, \\ x_q^{q+1} &= \frac{y^{q+1}}{y^q}, \ldots, x_q^n = \frac{y^n}{y^q}. \end{aligned} \tag{6}$$

Clearly the regions U_q, $q = 0, 1, \ldots, n$, cover the whole of projective n-space.

We next calculate the transition functions. For notational simplicity we do this for the particular pair U_0, U_1: the general formulae for the transition functions on $U_j \cap U_k$ can be obtained from those for $U_0 \cap U_1$ by the appropriate replacement of indices. Now the co-ordinates in U_0 are given by

$$x_0^1 = \frac{y^1}{y^0}, x_0^2 = \frac{y^2}{y^0}, \ldots, x_0^n = \frac{y^n}{y^0},$$

and in U_1 by

$$x_1^1 = \frac{y^0}{y^1}, x_1^2 = \frac{y^2}{y^1}, \ldots, x_1^n = \frac{y^n}{y^1}.$$

Hence in the region $U_0 \cap U_1$, where both $y^0, y^1 \neq 0$, the transition function from (x_0) to (x_1) is obviously

$$x_1^1 = \frac{1}{x_0^1}, x_1^2 = \frac{x_0^2}{x_0^1}, x_1^3 = \frac{x_0^3}{x_0^1}, \ldots, x_1^n = \frac{x_0^n}{x_0^1}. \tag{7}$$

(Note that $x_0^1 = y^1/y^0$ is non-zero on $U_0 \cap U_1$.) The Jacobian of this transition function is given by

$$J_{(x_0)\to(x_1)} = \det\begin{pmatrix} -\frac{1}{(x_0^1)^2} & 0 & \ldots & \ldots & 0 \\ -\frac{x_0}{(x_0^1)^2} & \frac{1}{x_0^1} & 0 & \ldots & 0 \\ \multicolumn{5}{c}{\cdots\cdots\cdots\cdots\cdots\cdots} \end{pmatrix} = -\frac{1}{(x_0^1)^{n+1}} \neq 0.$$

Since, as noted above, the general transition functions on $U_j \cap U_h$ are obtained similarly, it follows that $\mathbb{R}P^n$ with the U_q as local coordinate neighbourhoods is indeed a smooth manifold. The manifold $\mathbb{R}P^2$ (with $n=2$) is called the *projective plane*; in this case the region U_0 is called the *finite part of the projective plane*.

Finally we note that, as is easily shown, the one-to-one correspondences $S^1 \to \mathbb{R}P^1$ and $SO(3) \to \mathbb{R}P^3$ described in the above examples, are in fact diffeomorphisms.

We define *complex projective space* $\mathbb{C}P^n$ similarly; its points are the equivalence classes of non-zero vectors in $\mathbb{C}^{n+1}$ under the analogous equivalence relation (i.e. scalar multiples are identified), and the local co-ordinate neighbourhoods, with their co-ordinates, are defined as in the real case, making $\mathbb{C}P^n$ a $2n$-dimensional smooth manifold.

By way of an example, we consider in detail the complex projective line $\mathbb{C}P^1$. Its points are the equivalence classes of non-zero pairs (z^0, z^1) of complex numbers, where the equivalence is defined by $(z^0, z^1) \sim (\lambda z^0, \lambda z^1)$ for any non-zero complex number λ. Consider the (complex) function $w_0(z^0, z^1) = z^1/z^0$; this function is defined (and one-to-one) on all of $\mathbb{C}P^1$ except (the equivalence class of) $(0, 1)$: we shall formally define w_0 as taking the value ∞ at this point. Thus via the function w_0 the complex projective line $\mathbb{C}P^1$ becomes identified with the "extended complex plane" (i.e. the ordinary complex plane with an additional "point at infinity").

2.2.1. Theorem. *The complex projective line* $\mathbb{C}P^1$ *is diffeomorphic to the 2-dimensional sphere* S^2.

PROOF. On the region U_0 of the complex projective line consisting of all equivalence classes of non-zero pairs (i.e. non-zero pairs determined only up to scalar multiples) (z^0, z^1) with $z^0 \neq 0$, we introduce local co-ordinates u_0, v_0 defined by $u_0 + iv_0 = w_0 = z^1/z^0$. (These local co-ordinates may be regarded as defining a one-to-one map from U_0 onto the real plane $\mathbb{R}^2$.) Similarly, u_1, v_1, defined by $u_1 + iv_1 = w_1 = z^0/z^1$, will serve as co-ordinates on the region U_1 consisting of pairs (z^0, z^1) (up to scalar multiples) with $z^1 \neq 0$. Clearly the regions U_0 and U_1 cover $\mathbb{C}P^1$. The transition function from (u_0, v_0) to (u_1, v_1) on the region of intersection is given by

$$(u_1, v_1) = \left(\frac{u_0}{u_0^2 + v_0^2}, -\frac{v_0}{u_0^2 + v_0^2} \right),$$

or, in complex notation, by

$$u_1 + iv_1 = w_1 = \frac{1}{w_0} = \frac{u_0 - iv_0}{u_0^2 + v_0^2}.$$

Since this formula coincides with the formula (5) (in the case $n = 2$) for the transition functions for the stereographic co-ordinates on the sphere S^2, the theorem follows. □

It is on account of this result that the extended complex plane is often called the "Riemann sphere". Note that if $w = u + iv$ provides local co-ordinates u, v for the finite part of the extended complex plane (i.e. for the ordinary complex plane), then $1/w$ provides local co-ordinates of a (punctured) neighbourhood of the "point at infinity" ∞.

We now return to the consideration of the general complex projective space $\mathbb{C}P^n$. From each equivalence class of $(n+1)$-vectors we may choose as representative a vector whose tip lies on the unit sphere S^{2n+1}, i.e. satisfying

$$|z^0|^2 + \cdots + |z^n|^2 = 1,$$

by simply multiplying any vector $z = (z^0, \ldots, z^n)$ in the class by the scalar $\lambda = (\sum_{\alpha=0}^{n} |z^\alpha|)^{-1/2}$. The resulting vector (with tip on S^{2n+1}) is then clearly unique only up to multiplication by scalars of the form $e^{i\varphi}$, i.e. by complex numbers of modulus 1. We therefore conclude that:

Complex projective space $\mathbb{C}P^n$ can be obtained from the (unit) sphere $S^{2n+1} = \{z \mid \sum_{\alpha=0}^{n} |z^\alpha|^2 = 1\}$, by identifying all points $e^{i\varphi}z$ on the sphere (φ variable) with z, i.e. by identifying all points differing by a scalar factor of the form $e^{i\varphi}$.

Thus we have a map

$$S^{2n+1} \to \mathbb{C}P^n, \tag{8}$$

such that the pre-image of each point of $\mathbb{C}P^n$ is (topologically equivalent to) the circle $S^1 = \{e^{i\varphi}\}$. In particular, in view of Theorem 2.2.1, we obtain thence a map

$$S^3 \to S^2, \qquad (z^0, z^1) \mapsto w = \frac{z^1}{z^0} \qquad (|z^0|^2 + |z^1|^2 = 1).$$

2.3. Exercises

1. Prove that the odd-dimensional projective spaces $\mathbb{R}P^{2k+1}$ are orientable.
2. Prove that the connected component containing the identity element of a Lie group, is a normal subgroup.
3. Prove that a connected Lie group is generated by an arbitrarily small neighbourhood of the identity element.
4. Prove that every Lie group is orientable.
5. Prove that the projective spaces $\mathbb{R}P^n$ and $\mathbb{C}P^n$ are compact.
6. *Quaternion projective space* $\mathbb{H}P^n$ is defined as the set of equivalence classes of non-zero quaternion vectors in $\mathbb{H}^{n+1}$, where two $(n+1)$-tuples are equivalent if one is a

multiple of the other (by a non-zero quaternion). Define a manifold structure on $\mathbb{H}P^n$, and verify that $\mathbb{H}P^1$ is diffeomorphic to S^4.

7. Construct a mapping $S^{4n+3} \to \mathbb{H}P^n$, analogous to the mapping (8), and identify the complete inverse image of a point of $\mathbb{H}P^n$ under this map.

§3. Essential Facts from the Theory of Lie Groups

3.1. The Structure of a Neighbourhood of the Identity of a Lie Group. The Lie Algebra of a Lie Group. Semisimplicity

Every Lie group G (see Definition 2.1.6) has a distinguished point $g_0 = 1 \in G$ (the identity element), and, being by definition a smooth manifold, has a tangent space $T = T_{(1)}$ at that point. For each $h \in G$ the transformation $G \to G$, defined by $g \mapsto hgh^{-1}$, is called the *inner automorphism* of G determined by h. Any such transformation of G clearly fixes the identity element $g_0 = 1$ (since $hg_0h^{-1} = g_0$), and therefore the induced linear map of the tangent space T to G at the identity (see §1.2 above) is a linear transformation of T, denoted by

$$\mathrm{Ad}(h)\colon T \to T.$$

From the definitions of the inner automorphism determined by each element h, and the linear map of the tangent space T which it induces, it follows easily that $\mathrm{Ad}(h^{-1}) = [\mathrm{Ad}(h)]^{-1}$ and $\mathrm{Ad}(h_1 h_2) = \mathrm{Ad}(h_1)\,\mathrm{Ad}(h_2)$, for all h, h_1, h_2 in G. Hence the map $h \mapsto \mathrm{Ad}(h)$ is a linear representation (i.e. a homomorphism to a group of linear transformations) of the group G:

$$\mathrm{Ad}\colon G \to GL(n, \mathbb{R}),$$

where n is the dimension of G. (Note that for commutative groups G the representation Ad is trivial, i.e. $\mathrm{Ad}(h) = 1$ for all $h \in G$.)

We shall now express the group operation on a Lie group G in a neighbourhood of the identity, in terms of local co-ordinates on such a neighbourhood. We first re-choose co-ordinates in a neighbourhood of the identity element so that the identity element is the origin: $1 = g_0 = (0, \ldots, 0)$. We then express in functional notation the co-ordinates of the product $g_1 g_2$ (if it is still in the neighbourhood) of elements $g_1 = (x^1, \ldots, x^n)$ and $g_2 = (y^1, \ldots, y^n)$ by

$$\psi^\alpha(x, y) = \psi^\alpha(x^1, \ldots, x^n, y^1, \ldots, y^n), \qquad \alpha = 1, \ldots, n,$$

and the co-ordinates of the inverse g^{-1} of an element $g = (x^1, \ldots, x^n)$ by

$$\varphi^\alpha(x) = \varphi^\alpha(x^1, \ldots, x^n), \qquad \alpha = 1, \ldots, n.$$

The functions $\psi(x, y)$ $(=g_1g_2)$ and $\varphi(x)$ $(=g^{-1})$ obviously satisfy the following conditions (arising from the defining properties of a group):

(i) $\psi(x, 0)=\psi(0, x)=x$ (property of the identity);
(ii) $\psi(x, \varphi(x))=0$ (property of inverses);
(iii) $\psi(x, \psi(y, z))=\psi(\psi(x, y), z)$ (associative property).

Given sufficient smoothness of the function $\psi(x, y)$, it follows from condition (i) (and Taylor's theorem) that

$$\psi^\alpha(x, y)=x^\alpha+y^\alpha+b^\alpha_{\beta\gamma}x^\beta y^\gamma+(\text{terms of order} \geq 3) \tag{1}$$

(where of course $b^\alpha_{\beta\gamma}=\partial^2\psi^\alpha/\partial x^\beta\partial y^\gamma$ evaluated at the origin).

Now let ξ and η be tangent vectors to the group at the identity, i.e. elements of the space T, and as usual denote their components, in terms of our co-ordinates x^α, by ξ^α, η^α respectively. The *commutator* $[\xi, \eta]\in T$ of tangent vectors ξ, η is defined by

$$[\xi, \eta]^\alpha=(b^\alpha_{\beta\gamma}-b^\alpha_{\gamma\beta})\xi^\beta\eta^\gamma. \tag{2}$$

This commutator operation on T has the following three basic properties:

(a) $[\ ,\ ]$ is a bilinear operation on the n-dimensional linear space T (where n is the dimension of G);
(b) $[\xi, \eta]=-[\eta, \xi]$;
(c) $[[\xi, \eta], \zeta]+[[\zeta, \xi], \eta]+[[\eta, \zeta], \xi]=0$ ("Jacobi's identity"). (3)

(The first two of these properties are almost immediate from the definition of the commutator operation. Here is a sketch of the proof of (c) as a consequence of the associative law (iii) above: From (1) we obtain that

$$\psi^\alpha(\psi(x, y), z)=\psi^\alpha(x, y)+z^\alpha+b^\alpha_{\beta\gamma}\psi^\beta(x, y)z^\gamma$$
$$+(\text{terms of degree} \geq 3 \text{ in } \psi^i, z^j).$$

Substitution in this from (1) yields an expansion of $\psi^\alpha(\psi(x, y), z)$ in terms of x^μ, y^ν, z^γ in which the coefficient of $x^\mu y^\nu z^\gamma$ is $b^\alpha_{\beta\gamma}b^\beta_{\mu\nu}$. Repeating this procedure for $\psi^\alpha(x, \psi(y, z))$ and comparing the coefficient of $x^\mu y^\nu z^\gamma$ with that obtained in the case of $\psi^\alpha(\psi(x, y), z)$, we find that

$$b^\alpha_{\beta\gamma}b^\beta_{\mu\nu}=b^\alpha_{\mu\beta}b^\beta_{\nu\gamma} \tag{4}$$

On the other hand from the definition of the commutator we obtain

$$[[\xi, \eta], \zeta]^\alpha=(b^\alpha_{\beta\gamma}-b^\alpha_{\gamma\beta})[\xi, \eta]^\beta\zeta^\gamma=(b^\alpha_{\beta\gamma}-b^\alpha_{\gamma\beta})(b^\beta_{\mu\nu}-b^\beta_{\nu\mu})\xi^\mu\eta^\nu\zeta^\gamma.$$

It follows that Jacobi's identity is equivalent to

$$(b^\alpha_{\beta\gamma}-b^\alpha_{\gamma\beta})(b^\beta_{\mu\nu}-b^\beta_{\nu\mu})+(b^\alpha_{\beta\nu}-b^\alpha_{\nu\beta})(b^\beta_{\gamma\mu}-b^\beta_{\mu\gamma})+(b^\alpha_{\beta\mu}-b^\alpha_{\mu\beta})(b^\beta_{\nu\gamma}-b^\beta_{\gamma\nu})=0,$$

which is easily seen to be a consequence of (4), as required.)

Thus the tangent space to G at the identity is with respect to the commutator operation a *Lie algebra*; since it arises from G it is called the *Lie algebra of the Lie group G*. (Cf. §24.1 of Part I.)

If $e_1, \ldots, e_n$ are the standard basis vectors of T (in terms of the co-ordinates $x^1, \ldots, x^n$) then since $[e_\beta, e_\gamma]$ is again a vector in T, we may write

$$[e_\beta, e_\gamma] = c^\alpha_{\beta\gamma} e_\alpha,$$

whence by bilinearity

$$[\xi, \eta]^\alpha = c^\alpha_{\beta\gamma} \xi^\beta \eta^\gamma, \tag{5}$$

for all vectors ξ, η in T. The constants $c^\alpha_{\beta\gamma}$, which clearly determine the commutator operation on the Lie algebra, and which are skew-symmetric in the indices β, γ, are called the *structural constants of the Lie algebra.*

A *one-parameter subgroup* of a Lie group G is defined to be a parametrized curve $F(t)$ on the manifold G such that $F(0) = 1$, $F(t_1 + t_2) = F(t_1)F(t_2)$, $F(-t) = F(t)^{-1}$. (Thus the one-parameter subgroup is determined by a homomorphism $t \mapsto F(t)$ from the additive reals to G.)

(Before proceeding we note parenthetically that left multiplication by a fixed element h of an abstract Lie group G defines a diffeomorphism $G \to G$ $(g \mapsto hg)$; the induced map of tangent spaces is defined as before to send the tangent vector $\dot{g}(t)$ to a curve $g(t)$, to the tangent vector $(d/dt)(hg(t))$ to the curve $hg(t)$.) Now if $F(t)$ is a one-parameter subgroup of G, then

$$\frac{dF}{dt} = \left.\frac{dF(t+\varepsilon)}{d\varepsilon}\right|_{\varepsilon=0} = \frac{d}{d\varepsilon}(F(t)F(\varepsilon))|_{\varepsilon=0} = F(t)\left.\frac{dF(\varepsilon)}{d\varepsilon}\right|_{\varepsilon=0},$$

where the last equality follows from the preceding parenthetical definition of the tangent-space map induced by left multiplication on the group by the element $F(t)$. Hence $\dot{F}(t) = F(t)\dot{F}(0)$, or $F(t)^{-1}\dot{F}(t) = \dot{F}(0)$, i.e. the induced action of left multiplication by $F(t)^{-1}$ sends $\dot{F}(t)$ to $\dot{F}(0) = \text{const}$. Conversely, for each particular tangent vector A of T, the equation

$$F^{-1}\dot{F} = A \tag{6}$$

is satisfied by a unique one-parameter subgroup $F(t)$ of G; to see this note first that (6) is (when formulated in terms of the function $\psi(x, y)$ defining the multiplication of points x, y of G) a system of ordinary differential equations, and therefore by the appropriate existence and uniqueness theorem for the solutions of such systems, has, for some sufficiently small $\varepsilon > 0$ a unique solution $F(t)$ for $|t| < \varepsilon$. The values of $F(t)$ for all larger $|t|$ can then be obtained by forming long enough products of elements $F(\delta)$ with $|\delta| < \varepsilon$.

In the case that G is a matrix group it follows from (6) that $F(t) = \exp At$ (see §§14.2, 24.3 of Part I). We shall use this notation also for the one-parameter subgroup arising from A via (6) in the general case of an arbitrary Lie group.

Exercise

Let $F_1(t)$ and $F_2(t)$ be two one-parameter subgroups of a Lie group G with $A_1 = \dot{F}_1(0)$, $A_2 = \dot{F}_1(0)$, whence $F_1(t) = \exp A_1 t$, $F_2(t) = \exp A_2 t$. Prove that

$$t^2[A_1, A_2] = F_1(t)F_2(t)F_1^{-1}(t)F_2^{-1}(t) + O(t^3). \tag{7}$$

Let $F(t) = \exp At$ be a one-parameter subgroup of a Lie group G. For each t the inner automorphism $g \mapsto FgF^{-1}$ induces (as we saw above) a linear transformation $\operatorname{Ad} F(t)$ of the Lie algebra $T = T_{(1)}$, which, since T is n-dimensional, lies in $GL(n, \mathbb{R})$. It follows that $\operatorname{Ad} F(t)$ is a one-parameter subgroup of $GL(n, \mathbb{R})$, whence the vector $(d/dt) \operatorname{Ad} F(t)|_{t=0}$ lies in the Lie algebra of the group $GL(n, \mathbb{R})$, and so can be regarded as a linear operator.

EXERCISE

Prove that, as operator, $(d/dt) \operatorname{Ad} F(t)|_{t=0}$ is given by $B \mapsto [A, B]$ for B in the Lie algebra (which is identifiable with $\mathbb{R}^n$). (As in §24.1 of Part I, we denote the map $B \mapsto [A, B]$ by $\operatorname{ad} A: \mathbb{R}^n \to \mathbb{R}^n$.)

We next use the one-parameter subgroups to define canonical co-ordinates in a neighbourhood of the identity of a Lie group G. Let $A_1, \ldots, A_n$ form a basis for the Lie algebra T (which we may identify with $\mathbb{R}^n$), the tangent space to G at the point $g_0 = 1$. We saw above that to each vector $A = \sum A_i x^i$ in T there corresponds a one-parameter group $F(t) = \exp At$. To the point $F(1)$ (which it is natural to denote also by $\exp A$) we assign as co-ordinates the coefficients $x^1, \ldots, x^n$; in this way we obtain a system of co-ordinates (by "projecting down from the tangent space" as it were) in a sufficiently small neighbourhood of the identity element of G. (Verify this!) These are called *canonical co-ordinates of the first kind.*

Alternatively, writing $F_i(t) = \exp A_i t$, we have that each point g of a sufficiently small neighbourhood of the identity element can be expressed uniquely as

$$g = F_1(t_1) \ldots F_n(t_n),$$

for small $t_1, \ldots, t_n$. Assigning co-ordinates $t_1 = x_1, \ldots, t_n = x_n$ to the point g, we thus obtain the *co-ordinates of the second kind* in a neighbourhood of the identity.

EXERCISES

1. Given a curve in the form $g(\tau) = F_1(\tau t_1) \ldots F_n(\tau t_n)$, prove that

$$\left.\frac{dg}{d\tau}\right|_{\tau=0} = \sum_{i=1}^{n} t_i A_i.$$

2. Show that the "Euler angles" φ, ψ, θ (see §14.1 of Part I) constitute co-ordinates of the second kind on $SO(3)$.

Co-ordinates of the first kind are exploited in the proof of the following result.

3.1.1. Theorem. *If the functions $\psi^\alpha(x, y)$ defining the multiplication of points x, y of a Lie group G are real analytic (i.e. are representable by power series), then in some neighbourhood of $1 \in G$, the structure of the Lie algebra of G determines the multiplication in G.*

(Here the condition that ψ be real analytic (or, as they say, that G be analytic) is not crucial; however, the proof under weaker assumptions about ψ is more complicated.)

PROOF. Define auxiliary functions $v^\alpha_\beta(x)$ by

$$v^\alpha_\beta(x) = \left.\frac{\partial \psi^\alpha(x, y)}{\partial x^\beta}\right|_{y=\varphi(x)},$$

where as before $\varphi(x)$ is the function defining (in terms of co-ordinates) the inverse of x. It follows from properties (i), (ii) and (iii) of ψ and φ (towards the beginning of this section) that the functions $\psi^\alpha(x, y)$ satisfy the following system of partial differential equations in x:

$$v^\alpha_\beta(\psi(x, y))\frac{\partial \psi^\beta(x, y)}{\partial x^\gamma} = v^\alpha_\gamma(x), \tag{8}$$

with the initial conditions

$$\psi(0, y) = y.$$

(To see (8), note first that the left-hand side is

$$\left.\frac{\partial \psi^\alpha(x, y)}{\partial x^\beta}\right|_{\substack{x=\psi(x,y)\\ y=\varphi(\psi(x,y))}} \cdot \frac{\partial \psi^\beta(x, y)}{\partial x^\gamma},$$

which is the same as

$$\left.\frac{\partial \psi^\alpha(\psi(x, y), z)}{\partial x^\gamma}\right|_{z=\varphi(\psi(x,y))},$$

and then apply properties (i), (ii) and (iii) of φ and ψ.)

It can be shown that the system (8) has a solution precisely if

$$\frac{\partial v^\alpha_\beta}{\partial x^\gamma} - \frac{\partial v^\alpha_\gamma}{\partial x^\beta} = 2c^\alpha_{\mu\nu}v^\mu_\beta v^\nu_\gamma. \tag{9}$$

EXERCISE

Taking the invertibility of the matrix $(v^\alpha_\beta(\psi))$ into account, show that (9) is equivalent to

$$\frac{\partial^2 \psi}{\partial x^\alpha \partial x^\gamma} = \frac{\partial^2 \psi}{\partial x^\gamma \partial x^\alpha},$$

the condition for solubility of the system of "Pfaffian" equations (8) (cf. (5), (6) in §29.1).

Since (8) does indeed have a solution, namely the ψ defining the multiplication in G, it follows that equation (9) must hold.

On the other hand, if $x = x(t)$ represents the one-parameter subgroup determined by the initial velocity vector $A = (A^i)$, then, putting equation (6) into functional notation, we have

$$A^\alpha = \frac{d}{d\varepsilon} x^\alpha(\varepsilon)|_{\varepsilon=0} = \frac{d}{d\varepsilon} \psi^\alpha(x(\varepsilon+t), x(-t))|_{\varepsilon=0}$$

$$= \frac{\partial \psi^\alpha(x, y)}{\partial x^\beta}\bigg|_{y=\varphi(x)} \frac{dx^\beta(\varepsilon+t)}{d\varepsilon}\bigg|_{\varepsilon=0} = v^\alpha_\beta(x(t)) \frac{dx^\beta(t)}{dt}.$$

If we now take the x^α to be canonical co-ordinates of the first kind (in some neighbourhood of 1) then by definition of such co-ordinates, $x^\alpha(t) = A^\alpha t$, whence by the above

$$A^\alpha = v^\alpha_\beta(At)A^\beta,$$

yielding

$$x^\alpha = v^\alpha_\beta(x)x^\beta. \tag{10}$$

Our aim is to show that the functions $v^\alpha_\beta(x)$ are fully determined (in some neighbourhood of 1) by these canonical co-ordinates. Differentiating the last equation with respect to x^β, we obtain

$$\delta^\alpha_\beta = x^\gamma \frac{\partial v^\alpha_\gamma}{\partial x^\beta} + v^\alpha_\beta. \tag{11}$$

By multiplying equation (9) by x^β (and summing with respect to β), and then substituting from (10) and (11), we obtain

$$x^\beta \frac{\partial v^\alpha_\gamma}{\partial x^\beta} + v^\alpha_\gamma(x) = \delta^\alpha_\gamma + c^\alpha_{\mu\nu} x^\nu v^\mu_\gamma,$$

whence, on replacing x by At,

$$tA^\beta \frac{\partial v^\alpha_\gamma}{\partial x^\beta} + v^\alpha_\gamma(x) = \delta^\alpha_\gamma + c^\alpha_{\mu\nu} A^\nu t v^\mu_\gamma. \tag{12}$$

In terms of the new functions $w^\alpha_\gamma(t) = tv^\alpha_\gamma(At)$ (also dependent on A) the equations (12) take the form

$$\frac{dw^\alpha_\gamma}{dt} = \delta^\alpha_\gamma + c^\alpha_{\mu\nu} A^\nu w^\mu_\gamma, \tag{13}$$

which is a system of ordinary linear differential equations for the functions $w^\alpha_\gamma(t)$, with initial conditions $w^\alpha_\gamma(0) = 0$. Hence for each fixed A the functions $w^\alpha_\gamma(t)$ are uniquely determined by the Lie algebra structure (since the system (13) is determined by the structure constants $c^\alpha_{\mu\nu}$ (as well as A)). The w^α_γ in turn determine the functions $v^\alpha_\gamma(x)$, and thence the multiplication operation $\psi(x, y)$ as the solution of the system (8) with the given initial conditions. (It is here that the assumption of analyticity of the $\psi^\alpha(x, y)$ enters the picture, via for instance the Cauchy–Kovalevskaja theorem on (existence and) uniqueness of solutions of certain systems of partial differential equations.) This completes the proof of the theorem. □

3.1.2. Corollary. *If the Lie algebra of a connected analytic Lie group G is commutative* (i.e. $[A, B] \equiv 0$), *then the group G is commutative* (i.e. *abelian*).

PROOF. Setting $c^{\alpha}_{\mu\nu} = 0$ in (13), yields that $v^{\alpha}_{\beta}(x) = \delta^{\alpha}_{\beta}$, and then from (8) with the initial condition $\psi(0, y) = y$, that $\psi(x, y) = x + y$ on some neighbourhood. The corollary then follows from the connectedness of G, since this implies that G is generated as a group by the elements in any (arbitrarily small) neighbourhood of the identity. □

3.1.3. Definition. A Lie algebra $L = \{\mathbb{R}^n, c^i_{jk}\}$ is said to be *simple* if it is non-commutative and has no proper ideals (i.e subspaces $I \neq L, 0$, for which $[I, L] \subset I$), and *semisimple* if $L = I_1 \oplus \cdots \oplus I_k$ where the I_j are ideals which are simple as Lie algebras. (It follows that these ideals are pairwise commuting, i.e. $[I_i, I_l] = 0$ for $i \neq l$). A Lie group is defined to be *simple* or *semisimple* according as its Lie algebra is respectively simple or semisimple. The *Killing form* on an arbitrary Lie algebra L is defined by

$$\langle A, B\rangle = -\operatorname{tr}(\operatorname{ad} A \operatorname{ad} B), \tag{14}$$

where the operator ad A on L is in turn defined by

$$u \mapsto [A, u], \qquad u \in L. \tag{15}$$

Earlier in this section we defined for each g in a Lie group G an automorphism Ad(g) of the Lie algebra of G (namely, that induced by the inner automorphism of G determined by its element g); it is thus natural to call the automorphism Ad(g) an *inner automorphism* of the Lie algebra.

3.1.4. Theorem. (i) *If the Lie algebra L of a Lie group G is simple, then the linear representation* Ad: $G \to GL(n, \mathbb{R})$ *is irreducible* (i.e. *L has no proper invariant subspaces under the group of inner automorphisms* Ad(G)).
(ii) *If the Killing form of a Lie algebra is positive definite then the Lie algebra is semisimple.*

PROOF. (i) Suppose on the contrary that the representation Ad is reducible, and let I be a proper invariant subspace of L invariant under Ad G. Let X, Y be any elements of L, I, respectively, and let $x(\tau)$, $y(t)$ be the one-parameter subgroups determined by the tangent vectors X, Y. The invariance of I means in particular that for all τ, the vector

$$\operatorname{Ad}(x(\tau))(Y) = \frac{d}{dt}(x(\tau)y(t)x(\tau)^{-1})|_{t=0}$$

lies again in I. We shall use canonical co-ordinates of the first kind, so that in some neighbourhood of 1 we have $x(\tau) = X\tau$, $y(t) = Yt$. From (1) applied twice in succession it follows easily that the αth component of $x(\tau)y(t)x(\tau)^{-1} = x(\tau)y(t)x(-\tau)$ is given by

$$Y^{\alpha}t + [X, Y]^{\alpha}t\tau - b^{\alpha}_{\beta\gamma}X^{\beta}X^{\gamma}\tau^2 - b^{\alpha}_{\beta\gamma}Y^{\beta}Y^{\gamma}t^2 + \text{higher-order terms in } t, \tau.$$

(From (1) and the fact that $\psi(x(\tau), x(-\tau)) = 0 = \psi(y(t), y(-t))$, it follows that the two negative terms vanish.) On differentiating with respect to t and then setting $t = 0$, we obtain for all sufficiently small τ that

$$\mathrm{Ad}(x(\tau))(Y) = Y + [X, Y]\tau + O(\tau^2).$$

It follows that

$$[X, Y]\tau + O(\tau^2) \in I.$$

Dividing by τ and letting $\tau \to 0$, we get finally that $[X, Y] \in I$ (since with respect to the Euclidean norm every subspace of a finite-dimensional vector space is closed). Hence $[L, I] \subset I$, so that I is an ideal of L, contradicting the assumed simplicity of L.

(ii) Let I be any ideal of the Lie algebra L, and let J be the orthogonal complement of I in L with respect to the Killing form (i.e. J is the subspace of all vectors in L orthogonal to I). We first prove that J is also an ideal of L.

To this end let X, Y, Z be arbitrary elements of L, J, I, respectively; we wish to show that $[X, Y]$ is orthogonal to Z, i.e. that $\mathrm{tr}(\mathrm{ad}[X, Y]\mathrm{ad}\, Z) = 0$. Now it is easily verified from the Jacobi identity that

$$\mathrm{ad}[A, B] = \mathrm{ad}\, A\, \mathrm{ad}\, B - \mathrm{ad}\, B\, \mathrm{ad}\, A.$$

Hence

$$\mathrm{tr}(\mathrm{ad}[X, Y]\, \mathrm{ad}\, Z) = \mathrm{tr}(\mathrm{ad}\, X\, \mathrm{ad}\, Y\, \mathrm{ad}\, Z - \mathrm{ad}\, Y\, \mathrm{ad}\, X\, \mathrm{ad}\, Z),$$

and since a trace of a matrix product is invariant under cyclic permutations of the factors, it follows that

$$\begin{aligned}\mathrm{tr}(\mathrm{ad}[X, Y]\, \mathrm{ad}\, Z) &= \mathrm{tr}(\mathrm{ad}\, X\, \mathrm{ad}\, Y\, \mathrm{ad}\, Z - \mathrm{ad}\, X\, \mathrm{ad}\, Z\, \mathrm{ad}\, Y)\\ &= \mathrm{tr}(\mathrm{ad}\, X\, \mathrm{ad}[Y, Z]).\end{aligned}$$

Since $[Y, Z] \in I$ and $X \in J$, the final expression above is zero, as required.

The positive definiteness of the Killing form implies both that $L = I \oplus J$, and that no non-zero ideals of L can be commutative (since the restriction of the Killing form to a commutative ideal is zero). This completes the proof. □

Remark. There is a stronger result than (ii), due to Killing and E. Cartan: *A Lie algebra is semisimple if and only if its Killing form is non-degenerate.* In addition to the above argument, the proof of this stronger result uses the fact that the Killing form of a (non-commutative) simple Lie algebra cannot be identically zero. This is in turn a consequence of a theorem of Engel which states that the Killing form of a Lie algebra L is identically zero if and only if the Lie algebra is "nilpotent"; i.e. if there exists a positive integer k such that

$$[[\ldots[A_1, A_2], \ldots], A_k] = 0$$

for all $A_1, \ldots, A_k \in L$.

EXERCISES

1. (i) Prove that the isometries of a connected Riemannian manifold form a Lie group.
   (ii) Prove the analogous result for the group of all conformal transformations of a Riemannian manifold (see §15 of Part I).

2. Decide which of the Lie algebras encountered in Part I (see especially §24) are simple or semisimple.

3.2. The Concept of a Linear Representation. An Example of a Non-matrix Lie Group

We begin with a definition:

3.2.1. Definition. A (*linear*) *representation* of a group G is a homomorphism $\rho: G \to GL(n, \mathbb{R})$ or $\rho: G \to GL(n, \mathbb{C})$, from G to a group of real or complex matrices. Given a representation ρ of G, the map $\chi_\rho: G \to \mathbb{R}$ (or $G \to \mathbb{C}$) defined by $\chi_\rho(g) = \operatorname{tr} \rho(g)$, $g \in G$, is called the *character* of the representation ρ. As noted above, a representation ρ of G is said to be *irreducible* if the vector space $\mathbb{R}^n$ (or $\mathbb{C}^n$) contains no proper subspaces invariant under the matrix group $\rho(G)$.

3.2.2. Theorem ("Schur's Lemma"). *Let $\rho_i: G \to GL(n_i, \mathbb{R})$, $i = 1, 2$, be two irreducible representations of a group G. If $A: \mathbb{R}^{n_1} \to \mathbb{R}^{n_2}$ is a linear transformation changing ρ_1 into ρ_2 (i.e. satisfying $A\rho_1(g) = \rho_2(g)A$), then either A is the zero transformation or else a bijection (in which case of course $n_1 = n_2$).*

PROOF. If x is an element of the kernel of A, i.e. $Ax = 0$, then for all $g \in G$

$$A\rho_1(g)x = \rho_2(g)Ax = 0,$$

whence the kernel of A is invariant under $\rho_1(G)$ and so by the irreducibility of ρ_1 must be either the whole of $\mathbb{R}^{n_1}$ (in which case $A = 0$) or else the null space. Similarly the image space $A(\mathbb{R}^{n_1}) \subset \mathbb{R}^{n_2}$ is invariant under $\rho_2(G)$, and must therefore either be the null space or the whole of $\mathbb{R}^{n_2}$. This completes the proof. □

Note that if G is a Lie group and we have a representation $\rho: G \to GL(N, \mathbb{R})$ which is a smooth map, then the differential (i.e. induced map) ρ_* is a linear map from the Lie algebra $\mathfrak{g} = T_{(1)}$ to the space of all $N \times N$ matrices:

$$\rho_*: \mathfrak{g} \to M(N, \mathbb{R}).$$

We leave it to the reader to verify that ρ_* is actually a *representation of the Lie algebra* $\mathfrak{g}$, i.e. that it is a Lie algebra homomorphism: as well as being linear, it preserves commutators:

$$\rho_*[\zeta, \eta] = [\rho_*\zeta, \rho_*\eta].$$

(We note that it can be shown that if ρ is continuous, then it will automatically be smooth.)

A representation $\rho: G \to GL(N, \mathbb{R})$ (or $G \to GL(N, \mathbb{C})$) is called *faithful* if it is one-to-one, i.e. if its kernel is trivial: $\rho(g) \neq 1$ unless $g = 1$. A matrix Lie group trivially has a faithful Lie representation (i.e. a representation which is also a topological equivalence). However, as we shall now show (by means of an example) not every Lie group can be realized (i.e. has a faithful Lie representation) as a matrix Lie group. As our example we take the group $G = \widetilde{SL}(2, \mathbb{R})$ consisting of all transformations of the real line of the form

$$x \mapsto x + 2\pi a + \frac{1}{i} \ln \frac{1 - ze^{-ix}}{1 - \bar{z}e^{ix}}, \tag{16}$$

where $x \in \mathbb{R}$, $a \in \mathbb{R}$, $z \in \mathbb{C}$, $|z| < 1$, and ln denotes the main branch of the natural logarithmic function, i.e. the continuous branch determined by $\ln 1 = 0$. (Note that in (16) the argument of the function ln is a fraction whose numerator and denominator are complex conjugates; hence the fraction has modulus 1, so that its natural logarithm is either zero or purely imaginary, and therefore the image of x in (16) is indeed real (in fact between $-\pi$ and π).)

It is not difficult to see that the group $\widetilde{SL}(2, \mathbb{R})$ is a connected 3-dimensional Lie group (with the obvious co-ordinates a and the real and imaginary parts of z). The subgroup isomorphic to $\mathbb{Z}$ consisting of those transformations (16) with $a \in \mathbb{Z}$ and $z = 0$, is easily seen to be central in the whole group $\widetilde{SL}(2, \mathbb{R})$ (i.e. each of its elements commutes with all elements). (We shall see below that in fact it coincides with the centre of $\widetilde{SL}(2, \mathbb{R})$.) Note also for later use that the transformations (16) with $a \in \mathbb{R}$ and $z = 0$ form a one-parameter subgroup of $\widetilde{SL}(2, \mathbb{R})$.

Each transformation (16) has the property that if $x \mapsto y$ under the transformation, then $x + 2\pi k \mapsto y + 2\pi k$ for all $k \in \mathbb{Z}$. Hence each such transformation yields a transformation $w = e^{ix} \mapsto e^{iy}$ of the unit circle $|w| = 1$. It is easily verified that the latter transformation has the explicit form

$$w \mapsto \frac{w - z}{1 - \bar{z}w} e^{2\pi i a} \tag{17}$$

If one conjugates this by the linear fractional transformation $z = i[(1 - w)/(1 + w)]$, which maps the unit circle (with one point removed) to the real line (and the interior of the unit circle to the open upper half-plane) then one finds that the group of such transformations is isomorphic to $SL(2, \mathbb{R})/\{\pm 1\}$. (Alternatively one may use the results of §13.2 of Part I to get that the group of transformations (17) is isomorphic to $SU(1, 1)/\{\pm 1\} \simeq SL(2, \mathbb{R})/\{\pm 1\}$.) We thus have a homomorphism from our group $\widetilde{SL}(2, \mathbb{R})$ onto $SL(2, \mathbb{R})$ with kernel the above central subgroup isomorphic to $\mathbb{Z}$. Since $SL(2, \mathbb{R})/\{\pm 1\}$ has trivial centre, it follows that the centre of $\widetilde{SL}(2, \mathbb{R})$ is precisely that infinite cyclic subgroup.

3.2.3. Theorem. *The group $\widetilde{SL}(2, \mathbb{R})$ has no faithful Lie representation.*

PROOF. While the above-mentioned one-parameter subgroup (consisting of the transformations (16) with $a \in \mathbb{R}$ arbitrary and $z = 0$) clearly has infinite intersection with the centre of $\widetilde{SL}(2, \mathbb{R})$ (which consists, as just shown, of the transformations (16) with $a \in Z$ and $z = 0$), it is obviously not contained in the centre. We shall show that this is incompatible with the existence of a faithful Lie representation of $\widetilde{SL}(2, \mathbb{R})$.

Thus suppose that there is a subgroup G of $GL(n, \mathbb{C})$ which is identical with $\widetilde{SL}(2, \mathbb{R})$ as far as its Lie group structure is concerned. Denote by H the one-parameter subgroup of G corresponding to the one-parameter subgroup of $\widetilde{SL}(2, \mathbb{R})$ just mentioned. It follows from §14.2 of Part I and §3.1 above that H has the form $\{\exp tA \mid t \in \mathbb{R}\}$, where A is some fixed $n \times n$ matrix. By conjugating, if need be, by a suitable matrix from $GL(n, \mathbb{C})$ we may bring A into its Jordan canonical form; hence we may assume that A is in block diagonal form, with blocks each of the type

$$\begin{pmatrix} \lambda & a_1 & & 0 \\ & \ddots & \ddots & \\ & & \ddots & a_k \\ 0 & & & \lambda \end{pmatrix}, \tag{18}$$

where $a_i = 0$ or 1. (We are supposing here that different blocks correspond to different λ, i.e. that the degree of each block is equal to the multiplicity of the eigenvalue λ.) The matrix $\exp(tA)$ (see §14.2 of Part I) will then also be in block diagonal form with blocks of the same size as those of A, and with the block corresponding to (18) having the form $e^{\lambda t} B_\lambda(t)$ where

$$B_\lambda(t) = \begin{pmatrix} 1 & a_1 t & \frac{1}{2} a_1 a_2 t^2 & \frac{1}{6} a_1 a_2 a_3 t^3 & \cdots & \frac{1}{k!} a_1 \dots a_k t^k \\ 0 & 1 & a_2 t & \frac{1}{2} a_2 a_3 t^2 & \cdots & \frac{1}{(k-1)!} a_2 \dots a_k t^{k-1} \\ 0 & 0 & 1 & a_3 t & \cdots & \frac{1}{(k-2)!} a_3 \dots a_k t^{k-2} \\ \cdots & \cdots & \cdots & \cdots & \cdots & \cdots \\ 0 & \cdots & \cdots & \cdots & 0 & 1 \end{pmatrix}.$$

Since for infinitely many t the matrix with blocks $e^{\lambda t} B_\lambda(t)$ lies in the centre of G, it follows that every element of G also has the same block diagonal form as A, i.e. has blocks of the same degree in the same order. The set P of all $n \times n$ matrices (including the singular ones) which commute with every element of G, clearly forms a linear subspace of the vector space $\mathbb{C}^{n^2}$ of all $n \times n$ matrices; the intersection $P \cap G$ is the centre of G. For reasons similar to before, every element of P again has the same block diagonal form as the matrices in G.

The condition that any given $n \times n$ matrix of that block diagonal form lie in P is equivalent to the condition that its image in the quotient space $\mathbb{C}^{n^2}/P$ be zero, and is therefore expressible as a homogeneous system of linear equations in the entries in the diagonal blocks of the given matrix, which can be so arranged that any single equation involves only those entries in a single block. Applying this to the matrix $\exp tA$ with blocks $e^{\lambda t}B_\lambda(t)$, we obtain, on multiplying each equation by $e^{-\lambda t}$ for the appropriate λ, a system of polynomial equations in t. Since such a system is satisfied by either all or else only finitely many values of t, this contradicts the fact that the group $H = \{\exp tA\}$ is not contained in the centre of G but has infinite intersection with it. □

EXERCISES

1. Calculate the Lie algebra of the Lie group $\widetilde{SL}(2, \mathbb{R})$.

2. Verify that the above-described group homomorphism $\widetilde{SL}(2, \mathbb{R}) \to SL(2, \mathbb{R})/\{\pm 1\}$ is a local isomorphism in some neighbourhood of the identity.

§4. Complex Manifolds

4.1. Definitions and Examples

We now introduce the concept of a complex manifold.

4.1.1. Definition. A *complex analytic manifold* of *complex dimension n* is a manifold M of dimension $2n$, for which the charts $U_q (M = \bigcup_q U_q)$ with their local co-ordinate systems $z_q^\alpha = x_q^\alpha + iy_q^\alpha$, $\alpha = 1, \ldots, n$, are identifiable with regions of n-dimensional complex space $\mathbb{C}^n$. It is further required that on each region of intersection $U_q \cap U_p$, the transition functions from the co-ordinates z_q^α to the co-ordinates z_p^α and in the reverse direction, be complex analytic (see §12.1 of Part I):

$$\frac{\partial z_q^\alpha}{\partial \bar{z}_p^\beta} \equiv 0; \qquad \frac{\partial z_p^\beta}{\partial \bar{z}_q^\alpha} \equiv 0. \tag{1}$$

We define a *holomorphic map* between complex manifolds to be one which is complex analytic (in terms of the given complex local co-ordinates on the manifolds). Holomorphic maps from a complex manifold to the complex line $\mathbb{C}$ will be called *analytic* or *holomorphic functions* on the manifold. A bijection between complex manifolds will be said to be *biholomorphic* if both it and its inverse are holomorphic. If two complex manifolds are such that there exists a biholomorphic map between them, we shall say that they are *biholomorphically equivalent* or *complex diffeomorphic*.

One important property of a complex manifold is that it always comes with an orientation.

4.1.2. Theorem. *A complex analytic manifold is oriented.*

Proof. Let M be a complex manifold and let $z_q^\alpha = x_q^\alpha + iy_q^\alpha$, $z_p^\alpha = x_p^\alpha + iy_p^\alpha$ be local co-ordinates on charts U_q, U_p, respectively. By Lemma 12.2.2 of Part I, on the region of overlap of each such pair of neighbourhoods, the (real) Jacobian of the transition function from the co-ordinates x_q^α, y_q^α to the co-ordinates x_p^α, y_p^α, satisfies

$$J^{\mathbb{R}} = |J^{\mathbb{C}}|^2 = \left|\det\left(\frac{\partial z_q^\alpha}{\partial z_p^\beta}\right)\right|^2.$$

Since such Jacobians are therefore all positive, the theorem follows. □

The complex projective spaces $\mathbb{C}P^n$, introduced in §2.2 above, provide examples of complex analytic manifolds. Complex local co-ordinates on $\mathbb{C}P^n$ are defined as in the real case; the transition functions, exemplified by formula (7) of §2.2 (with the x_i^j replaced by the complex co-ordinates z_i^j) are clearly complex analytic. The manifolds $\mathbb{C}P^n$ are compact (see Exercise 5 of §2.3). It follows from the discussion in §2.2 that the complex manifold $\mathbb{C}P^1$ is biholomorphic to the extended complex plane with complex local co-ordinates $w = 1/z$ in the neighbourhood of the point at infinity (and with $w = 0$ at ∞).

The simplest examples of complex manifolds are furnished by regions of $\mathbb{C}^n$. Further important examples are provided by the non-singular complex surfaces in $\mathbb{C}^n$. Such a manifold is defined by a system of equations

$$\left.\begin{array}{c} f_1(z^1, \ldots, z^n) = 0 \\ \cdots\cdots\cdots\cdots\cdots\cdots \\ f_{n-k}(z^1, \ldots, z^n) = 0 \end{array}\right\}, \tag{2}$$

where the functions $f_1, \ldots, f_{n-k}$ are all complex analytic, and at every point the rank of the matrix $(\partial f_i/\partial z^j)$ is largest possible (namely $n-k$). The verification that a non-singular complex surface in $\mathbb{C}^n$ is indeed a complex analytic manifold is carried out in a manner analogous to that of the real case (see the proof of Theorem 2.1.1), with the aid of results from §12 of Part I.

In contrast with the real case (see §9), compact complex analytic manifolds are not realizable as non-singular complex surfaces in some $\mathbb{C}^n$. This is a consequence of the following theorem.

4.1.3. Theorem. *A holomorphic function on a compact, connected complex manifold is necessarily constant.*

Proof. If $f: M \to \mathbb{C}$ is a holomorphic function on a compact, connected complex manifold M, then it follows by means of a well-known argument (using the continuity of $|f|$ and the compactness of M) that $|f|$ attains a largest value; i.e. there is a point P_0 of M such that $|f(P)| \leq |f(P_0)|$ for all points P of M. The constancy of f on M follows from the connectedness assumption, and the following basic result of complex function theory (to state and prove which, we now interrupt the present proof).

4.1.4. Lemma ("Maximum-Modulus Principle"). *Let f be a function holomorphic on some region U of n-dimensional complex space $\mathbb{C}^n$. If the function $|f|$ has a local maximum at some point P_0 of U, i.e. if $|f(P)| \leq |f(P_0)|$ for all points P of U sufficiently close to P_0, then f is constant in some neighbourhood of P_0.*

PROOF. Since the function $|f|$ will clearly have a local maximum at P_0 on any complex line through P_0, it suffices to prove the lemma for the case $n = 1$. We may also assume without loss of generality that $P_0 = 0$, and that $f(0) \neq 0$ (since if $f(0) = 0$, the assertion of the lemma is trivial). By multiplying the function f, if necessary, by an appropriate complex number we may further suppose that $f(0)$ is a positive real number.

In §26.3 of Part I we gave a proof of the "Residue Theorem" of complex function theory; the well-known "Cauchy integral formula" for holomorphic functions of a complex variable is an almost immediate corollary of that result:

$$f(0) = \frac{1}{2\pi i} \oint_\gamma \frac{f(z)\,dz}{z},$$

where γ is any circle enclosing the origin. Putting $z = re^{i\varphi}$ where r is any constant small enough to ensure that both γ and its interior are contained in U, this becomes

$$f(0) = \frac{1}{2\pi} \int_0^{2\pi} f(re^{i\varphi})\,d\varphi, \tag{3}$$

which formula obviously must also hold if in it f is replaced by $\operatorname{Re} f$ (or $\operatorname{Im} f$).

The function $g(z) = \operatorname{Re}(f(0) - f(z))$ is non-negative on some neighbourhood of the origin, since, by hypothesis, for all z sufficiently close to 0 we have $f(0) - |f(z)| \geq 0$, and since also $|\operatorname{Re} f(z)| \leq |f(z)|$. On the other hand since formula (3) continues to hold with f replaced by g, it follows that

$$\int_0^{2\pi} g(re^{i\varphi})\,d\varphi = 0$$

for all sufficiently small r. Hence throughout some neighbourhood of the origin $g(z)$ must be identically zero, i.e. $\operatorname{Re} f(z) \equiv f(0)$. Since $|f(z)| \leq f(0)$ on some neighbourhood of the origin, we deduce that $f(z) = f(0)$ on some such neighbourhood, as required. □

The proof of Theorem 4.1.3 is now completed as follows. Recall that $|f(P_0)|$ is the maximum modulus of $f: M \to \mathbb{C}$ on the compact, connected complex manifold M. If we denote by M' the set of all points P of M such that $f(P) = P_0$, then by the maximum-modulus principle, M' is open. The set M' has no boundary, since by continuity any hypothetical boundary point would have to lie in M', contradicting the fact that M' is open. Hence the complement of M' is also open, whence M is the union of two disjoint open sets. Since M is connected this is impossible unless M' is empty (which it certainly is not) or the whole space M. This completes the proof of the theorem. □

4.1.5. Corollary. *Any complex analytic submanifold of $\mathbb{C}^n$, of dimension greater than zero, is non-compact.*

PROOF. Suppose that M is a compact complex analytic manifold which can be embedded in $\mathbb{C}^n$ for some n and let $f: M \to \mathbb{C}^n$ be an holomorphic embedding. Then in view of the above theorem, on each connected component of M, each co-ordinate function f^i of f, being an analytic function on that connected component, is constant. Hence f maps each connected component of M to a single point, which proves the corollary. □

The classical complex transformation groups constitute important examples of non-singular complex surfaces:

(1) $GL(n, \mathbb{C})$, the set of all non-singular, complex, $n \times n$ matrices, is an open region of the space $\mathbb{C}^{n^2} = \mathbb{R}^{2n^2}$ of all complex matrices;
(2) $SL(n, \mathbb{C})$, the surface in $\mathbb{C}^{n^2}$ of all unimodular complex $n \times n$ matrices (i.e. of determinant 1).
(3) $O(n, \mathbb{C})$, the surface in $\mathbb{C}^{n^2}$ whose points comprise all complex orthogonal matrices, i.e. complex matrices A satisfying $AA^{\mathrm{T}} = 1$.

The non-singularity of these surfaces is verified much as it was for their real analogues (see §14.1 of Part I).

Each of these manifolds is a Lie group (see Definition 2.1.6). In fact the maps ψ and φ defining the group structure:

$$\psi: G \times G \to G, \qquad \psi(g, h) = gh;$$

$$\varphi: G \to G, \qquad \varphi(g) = g^{-1},$$

are everywhere complex analytic (i.e. holomorphic). Thus the above groups are examples of matrix "complex Lie groups".

4.1.6. Definition. A Lie group G which is a complex analytic manifold, is called a *complex Lie group* if the above maps ψ and φ are complex analytic.

4.1.7. Theorem. *Every compact, connected, complex Lie group G is commutative.*

PROOF. As usual we denote by $\mathfrak{g}$ the Lie algebra of the group G. It is not difficult to see that the adjoint representation Ad: $G \to GL(n, \mathbb{C})$ is a complex analytic map (between complex manifolds). Since G is compact and connected, Theorem 4.1.3 implies that Ad is a constant map; since Ad is a homomorphism, we must in fact have that $\mathrm{Ad}(g) = 1$ for all $g \in G$.

If $g(t)$ is any (smooth) curve in G passing through the identity element, with $\dot{g}(0) = X$ say, and if Y is any element of $\mathfrak{g}$, then, as we showed in the proof of part (i) of Theorem 3.1.4,

$$\mathrm{Ad}(g(t))(Y) = Y + t[X, Y] + O(t^2).$$

Since $\mathrm{Ad}(g(t))(Y) = Y$, we conclude that $[X, Y] = 0$ for all $X, Y \in \mathfrak{g}$, whence by Corollary 3.1.2, G is commutative, as required. □

It can be shown that in fact the only compact, connected, complex Lie groups are the "complex tori", which we shall now consider. As usual let $e_1, \ldots, e_{2n}$ denote the standard basis vectors in $\mathbb{R}^{2n} = \mathbb{C}^n$. (In fact for our purposes any basis for $\mathbb{R}^{2n}$ will serve.) The *complex torus* T^{2n} is defined to have as its points the equivalence classes of vectors, where two vectors are equivalent if they differ by an integral linear combination of the given basis vectors:

$$z \sim z + \sum_{\alpha=1}^{2n} n_\alpha e_\alpha, \qquad n_\alpha \in \mathbb{Z}.$$

(Such integral linear combinations form a subgroup Γ of $\mathbb{C}^n$ called the *integral lattice* determined by the given basis $e_1, \ldots, e_{2n}$.) Thus T^{2n} is the quotient group of $\mathbb{C}^n$ by Γ:

$$T^{2n} = \mathbb{C}^n/\Gamma.$$

Obviously two integral lattices Γ and Γ', determined by bases $e_1, \ldots, e_{2n}$ and $f_1, \ldots, f_{2n}$, respectively, coincide precisely if the vectors f_i lie in Γ and the e_j in Γ':

$$f_i = n_i^j e_j, \qquad e_j = m_j^i f_i.$$

Since the matrices (n_i^j) and (m_j^i) have integer entries and are mutual inverses, it follows that their determinants are both ± 1.

Thus far we have defined the group structure of the torus T^{2n}. We now endow it with its manifold structure by taking as local (complex) co-ordinate neighbourhoods the images of appropriately chosen open subsets of $\mathbb{C}^n$ under the natural map

$$\mathbb{C}^n \to T^{2n} = \mathbb{C}^n/\Gamma,$$

where these open subsets are chosen on the one hand sufficiently small for the restriction to each of the natural map to be one-to-one, and on the other hand so that their images cover T^{2n}. We leave to the reader the details of the verification that with this manifold structure and the above abelian group structure, T^{2n} is indeed a complex Lie group.

Functions on T^{2n} may obviously be regarded as $2n$-fold periodic functions on $\mathbb{C}^n$:

$$f\left(z+\sum_{\alpha=1}^{2n} n_\alpha e_\alpha\right)=f(z).$$

It follows from Theorem 4.1.3 that: *A holomorphic 2n-fold periodic function on $\mathbb{C}^n$ is constant.*

By way of an interesting example we consider the special case $n=1$. A complex torus T^2 is determined by a basis for $\mathbb{R}^2=\mathbb{C}$, i.e. by a pair of non-zero complex numbers z_1, z_2 such that $z_1 \notin \mathbb{R}z_2$. Multiplying by z_1^{-1} we obtain a pair of the form $(1, \tau)$ where $\tau(=z_2/z_1 \in \mathbb{C})$ is non-real (since 1 and τ are linearly independent over $\mathbb{R}$). Since, as it is easy to see, the multiplications of $\mathbb{C}$ by z_1 and z_1^{-1} induce holomorphic maps between the tori determined by the pairs (z_1, z_2) and $(1, \tau)$, it follows that those tori are biholomorphically equivalent. Hence each one-dimensional torus is determined, at least up to biholomorphic equivalence, by a complex number τ with non-zero imaginary part.

4.1.8. Lemma. *If τ and τ' are two non-real complex numbers related by a linear-fractional transformation of the form*

$$\tau'=\frac{m\tau+n}{p\tau+q},$$

where the matrix $\begin{pmatrix} m & n \\ p & q \end{pmatrix}$ is integral and has determinant ± 1, then the tori determined by τ and τ' are biholomorphically equivalent.

PROOF. In view of the conditions on the coefficients m, n, p, q, the integral lattices determined by the pairs of vectors $(1, \tau)$ and $(p\tau+q\cdot 1, m\tau+n\cdot 1)$ coincide. The lemma now follows from the fact that the second pair defines a torus which, by the remark preceding the lemma, is biholomorphically equivalent to that determined by $(1, \tau')$. □

Remark. It can be shown (using the theory of elliptic functions) that tori determined by complex numbers τ, τ' sufficiently close to one another, are not biholomorphically equivalent.

Regarded as merely a (2-dimensional) real manifold, the torus T^2 is diffeomorphic to the familiar 2-dimensional real torus $S^1 \times S^1$, where one of the two circles is obtained by identifying points on the straight line determined by 0 and z_1 which differ by an integral multiple of z_1, and the other circle is obtained by carrying out a similar identification of points on the line through 0 and z_2. Similarly, the torus T^{2n} is diffeomorphic to the $2n$-dimensional real torus $S^1 \times \cdots \times S^1$ ($2n$ factors).

Returning to the complex case, suppose we have a torus T^{2n} determined by a basis $e_1, \ldots, e_{2n}$ (not necessarily standard) for $\mathbb{R}^{2n}=\mathbb{C}^n$. Among these $2n$

vectors there will be n which are linearly independent over $\mathbb{C}$; by re-indexing if necessary, we may suppose that $e_1, \ldots, e_n$ are linearly independent over $\mathbb{C}$. If we express the remaining vectors $e_{n+1}, \ldots, e_{2n}$ in terms of the first n, say

$$e_{n+k} = \sum_{j=1}^{n} b_{kj} e_j, \qquad k = 1, \ldots, n,$$

we obtain a complex matrix $B = (b_{kj})$, which (as in the particular case $n = 1$ examined above) determines the torus up to a biholomorphic equivalence. It is easy to see that the imaginary part of the matrix B must be non-singular, since otherwise the vectors $e_1, \ldots, e_{2n}$ would be linearly dependent over $\mathbb{R}$.

4.1.9. Definition. A complex torus T^{2n} is said to be *abelian* if for some basis $e_1, \ldots, e_{2n}$ of its integral lattice, the above-defined matrix $B = (b_{kj})$ is symmetric and its imaginary part $H = (h_{kj}) = (\operatorname{Im} b_{kj})$ is positive definite; i.e. if

$$b_{jk} = b_{kj} \quad \text{and} \quad h_{kj} \xi^k \xi^j > 0,$$

for all non-zero real vectors $(\xi^1, \ldots, \xi^n)$.

For example, the one-dimensional complex torus determined (up to a biholomorphic equivalence—see above) by a complex number τ with $\operatorname{Im} \tau > 0$, is abelian; since the tori determined by τ and $-\tau$ clearly coincide, it follows that in fact all one-dimensional complex tori are abelian. However even for $n = 2$ non-abelian tori exist.

EXERCISE

Show that almost all 2-dimensional complex tori T^4 are non-abelian.

On an abelian torus the *Jacobi-Riemann θ-function* $\theta(z_1, \ldots, z_n)$ of n complex variables, is defined by

$$\theta(z_1, \ldots, z_n) = \sum_{m_1, \ldots, m_n} \exp i \left\{ \frac{1}{2} \sum_{j,k} b_{kj} m_k m_j + \sum_k m_k z_k \right\}, \tag{4}$$

where the summation is over all n-tuples $(m_1, \ldots, m_n)$ of integers. The condition that the imaginary part of the matrix $B = (b_{kj})$ be positive definite, guarantees convergence of the series.

4.2. Riemann Surfaces as Manifolds

A *Riemann surface* is defined (cf. §12.3 of Part I) as a non-singular surface in $\mathbb{C}^2$ given by an equation of the form

$$f(z, w) = 0, \tag{5}$$

where $f(z, w)$ is an analytic function of z and w (for instance a polynomial in z and w). The condition for non-singularity, which makes the surface a one-dimensional complex manifold (i.e. complex curve), is as follows (see §12.3 of Part I, especially Theorem 12.3.1):

$$\operatorname{grad}_{\mathbb{C}} f = \left(\frac{\partial f}{\partial z}, \frac{\partial f}{\partial w}\right) \neq 0.$$

If we solve equation (5) for w it may happen that we obtain a multi-valued function; for instance:

(i) if $f(z, w) = w^2 - P_n(z)$ where $P_n(z)$ is a polynomial without multiple roots (see 12.3.2 of Part I), we obtain the two-valued function $w = \sqrt{P_n(z)}$ (a "hyperelliptic" Riemann surface);
(ii) if $f(z, w) = e^w - z$, we obtain $w = \ln z = \ln|z| + i \arg z + 2\pi i n$.

(The geometric meaning of multi-valuedness of $w(z)$ is that (some of) the surfaces $z = \text{const.}$ meet the surface $f(z, w) = 0$ in more than one point.)

Consider the case where $f(z, w)$ is a polynomial of degree n in the variables z, w. On making the substitution $z = y^1/y^0$, $w = y^2/y^0$, we obtain

$$f(z, w) = \frac{1}{(y^0)^n} Q_n(y^0, y^1, y^2),$$

where Q_n is a homogeneous polynomial in three variables. This furnishes a device for re-realizing our surface $f(z, w) = 0$ in $\mathbb{C}^2$, as the surface in the projective space $\mathbb{C}P^2$ given by the equation

$$Q_n(y^0, y^1, y^2) = 0, \tag{6}$$

except that the points of the latter surface for which $y^0 = 0$ correspond to "points at infinity" on the original Riemann surface (5). The adjunction of these points at infinity has compactified our surface:

4.2.1. Lemma. *The Riemann surface in complex projective space $\mathbb{C}P^2$ defined by equation* (6), *is compact.*

PROOF. The set of zeros of Q_n is clearly a closed set in $\mathbb{C}P^2$. Since $\mathbb{C}P^2$ is compact, and any closed subset of a compact space is compact, the lemma follows. □

Thus the original Riemann surface $f(z, w) = 0$ gives rise to a compact 2-dimensional real manifold. What do these manifolds actually look like in the case where $f(z, w) = w^2 - P_n(z)$, i.e. when $f(z, w)$ is as in (a) above? We first examine cases of low degree, and from these infer a general result. (It turns out that the points at infinity on such a surface are singular points, so that they may not all appear on the manifolds which we are about to construct (as realizations of such surfaces), while those which do appear should strictly speaking be removed from the manifolds.)

Examples. (a) Let $f(z, w) = w^2 - z$; then $Q_2(y^0, y^1, y^2) = (y^2)^2 - y^1 y^0$. We join the points $z = 0$, $z = \infty$ in the domain of the (extended) multiple-valued function $w = \sqrt{z}$ by a line segment α (or to put it more vividly, we "cut" the Riemann sphere S^2, diffeomorphic to $\mathbb{C}P^1$, to obtain the sphere with the segment α removed, depicted in Figure 4). It is not difficult to see intuitively that the restriction of the (extended) surface $f(z, w) = 0$ to those z of S^2 off the slit α, consists of two disjoint connected components, each of which can be seen, by projecting onto the extended z-plane, to be diffeomorphic to the sphere S^2 with the line segment α removed (see Figure 5). (These connected pieces are called the "branches" of the multi-valued function.) At the points 0 and ∞ the (extended) function $w = \sqrt{z}$ is single-valued. The desired surface is obtained by identifying the boundary segment α_1 of the component denoted I in Figure 5, with the boundary segment β_2 of the region II, and the boundary segment β_1 of region I with the boundary segment α_2 of region II of the surface (as indicated in Figure 5). It is intuitively plausible that as a result of this cutting and pasting, we obtain a manifold diffeomorphic to S^2.

(b) We next consider the case $f(z, w) = w^2 - P_2(z)$, where $P_2(z)$ is a polynomial of degree 2 with simple roots $z = z_1$, $z = z_2$, $z_1 \neq z_2$. Join the points z_1, z_2 by a straight line segment on the z-plane. For z outside that line segment the surface $f(z, w) = 0$ falls into two disjoint connected parts. If we adjoin a point at infinity to each of these connected parts, they will be as shown in Example (a), with the difference that here $z_1 \neq \infty$ (see Figure 6). As in that example, on identifying the appropriate boundary segments ($\alpha_1 \sim \beta_2$ and $\beta_1 \sim \alpha_2$), we see that the Riemannian manifold is in this case also diffeomorphic to S^2 (with two points removed).

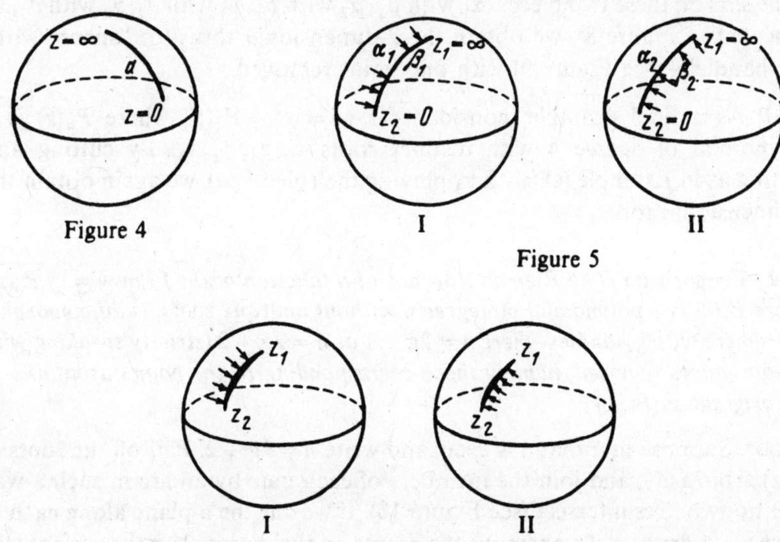

Figure 4

Figure 5

Figure 6

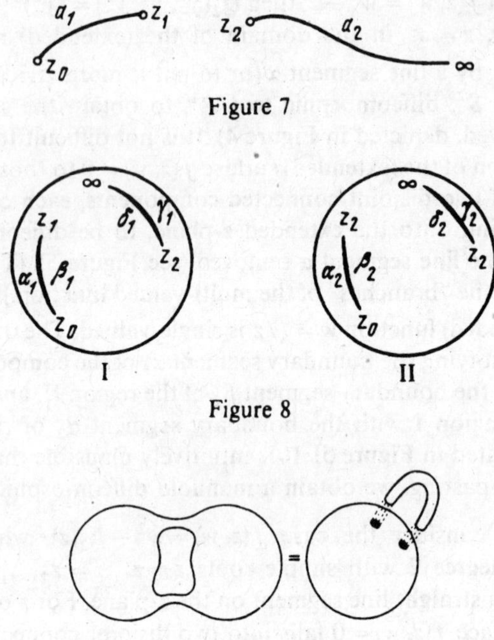

Figure 7

Figure 8

Figure 9

(c) Consider $f(z, w) = w^2 - P_3(z)$ where $P_3(z)$ is a polynomial of degree 3 with distinct roots z_0, z_1, z_2. Make cuts on S^2 as indicated in Figure 7; for z off these slits the extended surface $f(z, w) = 0$ again falls into two disjoint connected pieces, as shown in Figure 8. On identifying the appropriate edges of the slits on these two pieces (α_1 with β_2, α_2 with β_1, γ_1 with δ_2, γ_2 with δ_1, as indicated in Figure 8), we obtain the 2-dimensional torus (or "sphere-with-one-handle"—see Figure 9) with one point removed.

(d) As a final example, consider $f(z, w) = w^2 - P_4(z)$, where $P_4(z)$ is a polynomial of degree 4 with distinct roots z_0, z_1, z_2, z_3. By cutting and pasting as in Example (c) (with z_3 playing the role of ∞), we again obtain the 2-dimensional torus.

4.2.2. Proposition. *The Riemann surface of a function of the form $w = \sqrt{P_n(z)}$, where $P_n(z)$ is a polynomial of degree n without multiple roots, is diffeomorphic to a sphere with g handles where $n = 2g + 1$ or $n = 2g + 2$ (strictly speaking with certain points removed, namely those corresponding to the points at infinity of the original surface.)*

PROOF. Suppose first that n is even, and write $n = 2g + 2$. Pair off the roots of $P_n(z)$ arbitrarily, and join the members of each pair by an arc in such a way that no two arcs intersect (see Figure 10). If we cut the z-plane along each of these $g + 1$ arcs, i.e. if we remove the points on these arcs, then the surface falls into two disjoint connected parts U_1 and U_2. (If we move around any pair of

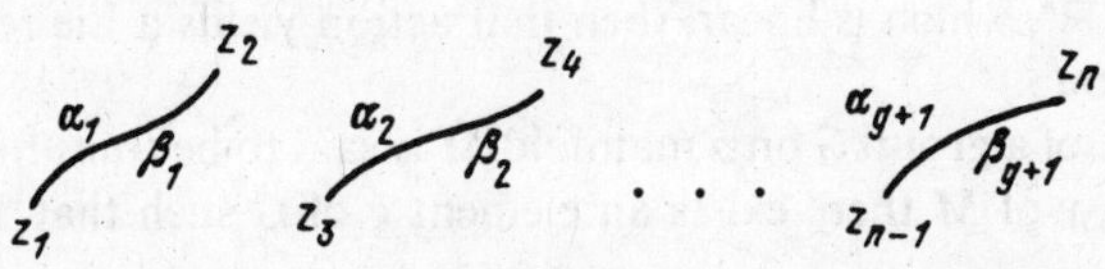

Figure 10

roots on the original surface, we stay on the same branch.) The edges α_i, β_i of the ith cut lie on (or rather are boundary segments of) different connected pieces U_1, U_2. We now glue these edges back together as follows:

$$(U_1, \alpha_i) \sim (U_2, \beta_i), \qquad (U_1, \beta_i) \sim (U_2, \alpha_i).$$

(This is justified by the fact that if on the original surface we move along the piece U_1 approaching the edge α_i, then on crossing it we pass smoothly over onto the branch U_2 (with corresponding edge β_i), and similarly if we approach on U_1 the edge β_i, we cross over onto U_2 (with corresponding edge α_i).) □

For odd n the construction is similar, with $z_{n+1} = \infty$ taken as the $(n+1)$st branch point.

§5. The Simplest Homogeneous Spaces

5.1. Action of a Group on a Manifold

We begin with the definition of such an action.

5.1.1. Definition. We say that a Lie group G (e.g. one of the matrix Lie groups considered in §14 of Part I) is *represented as a (Lie) group of transformations of a manifold* M (or *has a left (Lie)-action* on M) if there is associated with each of its elements g a diffeomorphism from M to itself

$$x \mapsto T_g(x), \qquad x \in M,$$

such that $T_{gh} = T_g T_h$ for all $g, h \in G$ (whence $T_1 = 1$), and if furthermore $T_g(x)$ depends smoothly on the arguments g, x (i.e. the map $(g, x) \mapsto T_g(x)$ should be a smooth map from $G \times M$ to M).

The Lie group G is said to have a *right action* on M if the above definition is valid with the property $T_g T_h = T_{gh}$ replaced by $T_g T_h = T_{hg}$.

If G is any of the Lie groups $GL(n, \mathbb{R})$, $O(n, \mathbb{R})$, $O(p, q)$, or $GL(n, \mathbb{C})$, $U(n)$, $U(p, q)$ (where $p + q = n$), then G acts in the obvious way on the manifold $\mathbb{R}^n$ or $\mathbb{R}^{2n} = \mathbb{C}^n$; moreover, in these cases the elements of G act as linear transformations. (Note that if, more generally, a Lie group has a Lie action on the manifold $\mathbb{R}^n$, which is linear, then that action yields a Lie representation of the Lie group.)

The action of a group G on a manifold M is said to be *transitive* if for every two points x, y of M there exists an element g of G such that $T_g(x) = y$.

5.1.2. Definition. A manifold on which a Lie group acts transitively is called a *homogeneous space* of the Lie group.

In particular, any Lie group G is a homogeneous space for itself under the action of left multiplication: $T_g(h) = gh$; in this context G is called the *principal (left) homogeneous space* (of itself). (The action $T_g(h) = hg^{-1}$ makes G into its own *principal right homogeneous space*.)

Let x be any point of a homogeneous space of a Lie group G. The *isotropy group* (or *stationary group*) H_x of the point x is the stabilizer of x under the action of G:

$$H_x = \{g \mid T_g(x) = x\}.$$

5.1.3. Lemma. *All isotropy groups H_x of points x of a homogeneous space, are isomorphic.*

PROOF. Let x, y be any two points of the homogeneous space and g be an element of the Lie group such that $T_g(x) = y$. It is then easy to check that the map $H_x \to H_y$ defined by $h \mapsto ghg^{-1}$ is an isomorphism (assuming a left action). □

5.1.4. Theorem. *There is a one-to-one correspondence between the points of a homogeneous space M of a group G, and the left cosets gH of H in G, where H is the isotropy group (and G is assumed to act on the left).*

PROOF. Let x_0 be any point of the manifold M. Then with each left coset gH_{x_0} we let correspond the point $T_g(x_0)$ of M. It is straightforward to verify that this correspondence is well defined (i.e. independent of the choice of representative of the coset), one-to-one, and onto. □

For right actions the analogous result holds with right cosets instead of left cosets.

Remark. It can be shown under certain general conditions that the isotropy group H is a closed subgroup of G, and that the set G/H of left cosets of H with the natural quotient topology can be given a unique (real) analytic manifold structure such that G is a Lie transformation group of G/H.

5.2. Examples of Homogeneous Spaces

(a) The group $O(n+1)$ clearly acts (in the natural way) of the sphere S^n (defined as the surface in Euclidean space $\mathbb{R}^{n+1}$ given by the equation $(x^1)^2 + \cdots + (x^{n+1})^2 = 1$). It is easy to see that this action is transitive, so that S^n is a homogeneous space for the Lie group $O(n+1)$ of orthogonal transformations of $\mathbb{R}^{n+1}$. The isotropy group of the point $x=(1,0,\ldots,0)\in S^n$ is comprised of all matrices of the form

$$\begin{pmatrix} 1 & 0 \\ 0 & A \end{pmatrix}, \qquad A \in O(n).$$

Hence by Theorem 5.1.4 above S^n can be identified with $O(n+1)/O(n)$ (where the quotient denotes merely the set of left cosets of the isotropy group, which is not normal in $O(n+1)$). In fact $S^n \cong O(n+1)/O(n)$, where $\cong$ denotes diffeomorphism (cf. the above remark).

The group $SO(n+1)$ is also transitive on S^n, and, analogously to the above, the isotropy group is isomorphic to $SO(n)$, so that we may identify S^n with $SO(n+1)/SO(n)$, and again $S^n \cong SO(n+1)/SO(n)$, as quotient space of $SO(n+1)$.

(b) From the definition of real projective space $\mathbb{R}P^n$ as consisting of the straight lines through the origin in $\mathbb{R}^{n+1}$, we obtain a transitive action of $O(n+1)$ on the manifold $\mathbb{R}P^n$. The subgroup of orthogonal transformations fixing the straight line through O with direction vector $(1,0,\ldots,0)$ is comprised of all matrices of the form

$$\begin{pmatrix} \pm 1 & 0 \\ 0 & A \end{pmatrix}, \qquad A \in O(n).$$

Hence the isotropy group is isomorphic to the direct product $O(1)\times O(n)$, and, again essentially by Theorem 5.1.4, we have

$$\mathbb{R}P^n \cong O(n+1)/O(1)\times O(n).$$

(c) The additive group of reals $\mathbb{R}$ acts (transitively) on the circle $S^1=\{e^{2\pi i\varphi}\}$ in the following way:

$$T_t(e^{2\pi i\varphi}) = e^{2\pi i(\varphi+t)}, \qquad t \in \mathbb{R}.$$

From the equality $e^{2\pi i}=1$ it follows that the isotropy group is exactly the group of integers.

More generally the group of all translations of $\mathbb{R}^n$ (which group it is natural to denote also by $\mathbb{R}^n$) acts transitively on the n-dimensional torus $T^n = (S^1)^n$, in the following way: if $y=(t_1,\ldots,t_n)\in\mathbb{R}^n$, and $z=(e^{2\pi i\varphi_1},\ldots,e^{2\pi i\varphi_n})$ is a point of the n-dimensional torus, define

$$T_y(z) = (e^{2\pi i(\varphi_1+t_1)},\ldots,e^{2\pi i(\varphi_n+t_n)}).$$

Clearly the isotropy group consists of all vectors y with integer components, i.e. the isotropy group of this homogeneous space is the integral lattice Γ of $\mathbb{R}^n$. Hence (cf. §4.1)

$$T^n \cong \mathbb{R}^n/\Gamma.$$

(d) *Stiefel manifolds.* For each n, k ($k \leq n$) the Stiefel manifold $V_{n,k}$ has as its points all orthonormal frames $x = (e_1, \ldots, e_k)$ of k vectors in Euclidean n-space (i.e. ordered sequences of k orthonormal vectors in Euclidean $\mathbb{R}^n$). Any orthogonal matrix A of degree n sends any such orthonormal frame x to another, namely $Ax = (Ae_1, \ldots, Ae_k)$; this defines an action of $O(n)$ on $V_{n,k}$ which is transitive. (Verify this!)

Each Stiefel manifold $V_{n,k}$ can be realized as a non-singular surface in the Euclidean space $\mathbb{R}^{nk}$ in the following way. Fix on an orthonormal basis for $\mathbb{R}^n$ (e.g. the standard basis), and introduce the following notation for the components with respect to this basis of any orthonormal k-frame $(e_1, \ldots, e_k)$ (i.e. point of $V_{n,k}$):

$$e_i = (x_{i1}, \ldots, x_{in}), \qquad i = 1, \ldots, k.$$

The nk quantities x_{ij}, $i = 1, \ldots, k; j = 1, \ldots, n$, (in lexicographic order, say) are now to be regarded as the co-ordinates of a point in nk-dimensional Euclidean space $\mathbb{R}^{nk}$, related by the following $k(k+1)/2$ equations:

$$\langle e_i, e_j \rangle = \delta_{ij} \Leftrightarrow \sum_{s=1}^{n} x_{is} x_{js} = \delta_{ij}, \qquad i, j = 1, \ldots, k, \quad i \leq j. \tag{1}$$

5.2.1. Lemma. *The Stiefel manifold $V_{n,k}$ is (embeddable as) a non-singular surface of dimension $nk - k(k+1)/2$ in $\mathbb{R}^{nk}$.*

PROOF. In view of the transitive action of the group $O(n)$ on $V_{n,k}$, it suffices to establish the non-singularity at any particular point. For convenience we choose the point $x_0 = (x_{ij})$ where $x_{ij} = \delta_{ij}$, $i = 1, \ldots, k; j = 1, \ldots, n$. Thus we wish to show that at x_0 the rank of the Jacobian matrix of the system of equations (1) is largest possible, namely $k(k+1)/2$, or, equivalently, that the tangent space at the point x_0, to the surface defined by that system, has dimension $nk - k(k+1)/2$.

To this end let $x_{ij} = x_{ij}(t)$ be a curve on the Stiefel manifold (as defined by (1)), passing through x_0 when $t = 0$:

$$\sum_{s=1}^{n} x_{is}(t) x_{js}(t) = \delta_{ij}, \qquad i, j = 1, \ldots, k;$$

$$x_{ij}(0) = \delta_{ij}, \qquad i = 1, \ldots, k; \quad j = 1, \ldots, n.$$

It follows that the components

$$\xi_{ij} = \frac{d}{dt} x_{ij}(t) \bigg|_{t=0}$$

of the velocity vector at the point x_0, satisfy

$$0 = \frac{d}{dt}\left(\sum_{s=1}^{n} x_{is}(t) x_{js}(t) \right)_{t=0} = \xi_{ij} + \xi_{ji}, \qquad i, j = 1, \ldots, k.$$

Hence the tangent space at the point x_0 to the surface $V_{n,k}$ consists of all nk-component vectors (ξ_{ij}), $i = 1, \ldots, k$; $j = 1, \ldots, n$, satisfying $\xi_{ij} = -\xi_{ji}$,

$i, j = 1, \ldots, k$. Since the dimension of this space is clearly $nk - k(k+1)/2$. the lemma is proved. □

Thus $V_{n,k}$ is indeed a smooth manifold. We now investigate the isotropy group of this homogeneous space. Take any orthonormal k-frame $e_1, \ldots, e_k$ and enlarge it to an orthonormal basis $e_1, \ldots, e_n$ for the whole of Euclidean n-space. Any orthogonal transformation fixing the vectors $e_1, \ldots, e_k$ must (relative to the above basis for $\mathbb{R}^n$) have the form

$$\begin{matrix} k\left\{ \vphantom{\begin{matrix}1\\ \\0\end{matrix}} \right. \\ n-k\{ \end{matrix} \begin{pmatrix} 1 & & 0 & \\ & \ddots & & 0 \\ 0 & & 1 & \\ & 0 & & A \end{pmatrix}, \qquad A \in O(n-k),$$

whence the isotropy group is isomorphic to $O(n, k)$, and $V_{n,k}$ can be identified with $O(n)/O(n-k)$. (In fact $V_{n,k} \cong O(n)/O(n-k)$.)

The Stiefel manifolds $V_{n,k}$ for $k < n$ are also homogeneous spaces for the group $SO(n)$. From this point of view the isotropy group is clearly (isomorphic to) $SO(n-k)$, and therefore also

$$V_{n,k} \cong SO(n)/SO(n-k).$$

In particular, we have

$$V_{n,n} \cong O(n), \qquad V_{n,n-1} \cong SO(n), \qquad V_{n,1} \cong S^{n-1}.$$

(e) *Grassmannian manifolds*. The points of the Grassmannian manifold $G_{n,k}$ are by definition the k-dimensional planes passing through the origin of n-dimensional Euclidean space. The usual action of the group $O(n)$ on $\mathbb{R}^n$ yields a transitive action of that group on the set of all k-dimensional planes through 0, i.e. on $G_{n,k}$. To find the (isomomorphism class of the) isotropy group, choose any k-dimensional plane π through 0, and then choose an orthonormal frame for $\mathbb{R}^n$ with its first k vectors in the plane π (whence the remaining $n-k$, however chosen, will be perpendicular to it). In terms of such a basis an orthogonal matrix fixing π (as a whole) will necessarily have the form

$$\begin{matrix} k \ \{ \\ n-k\{ \end{matrix} \begin{pmatrix} A & 0 \\ 0 & B \end{pmatrix}, \qquad A \in O(k), \qquad B \in O(n-k).$$

It follows that

$$G_{n,k} \cong O(n)/(O(k) \times O(n-k)).$$

Note finally that there is an obvious identification of the manifolds $G_{n,k}$ and $G_{n,n-k}$, and that, by its very definition $G_{n,1}$ is the same manifold as $\mathbb{R}P^{n-1}$.

(f) The following manifolds are homogeneous spaces for the unitary group $U(n)$:

(i) The odd-dimensional sphere S^{2n-1}, defined in n-dimensional complex space $\mathbb{C}^n$ by the equation

$$|z^1|^2 + \cdots + |z^n|^2 = 1.$$

It is not difficult to show that in this case

$$S^{2n-1} \cong U(n)/U(n-1) \cong SU(n)/SU(n-1).$$

(ii) The complex projective space $\mathbb{C}P^{n-1}$. In this case we have

$$\mathbb{C}P^{n-1} \cong U(n)/(U(1) \times U(n-1)).$$

(iii) The complex Grassmannian manifold $G^{\mathbb{C}}_{n,k}$ consisting of the k-dimensional complex planes in $\mathbb{C}^n$ passing through the origin. Here

$$G^{\mathbb{C}}_{n,k} \cong U(n)/(U(k) \times U(n-k)).$$

5.3. Exercises

1. Let M be a homogeneous space of a Lie group G, and let H be the isotropy group. Prove that the dimension of the manifold M is the difference in the dimensions of G and H:

$$\dim M = \dim G - \dim H.$$

Compute the dimension of the Grassmannian manifold $G_{n,k}$.

2. Prove the compactness of the manifolds $V_{n,k}$ and $G_{n,k}$.

3. Let $m = (m_1, \ldots, m_k)$ be a partition of the integer n, i.e.

$$n = m_1 + \cdots + m_k, \qquad m_i \geq 0.$$

A collection of linear subspaces $\pi_0, \pi_1, \ldots, \pi_k$ of the space $\mathbb{R}^n$ is called an *m-flag* if:

(i) $\dim \pi_i - \dim \pi_{i-1} = m_i$;
(ii) $\pi_0 = 0$, $\pi_k = \mathbb{R}^n$;
(iii) $\pi_{i-1} \subset \pi_i$.

Show how the totality of all *m-flags* $F(n, m)$ can be made to serve as a homogeneous space for the group $O(n)$, and calculate its isotropy group.

§6. Spaces of Constant Curvature (Symmetric Spaces)

6.1. The Concept of a Symmetric Space

Of great interest are those manifolds endowed with a metric g_{ab} whose curvature tensor (defined in §30.1 of Part I in terms of the connexion compatible with the metric) has identically zero covariant derivative (see §§28, 29 of Part I):

$$\nabla_s(R_{abcd}) = 0. \tag{1}$$

In any metrized space the components of the covariant derivative of the curvature tensor satisfy certain relations (namely Bianchi's identities—see Exercise 7, §30.5, Part I). However condition (1) places further severe restrictions on those components; thus, in particular, it follows from (1) that the scalar characteristics of the curvature are constant:

$$R = R^a_a = \text{const.}; \qquad R_{abcd}R^{abcd} = \text{const.}$$

It turns out also that under certain global conditions on a manifold, condition (1) implies the homogeneity of the metric g_{ab}; this is the case if the manifold is "simply connected" (see §17 below). Any manifold satisfying (1) is obtainable from some simply connected such manifold M as the quotient space (i.e. by identification) under some discrete group of motions. In this construction it can happen that the discrete group Γ in question is not central in the full isometry group of M, in which case the space M/Γ will not be homogeneous; such spaces are sometimes called "locally homogeneous" or "locally symmetric".

However our approach to symmetric spaces will be via the following definition.

6.1.1. Definition. A simply connected manifold M with a metric g_{ab} defined on it, is called a *symmetric space* (or *symmetric manifold*) if for every point x of M there exists an isometry (motion) s_x: $M \to M$ with the properties that x is an isolated fixed point of it, and that the induced map on the tangent space at x reflects (i.e. reverses) every tangent vector at x, i.e. $\xi \mapsto -\xi$. Such an isometry is called a *symmetry of M at the point x*.

The significance of the requirement in this definition that the manifold be simply connected will appear below (in §§17, 18). In the present section we shall not make use of the properties of simply connected manifolds; the reader not familiar with these properties may wish to attempt the appropriate exercises in §6.6 below, after he has studied the relevant sections of Chapter 4.

6.1.2. Lemma. *Every symmetric space satisfies condition* (1).

Proof. Let x be any particular point of the symmetric space M and let s_x be a symmetry of M at the point x. We can choose co-ordinates in some neighbourhood of x such that at x itself we have (see §29.3 of Part I)

$$x^\alpha = 0, \qquad g_{ab} = \delta_{ab}, \qquad \frac{\partial g_{ab}}{\partial x^\alpha} = 0.$$

(Here we are making the (inessential) assumption that the metric is Riemannian.)

Under the symmetry s_x, at the point x the tensor $\nabla_s R_{abcd}$ goes into its negative (and the tensors g_{ab} and R_{abcd} go into themselves) simply by virtue of the symmetry property of s_x and the transformation rule for tensors. On the other hand, since s_x is actually an isometry it is clear that $\nabla_s R_{abcd}$ must go into itself at x. The only way out is for $\nabla_s R_{abcd}$ to vanish at x, as required. □

Remark. The converse of this statement (under the assumption of simple-connectedness) is also true; however, in view of the greater technical complexity of its proof, we shall not prove it here. The following discussion may, however, be illuminating. In some neighbourhood of each point x of a Riemannian manifold we can define a *local symmetry* s_x as follows: Consider the pencil of geodesics on M passing through the point x (and assumed parametrized so that for each geodesic γ we have $\gamma(0)=x$); then for sufficiently small τ set

$$s_x(\gamma(\tau))=\gamma(-\tau).$$

(Recall (e.g. from §29.2 of Part I) that in a sufficiently small neighbourhood of x the geodesics through x will intersect nowhere else.)

EXERCISE

Prove that such local transformations s_x are for all x isometries of the manifold precisely if condition (1) holds on the manifold. (*Hint.* The simplest case is $n=2$, when the curvature tensor is determined by a single scalar R. For the general case it is easiest to deduce the preservation of the curvature tensor by the transformations s_x, from the preservation of Jacobi's equation along a geodesic.)

The existence of the "symmetries" s_x for all points x of a manifold M guarantees a sufficiently large supply of isometries for M to be (locally at least) homogeneous.

6.1.3. Lemma. *A symmetric space M is locally homogeneous; i.e. around each point $x\in M$, there is a neighbourhood such that for each point $\bar{x}$ in that neighbourhood (i.e. for each $\bar{x}$ sufficiently close to x) there exists an isometry of M sending x to $\bar{x}$. For each pair of points x, $y\in M$ which can be joined by a geodesic, there exists an isometry g of M such that $g(x)=y$.*

PROOF. The first statement of the lemma follows from the second since around every point x there is a neighbourhood with the property that every point $\bar{x}$ in that neighbourhood is joined to x by a geodesic, i.e. the geodesics through x sweep out the whole neighbourhood.

For the second statement, let γ be a geodesic arc joining the points x and y, and parametrized by the natural (length) parameter τ, with $0\le\tau\le T$, $\gamma(0)=x$, $\gamma(T)=y$. Let s_z be a symmetry of the manifold M at the point $z=\gamma(T/2)$. It follows from its symmetry property, together with the fact that it is an isometry of M, that s_z must send γ to γ, and therefore interchange x and y. (If the metric is pseudo-Riemannian of type $(1, n-1)$, and γ is an isotropic geodesic, then we may take as τ the "affine" (called also "natural" in Chapter 5 of Part I) parameter yielded in solving the equation for the geodesics (see §29.2 of Part I). □

Remark. Since any two points of a (connected) Riemannian manifold can be joined by a broken geodesic, it follows almost immediately that a symmetric Riemannian manifold is always homogeneous.

6.2. The Isometry Group of a Manifold. Properties of Its Lie Algebra

Henceforth in this section we shall consider only homogeneous, symmetric manifolds M (hence satisfying (1)), with metric g_{ab}. The Lie group of all isometries of M will be denoted by G, and the isotropy group by H, so that we may identify M with the set of left cosets of H in G (i.e. $M = G/H$, in the notation of the preceding section).

Let $\gamma = \gamma(\tau)$ be a geodesic on M parametrized by a natural parameter τ, and write $x_0 = \gamma(0)$. For each appropriate real T we define a map $f_{T,\gamma}: M \to M$ by setting

$$f_{T,\gamma} = s_{x_0}^{-1} s_x,$$

where $x = \gamma(-T/2)$ (see Figure 11). This map has the following three important properties:

(i) $f_{T,\gamma}$ moves each point of γ through a time-interval T along the geodesic (as indicated in Figure 11):

$$\gamma(T) \mapsto \gamma(\tau + T);$$

(ii) $f_{T,\gamma}$ parallel transports vectors along the geodesic;
(iii) for any fixed geodesic γ, the transformations $f_{T,\gamma}$ with T variable, form a one-parameter subgroup of the isometry group G:

$$f_{T_1+T_2,\gamma} = f_{T_1,\gamma} f_{T_2,\gamma},$$

$$f_{-T,\gamma} = (f_{T,\gamma})^{-1}.$$

From the last of these properties and §3.1 above, it follows that for each geodesic γ the one-parameter subgroup $f_{T,\gamma}$ of G has the form

$$f_{T,\gamma} = \exp(TB_\gamma),$$

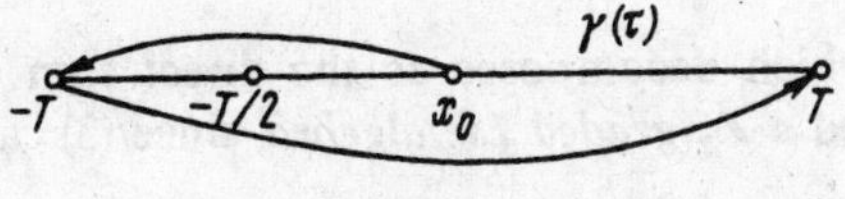

Figure 11

where B_γ is a certain vector in the Lie algebra $\mathfrak{g}$ of G (namely the tangent vector to the curve $f_{T,\gamma}$ at $T=0$). We denote by L^1 the linear subspace of the algebra $\mathfrak{g}$ spanned by the vectors $B_\gamma \in \mathfrak{g}$, where γ ranges over all geodesics through x_0, and by L^0 the Lie algebra of the isotropy group H_{x_0} of the point x_0. It follows (essentially from the fact that corresponding to each direction on the tangent plane to M at x_0 there is a geodesic through x_0 with that direction (see §29.2 of Part I)) that

$$\mathfrak{g} = L^0 + L^1 \quad \text{(direct sum of subspaces).} \tag{2}$$

If γ_1, γ_2 are two geodesics through the point x_0 then it can be shown that for small ε the product

$$f_{\varepsilon,\gamma_1} f_{\varepsilon,\gamma_2} f_{-\varepsilon,\gamma_1} f_{-\varepsilon,\gamma_2}$$

sends x_0 to a point whose distance from x_0 is of order ε^3. (This follows without difficulty from the properties of the Riemann curvature tensor. Verify!) We deduce from this (e.g. using formula (7) of §3.1) that we must have $[B_{\gamma_1}, B_{\gamma_2}] \in L_0$, whence $[L^1, L^1] \subset L^0$.

Suppose now that $g_T = \exp(TA)$ is a one-parameter subgroup of G leaving x_0 fixed (so that $A \in L_0$). Let γ be any geodesic through x_0, and denote by $\tilde{\gamma}$ the image of γ under the map g_ε. It then follows, again from the isometric property of the maps involved, that for small enough ε the map $g_\varepsilon f_{T,\gamma} g_{-\varepsilon}$ translates the points of the geodesic $\tilde{\gamma}$ along that geodesic (which of course also passes through x_0). Hence the tangent vector to the one-parameter subgroup $g_\varepsilon f_{T,\gamma} g_{-\varepsilon}$ (with parameter T), is in L_1. We now look for this tangent vector. From the basic facts about Lie algebras described in §3.1 it follows that for any two elements X, Y of the Lie algebra of a Lie group we have

$$\exp(tX)\exp(tY) = \exp\left(t(X+Y) + \frac{t^2}{2}[X, Y] + \text{higher-order terms}\right).$$

(This is a weak form of the "Campbell–Baker–Hausdorff formula".) It is an easy consequence of this that

$$\exp(tX)\exp(tY)\exp(-tX) = \exp(tY + t^2[X, Y] + \text{higher-order terms}).$$

Putting $t=1$, $X=\varepsilon A$, $Y = TB_\gamma$, we deduce that the desired tangent vector is $B_\gamma + \varepsilon[A, B_\gamma]$. Since $B_\gamma \in L^1$, it follows that $[A, B_\gamma] \in L^1$, whence $[L^0, L^1] \subset L^1$.

We include these facts in the following

6.2.1. Lemma. *With G and $\mathfrak{g} = L^0 + L^1$ as above, we have*

$$[L^0, L^0] \subset L^0, \qquad [L^1, L^0] \subset L^1, \qquad [L^1, L^1] \subset L^0. \tag{3}$$

A Lie algebra which decomposes as the direct sum of two subspaces satisfying (3) is called a *$\mathbb{Z}_2$-graded Lie algebra*, since (3) can be rewritten as

$$[L^i, L^j] \subset L^{(i+j) \bmod 2}. \tag{4}$$

6.2.2. Corollary. *With G as above, and* $\mathfrak{g} = L^0 + L^1$ *(direct) its Lie algebra, the linear operator*

$$\sigma\colon \mathfrak{g} \to \mathfrak{g}$$

whose restriction to L^0 *is the identity map* 1, *and whose restriction to* L^1 *is the reflection* -1, *is a Lie algebra automorphism (i.e. also preserves commutators). (The map* σ *is an "involution", i.e.* $\sigma^2 = 1$.)

(The converse is also true. To each involutary automorphism σ of a Lie algebra $\mathfrak{g}$, there corresponds a $\mathbb{Z}_2$-grading $\mathfrak{g} = L^0 + L^1$ (direct sum of spaces) of the Lie algebra, where L^0 is the set of elements fixed by σ and L^1 the set of elements which σ negates.)

In view of the homogeneity of the manifold M, the local geometry around any point x_0 is determined by a scalar product on the tangent space $\mathbb{R}^n_{x_0}$ at the point. (Here $n = \dim M$.) We now elicit a certain property (familiar from Part I) which this scalar product must have.

Note first that the tangent space $\mathbb{R}^n_{x_0}$ can be identified naturally with the space $L^1 \subset \mathfrak{g}$. Let A be any element of L^0, and consider the one-parameter subgroup $g_T = \exp(TA)$. For each T we have the map

$$\mathrm{Ad}(g_T)\colon \xi \mapsto \xi_T, \qquad \xi \in L^1.$$

As was shown in the course of proving part (i) of Theorem 3.1.4, we have

$$\xi_T = \mathrm{Ad}(g_T)(\xi) = \xi + T[A, \xi] + O(T^2),$$

whence

$$\left.\frac{d\xi_T}{dT}\right|_{T=0} = [A, \xi] = (\mathrm{ad}\, A)(\xi). \tag{5}$$

In view of the fact that g_T is an isometry of M the inner product on L^1 should be invariant under $\mathrm{Ad}(g_T)$, i.e.

$$\langle \xi_T, \eta_T \rangle = \langle \xi, \eta \rangle,$$

whence on differentiating with respect to T at $T = 0$, and using (5), we obtain

$$\langle [A, \xi], \eta \rangle + \langle \xi, [A, \eta] \rangle = 0. \tag{6}$$

This is the condition on the metric (i.e. scalar product) on $L^1 = \mathbb{R}^n_{x_0}$, that we were seeking. (Cf. the definition of a Killing metric in §24.4 of Part I.)

6.3. Symmetric Spaces of the First and Second Types

In the preceding subsection we obtained what might be called the algebraic model of a symmetric space. In principle all symmetric spaces can be classified (in the framework of the classification of compact Lie groups). In the present subsection we consider the most important examples of such spaces.

The simplest examples of (simply-connected) symmetric spaces are provided of course by the Euclidean spaces $\mathbb{R}^n$ and pseudo-Euclidean spaces $\mathbb{R}^n_{p,q}$ (where the curvature is zero). Here the group G is the full isometry group of $\mathbb{R}^n$ (or $\mathbb{R}^n_{p,q}$). (For the structure of some of these groups see §4 of Part I.) The isotropy group H is $O(n)$ (or $O(p, q)$), and the space $L^1 = \mathbb{R}^n$ consists of the translations. We have, as in the general case, $\mathfrak{g} = L^0 + L^1$, with the usual commutator relations between L^0 and L^1 (as in Lemma 6.2.1), and with the additional relation $[L^1, L^1] = 0$. Non-simply-connected symmetric spaces (earlier called "locally symmetric") can be obtained from these simply-connected ones by identification under various discrete groups Γ consisting of translations (and possibly also involving certain reflections, as for instance in the case of the Klein bottle—see §18 below).

In the remainder of our examples the group G will be semisimple. Recall from §3.1 that semisimplicity is equivalent to non-degeneracy of the Killing form on the Lie algebra $\mathfrak{g}$. Recall also that the Killing form $\langle\ ,\ \rangle$ on $\mathfrak{g}$ is defined by:

$$\langle A, B\rangle = -\operatorname{tr}(\operatorname{ad} A \operatorname{ad} B),$$

where $(\operatorname{ad} X)(\xi) = [X, \xi]$. (We shall also restrict G to being the connected component of the identity of the full isometry group.)

There are two distinct important types of simply-connected symmetric spaces with such G (even assuming the metric positive definite, i.e. Riemannian), namely:

Type I: the group G is compact and the Killing form on $\mathfrak{g}$ is positive definite. (We note here the result that a Lie group is compact if and only if it is a closed subgroup of some $O(m)$.)

Type II: the group G is non-compact and the Killing form on the Lie algebra $\mathfrak{g}$ is indefinite.

We consider the simplest (non-trivial) example of each type.

(a) The sphere S^2 is of type I. Here $G = SO(3)$, which is compact, and $H = SO(2)$.

(b) The Lobachevskian plane L^2 is of type II. Here G is the connected component of the identity of $SO(1, 2)$ (shown in §13.2 of Part I to be isomorphic to $SL(2, \mathbb{R})$), and $H = SO(2)$. The Lie algebra $\mathfrak{g}$ consists of all 2×2 matrices with zero trace, and the Killing form is given by

$$\langle A, B\rangle = -\operatorname{tr}(AB).$$

As a basis for the Lie algebra we may take

$$A_1 = \begin{pmatrix} 0 & 1 \\ 0 & 0 \end{pmatrix}, \quad A_2 = \begin{pmatrix} 0 & 0 \\ 1 & 0 \end{pmatrix}, \quad A_3 = \begin{pmatrix} -1 & 0 \\ 0 & 1 \end{pmatrix}.$$

We then have

$$A_1^2 = A_2^2 = 0, \qquad A_3^2 = 1,$$

and

$$\langle A_1, A_2 \rangle = -1, \quad \langle A_1, A_3 \rangle = \langle A_2, A_3 \rangle = 0,$$
$$\langle A_3, A_3 \rangle = -2, \quad \langle A_1, A_1 \rangle = \langle A_2, A_2 \rangle = 0.$$

It follows essentially from the fact that the matrices in $H = SO(2)$ have the form

$$\begin{pmatrix} \cos \varphi & \sin \varphi \\ -\sin \varphi & \cos \varphi \end{pmatrix},$$

that the subalgebra $L^0 \subset \mathfrak{g}$ is comprised of the matrices of the form $\lambda(A_1 - A_2)$. The subspace L^1 of $\mathfrak{g}$ is spanned by the vectors $A_1 + A_2, A_3$. It is easy to check that the inclusions (3) hold.

The restriction of the Killing form to the subspace L^1 is positive definite; this reflects the positive definiteness of the metric on the Lobachevskian plane.

EXERCISE
Investigate the general cases S^n and L^n.

6.4. Lie Groups as Symmetric Spaces

A Lie group Q endowed with a Riemannian metric invariant under left and right multiplications by group elements, can itself be regarded as a symmetric space. The isometry group of Q has a subgroup isomorphic to $Q \times Q$, whose action on Q is defined by

$$(g_1, g_2)\colon q \mapsto g_1 q g_2^{-1}.$$

The isotropy group of this action is clearly the diagonal subgroup $H = \{(q, q) | q \in Q\}$, which is isomorphic to Q: clearly $H(1) = 1$. For each q in Q the corresponding symmetry is defined by

$$s_q\colon x \mapsto q x^{-1} q.$$

(Verify that this does indeed define a symmetry.) In particular, $s_1(x) = x^{-1}$.

We shall examine in detail the case where Q is a compact connected subgroup of $SO(m)$, with the Euclidean metric

$$\langle A, B \rangle = \operatorname{tr}(AB^{\mathrm{T}}), \tag{7}$$

where B^{T} denotes the transpose of the matrix B. (Recall that it can be shown that a Lie group is compact if and only if it is a closed subgroup of some $O(m)$.)

EXERCISE
Show that the scalar product (7) is the Killing metric on $SO(m)$ determined by the Killing form on its Lie algebra (cf. §24.4 of Part I).

(The formula for the curvature of the Killing metric was derived in §30.3 of Part I. It follows from that formula that the Ricci tensor R_{ab} is positive definite. In the same subsection it was shown that the geodesics through the identity are precisely the one-parameter subgroups of Q.)

As noted in §24.4 of Part I, the group $SO(m)$ lies on the sphere of radius $\sqrt{m}$ (in the Euclidean space $\mathbb{R}^{m^2}$ of all $m \times n$ matrices with the metric (7)), since for $A \in SO(m)$, we have $AA^{\mathrm{T}} = 1$, whence $\langle A, A \rangle = m$; thus

$$SO(m) \subset S^{m^2-1}$$

6.4.1. Lemma. *The (Euclidean) scalar product (7) is invariant under right and left translations (i.e. multiplications) by elements of $SO(m)$.*

PROOF. Let $g \in SO(m)$, and let A, B be any $m \times m$ matrices. Then

$$\langle gA, gB \rangle = \mathrm{tr}(gAB^{\mathrm{T}}g^{\mathrm{T}}) = \mathrm{tr}(gAB^{\mathrm{T}}g^{-1}) = \mathrm{tr}(AB^{\mathrm{T}}) = \langle A, B \rangle,$$

and

$$\langle Ag, Bg \rangle = \mathrm{tr}(Agg^{\mathrm{T}}B^{\mathrm{T}}) = \mathrm{tr}(AB^{\mathrm{T}}) = \langle A, B \rangle,$$

whence the desired conclusion. □

6.4.2. Corollary. *The metric (7) restricted to any subgroup Q of $SO(m)$ is invariant under right and left multiplications $q \mapsto q_1 q q_2$.*

We call such a metric *bi-invariant* or *two-sided invariant*.

6.4.3. Lemma. *Every bi-invariant metric on a simple Lie group is proportional (with constant proportionality factor) to the Killing metric.*

PROOF. Let Q be a simple Lie group with bi-invariant metric $\langle\ ,\ \rangle$. The bi-invariance implies that for all elements A, B, C of the Lie algebra L of Q, and all $g_T = \exp(AT)$, we have

$$\langle \mathrm{Ad}(g_T)(B), \mathrm{Ad}(g_T)(C) \rangle = \langle B, C \rangle, \tag{8}$$

whence it follows, just as in the derivation of (6) above, that

$$\langle [A, B], C \rangle + \langle B, [A, C] \rangle = 0. \tag{9}$$

Now let $g_{ab}, \bar{g}_{ab}$ be two metrics on Q satisfying (8), (9). Then any linear combination $g_{ab} - \lambda \bar{g}_{ab}$ will also be Ad-invariant (or equivalently skew ad-invariant). Let λ_1 be any eigenvalue of the pair of quadratic forms $g_{ab}, \bar{g}_{ab}$, i.e. $\det(g_{ab} - \lambda_1 \bar{g}_{ab}) = 0$. (The symmetry of $g_{ab}, \bar{g}_{ab}$ implies that λ_1 is real.) The subspace R_{λ_1} of all eigenvectors corresponding to the eigenvalue λ_1 is easily seen (from (9)) to be a (non-zero) ideal of L, whence by the assumed simplicity of L, we must have $R_{\lambda_1} = L$. Hence $g_{ab} = \lambda_1 \bar{g}_{ab}$. Since the Killing metric satisfies (9), the desired result follows. □

6.4.4. Corollary. *Every simple subgroup Q of the group $SO(m)$ endowed with the Killing metric, can be isometrically embedded in the sphere S^{m^2-1} endowed with a metric proportional to the usual metric on the sphere.*

6.4.5. Corollary. *Since the Ricci tensor R_{ab} (determined by the Killing metric g_{ab} on a group) also satisfies (8), (9), it follows that for simple groups, $R_{ab} = \lambda g_{ab}$ where $\lambda = \text{const}$.*

Note finally that, as remarked above, it follows essentially from §30.3 of Part I that R_{ab} is positive definite for compact connected Lie groups. This is true also for semisimple groups, since each such group is (locally) a direct product $G_1 \times \cdots \times G_k$ of simples, and the sign of λ_i is easily determined for each of the simple factors G_i.

6.5. Constructing Symmetric Spaces. Examples

We now return to general symmetric spaces. In the notation of §6.2 above write

$$M = G/H, \qquad \mathfrak{g} = L^0 + L^1 \quad \text{(direct sum of subspaces)},$$

where M is a given symmetric space, G is its isometry group, L^0 is the Lie algebra of the isotropy group H, and L^1 is identifiable with the tangent space $\mathbb{R}^n_{x_0}$ to M at the point x_0 (fixed by H, i.e. $H = Hx_0$). Recall also that by virtue of the homogeneity of M, its metric is determined locally by a metric on L^1 satisfying (6); in what follows we shall assume the metric on M to be obtained from the Killing form on $\mathfrak{g}$ (see below).

6.5.1. Lemma. *The subspaces L^0 and L^1 of the Lie algebra $\mathfrak{g}$ are orthogonal with respect to the Killing form.*

PROOF. From Lemma 6.2.1, it is immediate that for all $A \in L^0$, $B \in L^1$ we have

$$\text{ad}\, A(L^0) \subset L^0, \qquad \text{ad}\, A(L^1) \subset L^1,$$
$$\text{ad}\, B(L^1) \subset L^0, \qquad \text{ad}\, B(L^0) \subset L^1.$$

It follows readily (using a basis of $\mathfrak{g}$ which is the union of bases for L^0 and L^1) that $\text{tr}(\text{ad}\, A\ \text{ad}\, B) = 0$, as required. □

We deduce at once from this that in terms of a basis for $\mathfrak{g}$ of the kind just mentioned, the Killing form on $\mathfrak{g}$ has the form

$$(g_{ab}) = \begin{pmatrix} g^{(0)}_{\alpha\beta} & 0 \\ 0 & g^{(1)}_{\gamma\delta} \end{pmatrix}, \tag{10}$$

where α, β range over the indices of the basis for L^0, and γ, δ over the indices of the basis for L^1. (The form $g^{(1)}_{\gamma\delta}$ is often called the *Killing form of the symmetric space M*.)

Since the Killing form (10) on $\mathfrak{g}$ satisfies (9), so also does the form $g^{(0)}_{\alpha\beta}$ on L^0. Hence by the proof of Lemma 6.4.3, if L^0 is simple then the form $g^{(0)}_{\alpha\beta}$ will be a constant multiple of the Killing form on the algebra L^0. However, in the important examples the algebra L^0 is not simple, but rather semisimple of the form $L^0 = L^0_1 \oplus L^0_2$ where L^0_1 and L^0_2 are simple. From Lemma 6.5.1 (with now L^0 in the role of $\mathfrak{g}$), we see that in this situation the restrictions of the form $g^{(0)}_{\alpha\beta}$ to the factors L^0_1, L^0_2 will be constant multiples (by λ_1, λ_2 say) of the Killing forms on those factors.

It can be seen that if H is compact and the metric $g^{(1)}_{\gamma\delta}$ is positive definite, then with respect to suitable bases for L^0 and L^1 the matrices ad A, $A \in L^0$, are skew-symmetric. Hence $\langle A, A\rangle = -\operatorname{tr}(\operatorname{ad} A)^2$ is positive, and therefore in view of

$$-\operatorname{tr}(\operatorname{ad} A)^2 = -[\operatorname{tr}(\operatorname{ad} A)^2_{L^0} + \operatorname{tr}(\operatorname{ad} A)^2_{L^1}],$$

it follows that

$$\langle A, A\rangle_{\mathfrak{g}} > \langle A, A\rangle_{L^0}. \tag{11}$$

Hence for compact H (and positive-definite metric on the symmetric space M) the restriction to L^0 (namely $g^0_{\alpha\beta}$) of the Killing form on the Lie algebra $\mathfrak{g}$, is positive definite. (Cf. the fact that the Killing form on the Lie algebra of a compact Lie group (e.g. on the Lie algebra L^0 of H) is non-negative.)

We see that in order to construct a symmetric space it essentially suffices to choose a suitable subalgebra L^0 of $\mathfrak{g}$ on which the restriction of the Killing form of the enveloping algebra $\mathfrak{g}$ is non-degenerate; then L^1 is defined as the orthogonal complement of L^0 in $\mathfrak{g}$. However the inequality (11) greatly restricts the choice of L^0: If the Killing form on $\mathfrak{g}$ is indefinite (type II) then for symmetric spaces with Riemannian metric the subalgebra $L^0 \subset \mathfrak{g}$ must be such that the restriction of the Killing form to its orthogonal complement is either positive or negative definite, and at the same time L^0 must be the Lie algebra of a compact group, and therefore of a subgroup of $SO(m)$.

Remark. A given symmetric space can be realized as a submanifold of the group G in such a way that the geodesics of M are geodesics also in the manifold G. This embedding can be obtained in any one of the following three (equivalent) ways:

(i) by considering all one-parameter subgroups of G emanating from the identity in the direction of vectors $B \in L^1$ (show that these geodesics sweep out a submanifold of G diffeomorphic to M);
(ii) via the map $\varphi: M \to G$, defined by $\varphi(x) = s_{x_0}^{-1} s_x$ (where s_{x_0}, s_x are the appropriate symmetries);
(iii) by means of an "involution" $\bar{\sigma}: G \to G$ (by which we mean an anti-automorphism of the group $(\bar{\sigma}(g_1 g_2) = \bar{\sigma}(g_2)\bar{\sigma}(g_1))$ such that the map

induced on $\mathfrak{g}$ is the identity map on L^0 and the negative of the identity map on L^1); $M \subset G$ is then the image under the map $g \mapsto g\bar{\sigma}(g^{-1})$.

Exercise

Show the equivalence of these embeddings.

What follows is a list of the basic examples of connected symmetric spaces of type I. (As an exercise, find in each case the corresponding direct decomposition $\mathfrak{g} = L^0 + L^1$.)

(1) $SO(2n)/U(n)$.
(2) $SU(n)/SO(n)$.
(3) $SU(2n)/Sp(n)$.
(4) $Sp(n)/U(n)$.
(5) $SO(p+q)/(SO(p) \times SO(q))$.
(6) $SU(p+q)/(SU(p) \times U(q))$.
(7) $Sp(p+q)/(Sp(p) \times Sp(q))$.

(5)–(7): Grassmannian manifolds (including the projective spaces and spheres).

The following are examples of symmetric spaces of type II (with positive-definite metric). (The simply-connected ones among such spaces turn out to have the topology of Euclidean space $\mathbb{R}^n$.)

(1) $SO(p, q)/(SO(p) \times SO(q))$. (For $q = 1$ this is the Lobachevsky space L^p.)
(2) $SU(p, q)/(U(p) \times SU(q))$. (For $q = 1$ this is the unit ball in $\mathbb{C}^p$, as a complex manifold; if also $p = 1$, this manifold is identifiable with $L^2 \cong SU(1, 1)/U(1)$.)
(3) $Sp(p, q)/(Sp(p) \times Sp(q))$.
(4) $SL(n, \mathbb{R})/SO(n)$.
(5) $SL(n, \mathbb{C})/SU(n)$.
(6) $SO(n, \mathbb{C})/SO(n, \mathbb{R})$.

We conclude with a list of symmetric spaces of dimension 4 with metric of signature $(+ - - -)$. (These spaces are of potential importance for the general theory of relativity since (by Corollary 6.4.5) the metric g_{ab} satisfies the equation $R_{ab} - \lambda g_{ab} = 0$ (see §37.4 of Part I).

I. Spaces of constant curvature with isotropy group $H = SO(1, 3)$:

(1) Minkowski space $\mathbb{R}^4_{1,3}$.
(2) The de Sitter space $S_+ = SO(1, 4)/SO(1, 3)$; note that S_+ is homeomorphic to $\mathbb{R} \times S^3$. Here the curvature tensor R is the identity operator on the space of bivectors $\Lambda^2(\mathbb{R}^4)$: $R = 1$.
(3) The de Sitter space $S_- = SO(2, 3)/SO(1, 3)$; this space is homeomorphic to $S^1 \times \mathbb{R}^3$, and its "universal covering space" $\tilde{S}_- = \widetilde{SO}(2, 3)/\widetilde{SO}(1, 3)$ (see §18) is homeomorphic to $\mathbb{R}^4$. Here the curvature tensor $R = -1$.

II. Reducible spaces (products of spaces of constant curvature):

(1) $H = SO(3)$; $M = \mathbb{R}_+ \times M^3_{---}$, where M^3_{---} is a space of constant curvature, with signature $(- - -)$.

(2) $H = SO(1, 2)$; $M = \mathbb{R}_- \times M^3_{+--}$, where M^3_{+--} is a space of constant curvature with signature $(+ - -)$.

(3) $H = SO(2) \times SO(1, 1)$; $M = M^2_{--} \times M^2_{+-}$, the product of two 2-dimensional spaces of constant curvatures.

III. The symmetric spaces M_t of plane waves. (For these the isotropy group is abelian, and the isometry group is soluble.) In terms of a certain system of global co-ordinates the metric has the form

$$dl^2 = 2\,dx_1\,dx_4 + \underbrace{[(\cos t)x_2^2 + (\sin t)x_3^2]}_{K}\,dx_4^2 + dx_2^2 + dx_3^2,$$

$$\cos t \geq \sin t.$$

In terms of the tetrad (see §30.1 of Part I) given by the 1-forms

$$p = dx_1, \qquad q = dx_1 + K\,dx_4, \qquad x = dx_3, \qquad y = dx_4,$$

the curvature tensor is constant, of the form

$$R = -4[\cos t(p \wedge x) \otimes (p \wedge x) + \sin t(p \wedge y) \otimes (p \wedge y)].$$

Remarks. 1. A simply connected symmetric space is uniquely determined by its curvature tensor at a point. To see this let $R: \Lambda^2(V) \to \Lambda^2(V)$ be the curvature tensor, and denote by $\mathfrak{h}$ the Lie algebra of skew-symmetric linear operators on the space V generated by those operators of the form $R(x, y)$, $x, y \in V$. (Then $\mathfrak{h}$ is the Lie algebra of the isotropy group (previously denoted by L^0).) Let $\mathfrak{g}$ denote the Lie algebra $V + \mathfrak{h}$, where the commutator operation on this direct sum of spaces is defined by

$$[(u, a), (v, b)] = (av - bu, [a, b] + R(u, v)).$$

Then in terms of the pair $\mathfrak{g}$, $\mathfrak{h}$ the structure of the original symmetric space is naturally reproduced on the symmetric space $M = G/H$.

2. The problem of classifying all curvature tensors of symmetric spaces with a given isotropy group H reduces to that of finding the H-invariant tensors R of the type of the curvature tensor, for which $R(x, y)$ belongs to the Lie algebra of H for all x, y in V.

6.6. Exercises

1. Show that for symmetric spaces of type II with positive-definite metric, the dimension of the subalgebra L^0 of $\mathfrak{g}$ is equal to the number of positive squares in the Killing form on $\mathfrak{g}$.

2. Show that in the complex case (e.g. where $G = SL(n, \mathbb{C})$ or $SO(n, \mathbb{C})$) the numbers of negative and positive square terms in the complex Lie algebra $\mathfrak{g}$ are equal, and $\dim L^0 = \frac{1}{2} \dim \mathfrak{g}$. Find the subalgebra L^0 of the Lie algebra $\mathfrak{g}$ of the group $G = SL(n, \mathbb{C})$.

3. Show that for symmetric spaces of type II with positive-definite metric one always has $M \cong G/H$, where H is a maximal compact subgroup of G. Investigate the particular cases $SL(n, \mathbb{R})/SO(n)$, $SL(n, \mathbb{C})/SU(n)$.

4. Show that a simply-connected, symmetric space of type II always has the topology of Euclidean $\mathbb{R}^n$.

For the next few exercises, note that, as for Lie groups (see §30.3 of Part I), so also for symmetric spaces do we have

$$\langle R(\xi, \eta)\zeta, \tau\rangle|_{x_0} = \tfrac{1}{4}\langle[\xi, \eta], [\zeta, \tau]\rangle_{L^0}, \qquad \xi, \eta, \zeta, \tau \in \mathbb{R}^n_{x_0} = L^1.$$

5. Show that for spaces of type I, the Ricci tensor R_{ab} is positive definite, and the "sectional curvature" $\langle R(\xi, \eta)\xi, \eta\rangle$ (where ξ, η span a parallelogram of unit area) is non-negative.

6. Show that for spaces of type II the sectional curvature is non-positive. Deduce from this that a simply-connected, symmetric space of type II is topologically the same as $\mathbb{R}^n$ (assuming the metric Riemannian).

7. Decide which of the 7 symmetric spaces of type I and 6 spaces of type II listed above have non-vanishing sectional curvature. Investigate the spaces S^n, $\mathbb{C}P^n$, $\mathbb{H}P^n$ of type I, and the spaces L^n, $SU(n, 1)/U(n)$, $SL(n, \mathbb{R})/SO(n)$, $SL(n, \mathbb{C})/SU(n)$ of type II.

8. Prove that in dimensions $n = 2, 3$ the only simply-connected, symmetric spaces with positive-definite metric are L^n, S^n, $\mathbb{R}^n$. (*Hint*. Show that the isotropy group $H \subset G$ must be $SO(n)$ ($n = 2, 3$), and thence deduce (for $n = 3$) the constancy of all sectional curvatures.)

9. Prove that a simply-connected symmetric space M with semisimple $G = G_1 \times \cdots \times G_k$ (where the G_i are simple) has the form
$$M = (G_1/H_1) \times \cdots \times (G_k/H_k),$$
with the metric decomposing as a direct product of metrics on the factors $M_i = G_i/H_i$, each of which is proportional to the Killing metric on the subspace L_i^1 of the corresponding Lie algebra $\mathfrak{g}_i = L_i^0 + L_i^1$.

§7. Vector Bundles on a Manifold

7.1. Constructions Involving Tangent Vectors

From any n-dimensional manifold M we can construct a $2n$-dimensional manifold, called the *tangent bundle* $L(M)$ of M as follows. The points of the manifold $L(M)$ are defined to be the pairs (x, ξ) where x ranges over the points of M and ξ ranges over the tangent space to M at x. Local co-ordinates are introduced on $L(M)$ in the following way. Let U_q be a chart of M with local co-ordinates x_q^α. Then in terms of the usual standard basis $e_\alpha = \partial/\partial x_q^\alpha$ (of operators on real-valued functions on M), any vector ξ in the tangent space to M at a point x of U_q can be written in terms of components as $\xi = \xi_q^\alpha e_\alpha$. As a typical chart U_q^L of $L(M)$ we take the set of all pairs (x, ξ) where x ranges over U_q, with local co-ordinates

$$(y_q^1, \dots, y_q^{2n}) = (x_q^1, \dots, x_q^n, \xi_q^1, \dots, \xi_q^n).$$

The transition functions on the region of intersection of two charts U_q^L and U_p^L (with co-ordinates x_p^β) are then of the form

$$(y_p^1, \dots, y_p^{2n}) = (x_p^\beta, \xi_p^\beta) = \left(x_p^\beta(x_q^1, \dots, x_q^n), \frac{\partial x_p^\beta}{\partial x_q^\alpha} \xi_q^\alpha \right).$$

The Jacobian matrix of such a transition function is then clearly

$$\left(\frac{\partial y_p^i}{\partial y_q^j}\right) = \begin{pmatrix} A & 0 \\ H & A \end{pmatrix}, \qquad A = \left(\frac{\partial x_p^\beta}{\partial x_q^\alpha}\right), \qquad H = \left(\frac{\partial^2 x_p^\beta}{\partial x_q^\alpha \partial x_q^\gamma} \xi_q^\alpha\right),$$

whence the Jacobian is $(\det A)^2 > 0$. This gives immediately the

7.1.1. Proposition. *The tangent bundle $L(M)$ is a smooth, oriented $2n$-dimensional manifold.*

Note by way of an example that the tangent bundle on a region U of Euclidean space $\mathbb{R}^n$ is diffeomorphic to the direct product $U \times \mathbb{R}^n$.

If the manifold M comes with a Riemannian metric, then we can delineate in $L(M)$ a submanifold, the *unit tangent bundle* $L_1(M)$, consisting of those points (x, ξ) with $|\xi| = 1$. The dimension of L_1 is $2n - 1$. (It is defined in $L(M)$ by the single non-singular equation $f(x, \xi) = g_{\alpha\beta}\xi^\alpha\xi^\beta = 1$.)

Example. Each tangent vector ξ at a point of the n-sphere S^n (defined in Euclidean space $\mathbb{R}^{n+1}$ by the equation $\sum_{\alpha=0}^n (x^\alpha)^2 = 1$) is perpendicular to the radius vector x from the origin to the point x. Hence in the case $M = S^n$, the unit tangent bundle $L_1(M)$ of pairs (x, ξ) with $|\xi| = 1$, is (intuitively) identifiable with the Stiefel manifold $V_{n+1,2}$ (see §5.2). In particular for $n = 2$, the unit tangent bundle $L_1(S^2)$ is identifiable with $V_{3,2} \cong SO(3)$ (which is in turn diffeomorphic to $\mathbb{R}P^3$—see §2.2).

A smooth map $f: M \to N$ from a manifold M to a manifold N, determines a smooth map of the corresponding tangent bundles:

$$L(M) \to L(N), \qquad (x, \xi) \mapsto (f(x), f_*\xi),$$

where f_* is the induced map of the tangent spaces (see §1.2).

We note briefly a few other constructions similar to that of the tangent bundle.

(i) One often meets with the manifold $L_p(M)$ whose points are the pairs (x, τ) where τ ranges over the straight lines through the origin, in the tangent space $\mathbb{R}^n$ to M at the point $x \in M$.
(ii) Given any n-dimensional manifold M, we may construct from it the *tangent n-frame bundle* $E = E(M)$ having as points the pairs (x, τ) with $x \in M$ and $\tau = (\xi_1, \ldots, \xi_n)$ any ordered basis (i.e. frame) for the tangent space to M at x.
(iii) If M is oriented then $\tilde{E} = \tilde{E}(M)$ is defined as in (ii) except that the frames τ are required to be in the orientation class determining the orientation of M.
(iv) If M is a Riemannian manifold, then $E_0 = E_0(M)$ is defined as in (ii) with the frames τ restricted to being orthogonal.

Further examples of such constructions will be considered in Chapter 6.

We now define the *cotangent bundle* $L^*(M)$ on a manifold M. The points of $L^*(M)$ are taken to be the pairs (x, p) where p is a covector (i.e. 1-form on M) at the point x. Local co-ordinates x_p^α on a chart U_p of M determine the local co-ordinates $(x_p^\alpha, p_{p\alpha})$ on the corresponding chart of $L^*(M)$, where the $p_{p\alpha}$ are defined by

$$p = p_{p\alpha}\, dx_p^\alpha$$

(i.e. they are the components of p with respect to the standard dual basis of 1-forms on U_p).

The transition functions from co-ordinates $(x_p^\alpha, p_{p\alpha})$ to co-ordinates $(x_q^\beta, p_{q\beta})$ on $U_p \cap U_q$ are as follows:

$$(x_q^\beta, p_{q\beta}) = \left(x_q^\beta(x_p^1, \ldots, x_p^n), \frac{\partial x_p^\alpha}{\partial x_q^\beta} p_{p\alpha} \right). \tag{1}$$

The Jacobian matrix is then

$$\begin{pmatrix} A^{-1} & 0 \\ \tilde{H} & A \end{pmatrix}, \qquad A = \left(\frac{\partial x_p^\alpha}{\partial x_q^\beta} \right), \qquad \tilde{H} = \left(\frac{\partial^2 x_p^\alpha}{\partial x_q^\beta \partial x_q^\gamma} p_{p\alpha} \right),$$

whence the Jacobian is 1, and the manifold $L^*(M)$ is oriented.

The existence of a metric $g_{\alpha\beta}$ on the manifold M gives rise to a map $L(M) \to L^*(M)$, defined by

$$(x^\alpha, \xi^\alpha) \mapsto (x^\alpha, g_{\alpha\beta}(x)\xi^\beta),$$

i.e. by means of the tensor operation of lowering of indices (see §19.1 of Part I).

Since the expression $\omega = p_\alpha\, dx^\alpha$ (a differential form on M) is invariant under transformations of the form (1), it can be regarded as a differential form on $L^*(M)$. Its differential $\Omega = d\omega = \sum_{\alpha=1}^n dp_\alpha \wedge dx^\alpha$ (see §25.2 of Part I) is a non-degenerate (skew-symmetric) 2-form on $L^*(M)$, which is, obviously, closed, i.e. $d\Omega = 0$. We conclude that: *The manifold $L^*(M)$ is symplectic.* (Recall that in Part I we defined a symplectic space to be one equipped with a closed (skew-symmetric) 2-form.)

7.2. The Normal Vector Bundle on a Submanifold

Let M be an n-dimensional Riemannian manifold with metric $g_{\alpha\beta}$, and let N be a smooth k-dimensional submanifold of M. The *normal (vector) bundle* $\nu_M(N)$ on the submanifold N in M, is defined to have as its points the pairs (x, v) where x ranges over the points of N, and v is a vector tangent to M at the point x, and orthogonal to N at x (i.e. orthogonal to the tangent space to N at x, which is a subspace of the tangent space to M at x). Assuming (as always—see §1.3) that the submanifold N is defined by a non-singular system of $(n-k)$ equations, then (as noted in §1.3) in terms of suitable local co-ordinates $y^1, \ldots, y^n$ on M these equations take the simple form $y^{k+1}=0, \ldots, y^n=0$, and $y^1, \ldots, y^k$ serve as local co-ordinates on N. In terms of such local co-ordinates $y^1, \ldots, y^n$ on M, the normal bundle $\nu_M(N)$ is then determined as a submanifold of $L(M)$ by the system of equations

$$y^{k+1}=0, \ldots, y^n=0, \qquad g_{\alpha\beta}(y)v^\beta=0, \qquad \alpha=1, \ldots, k.$$

Since this system is non-singular (verify it!), it follows that $\nu_M(N)$ is an n-dimensional submanifold of $L(M)$.

Examples. (a) Let M be Euclidean n-space $\mathbb{R}^n$, and suppose N is defined by the non-singular system of $(n-k)$ equations

$$f_1(y)=0, \ldots, f_{n-k}(y)=0, \qquad y=(y^1, \ldots, y^n),$$

where $y^1, \ldots, y^n$ are Euclidean co-ordinates on $\mathbb{R}^n$. Then the vectors grad $f_1, \ldots,$ grad f_{n-k} are at each point of N perpendicular to the surface N and linearly independent (see §7.2 of Part I). Hence we see that $\nu_{\mathbb{R}^n}(N)$ has the structure of a direct product:

$$\nu_{\mathbb{R}^n}(N) \cong N \times \mathbb{R}^{n-k}.$$

More generally if N is defined as a submanifold of a manifold M by a non-singular system of equations

$$f_1(x)=0, \ldots, f_{n-k}(x)=0,$$

then at each point x of N the vector fields

$$e_i(x) = \text{grad } f_i(x) = g_{ij}\frac{\partial f_i}{\partial x^j}, \qquad i=1, \ldots, n-k,$$

are orthogonal to N and linearly independent, whence any vector normal to N at $x \in N$ has the form $v = v^i e_i(x)$. The correspondence $(x, v) \leftrightarrow (x, v^1, \ldots, v^k)$ is then a diffeomorphism:

$$\nu_M(N) \cong N \times \mathbb{R}^{n-k}.$$

An important special case arises from the consideration of a manifold A with boundary, defined by an inequality $f(x) \leq 0$ in M. Here N is the boundary ∂A of A defined by the single equation $f(x)=0$, and of dimension

$n-1$. The normal bundle to the boundary then decomposes as a direct product:

$$\nu_M(\partial A) = \partial A \times \mathbb{R}.$$

(b) Suppose $M = N \times N$ where N is a Riemannian manifold. A typical tangent vector to M at a point is then a pair (ξ, η) of tangent vectors to N. Define a Riemannian metric on M by setting

$$\langle(\xi_1, \eta_1), (\xi_2, \eta_2)\rangle = \langle\xi_1, \xi_2\rangle + \langle\eta_1, \eta_2\rangle.$$

Consider the diagonal $\Delta = \{(x, x) | x \in N\}$ of M; this is a submanifold of M manifestly identifiable with N. The tangent vectors to Δ at any point will have the form (ζ, ζ); hence a tangent vector $v = (\xi, \eta)$ will be perpendicular to Δ precisely if

$$0 = \langle(\zeta, \zeta), (\xi, \eta)\rangle = \langle\zeta, \xi + \eta\rangle$$

for all tangent vectors ζ to N. since this is possible if and only if $\xi = -\eta$, it follows that the vectors normal to the diagonal $\Delta \cong N$ have the form $v = (\xi, -\xi)$. Hence we conclude that:

$$\nu_{N \times N}(\Delta) \cong L(N).$$

(c) Let $\nu_M(N)$ be the normal bundle on the submanifold N of the Riemannian manifold M. We define a map h, the *geodesic map* from $\nu_M(N)$ to M as follows. Let (x, v) be any point of $\nu_M(N)$ and let $\gamma(t)$ be the geodesic of M emanating from x with initial velocity vector v; thus $\dot{\gamma}(0) = v$. Then define h by $h(x, v) = \gamma(1)$.

7.2.1. Lemma. *The Jacobian of the map h is non-zero at every point of $\nu_M(N)$ of the form* $(x, 0)$.

PROOF. We give the proof only for the case when M is the space $\mathbb{R}^n$ with the usual Euclidean metric, and N is a hypersurface in $\mathbb{R}^n$ given (locally) by parametric equations $x^i = x^i(u^1, \ldots, u^{n-1})$, $i = 1, \ldots, n$. Then as local co-ordinates for the points $(x, v) \in \nu_{\mathbb{R}^n}(N)$ we may take the n-tuples $(u^1, \ldots, u^{n-1}, t)$, where $x = x(u)$, $v = tn(u)$; here $n(u)$ is the unit normal to the surface N at the point $x(u)$. In terms of these co-ordinates the geodesic map h is clearly given by

$$h(u^1, \ldots, u^{n-1}, t) = x(u) + tn(u).$$

Hence its partial derivatives are as follows:

$$\frac{\partial h}{\partial u^i} = \frac{\partial x}{\partial u^i} + t\frac{\partial n}{\partial u^i}, \qquad \frac{\partial h}{\partial t} = n.$$

On putting $t = 0$ we obtain the non-singular Jacobian matrix $(\partial h/\partial u, \partial h/\partial t) = (\partial x/\partial u, n)$, whence the lemma. □

7.2.2. Corollary. *Suppose that the submanifold N is compact. Then the geodesic map h maps the region*

$$v_M^\varepsilon = \{(x, v) \mid |v| < \varepsilon\}$$

diffeomorphically onto some neighbourhood $U_\varepsilon(N)$ of N in M.

PROOF. In view of the preceding lemma the map h is a diffeomorphism on some neighbourhood of any point of $v_M(N)$ of the form $(x, 0)$. Since N is compact, some finitely many of these neighbourhoods cover the subset $(N, 0)$ of $v_M(N)$. Then the union of these finitely many neighbourhoods contains some ε-neighbourhood $v_M^\varepsilon(N)$ of $(N, 0)$, and on this neighbourhood h is diffeomorphic. □

Remark. Let $U_\varepsilon(N)$ be, as in the corollary, the (diffeomorphic) image of $v_M^\varepsilon(N)$ under h. Then emanating from each point x in $U_\varepsilon(N)$ there is a (locally unique) "perpendicular geodesic" arc γ to N. We shall call the length of this "perpendicular" the *distance* from $x \in U_\varepsilon(N)$ to the submanifold N, and denote it by $\rho(x, N)$. Clearly the function $\rho(x, N)$ depends smoothly on the points x of the region $U_\varepsilon(N)$ of M.

7.2.3. Theorem. *If M is a compact, two-sided hypersurface in Euclidean $\mathbb{R}^n$ (see §2.1), then M is given by a single non-singular equation $f(x) = 0$.*

PROOF. Let $\varphi(t)$ be a smooth function with graph something like that shown in Figure 12. Define a function $f: \mathbb{R}^n \to \mathbb{R}$ by:

$$f(x) = \begin{cases} \pm\varepsilon & \text{if } x \notin U_\varepsilon(M), \\ \varphi(\pm\rho(x, M)) & \text{if } x \in U_\varepsilon(M), \end{cases}$$

where $U_\varepsilon(M)$ is the region of M appearing in the corollary, and where the plus sign is taken if x lies in a particular one of the two disjoint connected regions comprising $\mathbb{R}^n - M$, and the minus sign if x is in the other. (It is here that we are using the two-sidedness of M in $\mathbb{R}^n$.) Then M is defined in $\mathbb{R}^n$ by the equation $f(x) = 0$. □

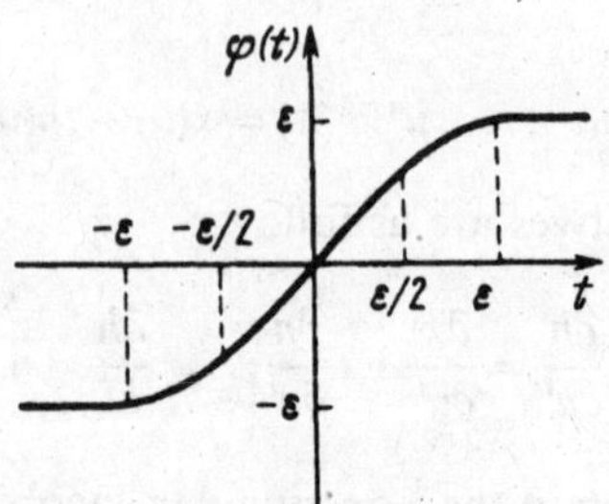

Figure 12

CHAPTER 2

Foundational Questions. Essential Facts Concerning Functions on a Manifold. Typical Smooth Mappings.

The present chapter is devoted to foundational questions in the theory of smooth manifolds. The proofs of the theorems will play no role whatever in the development of the basic topology and geometry of manifolds contained in succeeding chapters. Consequently in this chapter the reader may, if he wishes, acquaint himself with the definitions and statements of results only, without thereby sacrificing anything in the way of comprehension of the later material.

The subject matter of the chapter falls into two parts. In the first part "partitions of unity", so-called, are constructed, and then used in proving various "existence theorems" (which are in many concrete instances self-evident): the existence of Riemannian metrics and connexions on manifolds, the rigorous verification of the general Stokes formula, the existence of a smooth embedding of any compact manifold into a suitable Euclidean space, the approximability of continuous functions and mappings by smooth ones, and the definition of the operation of "group averaging" of a form or metric on a manifold with respect to a compact transformation group.

The second part, beginning with "Sard's theorem", is concerned with making precise ideas of the "typical" singularities of a function or mapping. This part will be found very useful in subsequent concrete topological constructions, so that the definitions and statements of results contained in it merit closer study.

§8. Partitions of Unity and Their Applications

We first introduce some notation. The space of all (real-valued) functions on a manifold M, with continuous partial derivatives of all orders, will be denoted by $C^\infty(M)$ (these will be our "smooth" functions); the supremum (i.e. least upper bound) of the values $f(x)$ taken by a function f will be denoted by $\sup f(x)$; and $\operatorname{supp} f$ will denote the support of f, i.e. the closure of the set of all points x at which $f(x) \neq 0$.

8.1. Partitions of Unity

We begin with a lemma concerning Euclidean space $\mathbb{R}^n$.

8.1.1. Lemma. *Let A, B be two non-intersecting, closed subsets of Euclidean space $\mathbb{R}^n$, with A bounded. Then there exists a C^∞-function φ on $\mathbb{R}^n$ such that $\varphi(x) \equiv 1$ on A and $\varphi(x) \equiv 0$ on B (see Figure 13). Moreover such a φ can be found satisfying $0 \leq \varphi(x) \leq 1$.*

PROOF. Let a, b be two real numbers with $0 < a < b$. It is easy to verify that the function on $\mathbb{R}^1$ defined by

$$f(x) = \begin{cases} \exp\left(\dfrac{1}{x-b} - \dfrac{1}{x-a}\right) & \text{for } a < x < b, \\ 0 & \text{for all other } x, \end{cases}$$

is smooth (i.e. is C^∞). (Verify it!) In terms of f we define a new smooth function F by

$$F(x) = \left(\int_x^b f(t)\,dt\right) \bigg/ \int_a^b f(t)\,dt.$$

It is readily seen that this smooth function F has the following properties:

$$F(x) \begin{cases} = 0 & \text{for } x \geq b, \\ = 1 & \text{for } x \leq a, \\ \text{decreases from 1 to 0} & \text{for } a \leq x \leq b. \end{cases}$$

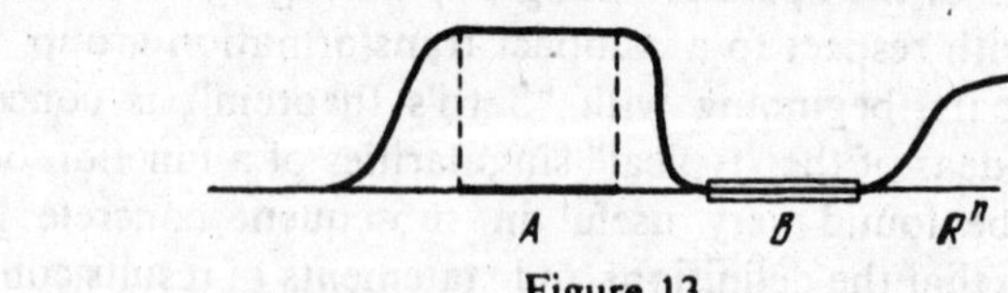

Figure 13

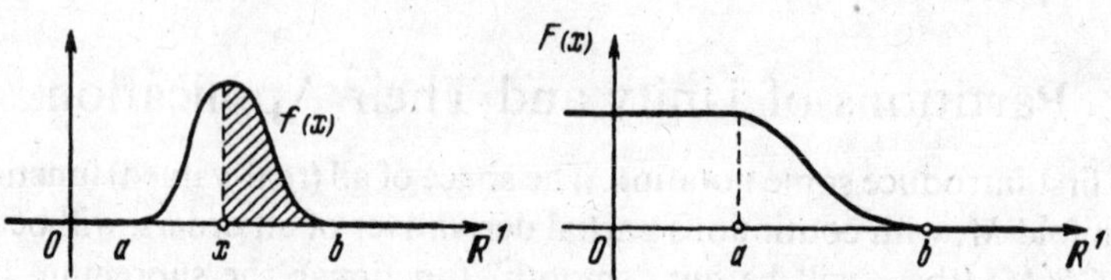

Figure 14

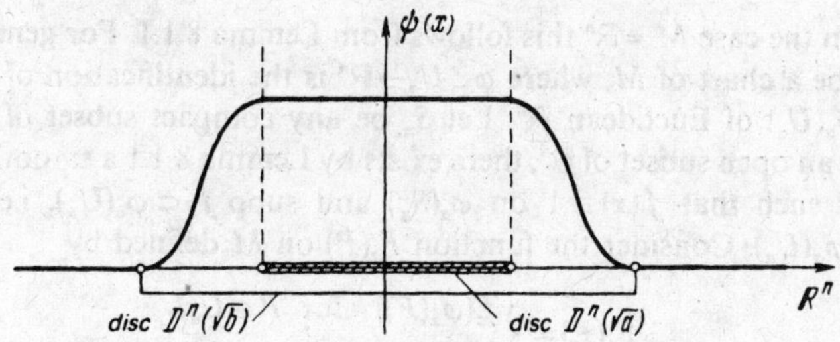

Figure 15

We next define a function ψ on $\mathbb{R}^n$, by the formula

$$\psi(x^1, \ldots, x^n) = F((x^1)^2 + \cdots + (x^n)^2) = F\left(\sum_{i=1}^{n} (x^i)^2\right).$$

It is again clear that ψ is a smooth function with the following properties (see Figure 15):

$$\psi(x) \begin{cases} = 0 & \text{for } r^2 \geq b, \\ = 1 & \text{for } r^2 \leq a, \\ \text{decreases from 1 to 0} & \text{for } a \leq r^2 \leq b. \end{cases}$$

(Here of course $r^2 = \sum_{i=1}^n (x^i)^2$.) We have thus shown that, given any two concentric spheres S and S' in $\mathbb{R}^n$, with S the larger, there exists a C^∞-function ψ which vanishes identically outside S, and is identically 1 on the ball bounded by S'.

Consider now the sets A, B (as in the lemma). Since A is compact, B closed, and $A \cap B = \varnothing$, there exists a finite collection of spheres S_i $(1 \leq i \leq m)$ such that the open balls D_i which they bound ($\partial \bar{D}_i = S_i$, where the bar denotes the closure operation), cover the set A (i.e. $A \subset \bigcup_{i=1}^m D_i$), and have the further property that $\bar{D}_i \cap B = \varnothing$ for all i. It is clear that for each i we can find a strictly smaller S_i' concentric with S_i such that the open balls D_i' which they bound still cover A (i.e. $A \subset \bigcup_{i=1}^m D_i'$). For each $i = 1, \ldots, m$, let ψ_i be a function in $C^\infty(\mathbb{R}^n)$ such that $0 \leq \psi_i(x) \leq 1$ and

$$\psi_i(x) = \begin{cases} 1 & \text{on } D_i', \\ 0 & \text{outside } D_i, \end{cases}$$

and set $\varphi(x) = 1 - \prod_{i=1}^m (1 - \psi_i(x))$. It is then immediate that $\varphi(x) \in C^\infty(\mathbb{R}^n)$, and that $\varphi(x) \equiv 1$ on A and $\varphi(x) \equiv 0$ on B, completing the proof. □

8.1.2. Lemma. *Let C be a compact subset of a smooth manifold M, and let V be any open subset of M containing C. Then there exists a function $\varphi \in C^\infty(M)$ such that $0 \leq \varphi(x) \leq 1$ on M, $\varphi(x) \equiv 1$ on C, and $\varphi(x) \equiv 0$ outside V.*

PROOF. In the case $M=\mathbb{R}^n$ this follows from Lemma 8.1.1. For general M, let $(U_\alpha, \varphi_\alpha)$ be a chart of M, where $\varphi_\alpha\colon U_\alpha\to\mathbb{R}^n$ is the identification of U_α with a region $\varphi_\alpha(U_\alpha)$ of Euclidean $\mathbb{R}^n$. Let S_α be any compact subset of U_α. Since $\varphi_\alpha(U_\alpha)$ is an open subset of $\mathbb{R}^n$, there exists by Lemma 8.1.1 a smooth function f_α on $\mathbb{R}^n$ such that $f(x)\equiv 1$ on $\varphi_\alpha(S_\alpha)$ and supp $f_\alpha\subset\varphi_\alpha(U_\alpha)$, i.e. $f_\alpha(x)\equiv 0$ outside $\varphi_\alpha(U_\alpha)$. Consider the function $F_\alpha(P)$ on M defined by

$$F_\alpha(P)=\begin{cases} f_\alpha(\varphi_\alpha(P)) & \text{for } P\in U_\alpha,\\ 0 & \text{for } P\notin U_\alpha.\end{cases}$$

Clearly $F_\alpha\in C^\infty(M)$, $F_\alpha(P)\equiv 1$ on S_α, and $F_\alpha(P)\equiv 0$ outside U_α.

We are now ready to turn our attention to the compact subset C of M contained in the open subset V (as in the lemma). In view of the compactness of C we can find a finite collection of (possibly new) local co-ordinate neighbourhoods $U_1,\ldots,U_N$ and compact subsets $S_1,\ldots,S_N$, such that

$$S_\alpha\subset U_\alpha,\qquad C\subset\bigcup_{\alpha=1}^{N}S_\alpha,\qquad \bigcup_{\alpha=1}^{N}U_\alpha\subset V.$$

By what we have just shown, for each $\alpha=1,\ldots,N$ there exists a function $F_\alpha\in C^\infty(M)$ such that $F_\alpha\equiv 1$ on S_α and $F_\alpha\equiv 0$ outside u_α. The function $F=1-\prod_{\alpha=1}^{N}(1-F_\alpha)$ then belongs to $C^\infty(M)$, is identically 1 on C, and vanishes outside $\bigcup_{\alpha=1}^{N}U_\alpha$, so that certainly $F(P)\equiv 0$ outside V. □

8.1.3. Theorem (Existence of "Partitions of Unity"). *Let M be a compact, smooth manifold and let $\{U_\alpha\}$ $(1\le\alpha\le N)$ be an arbitrary finite covering of M by local co-ordinate regions (for instance by open balls). Then there exists a family of functions $\varphi_\alpha\in C^\infty(M)$ with the following properties:*

(i) supp $\varphi_\alpha\subset U_\alpha$ *for all* α;
(ii) $0\le\varphi_\alpha(x)\le 1$ *for all* $x\in M$;
(iii) $\sum_\alpha\varphi_\alpha(x)\equiv 1$ *for all* $x\in M$.

PROOF. There always exists a "constricted" family of open sets V_α, $1\le\alpha\le N$, such that $\bar{V}_\alpha\subset U_\alpha$ and $\{V_\alpha\}$ still covers M. By Lemma 8.1.2 applied to each pair U_α, V_α, there exists a function $\psi_\alpha\in C^\infty(M)$ such that $0\le\psi_\alpha(x)\le 1$ on M, $\psi_\alpha(x)\equiv 1$ on $\bar{V}_\alpha$, and $\psi_\alpha(x)\equiv 0$ outside U_α. It is immediate that the function $\psi=\sum_{\alpha=1}^{N}\psi_\alpha$ belongs to $C^\infty(M)$ and is positive on M, i.e. $\psi(x)>0$ for all $x\in M$. If we take $\varphi_\alpha=\psi_\alpha/\psi$, then these φ_α satisfy the requirements of the theorem. This completes the proof. □

The family of functions φ_α is called a *partition of unity subordinate to the covering* $\{U_\alpha\}$.

Remark. The assumption that the manifold M be compact is not essential. It is readily seen that the proof of the existence of partitions of unity carries over to manifolds having suitable "locally finite" coverings (such a covering being one for which there is a neighbourhood of each point intersecting only finitely many regions of the covering). Recall that a Hausdorff topological space is called *paracompact* if every open covering has a locally finite refinement which covers the space. Thus the above proof of the existence of partitions of unity works more generally for any manifold which is paracompact.

8.2. The Simplest Applications of Partitions of Unity. Integrals Over a Manifold and the General Stokes Formula

The theorem on the existence of partitions of unity has useful consequences; we shall now consider some of these. For the sake of simplicity we shall assume throughout that the manifolds we deal with are compact.

8.2.1. Corollary. *On any compact manifold a Riemannian metric can be defined.*

PROOF. Let $\{U_\alpha\}$, $1 \leq \alpha \leq N$, be any finite covering of a compact manifold M by open balls U_α with local co-ordinates x_α^i. In each U_α take any Riemannian metric $(g_{ab}^{(\alpha)})$ (e.g. $g_{ab}^{(\alpha)} = \delta_{ab}$); we then need somehow to combine the $g_{ab}^{(\alpha)}$ to obtain a metric on M. This is done by defining

$$g_{ab} = \sum_{\alpha=1}^{N} g_{ab}^{(\alpha)}(x)\, \psi_\alpha(x),$$

where $\{\psi_\alpha\}$ is a partition of unity subordinate to the covering $\{U_\alpha\}$. Clearly the g_{ab} are smooth. Since $\psi_\alpha(x) \geq 0$ for all x, and since the set of Riemannian metrics on any space forms a "convex cone" (i.e. for any Riemannian metrics g_1, g_2 and any positive reals c, d, the linear combination $cg_1 + dg_2$ is again a Riemannian metric), it follows that (g_{ab}) is indeed a Riemannian metric. □

It follows immediately that

8.2.2. Corollary. *On any compact manifold there exists a Riemannian connexion.*

The existence of partitions of unity is similarly exploited in defining the integral of an exterior form ω of degree $n = \dim M$ over a manifold M. As before let $\{U_\alpha\}$, $\alpha = 1, \ldots, N$, be a finite covering of the (compact) manifold M by charts U_α with local co-ordinates $x_\alpha^1, \ldots, x_\alpha^n$. In terms of these local co-ordinates the form $\omega^{(n)}$ can in each U_α be written as

$$\omega^{(n)}(x) = a_{1\ldots n}(x)\, dx_\alpha^1 \wedge \cdots \wedge dx_\alpha^n,$$

and the integral of $\omega^{(n)}$ over the region U_α is, as usual, just the multiple integral:

$$\int_{U_\alpha} \omega^{(n)} = \int_{U_\alpha} a_{1\ldots n}(x)\, dx_\alpha^1 \wedge \cdots \wedge dx_\alpha^n.$$

To define the integral over the whole of $M=M^n$ we need to piece these integrals together. With this in view, we take a partition of unity $\{\psi_\alpha\}$ subordinate to $\{U_\alpha\}$. The desired integral is then defined by:

$$\int_{M^n} \omega^{(n)} = \int_{M^n} \left(\sum_{\alpha=1}^{N} \psi_\alpha(x) \right) \omega^{(n)}(x) = \sum_{\alpha=1}^{N} \int_{U_\alpha} \psi_\alpha(x)\, \omega^{(n)}(x).$$

(Recall that $\psi_\alpha(x) \equiv 0$ outside U_α.) The verification that this definition is independent of the particular finite covering $\{U_\alpha\}$ and the partition of unity, presents no essential difficulty, and we omit the details.

As our next application of the existence of partitions of unity we give a rigorous proof of the general Stokes formula. Let $D \subset \mathbb{R}^n$ be a bounded region with smooth boundary ∂D, given in terms of Euclidean co-ordinates $x^1, \ldots, x^n$ by an equation $f(x^1, \ldots, x^n) \geq 0$, where $\operatorname{grad} f|_{\partial D} \neq 0$; thus the boundary of D is a smooth, non-singular hypersurface in $\mathbb{R}^n$. An orientation of $\mathbb{R}^n$ determines the order of the co-ordinates $x^1, \ldots, x^n$ (up to an even permutation), since the orientation is prescribed by the frame (i.e. ordered basis for the tangent space) $(e_1, \ldots, e_n)$ consisting of the standard basis vectors in the natural order, which frame moves smoothly from point to point in $\mathbb{R}^n$. For each point P in ∂D, denote by $n(P)$ the outward normal to ∂D. In some neighbourhood of each point P of ∂D we can define smooth local co-ordinates $y^1, \ldots, y^{n-1}$, which can be ordered so as to define an orientation of ∂D; recall that this orientation is said to be *induced by the orientation on D* if at each point of ∂D the frame $(\partial/\partial y^1, \ldots, \partial/\partial y^{n-1}, n(P))$ is obtained from the frame $(e_1, \ldots, e_n)$ by means of a linear transformation with positive determinant.

8.2.3. Theorem. *Let ω be an exterior differential form of degree $n-1$ on the region D of $\mathbb{R}^n$. Then*

$$\int_D d\omega = \int_{\partial D} i^*(\omega),$$

where $i: \partial D \to D$ is the embedding, $i^*(\omega)$ is the restriction of the form ω to the boundary ∂D of D (see §22.1 *of Part* I), *and the orientation on ∂D is that induced by the orientation on D.*

(Note that the orders of the co-ordinates $x^1, \ldots, x^n$ and $y^1, \ldots, y^{n-1}$, which are determined (up to even permutations) by the orientation, must be stipulated in calculating integrals of forms, since the order determines the sign of the integral.)

PROOF. Let $\{U_\alpha\}$, $1 \leq \alpha \leq N$, be a finite covering of the region D by open bails, and let $h_\alpha: B^n \to \mathbb{R}^n$; $h_\alpha(B^n) = U_\alpha$, be fixed co-ordinate maps, where B^n is the unit open ball in $\mathbb{R}^n$ (with fixed co-ordinates $x^1, \ldots, x^n$). Thus h_α assigns co-ordinates to the chart U_α. By choosing the U_α sufficiently small and arranging the co-ordinatization appropriately, we may assume (by virtue of the Implicit Function Theorem) that every intersection $\partial D \cap U_\alpha$ which is non-empty is given by the equation $x_\alpha^n=0$, where $x_\alpha^1, \ldots, x_\alpha^n$ are the local co-ordinates on U_α.

Now let $\{\varphi_\alpha\}$ be a partition of unity subordinate to the covering $\{U_\alpha\}$; thus $\{\varphi_\alpha\}$ has the following properties:

(i) $\operatorname{supp} \varphi_\alpha \subset U_\alpha$ for all α;
(ii) $\varphi_\alpha(x) \geq 0$ for all $x \in \bigcup_\alpha U_\alpha$;
(iii) $\sum_\alpha \varphi_\alpha(x) \equiv 1$ for all $x \in \bigcup_\alpha U_\alpha$.

From (iii), and since the φ_α are scalars, we have in view of the linearity of integrals that

$$\int_{\partial D} i^*(\omega) = \sum_\alpha \int_{\partial D} i^*(\varphi_\alpha \omega),$$

$$\int_D d\omega = \sum_\alpha \int_D d(\varphi_\alpha \omega).$$

Hence it suffices to show that for each α $(1 \leq \alpha \leq N)$,

$$\int_{\partial D} i^*(\varphi_\alpha \omega) = \int_D d(\varphi_\alpha \omega). \tag{1}$$

If in terms of the local co-ordinates $x_\alpha^1, \ldots, x_\alpha^n$ on U_α, we write

$$\varphi_\alpha \omega = \tilde{\omega}_\alpha = \sum_{k=1}^n (-1)^{k-1} a_k(x)\, dx_\alpha^1 \wedge \cdots \wedge \widehat{dx_\alpha^k} \wedge \cdots \wedge dx_\alpha^n \tag{2}$$

(where $a_k(x) \in C^\infty(D)$, and the hatted symbol is understood as omitted), then (see §25.2 of Part I)

$$d\tilde{\omega}_\alpha = \left(\sum_{k=1}^n \frac{\partial a_k(x)}{\partial x_\alpha^k}\right) dx_\alpha^1 \wedge \cdots \wedge dx_\alpha^n. \tag{3}$$

First case: $U_\alpha \cap \partial D = \varnothing$. Since $\operatorname{supp} \varphi_\alpha \subset U_\alpha$, it follows that $\operatorname{supp}(\varphi_\alpha \omega) \subset U_\alpha$; hence if $U_\alpha \cap \partial D = \varnothing$, then $\varphi_\alpha(x) \equiv 0$ on ∂D, whence $\int_{\partial D} i^*(\varphi_\alpha \omega) = 0$. We therefore wish to show that also $\int_D d(\varphi_\alpha \omega) = 0$.

Since $U_\alpha \cap \partial D = \varnothing$, we must have either $U_\alpha \subset D$ or $U_\alpha \subset \mathbb{R}^n - D$. In the latter case certainly $\int_D d(\varphi_\alpha \omega) = 0$, so we may suppose $U_\alpha \subset D$. Our problem is then to show that (see (3))

$$\int_{U_\alpha} \left(\sum_{k=1}^n \frac{\partial a_k}{\partial x_\alpha^k}\right) dx_\alpha^1 \wedge \cdots \wedge dx_\alpha^n = 0.$$

Via the co-ordinate function h_α we may identify U_α with the unit open ball $B^n \subset \mathbb{R}^n$. With this understood, we extend the region of definition of the integrand in the integral

$$\int_{U_\alpha = B^n} \left(\sum_{k=1}^n \frac{\partial a_k}{\partial x_\alpha^k}\right) dx_\alpha^1 \wedge \cdots \wedge dx_\alpha^n,$$

to the whole of $\mathbb{R}^n$ by defining it to be zero outside B^n. (Recall that $\operatorname{supp} a_k \subset U_\alpha = B^n$.) Let C^n be the cube of side $2R$ in $\mathbb{R}^n$ defined by

$$C^n = \{(x^1, \ldots, x^n) \,\|x^k| \leq R, 1 \leq k \leq n\},$$

large enough to contain B^n. Then

$$\int_{B^n} \left(\sum_{k=1}^{n} \frac{\partial a_k}{\partial x_\alpha^k} \right) dx_\alpha^1 \wedge \cdots \wedge dx_\alpha^n = \sum_{k=1}^{n} \int_{C^n} \frac{\partial a_k}{\partial x_\alpha^k} dx_\alpha^1 \wedge \cdots \wedge dx_\alpha^n$$

$$= \sum_{k=1}^{n} \int_{C^{n-1}} (-1)^{k-1} \left(\int_{-R}^{R} \frac{\partial a_k}{\partial x_\alpha^k} dx_\alpha^k \right) dx_\alpha^1 \wedge \cdots \wedge \widehat{dx_\alpha^k} \wedge \cdots \wedge dx_\alpha^n.$$

(Here C^{n-1} denotes the appropriate $(n-1)$-dimensional cube.) Up to sign, the kth term of this sum can be evaluated as follows:

$$\int_{C^{n-1}} \left(\int_{-R}^{R} \frac{\partial a_k}{\partial x_\alpha^k} dx_\alpha^k \right) dx_\alpha^1 \wedge \cdots \wedge \widehat{dx_\alpha^k} \wedge \cdots \wedge dx_\alpha^n$$

$$= \pm \int_{C^{n-1}} \{a_k(x_\alpha^1, \ldots, x_\alpha^{k-1}, R, x_\alpha^{k+1}, \ldots, x_\alpha^n)$$

$$- a_k(x_\alpha^1, \ldots, x_\alpha^{k-1}, -R, x_\alpha^{k+1}, \ldots, x_\alpha^n)\} \, dx_\alpha^1 \wedge \cdots \wedge \widehat{dx_\alpha^k} \wedge \cdots \wedge dx_\alpha^n$$

$$= 0,$$

since $a_k(x_\alpha^1, \ldots, \pm R, \ldots, x_\alpha^n) = 0$.

Second case: $U_\alpha \cap \partial D \neq \varnothing$. We wish to establish (1). In view of the supports of the integrands it suffices to verify that

$$\int_{\partial D \cap U_\alpha} i^*(\tilde{\omega}_\alpha) = \int_{U_\alpha} d\tilde{\omega}_\alpha. \tag{4}$$

From (2) and our initial provision that $\partial D \cap U_\alpha$ be given by the equation $x_\alpha^n = 0$, it follows that

$$i^*(\tilde{\omega}_\alpha) = (-1)^{n-1} a_n \, dx_\alpha^1 \wedge \cdots \wedge dx_\alpha^{n-1}.$$

Thus the equality we seek to establish, namely (4), becomes

$$\int_{\partial D \cap U_\alpha} (-1)^{n-1} a_n \, dx_\alpha^1 \wedge \cdots \wedge dx_\alpha^{n-1} = \sum_{k=1}^{n} \int_{U_\alpha} \frac{\partial a_k}{\partial x_\alpha^k} dx_\alpha^1 \wedge \cdots \wedge dx_\alpha^n. \tag{5}$$

As in the first case we now identify U_α with the unit open ball B^n, and extend the domain of the a_k to all of $\mathbb{R}^n$ by defining them to be zero outside B^n. Then with the cube C^n as before, the right-hand side of (5) becomes

$$\sum_{k=1}^{n} \int_{C^n} \frac{\partial a_k}{\partial x_\alpha^k} dx_\alpha^1 \wedge \cdots \wedge dx_\alpha^n. \tag{6}$$

For $k \neq n$, certainly $\partial a_k/\partial x_\alpha^k$ is a continuous function of x_α^k, so that by the "Fundamental Theorem of Calculus"

$$\int_{C^n} \frac{\partial a_k}{\partial x_\alpha^k} dx_\alpha^1 \wedge \cdots \wedge dx_\alpha^n$$
$$= \int_{C^{n-1}} \left(\int_{-R}^{R} \frac{\partial a_k}{\partial x_\alpha^k} dx_\alpha^k \right) dx_\alpha^1 \wedge \cdots \wedge \widehat{dx_\alpha^k} \wedge \cdots \wedge dx_\alpha^n = 0,$$

since $a_k(x_\alpha^1, \ldots, \pm R, \ldots, x_\alpha^n) = 0$. On the other hand the nth summand in (6) is

$$\int_{C^n} \frac{\partial a_n}{\partial x_\alpha^n} dx_\alpha^1 \wedge \cdots \wedge dx_\alpha^n = (-1)^{n-1} \int_{C^{n-1}} \left(\int_{-R}^{R} \frac{\partial a_n}{\partial x_\alpha^n} dx_\alpha^n \right) dx_\alpha^1 \wedge \cdots \wedge dx_\alpha^{n-1} \tag{7}$$

Now as a function of x^n alone (i.e. for any particular fixed values of $x^1, \ldots, x^{n-1}$) a_n is continuous on each of the intervals $-R \leq x^n < 0$ and $0 < x^n \leq R$ (with a possible jump discontinuity at $x^n = 0$); hence it follows by integrating over each of these intervals and adding that

$$\int_{-R}^{R} \frac{\partial a_n}{\partial x_\alpha^n} dx^n = a_n|_{\partial D}.$$

Substituting from this in the right-hand side of (7) we get finally

$$\int_{B^n} d\tilde{\omega}_\alpha = \int_{C^{n-1}} (-1)^{n-1} a_n \, dx_\alpha^1 \wedge \cdots \wedge dx_\alpha^{n-1},$$

as required. This completes the proof in the second case, and thereby the proof of the theorem. □

Remark. The fact that the orientation on ∂D was taken to be that induced by the given orientation of D, was used in applying the "Fundamental Theorem of Calculus" in the form $\int_a^b df(x) = f(b) - f(a)$, with $b > a$, which inequality was determined by the direction of the outward normal $n(P)$ to ∂D; if we had used instead the inward normal we would have obtained the negative of the integral in question. For fixed $x_\alpha^1, \ldots, x_\alpha^{n-1}$, the function $a_n(x_\alpha^n)$ has graph something like that shown in Figure 16.

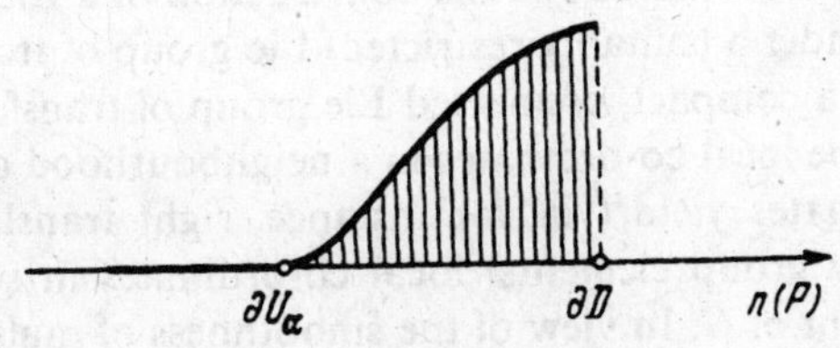

Figure 16

Exercise
Prove the general Stokes formula for compact manifolds M with boundary (see §1.3):

$$\int_M d\omega = \int_{\partial M} \omega.$$

(Here the orientation on the boundary ∂M of M is again chosen to be that induced by the given orientation of M.)

8.3. Invariant Metrics

We shall now show that the existence of partitions of unity allows the construction of a Riemannian metric on a manifold, invariant under the action of a given compact group of transformations.

We begin with the case of a finite group acting on a smooth, closed (i.e. compact and without boundary) manifold.

8.3.1. Theorem. *Given a smooth closed manifold M and a finite group G of transformations of M, there exists a Riemannian metric on M invariant under G.*

Proof. We have already shown (Corollary 8.2.1) that, as a consequence of the existence of partitions of unity, there exists a Riemannian metric $g_{ab}(x)$ say, on M. Denote by $\langle\ ,\ \rangle_x$ the scalar product on T_x (the tangent space to M at each point x), defined by the metric $g_{ab}(x)$, and denote by N the order of the finite group G. We define a new scalar product $(\ ,\ )_x$ (and thereby a new Riemannian metric on M), by means of the procedure of "group averaging" of the old metric, with respect to the group G:

$$(\xi, \eta)_x = \frac{1}{N} \sum_{g \in G} \langle g_*(\xi), g_*(\eta) \rangle_{g(x)}.$$

Here ζ, η are arbitrary vectors in T_x, and g_* is the map of tangent spaces induced by g. It is clear that this new metric is invariant under the action of G, i.e. that

$$(g_*(\xi), g_*(\eta))_{g(x)} = (\xi, \eta)_x,$$

for all $x \in M$, $\xi, \eta \in T_x$, $g \in G$. This completes the proof. □

An analogous procedure allows the construction of a Riemannian metric on M invariant under a (suitably restricted) Lie group of transformations of M. Thus let G be a compact, connected Lie group of transformations of M, and let $t^1, \ldots, t^m$ be local co-ordinates in a neighbourhood of the identity of G. These co-ordinates yield (via, for instance, right translations, i.e. right multiplications by group elements) local co-ordinates in some neighbourhood of every point of G. In view of the smoothness of multiplication on G, this collection of co-ordinatized neighbourhoods forms an atlas on the manifold G. In this sense the local co-ordinates $t^1, \ldots, t^m$ serve (via right translations) for all of G.

8.3.2. Lemma. *On a compact, connected Lie group G, there exists a "volume element" invariant under right translation, and any such form is expressible as $d\mu(\alpha) = \Omega\, dt^1 \wedge \cdots \wedge dt^m$, where $\alpha \in G$, Ω is constant, and $t^1, \ldots, t^m$ are local co-ordinates in a neighbourhood of α, obtained by means of right translations from a system of co-ordinates for a neighbourhood of the identity.*

(Note that such a differential form is often said to define a "right-invariant measure" on G.)

PROOF. Consider the volume m-form at the identity of G defined in the usual way (i.e. the form whose value at an ordered m-tuple of tangent vectors is the determinant of the matrix whose rows are made up of the components of those vectors). This form is then extended to the whole group by means of right translations. This shows the existence of a right-invariant volume element. That any such volume element must be expressible as indicated is immediate from the fact that up to a constant factor there is only one skew-symmetric form of rank m on m-dimensional space (in this case on $T_1(G)$). This proves the lemma. □

(Note that a change of variables leads to a multiplication of the volume element by the Jacobian of the change, and also that a left-invariant measure on G can be constructed similarly.)

There is a standard notation used to express right-invariance of the measure, namely $d\mu(gg_0) = d\mu(g)$. In terms of integrals the right-invariance of the measure $d\mu(\alpha)$ is expressed by

$$\int_G f(gg_0)\, d\mu(g) = \int_G f(g)\, d\mu(g),$$

where $f(g)$ is any function on G for which the integrals exist.

Suppose now that G is a Lie group of transformations of a manifold M.

8.3.3. Theorem. *Let G be a compact, connected Lie group of transformations of a smooth, closed manifold M. Then there is a Riemannian metric on M invariant under G.*

(We note that if G is not transitive on M, then there will in general exist many such G-invariant metrics on M.)

PROOF. The construction of the desired metric is again carried out by means of the procedure of group averaging of a given metric with respect to the action of the group G. As before let $g_{ab}(x)$ be any Riemannian metric on M (guaranteed by Corollary 8.2.1), with corresponding scalar product $\langle\ ,\ \rangle_x$ on T_x, the tangent space to M at x. Define a new scalar product $(\ ,\ )_x$ on T_x (and thereby a new metric on M) by setting

$$(\xi, \eta)_x = \frac{1}{\mu(G)} \int_G \langle g_*(\xi), g_*(\eta) \rangle_{g(x)} \, d\mu(g),$$

where $x \in M$, $\xi, \eta \in T_x$, $g \in G$, g_* is the induced map of tangent spaces, $d\mu(g)$ is a right-invariant measure on G, and $\mu(G)$ is the volume of the whole group G. We then have

$$(g_{0*}(\xi), g_{0*}(\eta))_{g_0(x)} = \frac{1}{\mu(G)} \int_G \langle (gg_0)_*(\xi), (gg_0)_*(\eta) \rangle_{gg_0(x)} \, d\mu(gg_0)$$

$$= \frac{1}{\mu(G)} \int_G \langle g'_*(\xi), g'_*(\eta) \rangle_{g'(x)} \, d\mu(g') = (\xi, \eta)_x,$$

which shows the G-invariance of the new metric under the action of G. □

§9. The Realization of Compact Manifolds as Surfaces in $\mathbb{R}^N$

Let M and N be smooth manifolds of dimensions n and p respectively. Recall (from §1.3) that a smooth map $f\colon M \to N$ is called an "immersion" if the rank of the induced map $df|_x\colon T_x \to T_{f(x)}$ at each point x, is equal to n, i.e. if the induced map of tangent spaces is at each point an embedding. (Hence in particular we must then have $p \geq n$.) It follows from the Implicit Function Theorem that around each point x there is a neighbourhood $U(x)$ on which the restriction of f is a diffeomorphism from $U(x)$ to its image $f(U(x)) \subset N$. It is easy to give examples of immersions which are, however, not "globally" one-to-one. Recall that a one-to-one immersion is called an "embedding".

Theorem. *Any compact, smooth manifold M can be smoothly embedded in Euclidean space $\mathbb{R}^k$ for sufficiently large k.*

PROOF. Let $\{U_i\}$, $1 \leq i \leq s$, be a fixed finite covering of the manifold M by open neighbourhoods diffeomorphic to $\mathbb{R}^n$ (where $n = \dim M$), and for each i let $\varphi_i\colon M \to S^n \subset \mathbb{R}^{n+1}$, be a map sending $M \setminus U_i$ to a single point, and filling the rest of S^n out with U_i, i.e. mapping U_i onto the sphere S^n with the point removed. It is intuitively clear (especially in the cases $n = 1, 2$) that the map φ_i can be so constructed that its restriction to U_i is an embedding of manifolds. We now define $\varphi\colon M \to \mathbb{R}^k$, where $k = s(n+1)$, by

$$\varphi(x) = (\varphi_1(x), \ldots, \varphi_s(x)).$$

The fact that φ is an embedding now follows from the construction of each $\varphi_i|_{U_i}$ as an embedding of U_i, and the fact that the U_i cover M. □

We note in conclusion that the dimension k of the containing Euclidean space $\mathbb{R}^k$ can be reduced to $2n+1$ (see §11.1). It can, moreover, be shown that any continuous map $M \to \mathbb{R}^{2n+1}$ is approximable by smooth embeddings (see below); for $n=1$ this is intuitive.

§10. Various Properties of Smooth Maps of Manifolds

10.1. Approximation of Continuous Mappings by Smooth Ones

We first establish, on the basis of a result concerning functions on regions of $\mathbb{R}^n$, that any continuous map between manifolds can be approximated (in a precise sense which we shall now define) by smooth maps. Mainly for the sake of simplicity, we shall assume our (smooth) manifolds to be connected and closed (i.e. compact and without boundary). Let M and N be two such manifolds. As we saw above (in §8.2) on such manifolds there exist Riemannian metrics; let ρ denote the distance function on N defined (as in §1.3) by any particular Riemannian metric on N. In terms of ρ we can define the distance between any two continuous maps $f, g\colon M \to N$ by

$$\rho(f, g) = \max_{x \in M} \rho(f(x), g(x)).$$

This turns the space of all continuous functions from M to N into a metric space.

As noted above, we shall need the following result, perhaps familiar from courses in analysis. We omit its proof.

10.1.1. Proposition. *Let $f(x^1, \ldots, x^n)$ be a continuous function on the (open) region U of $\mathbb{R}^n$. Then corresponding to any $\varepsilon > 0$, and any open set $V \subset U$ such that also $\bar{V} \subset U$, there exists a continuous function $g(x^1, \ldots, x^n)$ with the following four properties:*

(i) *the function g is smooth on V;*
(ii) $g|_{U \setminus V} = f|_{U \setminus V}$;
(iii) $\max_{x \in \bar{V}} |f(x) - g(x)| \le \varepsilon$;
(iv) *the function g is smooth at every point where f is smooth.*

We can now proceed to the following important approximation theorem.

10.1.2. Theorem. *Let M, N be connected, closed smooth manifolds. Then any continuous map $f\colon M \to N$, can be approximated arbitrarily closely by smooth maps; i.e. for each $\varepsilon > 0$ there exists a smooth map $g\colon M \to N$, such that $\rho(f, g) < \varepsilon$.*

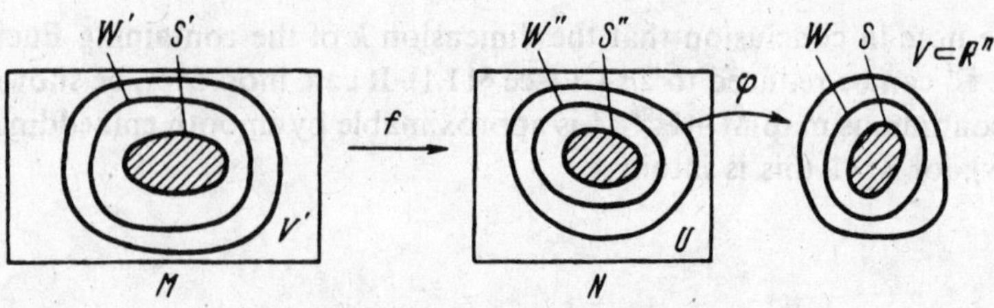

Figure 17

PROOF. Let $U \subset N$ be an open set homeomorphic to a region V of Euclidean space $\mathbb{R}^n$ (where $n = \dim N$), and let $\varphi\colon U \to V$ be a homeomorphism (for instance, we may take U to be any local co-ordinate neighbourhood of the manifold N, and φ to be the co-ordinate map identifying U with a region V of $\mathbb{R}^n$). Let S, W be open sets such that $S \subset \bar{S} \subset W \subset \bar{W} \subset V \subset \mathbb{R}^n$, and set $W'' = \varphi^{-1}(W)$, $S'' = \varphi^{-1}(S)$, $V' = f^{-1}(U)$, $W' = f^{-1}(W'')$, and $S' = f^{-1}(S'')$ (see Figure 17). Since $\bar{S}'' \subset W'' \subset \bar{W}'' \subset U$, there exists a positive number $\eta < \varepsilon$ such that $\rho(W'', N \setminus U) > \eta > 0$, and $\rho(S'', N \setminus W'') > \eta > 0$. By applying Proposition 10.1.1 to the co-ordinate functions of the map $\varphi \circ f\colon V' \to \mathbb{R}^n$, we deduce the existence of a continuous map $\tilde{g}\colon V' \to \mathbb{R}^n$, which is smooth on S' and at all points of V' where f is smooth, and which has the further properties that

$$\tilde{g}|_{V' \setminus W'} \equiv (\varphi \circ f)|_{V' \setminus W'}, \qquad \rho(f, \varphi^{-1} \circ \tilde{g}) \leq \eta < \varepsilon.$$

It follows that $(\varphi^{-1} \circ \tilde{g})(V') \subset U$. Hence we have a continuous map $g' = \varphi^{-1} \circ \tilde{g}\colon V' \to U$, which is smooth on S' and agrees with f on $V' \setminus W'$. We deduce that the map $g\colon M \to N$, defined by: $f = g$ on $M \setminus W'$, and $g = g'$ on V' (i.e. the extension of g' to all of M obtained by defining it to agree with f outside W'), is continuous, is smooth on S' and at all points where f is smooth, and satisfies $\rho(f, g) < \varepsilon$. By covering the manifold N suitably with finitely many open sets homeomorphic to regions of $\mathbb{R}^n$, and approximating the given continuous function f successively on the inverse image of each open set of the cover, in the manner described above, we arrive at the desired conclusion. □

Remarks. 1. The proof of the theorem brings out the local character of the approximation: the map f is successively approximated by functions smooth on (large) open subsets of the charts of M. Hence if the given map f happened to be already smooth on some open region $U \subset M$, then for any closed subset $V \subset U$ we could arrange that $g \equiv f$ on V.

2. In the sequel (§12.1 below) it will be shown that if M, N are connected, closed smooth manifolds, then there exists an $\varepsilon_0 > 0$ such that the inequality $\rho(f, g) < \varepsilon_0$ (where $f, g\colon M \to N$ are continuous maps) implies that the maps f and g are "homotopic" (see §12.1 for the definition of a homotopy between maps). In view of this we may assume (by restricting ε) that in the above theorem the smooth map g approximating f is homotopic to f.

10.2. Sard's Theorem

For each smooth map $f\colon M \to N$, between smooth manifolds M, N, let $C = C(f) \subset M$ denote the set of all $x \in M$ at which the differential $df_x\colon T_x \to T_{f(x)}$ has rank strictly less than $n = \dim N$. The set C is called the *set of critical points* of the map f, and $f(C)$ the *set of critical values* of f.

Before stating the theorem, we recall the concept of a set of "measure zero". A subset $B \subset \mathbb{R}^n$ is said to have (n-dimensional) *measure zero* if for each $\varepsilon > 0$ it can be covered by a countable collection of n-dimensional cubes, the sum of whose volumes is less than ε. (It is usually shown in courses in analysis that then the complementary set $\mathbb{R}^n \backslash B$ is an everywhere dense subset of $\mathbb{R}^n$.) This definition is extended in the following way to subsets of an n-dimensional manifold N: a subset B of N has *measure zero* if for each co-ordinate map $\varphi\colon U \to \mathbb{R}^n$, where U ranges over the charts of N, the set $\varphi(U \cap B)$ has measure zero in $\mathbb{R}^n$.

10.2.1. Theorem (Sard). *Let $f\colon M \to N$ be a smooth (i.e. of class C^∞) map between smooth manifolds M and N. Then the set $f(C)$ of critical values of f has measure zero in N.*

Proof. In view of the countability requirement on the collection of charts making up an atlas for a manifold, it suffices to prove that if U is any (open) region of $\mathbb{R}^m$ (identifiable with a chart of M) and $f\colon U \to \mathbb{R}^n$ is any smooth map, then the set $f(C)$ has measure zero in $\mathbb{R}^n$. The proof will be by induction on the sum $m+n$ of the dimensions m, n of the manifolds M, N. Since the theorem is obvious if either $m=0$ or $n=0$, we may assume $m, n \geq 1$.

Denote by C_i the subset of M consisting of those points x of M at which ail of the partial derivatives of f of order $\leq i$ are zero. We then have a descending sequence of closed subsets of M:

$$C \supset C_1 \supset C_2 \supset \dots.$$

We split the proof up into three lemmas.

10.2.2. Lemma. *The set $f(C \backslash C_1)$ has measure zero.*

Proof. Since clearly $C = C_1$ for $n=1$, we may assume that $n \geq 2$. We shall need the following special case of "Fubini's theorem" (whose proof can be found in most analysis textbooks): a set $A \subset \mathbb{R}^n = \mathbb{R}^1 \times \mathbb{R}^{n-1}$ has zero n-dimensional measure if it intersects each hyperplane $q \times \mathbb{R}^{n-1}$ ($q \in \mathbb{R}^1$) in a set of zero $(n-1)$-dimensional measure.

Our aim is to find, for each $x' \in C \backslash C_1$, an open set $V \subset \mathbb{R}^m$, containing x', such that $f(V \cap C)$ has zero measure. The lemma will then follow since $C \backslash C_1$ can be covered by countably many such V.

Since $x' \notin C_1$, at least one of the first partial derivatives of f at x' must be non-zero; without loss of generality we may suppose $\partial f_1/\partial x^1 \neq 0$ at x', where f_1 is the first co-ordinate function of the map f. Define a map $h\colon U \to \mathbb{R}^m$ by

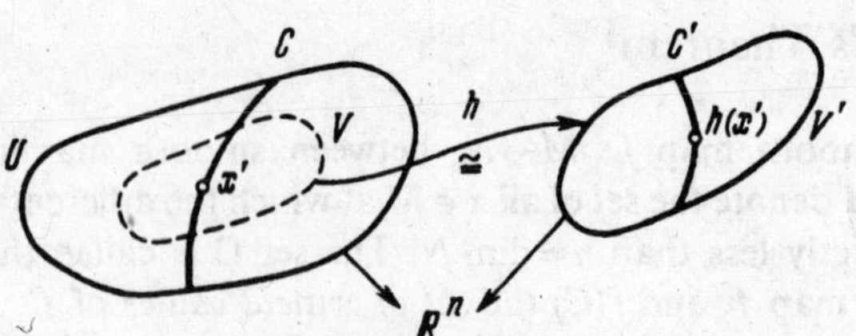

Figure 18

$h(x)=(f_1(x), x^2, \ldots, x^m)$. Since the rank of dh at x' is clearly m, it follows (from the Implicit Function Theorem) that there is a neighbourhood $V=V(x')\subset U$, such that the restriction of h to V is a diffeomorphism from V onto some open neighbourhood V' of the point $h(x')$. Hence the set C' of critical points of the composite mapping $g=f\circ h^{-1}\colon V'\to\mathbb{R}^n$ is just $h(V\cap C)$, i.e. $g(C')=f(V\cap C)$ is the set of critical values of g (see Figure 18).

From the definition of the map g it follows that the image under g of each point of V' of the form $(t, x^2, \ldots, x^m)$ lies in the hyperplane $t\times\mathbb{R}^{n-1}$, so that g sends hyperplanes in V' to hyperplanes in $\mathbb{R}^n$. (Incidentally, rather than introduce the map h, we could have worked directly with curved hypersurfaces.)

Consider the family of smooth maps

$$g^t\colon (t\times\mathbb{R}^{m-1})\cap V'\to t\times\mathbb{R}^{n-1},$$

obtained by restricting g appropriately. Then since

$$\left(\frac{\partial g_i}{\partial x^j}\right)=\begin{pmatrix}1 & 0\\ * & \dfrac{\partial g_i^t}{\partial x^j}\end{pmatrix}$$

at each point $\alpha=(t, x^2, \ldots, x^m)$ of V', it follows that α is a critical point of g if and only if it (or rather the $(m-1)$-tuple $(x^2, \ldots, x^m)$) is a critical point of g^t. By the inductive hypothesis the set of critical values of g^t has measure zero in $t\times\mathbb{R}^{n-1}$; hence the intersection $g(C')\cap(t\times\mathbb{R}^{n-1})$ has zero $(n-1)$-dimensional measure. It now follows from the special case of Fubini's theorem stated at the beginning of the proof, that $g(C')$ has measure zero, as required. □

10.2.3. Lemma. *For all $i\geq 1$ the sets $C_i\backslash C_{i+1}$ have measure zero.*

Proof. The argument will in part resemble that of the previous lemma. Thus let $x'\in C_i\backslash C_{i+1}$; then at the point x' all partial derivatives of orders $\leq i$ of the co-ordinate functions of the mapping f vanish, while for some family of indices $r; s_1, s_2, \ldots, s_{i+1}$, we have

$$\frac{\partial^{i+1} f_r}{\partial x_{s_1}\cdots\partial x_{s_{i+1}}}\neq 0 \quad\text{at } x'.$$

Denoting the function $\partial^i f_r/\partial x_{s_2}\cdots\partial x_{s_{i+1}}$ by W, we have

$$W(x')=0, \qquad \left.\frac{\partial W}{\partial x_{s_1}}\right|_{x'} \neq 0.$$

We may without loss of generality assume that $s_1=1$. The map $h\colon U\to\mathbb{R}^m$ defined by $h(x)=(W(x), x^2,\dots, x^m)$, is (for the usual reasons) when restricted to some neighbourhood V of x', a diffeomorphism from that neighbourhood onto an open set $V'\subset\mathbb{R}^m$. Consider now the image of the set $C_i\cap V$ under the mapping h. Since the first co-ordinate function of h is $W(x)$, which is a partial derivative of f of order i, and since all such partial derivatives vanish on the set C_i, it follows that h sends $C_i\cap V$ to the hyperplane $0\times\mathbb{R}^{m-1}$.

As in the preceding lemma we now form the composite map $g=f\circ h^{-1}$: $V'\to\mathbb{R}^n$. By the inductive hypothesis the set of critical values of its restriction $g'\colon(0\times\mathbb{R}^{m-1})\cap V'\to\mathbb{R}^n$, has measure zero in $\mathbb{R}^n$. Now all points of the set $h(C_i\cap V)$ (which is contained in the domain of g') are critical for g' since the set $C_i\cap V$ consists of points at which all partial derivatives of order $\le i$ of f vanish, so that (by the chain rule) all partial derivatives of g' of order $\le i$ vanish at all points of $h(C_i\cap V)$ (or for the simpler reason that the rank of f is less than n). Hence the set $g'\circ h(C_i\cap V)=f(C_i\cap V)$ has measure zero in $\mathbb{R}^n$. The lemma now follows (as in the proof of the previous one) by covering $C_i\backslash C_{i+1}$ by countably many such open sets V.

(Note that the difference in the proofs of the above two lemmas resides in the fact that in the first lemma we were, generally speaking, unable to arrange (via a suitable diffeomorphism h) for $C\cap V$ to lie in the hyperplane $0\times\mathbb{R}^{m-1}$, since the definition of C as the set of points at which f has rank $< n$, did not allow the selection of any such hyperplane in the way this was effected in the proof of the second lemma.) □

10.2.4. Lemma. *For sufficiently large k the set $f(C_k)$ has measure zero.*

Proof. Cover C_k with a countable collection of cubes of (small) side δ. Let I^m be any one of these cubes contained in U; we shall show that for k sufficiently large the set $f(C_k\cap I^m)$ has measure zero. By Taylor's theorem and the definition of C_k we have that for $x\in C_k\cap I^m$, $x+h\in I^m$,

$$f(x+h)=f(x)+R(x,h),$$

with $|R(x,h)|\le\alpha|h|^{k+1}$, where α depends only on f and I^m, and $|\ |$ denotes Euclidean length of vectors. Subdivide I^m into r^m subcubes each of side δ/r, and let I_1 denote a cube of the subdivision containing the point x. Then every point of the cube I_1 has the form $x+h$, where $|h|\le\sqrt{m}\,\delta/r$. Consequently $f(I_1)$ is contained in the cube of side a/r^{k+1} with centre at the point $f(x)$, where $a=2\alpha(\sqrt{m}\,\delta)^{k+1}$. Hence $f(C_k\cap I^m)$ is contained in the union of r^m cubes whose total volume is at most $r^m(a/r^{k+1})^n=a^n r^{m-n(k+1)}$. If $k+1>m/n$ then this volume approaches zero as $r\to\infty$. This proves the lemma. □

As noted earlier, Sard's theorem now follows from the above three lemmas.

10.2.5. Corollary. *Let $f: M \to N$ be a smooth map of smooth manifolds M, N, with set of critical points C. Then the set $N \backslash f(C)$ is everywhere dense in N.*

10.2.6. Corollary. *If $f: M \to N$ is a smooth map of smooth manifolds M, N, where $\dim M < \dim N$, then the image $f(M)$ has measure zero in N. Hence, in particular, f cannot map M onto N, i.e. $f(M)$ is properly contained in N.*

We shall later on (in §11.1) use Sard's theorem to prove "Whitney's theorem" on embeddings and immersions of smooth manifolds. For the present we turn our attention to the non-critical or "regular" points of a mapping.

10.2.7. Definition. A point $x \in M$ is called a *regular point* of a smooth map $f: M \to N$, if it is not critical, i.e. if the rank of the map df_x is equal to $n = \dim N$. A point $y \in N$ is called a *regular value* of a smooth map $f: M \to N$, if all of its preimages are regular points of f (so that, in particular, y will be a regular value of f if $f^{-1}(y)$ is empty). If y is a regular value of the map f, then we shall say also that the map f is *regular with respect to the point* y.

Thus given a smooth map $f: M \to N$, the sets of regular points and critical points are complementary in M, while the sets of regular values and critical values are complementary subsets of N.

Note that if $f: M \to N$ is a smooth map of smooth manifolds, and $y \in N$ is a regular value of f, then it follows from the Implicit Function Theorem that $f^{-1}(y)$ is a smooth submanifold of M.

In the sequel we shall find useful the following easy consequence of Sard's theorem.

10.2.8. Corollary. *Let M, N be smooth manifolds and let y be a (fixed) point of N. Then the set of smooth maps $f: M \to N$, for which y is a regular value, is everywhere dense in the space of all smooth maps from M to N.*

PROOF. We wish to show that given any smooth map $g: M \to N$, there is arbitrarily close to it a smooth map $f: M \to N$ for which y is a regular value. By Sard's theorem the set of regular values of the map g is everywhere dense in N, so that a regular value y' of g can be found in every open neighbourhood $U \subset N$ of the point y. We may choose U to be diffeomorphic to the open ball B^n (or disc D^n), with co-ordinate map $\varphi: U \to B^n$. Write $z = \varphi(y)$, $z' = \varphi(y')$. It is intuitively clear that there exists a diffeomorphism $h: B^n \to B^n$, which maps points near the boundary identically, and has the further properties that $h(z') = z$, and $|h(t) - t| < \varepsilon = \rho(z, z')$ for $t \in B^n$. Extend h to a diffeomorphism $h': N \to N$ of the whole of N (identifying B^n with U) by defining h' to be the identity outside U. Then the map $f = h' \circ g$ has the desired properties. □

For our later purposes (hinted at prior to the corollary) we shall actually require a somewhat stronger result than this one. We shall need the fact that the diffeomorphism $h'\colon N \to N$ obtained in the course of the above proof, can be chosen so that not only is the map $f = h' \circ g$ close to g in the former sense (namely $\rho(f, g) = \max \rho(f(x), g(x)) < \varepsilon$), but also each pair of corresponding first derivatives are close. We place on the reader (as an exercise) the onus of proving the existence of such a diffeomorphism.

10.3. Transversal Regularity

We now turn to the important concept of t-regularity (t for "transversal").

10.3.1. Definition. Let P be a smooth submanifold of a smooth manifold N, of codimension k (where by "codimension" is meant simply $\dim N - \dim P$). Then a smooth map $f\colon M \to N$ from a smooth manifold M to N is said to be *transversally regular on P* if the rank of the map

$$df\colon T_x \to T_{f(x)}N/T_{f(x)}P$$

is k whenever $f(x) \in P$, i.e. if the rank of the induced map of tangent spaces is, modulo the tangent space $T_{f(x)}P$ to P at $f(x)$, equal to k, or, in yet other words, if the subspaces $df(T_x)$ and $T_{f(x)}P$ together span the whole of the tangent space $T_{f(x)}N$ to N at $f(x)$, or if, as they say (somewhat imprecisely) "the image $f(M)$ is transverse to P at the point $f(x)$" (see Figure 19).

We note an important property of t-regular mappings: the complete inverse image $f^{-1}(P) \subset M$ is a smooth submanifold of M of the same codimension k as P, i.e. of dimension $m+p-n$; this follows from the Implicit Function Theorem. (Cf. the special case when P is a single point representing a regular value of f.)

10.3.2. Theorem. *Let M, N be smooth manifolds, and $P \subset N$ a smooth submanifold. Then the set of maps $g\colon M \to N$ which are t-regular on P, is everywhere dense in the space of all smooth maps from M to N; i.e. in every neighbourhood of any smooth map $M \to N$, there exists a map t-regular on P.*

PROOF. We wish to show that given any smooth map $f\colon M \to N$, we can find arbitrarily close to f a map g which is t-regular on P.

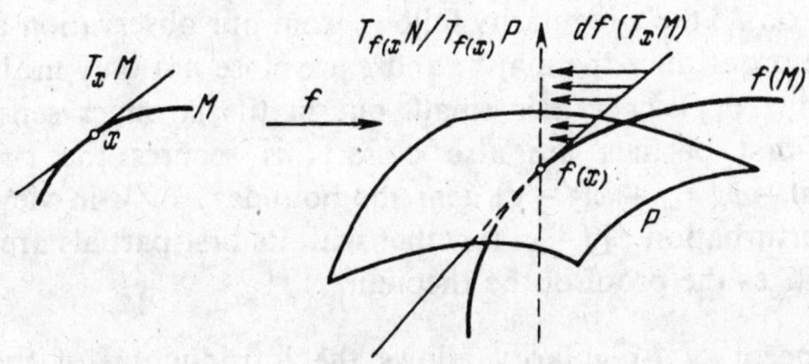

Figure 19

We begin with the following observation, whose validity is essentially immediate from the definition of t-regularity: Let $\rho: M \to N$ be a smooth map, x a point of M, U an (open) neighbourhood of x and $V \supset f(U)$ an open neighbourhood of $\rho(x)$ in N, and let P be a submanifold of N; then if the restriction $\rho: U \to V$ is t-regular on P (i.e. on $P \cap V$), then this will be true also of all smooth maps which, together with all their first partial derivatives, are sufficiently close to ρ and its respective first partial derivatives. (This in essence follows from the fact that ρ and its first derivatives at x determine how $d\rho(T_x)$ is situated in $T_{\rho(x)}N$.)

In view of this observation it is appropriate (as we shall see at the end of the proof) to first prove the theorem locally, i.e. for each open neighbourhood U (which we may identify with $\mathbb{R}^m$), corresponding V (identified with $\mathbb{R}^n$), and P (identified with $\mathbb{R}^p \subset \mathbb{R}^n$, where $\mathbb{R}^p$ consists of the points of $\mathbb{R}^n$ with their last $(n-p)$ co-ordinates zero). Then writing f in terms of its co-ordinate functions

$$f(x^1, \ldots, x^m) = (f_1(x), \ldots, f_p(x), f_{p+1}(x), \ldots, f_n(x)),$$

the t-regularity of f along P at the original point of interest, becomes equivalent to the condition that the point 0 be a regular value of the map $\alpha: \mathbb{R}^m \to \mathbb{R}^{n-p}$, defined by $\alpha(x) = (f_{p+1}(x), \ldots, f_n(x))$.

By Corollary 10.2.8 above (or rather by the remark immediately following its proof) the set of smooth maps $\mathbb{R}^m \to \mathbb{R}^{n-p}$ for which 0 is a regular value is everywhere dense (in the stronger sense of that remark) in the space of all such smooth maps. Hence there exists a smooth map $\alpha': \mathbb{R}^m \to \mathbb{R}^{n-p}$, such that α' and its first partial derivatives are arbitrarily close to α and its first partials, and such that also (denoting by $g_{p+1}, \ldots, g_n$ the co-ordinate functions of α') the map

$$g(x^1, \ldots, x^m) = (f_1(x), \ldots, f_p(x), g_{p+1}(x), \ldots, g_n(x))$$

is t-regular on P. Assuming that the original function $f: \mathbb{R}^m \to \mathbb{R}^n$ has already been made t-regular on P at (relevant) points near the boundary ∂V of V, let φ denote a suitable smooth function on V with the property

$$\varphi = \begin{cases} 0 & \text{on } \partial V, \\ 1 & \text{on } K, \end{cases}$$

where K is suitable compact subset of V, and put $\rho = f(1-\varphi) + \varphi \cdot g$. The map ρ is then t-regular on P and has the properties that $\rho(x) \equiv f(x)$ on ∂V and $\rho(x) = g(x)$ on K; the t-regularity follows from our observation at the beginning of the proof since the maps f and g are close not only in the sense that $\rho(f, g) = \max \rho(f(x), g(x))$ is small, but in the stronger sense that their respective first partials are also close: this secures the t-regularity of $\rho = f(1-\varphi) + \varphi g = f + \varphi(f-g)$ near the boundary of V in view of the fact that the perturbation $\varphi(f-g)$ together with its first partials are small there. This completes the proof of the theorem. □

The concept of t-regularity allows the introduction of the important concept of "transversely intersecting submanifolds". Let M and P be two smooth submanifolds of a smooth manifold N; we shall say that M and P *intersect transversely* if the inclusion map $i_M: M \to N$ is t-regular on P. This means that at each point x of the intersection $M \cap P$ the tangent spaces T_xM and T_xP together span the whole of T_xN. (Note that the relation of intersecting transversely is symmetric; i.e. in the above definition the map i_M might equivalently have been replaced by the inclusion $i_P: P \to N$ (verify this!) Note also that the intersection $M \cap P$ of a pair of transversely intersecting smooth submanifolds is a smooth submanifold.)

The preceding theorem (which ultimately derives from Sard's theorem) tells us that transversally regular maps are so abundant that they can be found in arbitrarily small neighbourhoods of any smooth map; in this sense we may perhaps say that t-regular maps are "typical" among smooth maps. Thus Sard's theorem permits us, by means of an arbitrarily small perturbation (within the class of all smooth maps) of a given smooth map, to bring it into "general position" relative to the submanifold P (i.e. make it t-regular); theorems of the above type form the basis for the kind of procedure referred to generally as *bringing a map into general position.*

(The last several results have been concerned with the existence of what might be called "small perturbations" in the class of all smooth mappings between two manifolds; we note that it is sometimes of use to be able to bring a map into general position by means of variations (i.e. perturbations) within a narrower class of mappings.)

The following result will prove useful to us.

10.3.3. Theorem. *Let A, M, N be smooth manifolds, let P be a submanifold of N, and let $f: A \times M \to N$ be a smooth mapping which is t-regular on P. Then the set of all points a of A for which $f_a = f(a, x): M \to N$, is t-regular on P, is everywhere dense in A.*

(The manifold A may be regarded as an auxiliary "manifold of parameters" by the aid of which an initial map $f(a_0, x): m \to N$, can be brought into general position.)

PROOF. Denote the manifold $f^{-1}(P)$ by Q. If a layer (or "fibre") $a \times M$ intersects Q transversely then (by definition) we have $T(A \times M) = T(Q) + H$, where $H \subset T(a \times M)$ (see Figure 20). (Here $T(\ )$ denotes the tangent space at a

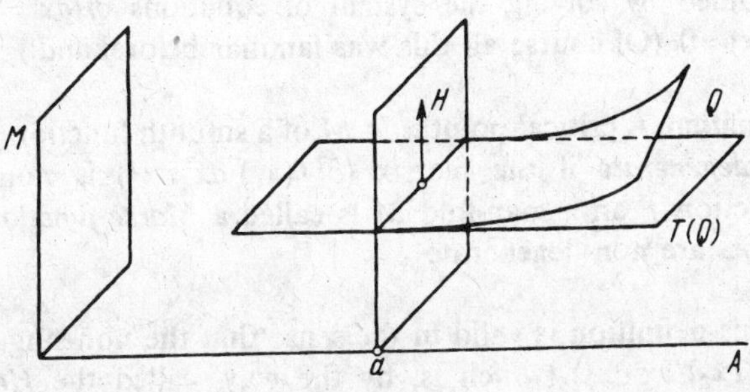

Figure 20

relevant point of the bracketed manifold.) It follows from this and the assumption that $f: A \times M \to N$, is t-regular on P, that df sends H to a subspace of $T(N)$ supplementing $T(P)$, i.e. that f_a is t-regular on P. The converse of this is clearly also true: if f_a is t-regular on P, then the manifolds $a \times M$ and Q intersect transversely.

Now a fibre $a \times M$ intersects Q transversely if and only if the point $a \in A$ is a regular value of the restriction to $Q \subset A \times M$ of the projection map $p: A \times M \to A$. Since by Corollary 10.2.5 of Sard's theorem the set of such regular values is everywhere dense in A, we have reached the desired conclusion. □

Remarks. 1. This theorem is in fact equivalent to Corollary 10.2.5. The above proof shows it to be a consequence of that corollary. For the converse implication, let $f: M \to N$ be a smooth map, and consider the map $F: N \times M \to N \times N$, defined by $F(x, y) = (x, f(y))$. It is easily seen that $x \in N$ is a regular value of f if and only if F is t-regular on the submanifold $N \times x$, whence by the above theorem (with N in the role of A) the set of regular values of f is everywhere dense in N.

2. An important case of transverse intersection is that where the submanifolds M and P of N have "complementary dimensions", i.e. $m + p = n$, where $m = \dim M$, $p = \dim P$, $n = \dim N$. Then from the very definition of transverse intersection of submanifolds it follows that $M \cap P$ consists of isolated points only. Hence if N is embedded in a manifold N_1 of dimension $> n$, then M and P can be brought into general position in N_1 (i.e. perturbed by an arbitrarily small amount) so that their intersection is empty; i.e. M and P can be separated in N_1.

10.4. Morse Functions

In §10.2 we introduced the important concept of a critical point of a smooth map $f: M \to N$ of smooth manifolds. Consider now the special case when $N = \mathbb{R}^1$, the real line, so that f is just a smooth (scalar) function on M. Since $\dim T_{f(x)} \mathbb{R}^1 = 1$, a point x of M is critical (i.e. rank df_x is less than 1) if and only if $df_x = 0$. Thus the critical points of a smooth function $f(x)$ on M are obtained by solving the system of equations $\partial f/\partial x^i = 0$, $1 \le i \le m$, i.e. $\operatorname{grad} f(x) = 0$. (Of course all this was familiar beforehand!)

10.4.1. Definition. A critical point $x_0 \in M$ of a smooth function $f(x)$ on M is called *non-degenerate* if the matrix $(\partial^2 f(x_0)/\partial x^i \partial x^j)$ is non-singular. A smooth function f on a manifold M is called a *Morse function* if all of its critical points are non-degenerate.

Remark. This definition is valid in the sense that the non-singularity of the matrix $(\partial^2 f(x_0)/\partial x^i \partial x^j)$ (which is, by the way, called the *Hessian* of the function f at the point x_0) is independent of the choice of local co-ordinates in a neighbourhood of the critical point x_0. This is a consequence of the fact that the Hessian is the matrix of the symmetric bilinear form $d^2 f$ on $T_{x_0}M$ defined as follows. Let $\xi, \eta \in T_{x_0}M$; we may suppose that ξ, η are, respectively, the values at x_0 of smooth vector fields X, Y defined on M. Then set $d^2 f(\xi, \eta) = \partial_X \partial_Y (f)_{x_0}$, where ∂_X, ∂_Y denote the operations of taking the directional derivatives of functions in the directions X, Y. It is then easily verified that $d^2 f$ is indeed a symmetric bilinear form, and that its matrix relative to the basis $e_1 = \partial/\partial x^1, \ldots, e_m = \partial/\partial x^m$ of $T_{x_0}M$ is precisely $(\partial^2 f(x_0)/\partial x^i \partial x^j)$.

10.4.2. Definition. By the *index* of a non-degenerate critical point x_0 of a function f on a manifold M, we shall mean the largest dimension attained by subspaces $V \subset T_{x_0}M$ on which the Hessian $d^2 f$ is negative definite, i.e. the number of negative squares in the bilinear form $d^2 f$ after it has been brought into canonical (diagonal) form.

The question now naturally arises as to whether Morse functions always exist on a manifold M, and if so, how many, i.e. whether they are everywhere dense in the space of all smooth functions on M. The answers to these questions are in the affirmative (for compact M): this forms the substance of the next theorem. As before we may express this everywhere-denseness of Morse functions by saying that an arbitrary smooth function can be changed to a Morse function by bringing it into "general position", or again that Morse functions are "typical" among smooth functions.

10.4.3. Theorem. (i) *On any compact smooth manifold M there exist Morse functions.*

(ii) *The Morse functions on M are everywhere dense in the space of all smooth functions on M.*

(iii) *Every Morse function on M has only finitely many critical points $x_1, \ldots, x_n$ (which are therefore certainly isolated points of M).*

(iv) *The subset of Morse functions f on M with the property that each critical value of f corresponds to exactly one critical point on M (i.e. such that for any pair of distinct critical points $x, y \in M$ of f, we have $f(x) \neq f(y)$) is everywhere dense in the set of all Morse functions on M.*

PROOF. It is almost immediate from the definition that the non-degenerate critical points of a smooth function f on M are isolated (since if x_0 is a non-isolated critical point, then by choosing local co-ordinates suitably in a neighbourhood of x_0 it can be arranged that, for instance, $\partial^2 f(x_0)/\partial x^1 \partial x^i = 0$ for all i). The compactness of M then implies that they are finite in number. This proves (iii).

We now turn to (i) and (ii). Thus we wish to show that given any smooth function f on M we can find a Morse function g arbitrarily close to it (i.e. such that $\max |f(x) - g(x)| < \varepsilon$ for arbitrarily prescribed $\varepsilon > 0$); hence we shall consider the "small perturbations" of f, and endeavour to detect among them a Morse function.

Consider the map $\alpha_f: M \to T^*M$, defined by $\alpha_f(x) = df_x$, where T^*M denotes the cotangent bundle on the manifold M, i.e. the $2m$-dimensional smooth manifold consisting of all pairs (x, ξ) with $x \in M$ and ξ a covector at the point x (i.e. $\xi \in T_x^*M$) (see §7.1 above). (Note that the linear functional df_x belongs to T_x^*M; strictly speaking $\alpha_f(x) = (x, df_x)$.) We shall now describe a construction whereby α_f is included as a member of an s-parameter family A of maps which are perturbations of α_f. To this end we first cover M with a finite collection $\{U_j\}$, $1 \le j \le k$, of open balls, and enclose each of these balls U_j in a larger one V_j such that $\bar{U}_j \subset V_j$. We then choose on each V_j a set of m independent linear functions $l^1_{V_j}(x), \ldots, l^m_{V_j}(x)$ of local co-ordinates on V_j; e.g. the local co-ordinates $x^1, \ldots, x^m$ themselves will serve. By Lemma 8.1.2, corresponding to each pair V_j, U_j there exists a smooth function φ_j on M, such that $\varphi_j(x) \equiv 1$ on $\bar{U}_j$, $\varphi_j(x) \equiv 0$ outside V_j, and $0 \le \varphi_j(x) \le 1$ on $V_j \backslash U_j$. The existence of such a function φ_j for each j allows us to define new smooth functions $\bar{l}^i_{V_j}$, $i = 1, \ldots, m$, which are defined on the whole of M and agree with $l^i_{V_j}$ on $\bar{U}_j$, by

$$\bar{l}^i_{V_j}(x) = \begin{cases} l^i_{V_j}(x)\,\varphi_j(x) & \text{for } x \in V_j, \\ 0 & \text{for } x \in M \backslash V_j. \end{cases}$$

Denote by A the linear space of smooth functions on M of the form

$$g(x, a) = f(x) + \sum_{V_j, i} a^i_{V_j}\, \bar{l}^i_{V_j}(x),$$

where $f(x)$ is our initially given smooth function on M, and the $a^i_{V_j}$ are real; clearly $\dim A = mk$ where k is the number of open balls U_j in our covering of M, and we may take the $a^i_{V_j}$, $1 \le i \le m$, $1 \le j \le k$, as co-ordinates for A. Consider the map $\psi: M \times A \to T^*M$, defined by $\psi(x, g) = dg_x$; clearly this map is smooth. The n-dimensional submanifold P of T^*M, defined by $P = \{(x, 0) | x \in M\}$ (the so-called "zero-th cross-section of the cotangent bundle") is obviously diffeomorphic to M. We claim that the map $\psi: M \times A \to T^*M$, is t-regular on P. To see this, note first that a typical element of $T_{(x,0)}(M \times A)$ has the form $(\zeta, h(x))$, where ζ is an arbitrary tangent vector M at x, and $h(x)$ is an arbitrary linear combination of the functions $\bar{l}^i_{V_j}$. (Recall that we may use the $a^i_{V_j}$ as co-ordinates for A; the partial derivatives of $g(x, a)$ with respect to the $a^i_{V_j}$ are just the $\bar{l}^i_{V_j}$.) It follows that a typical element of $d\psi T_{(x,0)}(M \times A)$ has the form (η, l), where η is an arbitrary tangent vector to M at x and l is an arbitrary linear combination of the gradients of the $\bar{l}^i_{V_j}$. Since the $l^i_{V_j}$ were chosen to be linearly independent on V_j, we deduce that $d\psi T_{(x,0)}(M \times A)$ does indeed supplement in T^*M the tangent space to the submanifold P, i.e. ψ is t-regular on P, as claimed (see Figure 21).

It now follows from Theorem 10.3.3 that the set of $a_0 \in A$ for which the map $\psi(x, a_0): M \to T^*M$ is t-regular on P, is everywhere dense in A. Hence arbitrarily close to the original map $\alpha_f: M \to T^*M$ (which is equal to $\psi(x, 0)$ since $f(x) = g(x, 0)$ corresponds to the zero values of the parameters $a^i_{V_j}$), there exists a map $\psi(g, x) = dg_x = \alpha_g(x)$, which is t-regular on P. The function g then

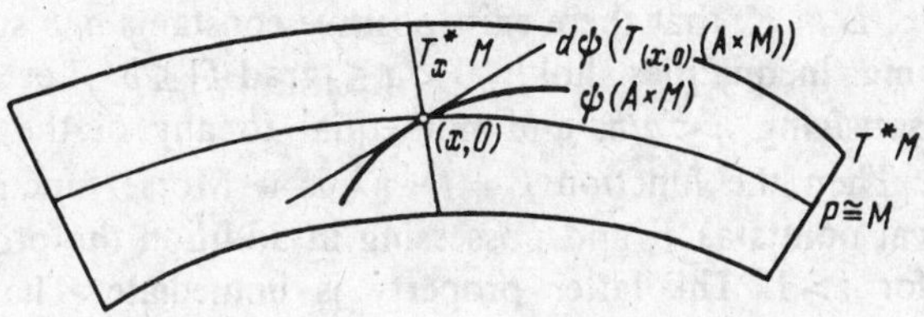

Figure 21

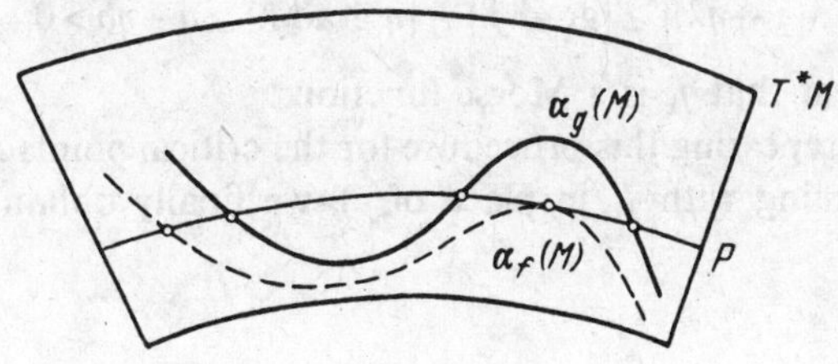

Figure 22

represents a small shift in the original function f by means of linear functions, having the property that α_g: $M \to T^*M$ is t-regular on P (see Figure 22).

Now t-regularity of α_g on P means precisely that $\alpha_g(M)$ and P intersect transversely at each point $(x_0, 0)$ of their intersection $\alpha_g(M) \cap P$, i.e. the tangent space to T^*M at $(x_0, 0)$ is the sum of the tangent spaces to P and to $\alpha_g(M)$. Now a typical element of $\alpha_g(M)$ has the form (x, dg_x) where dg_x can be written as $(\partial g/\partial x^1, \ldots, \partial g/\partial x^m)$; hence a typical tangent vector to $\alpha_g(M)$ at (x, dg_x), will have the form $(\xi, (d/dt)(\partial g(x(t))/\partial x)|_{t=0})$, where ξ is a tangent vector to M at x, and $x(t)$ is a curve in M with $x(0) = x$. It follows that tangent vectors to $\alpha_g(M)$ are of the form $(\xi, (\partial^2 g/\partial x^i \, \partial x^j)\eta)$, where ξ, η are arbitrary tangent vectors to M at x. Now since, as just observed, the tangent space to T^*M at $(x_0, 0)$ is the sum of the tangent spaces to P and to $\alpha_g(M)$, it follows that the matrix $(\partial^2 g(x_0)/\partial x^i \, \partial x^j)$ must define an isomorphism, i.e. must be non-singular at x_0. Since the x_0 such that $(x_0, 0) \in \alpha_g(M) \cap P$ are precisely the critical points of g (satisfying $dg_{x_0} \equiv 0$), we infer that the critical points of g are non-degenerate, i.e. that g is a Morse function.

Thus we have established parts (i), (ii) and (iii) of the theorem. It remains to prove (iv), i.e. that the set of Morse functions having precisely one critical point on each critical "fibre" (i.e. in each complete inverse image of a critical value) is everywhere dense in the set of all Morse functions.

Thus let $f(x)$ denote an arbitrary Morse function on M with critical points $x_1, \ldots, x_N$. Let U, W be two open neighbourhoods of the point x_1 with the properties that $\bar{U} \subset W$, $\bar{U}$ and $\bar{W}$ are compact, and $x_i \notin \bar{W}$ for $i > 1$. By Lemma 8.12 there exists a smooth function $\lambda(x)$ on M such that $\lambda(x) \equiv 1$ on $\bar{U}$, $\lambda(x) \equiv 0$ outside W, and $0 \le \lambda(x) \le 1$ on $W \backslash U$. The subset $\operatorname{supp} \lambda \cap \operatorname{supp}(1-\lambda) = K$, being a closed subset (by definition) of the compact space $\bar{W}$, is itself compact; it follows from this and the fact that none of the critical points $x_1, \ldots, x_N$ is in K, that there exist positive constants a, b such that on K the following inequalities hold: $0 < a \leq |\text{grad } f| \leq b$. Let η be any positive real satisfying $\eta < a/b$, and not equal to any of the differences $|f(x_1) - f(x_i)|$. Then the function $f_1 = f + \eta\lambda$ is a Morse function having the same critical points as f, and possessing in addition the property that $f_1(x_i) \neq f_1(x_1)$ for $i > 1$. The latter property is immediate. That f and f_1 have the same critical values follows from the facts that grad $\lambda \equiv 0$ outside K, that $f_1 = f + \eta$ (where η is constant) on U, and that on K

$$|\text{grad}(f + \eta\lambda)| \geq |\text{grad } f| - |\eta \text{ grad } \lambda| \geq a - \eta b > 0.$$

It is then also clear that f_1 is a Morse function.

Essentially by repeating this procedure for the critical points $x_2, \ldots, x_N$ in succession (beginning with f_1 in place of f), we finally obtain the desired Morse function. □

§11. Applications of Sard's Theorem

11.1. The Existence of Embeddings and Immersions

In §9 above we showed that any smooth, closed manifold can be smoothly embedded in Euclidean space $\mathbb{R}^N$ for N sufficiently large. Our next result, the so-called "weak form of Whitney's theorem", gives us an idea of how large an N will suffice.

11.1.1. Theorem (Whitney). *Any smooth, connected, closed manifold M of dimension n can be smoothly embedded in $\mathbb{R}^{2n+1}$, and smoothly immersed in $\mathbb{R}^{2n}$. Every continuous map $M \to \mathbb{R}^{2n+1}$ (resp. $M \to \mathbb{R}^{2n}$) can be approximated arbitrarily closely by smooth embeddings (resp. smooth immersions).*

PROOF. We shall prove only the first statement of the theorem. By the theorem of §9 we can regard M as a submanifold of $\mathbb{R}^N$ for some sufficiently large N. The basic idea of the following proof consists in the construction of successive projections of the submanifold $M \subset \mathbb{R}^n$ onto hyperplanes, at each step reducing by 1 the dimension of the ambient Euclidean space. Recall from §2.2 that the straight lines in $\mathbb{R}^N$ passing through the origin O are by definition the "points" of the $(N-1)$-dimensional projective space $\mathbb{R}P^{N-1}$. For each $l \in \mathbb{R}P^{N-1}$, let π_l be the orthogonal projection of $\mathbb{R}^N$ onto the hyperplane $\mathbb{R}_l^{N-1}$ orthogonal to l and containing the point O.

Our aim is in essence to find a straight line l such that the projection $\pi_l(M)$ of M is again a smooth manifold (now in $\mathbb{R}_l^{N-1}$). Considering first the case of immersions, we seek projections π_l with the property that for every $x \in M$ the differential $d\pi_l$: $T_xM \to \mathbb{R}_l^{N-1}$, has zero kernel. Those directions $l \in \mathbb{R}P^{N-1}$ for which $d\pi_l$ has non-zero kernel for some x, we shall call *prohibited directions of*

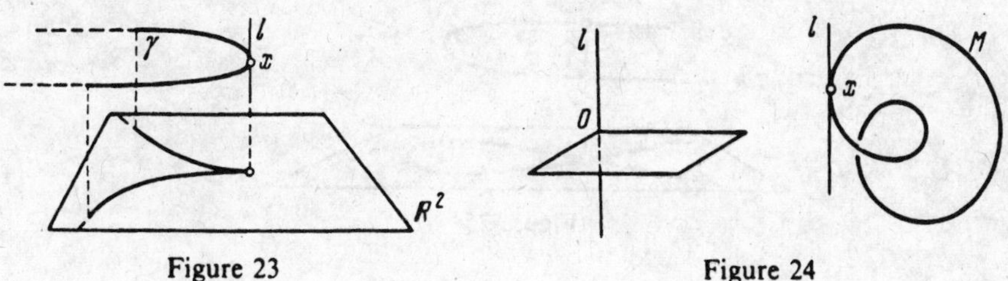

Figure 23

Figure 24

the first kind. (If, for example, a smooth curve $\gamma \subset \mathbb{R}^3$ is projected onto a 2-dimensional plane $\mathbb{R}^2$ along a prohibited direction, the result of the projection will be a curve having singularities of the form shown in Figure 23, usually called "cusps".) It is clear that the prohibited directions l are characterized by their having the property that for at least one point $x \in M$ some parallel translate of l is in T_xM (see Figure 24). In terms of the set of prohibited directions we define a smooth manifold Q as follows: the points of Q are the ordered pairs (x, l) where $x \in M$ and l is any straight line (in $\mathbb{R}P^{N-1}$) parallel to a vector in T_xM. (Show that Q is indeed a smooth manifold!) Since the x need n parameters for their determination and the l need $n-1$, it follows that Q is $(2n-1)$-dimensional. In view of the smoothness of the embedding $M \subset \mathbb{R}^N$, the map $\alpha\colon Q \to \mathbb{R}P^{N-1}$, defined by $\alpha(x, l) = l$, is also smooth. By Sard's theorem the set $\alpha(Q)$ consisting precisely of the prohibited directions, has $(N-1)$-dimensional measure zero if $N-1 > 2n-1$, i.e. if $N > 2n$. Hence if $2n < N$, then the set $\alpha(Q)$ of all prohibited directions certainly fails to exhaust $\mathbb{R}P^{N-1}$, so that there exists an element l_0 of $\mathbb{R}P^{N-1}$ outside $\alpha(Q)$. Hence in this case we have a smooth immersion (namely $\pi_{l_0}\colon M \to \mathbb{R}^{N-1}_{l_0}$) of M into $\mathbb{R}^{N-1}_{l_0}$.

Clearly this argument remains valid if from the beginning we assume only that M is immersed (rather than embedded) in $\mathbb{R}^N$; hence the above procedure of projecting M onto a hyperplane can be re-applied in the new situation where M is immersed in $\mathbb{R}^{N-1}$, and in fact successively repeated k times provided $2n \le N-k$. At the commencement of the final (say $(k+1)$st) step we shall have M immersed in $\mathbb{R}^{N-k}$ where $N-k = 2n+1$; a final projection will then achieve an immersion of M in $\mathbb{R}^{2n}$, after which the above procedure is no longer applicable. This concludes the proof that M can be immersed in $\mathbb{R}^{2n}$.

We now turn to the problem of embedding M in $\mathbb{R}^{2n+1}$. Thus we seek a straight line l in $\mathbb{R}P^{N-1}$ such that the projection π_l, in addition to being an immersion of M, is one-to-one on M, i.e. $\pi_l(M)$ involves no self-intersections of M. A straight line $l \in \mathbb{R}P^{N-1}$ will be called a *prohibited direction of the second kind* if the corresponding projection $\pi_l\colon M \to \mathbb{R}^{N-1}_l$ is not one-to-one. (If, for example, a smooth curve $\gamma \subset \mathbb{R}^3$ is projected onto a 2-dimensional plane $\mathbb{R}^2$ along a prohibited direction of the second kind, then the image will have self-intersections as illustrated in Figure 25.)

It is clear that the prohibited directions of the second kind are just those $l \in \mathbb{R}P^{N-1}$ some parallel translate of which intersects M in at least two points. The set of all prohibited directions gives rise to a $2n$-dimensional, smooth

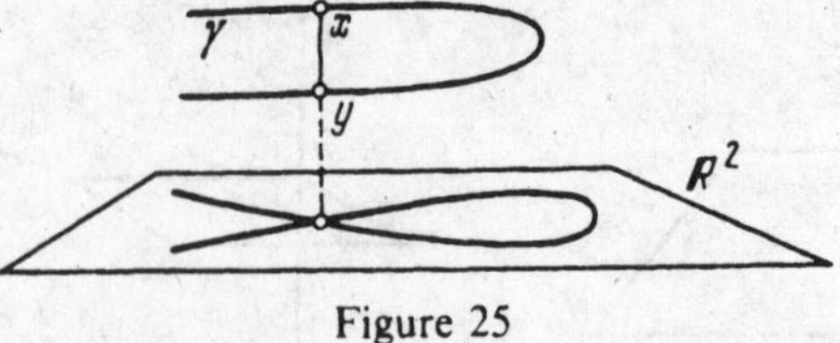

Figure 25

manifold P consisting of all ordered pairs (x, y), $x, y \in M$, where $x \neq y$; i.e. $P = (M \times M)\backslash\Delta$, where $\Delta = \{(x, x) | x \in M\}$ is the diagonal subset of $M \times M$. The map $\beta\colon P \to \mathbb{R}P^{N-1}$, where $\beta(x, y)$ is defined to be the straight line through O parallel to the straight line segment joining x and y, is easily seen to be smooth. Hence Sard's theorem tells us that if $2n < N-1$, then the set $\beta(P) \subset \mathbb{R}P^{N-1}$, has $(N-1)$-dimensional measure zero.

Hence the union $\alpha(Q) \cup \beta(P)$ (which incidentally coincides with the closure $\overline{\beta(P)}$ of $\beta(P)$, since $\alpha(Q)$ consists essentially of tangent lines to M, while $\beta(P)$ consists of chords) has $(N-1)$-dimensional measure zero. Thus if $N > 2n+1$ there certainly exist straight lines in $\mathbb{R}P^{N-1}$ which are not prohibited directions of either kind. If l_0 is any such straight line, then the projection $\pi_{l_0}\colon M \to \mathbb{R}^{N-1}_{l_0}$, will be a smooth embedding of M in $\mathbb{R}^{N-1}_{l_0}$. As in the case of immersions, we iterate this procedure of projecting M onto a hyperplane until the argument just given no longer applies; with the final iteration we shall have achieved an embedding of M into $\mathbb{R}^{2n+1}$, as required. □

Remark. In connexion with this proof we note that there is in general no possibility of refining the argument to produce further projections onto hyperplanes, which are still embeddings of M. For suppose that M is the circle S^1 initially embedded in $\mathbb{R}^3$ as a non-trivial knot (as, for example, in Figure 26); here $n = 1$, $N = 3 = 2n+1$. It is intuitively clear that all projections of this knotted circle onto 2-dimensional planes in $\mathbb{R}^3$ will result in self-intersecting curves. This counterexample shows that any attempt to reduce the dimension of the ambient Euclidean space which uses the "projection method" of the above proof, will necessarily be unavailing. Nonetheless by means of a subtler argument the dimension can be reduced: there is a more difficult result (whose proof we omit) to the effect that any compact n-manifold M can be smoothly embedded in $\mathbb{R}^{2n}$; however the embeddings

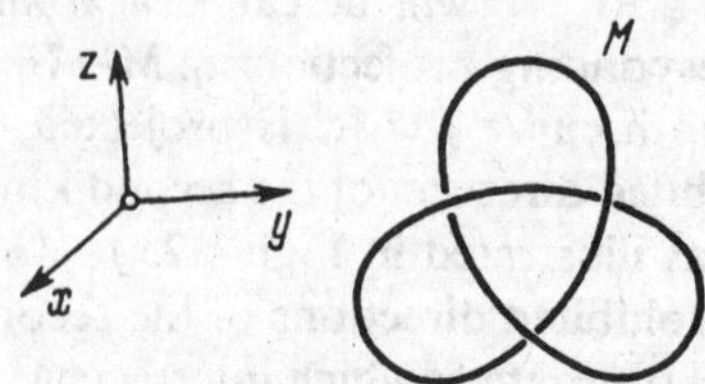

Figure 26

$M \to \mathbb{R}^{2n}$, are no longer everywhere dense in the space of all smooth maps $M \to \mathbb{R}^{2n}$. In general no further reduction of the dimension is possible: we invite the reader to prove that a non-orientable, closed 2-manifold is not embeddable in $\mathbb{R}^3$ (see §2.1). However in particular cases, i.e. for more restricted classes of manifolds, the dimension can be further reduced; for instance, it can be shown on the basis of the known classification of compact, orientable, 2-dimensional manifolds, that any such manifold can be embedded in $\mathbb{R}^3$.

11.2. The Construction of Morse Functions as Height Functions

With the aid of an embedding $M \to \mathbb{R}^N$, we shall now give a different proof of the existence of Morse functions on a compact smooth manifold M. We shall show that Morse functions can be found in a rather narrow class of functions on manifolds, called "height functions", which we shall now define.

Let M be smoothly embedded in $\mathbb{R}^N$, and let (in distinction to the previous subsection) $\xi_l(t)$ denote a (parametrized) straight line through the origin of $\mathbb{R}^N$ with direction vector l. Define a function h_l on M by setting $h_l(x)$ equal to t_x where t_x is that value of the parameter t at the point of orthogonal projection $\xi_l(t_x)$ of $x \in M$ onto the line $\xi_l(t)$. (Thus if t is distance (positive or negative) on $\xi_l(t)$ from the origin, then $h_l(x)$ is the distance of $x \in M$ from the origin (or "height above the origin") in the direction l. Such functions h_l are called *height functions* on $M \subset \mathbb{R}^N$.

The reader will easily verify the following two properties of height functions:

(i) The set of height functions on M is (neglecting the particular parametrizations of the $\xi_l(t)$) in one-to-one correspondence with the set of pairs of diametrically opposite points of the (unit) sphere S^{N-1}, or, equivalently, the points of $\mathbb{R}P^{N-1}$;

(ii) A point x_0 of M is a critical point of a height function h_l if and only if the vector l is orthogonal to the manifold M at the point x_0 (i.e. $l \perp T_{x_0}M$).

Our next task is to discover conditions under which a critical point $x_0 \in M$ of a height function h_l is non-degenerate. We shall at this point restrict our considerations to manifolds M embedded in $\mathbb{R}^{m+1}$, where $m = \dim M$, i.e. to hypersurfaces in $\mathbb{R}^{m+1}$ (but see the "Important Remark" below). Recall (from §26.2 of Part I) that for hypersurfaces $M \subset \mathbb{R}^{m+1}$ the *Gauss map* $r\colon M \to S^m \to \mathbb{R}P^m$ is defined by $r(x) = n(x)$, where $n(x)$ is the unit normal vector to M at the point x (translated back to the origin, i.e. with its tail at the origin).

11.2.1. Lemma. *A point x_0 of the hypersurface $M \subset \mathbb{R}^{m+1}$ is a non-degenerate critical point of a height function h_l if and only if it is a regular (i.e. non-critical) point of the Gauss map $r\colon M \to \mathbb{R}P^m$ (and of course also $l \perp T_{x_0}M$).*

PROOF. By means of a suitable orthogonal change of co-ordinates on $\mathbb{R}^{m+1}$ we can arrange that the x^{m+1}-axis is parallel to l (so that the other m-axes are perpendicular to l). The hyperplane $\mathbb{R}^m(x^1, \ldots, x^m)$ can then be regarded as the tangent space to M at x_0, and in some sufficiently small neighbourhood of x_0 the manifold M will be given by an equation of the form $x^{m+1} = \varphi(x^1, \ldots, x^m)$, with $d\varphi|_{x_0} = 0$, and $x^1, \ldots, x^m$ will serve as local co-ordinates. Then on that neighbourhood the "height" $h_l(x)$ is just the function $\varphi(x^1, \ldots, x^m) = x^{m+1}$. By defining in terms of $x^1, \ldots, x^{m+1}$, as in the proof of Theorem 26.2.2 of Part I, suitable local co-ordinates $y^1, \ldots, y^m$ on the sphere S^m in a neighbourhood of the image point $r(x_0)$, and repeating the argument given there (with $m+1$ in place of 3) leading to the formula $K\, d\sigma = r^*(\Omega)$, where K is the Gaussian curvature of M (which may be taken to be defined by the equation below), and $d\sigma$, Ω are the (induced) volume elements on M and S^m, respectively, we obtain the following equations:

$$\left(\frac{\partial^2 h_l(x_0)}{\partial x^\alpha\, \partial x^\beta}\right) = \left(\frac{\partial^2 \varphi(x_0)}{\partial x^\alpha\, \partial x^\beta}\right) = \left(\frac{\partial y^\alpha}{\partial x^\beta}\bigg|_{x_0}\right);$$

$$\det\left(\frac{\partial^2 h_l(x_0)}{\partial x^\alpha\, \partial x^\beta}\right) = K,$$

where here K is the Gaussian curvature of M at x_0, and $y^\alpha = r^\alpha(x^1, \ldots, x^m)$, $1 \le \alpha \le m$, is the Gauss map expressed in terms of the local co-ordinates $x^1, \ldots, x^m$; $y^1, \ldots, y^m$. It is immediate from the first of the above equations that the condition that the Gauss map r be regular at x_0, namely $\det(\partial y^\alpha / \partial x^\beta|_{x_0}) \neq 0$, is equivalent to the defining condition for non-degeneracy of the critical point x_0 of h_l, namely $\det(\partial^2 h_l(x_0)/\partial x^\alpha\, \partial x^\beta) \neq 0$. This completes the proof of the lemma. □

11.2.2. Theorem. *A height function h_l on a hypersurface $M \subset \mathbb{R}^{m+1}$ ($m = \dim M$) is a Morse function precisely if the point $\{\pm l\}$ of $\mathbb{R}P^m$ is a regular value of the Gauss map $r\colon M \to \mathbb{R}P^m$. It follows that the Morse functions are everywhere dense in the set of height functions on M.*

PROOF. The first statement is immediate from the lemma. The second follows from Sard's theorem, which tells us that the regular values of the Gauss map are everywhere dense in $\mathbb{R}P^m$. □

Important Remark. This theorem is valid without the restriction that M be a hypersurface in $\mathbb{R}^m$, i.e. it holds quite generally for any smooth embedding $M \subset \mathbb{R}^q$ ($q \ge m+1$). We now sketch an argument allowing us to deduce this general result from Lemma 11.2.1. Thus suppose $M \subset \mathbb{R}^q$ with $q > m+1$, where $m = \dim M$. We shall now define a $(q-1)$-dimensional manifold N, as the boundary of a so-called "tubular neighbourhood" of the manifold M. To this end consider the set of $(q-m)$-dimensional discs D_x^{q-m} in $\mathbb{R}^q$ (i.e. balls) orthogonal to the submanifold M, of radius $\varepsilon > 0$, and with centres at the points x of M. For sufficiently small $\varepsilon > 0$, the union of these discs is a q-

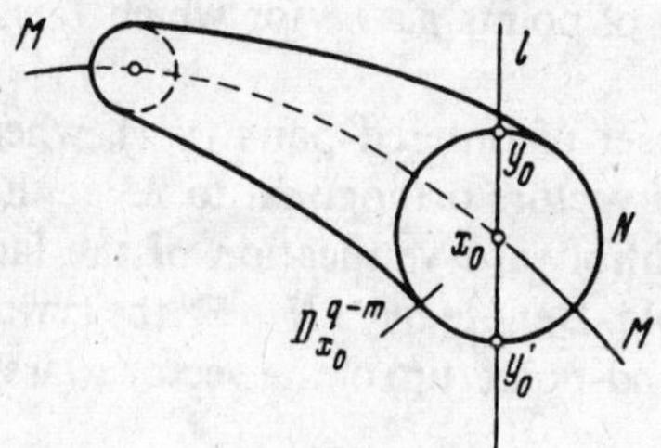

Figure 27

dimensional submanifold of $\mathbb{R}^q$ (a "tubular neighbourhood" of M), whose boundary we denote by N (see Corollary 11.3.3 below). Since N has dimension $q-1$, it can be mapped via the Gauss map r to S^{q-1}.

Now let $h_l(x)$ be any height function defined for points x in M or N (i.e. on $M \cup N$). It is intuitively clear that with each critical point $x_0 \in M$ of the function $h_l(x)$ there are associated exactly two critical points y_0 and y_0' of h_l restricted to N, namely the two points of intersection with N of the straight line parallel to l passing through x_0 and orthogonal to M at x_0 (see Figure 27).

It may be shown that a critical point $x_0 \in M$, of the function h_l restricted to M, is non-degenerate precisely if the points y_0, y_0' are non-degenerate for h_l restricted to N. (They are always either both degenerate or both non-degenerate.) Hence if h_l is a Morse function on N, then it is also a Morse function when restricted to M. The existence and everywhere denseness (in the set of all height functions on N) of Morse functions on N, guaranteed by Theorem 11.2.2, therefore implies the analogous conclusion for M. We have thus shown that the restriction that M be a hypersurface in $\mathbb{R}^{m+1}$ was inessential.

11.3. Focal Points

There exist yet other simple methods of obtaining a plentiful supply of Morse functions on a smooth manifold. Without going into great detail, we describe one such method.

Consider a smooth m-manifold M smoothly embedded as a submanifold of $\mathbb{R}^q$. With each point $p \in \mathbb{R}^q$, we associate a smooth function L_p on the submanifold M, defined by $L_p(x)=|p-x|^2$, where $x \in M$ and $|p-x|$ is the Euclidean distance between the points $p, x \in \mathbb{R}^q$. We shall show that for almost all points $p \in \mathbb{R}^q$ (i.e. for all points except those in a set of measure zero) the function L_p is a Morse function on M. (Clearly the set of functions of the form $L_p(x)$ is different from the set of height functions $h_l(x)$.)

For the proof we first define the "focal points" for $M \subset \mathbb{R}^q$; our ultimate aim will then be to show that the set of focal points, which has measure zero, coincides with the set of points $p \in \mathbb{R}^q$ for which L_p is not a Morse function on M.

Denote by N the set of ordered pairs (x, v), where $x \in M$, and v ($\in \mathbb{R}^q$) ranges over the set of vectors orthogonal to M at the point x. We leave to the reader the straightforward verification of the fact that N is a smooth q-dimensional manifold. Denote by $f: N \to \mathbb{R}^q$ the (smooth) map sending each pair $(x, v) \in N$ to the end-point (tip) of the vector v (assumed to have its tail at x).

11.3.1. Definition. A point P of $\mathbb{R}^q$ is called a *focal point of multiplicity* μ for M if for some $(x, v) \in N$, we have $P = f(x, v)$ and the Jacobian of the map f has rank $q - \mu < q$ at (x, v).

It is immediate from Sard's theorem that the set of focal points P for $M \subset \mathbb{R}^q$ has measure zero in $\mathbb{R}^q$.

We next define (analogously to the definition in the case $q=3$, $m=2$, given in §8.1 of Part I) the "second fundamental form" of the submanifold $M \subset \mathbb{R}^q$. Let $x_0 \in M$. We may suppose that in some neighbourhood of x_0 the submanifold $M \subset \mathbb{R}^q$ is defined by a parametric system $x = x(u^1, \ldots, u^m)$ of equations, where the u^α are local co-ordinates on M and $x = (x^1, \ldots, x^q)$. Write $x_{ij} = \partial^2 x / \partial u^i \, \partial u^j$, denote by v a unit normal to M at the point x_0, and consider the (Euclidean) scalar product (at x_0) $\langle v, x_{ij} \rangle = \langle v, n_{ij}(x_0) \rangle$, where $n_{ij}(x_0)$ is the component of the vector x_{ij} in the direction of the normal v at the point x_0. The matrix Q_v with entries $\langle v, x_{ij} \rangle$ is then by definition the matrix of the *second fundamental form* of M with respect to the normal v. Let $G = (g_{ij})$ be the matrix of the first fundamental form of $M \subset \mathbb{R}^q$, i.e. of the metric on M induced by the Euclidean metric on $\mathbb{R}^q$. We may suppose the local co-ordinates $u^1, \ldots, u^m$ so chosen that $G(x_0)$ is the identity matrix, so that at the particular point x_0 we have $G^{-1}Q_v = Q_v$. Analogously to the definition given in §8.3 of Part I, we define the *principal curvatures of M at the point x_0 in the direction v* to be the eigenvalues λ_α of the matrix $G^{-1}Q_v$ (which in terms of our specially chosen local co-ordinates is just Q_v at x_0).

11.3.2. Lemma. *The focal points on the straight line $x_0 + tv$ (through the point x_0 and parallel to v) in $\mathbb{R}^q$, are precisely the points where $t = \lambda_\alpha^{-1}$, $1 \le \alpha \le m$, i.e. where t takes on as values the reciprocals of the principal curvatures of $M \subset \mathbb{R}^q$ at the point x_0, in the direction v.*

PROOF. Since N has the same local manifold structure as the product $M \times \mathbb{R}^{q-m}$, we may introduce on some neighbourhood of the point (x_0, v) local co-ordinates of the form $u^1, \ldots, u^m, t^1, \ldots, t^{q-m}$, where $u^1, \ldots, u^m$ are local co-ordinates on M. In terms of such co-ordinates f has the form

$$f(x(u), t) = x(u) + \sum_\alpha t^\alpha a_\alpha(u),$$

where $(a_\alpha(u))$ is a frame for the space of all vectors orthogonal to M at the point $x(u)$, which frame is assumed to depend smoothly on u. Hence in terms of these co-ordinates df is given by the $q \times q$ matrix

$$\begin{pmatrix} \dfrac{\partial f}{\partial u^i} \\ \dfrac{\partial f}{\partial t^\alpha} \end{pmatrix} = \begin{pmatrix} \dfrac{\partial x}{\partial u^i} + \sum_{\alpha=1}^{q-m} t^\alpha \dfrac{\partial a_\alpha(u)}{\partial u^i} \\ a_\alpha(u) \end{pmatrix}. \tag{1}$$

The effect of df on a tangent vector to N is then computed by premultiplying the vector by this matrix. Since for x near x_0 we may take as a basis for T_xM the set of vectors $\partial x/\partial u^\alpha$, $1 \le \alpha \le m$, the set

$$\left\{\frac{\partial x}{\partial u^1}, \dots, \frac{\partial x}{\partial u^m}; a_1(u), \dots, a_{q-m}(u)\right\}$$

forms a basis for the tangent space to N at each point in the neighbourhood of (x_0, v). In view of (1), the image under df of this basis is given by the columns of the matrix

$$\begin{pmatrix} \left\langle \dfrac{\partial x}{\partial u^i}, \dfrac{\partial x}{\partial u^j} \right\rangle + \sum_{\alpha=1}^{q-m} t^\alpha \left\langle \dfrac{\partial a_\alpha}{\partial u^i}, \dfrac{\partial x}{\partial u^j} \right\rangle & \sum_{\alpha=1}^{q-m} t^\alpha \left\langle \dfrac{\partial a_\alpha}{\partial u^i}, a_\beta \right\rangle \\ 0 & 1 \end{pmatrix}, \tag{2}$$

where here 1 denotes the $(q-m) \times (q-m)$ identity matrix. The rank of this matrix will therefore be equal at each point to the rank of the map f, i.e. the rank of the Jacobian matrix (1) of f.

Now clearly the rank of the matrix (2) is equal to $(q-m) + \operatorname{rank} A$, where

$$A = \left(\left\langle \frac{\partial x}{\partial u^i}, \frac{\partial x}{\partial u^j} \right\rangle + \sum_\alpha t^\alpha \left\langle \frac{\partial a_\alpha}{\partial u^i}, \frac{\partial x}{\partial u^j} \right\rangle\right).$$

Since $\langle a_\alpha, \partial x/\partial u^j \rangle = 0$, we have

$$0 = \frac{\partial}{\partial u^i}\left\langle a_\alpha, \frac{\partial x}{\partial u^j}\right\rangle = \left\langle \frac{\partial a_\alpha}{\partial u^i}, \frac{\partial x}{\partial u^j}\right\rangle + \left\langle a_\alpha, \frac{\partial^2 x}{\partial u^i \partial u^j}\right\rangle,$$

whence A can be rewritten as

$$A = \left(g_{ij} - \sum_\alpha t^\alpha \left\langle a_\alpha, \frac{\partial^2 x}{\partial u^i \partial u^j}\right\rangle\right).$$

Thus for points of the form (x_0, tv) we have

$$A = \left(g_{ij} - t\left\langle v, \frac{\partial^2 x}{\partial u^i \partial u^j}\right\rangle\right) = (g_{ij} - t\langle v, x_{ij}\rangle).$$

Hence at such points the rank of f is the same as the rank of the matrix $(g_{ij} - t\langle v, x_{ij}\rangle)$, whence the lemma is immediate. □

With this lemma we can now quickly reach our objective. Thus consider the function $L_p(x) = |x-p|^2 = \langle x-p, x-p\rangle$, where $p \in \mathbb{R}^q$. Then

$$\frac{\partial L_p(x)}{\partial u^i} = 2\left\langle \frac{\partial x}{\partial u^i}, x \right\rangle - 2\left\langle \frac{\partial x}{\partial u^i}, p \right\rangle = 2\left\langle \frac{\partial x}{\partial u^i}, x-p \right\rangle.$$

Hence the critical points of the function L_p on M are precisely those points $x \in M$ such that $\langle \partial x/\partial u^i, p-x \rangle = 0$, i.e. such that as a vector, $x-p$ is perpendicular to T_xM. Thus if x is critical, p has the form $x+tv$ for some (unit) normal vector to T_xM. To examine non-degeneracy we need to consider the second partial derivatives:

$$\frac{\partial^2 L_p(x)}{\partial u^i\, \partial u^j} = \left\langle \frac{\partial^2 x}{\partial u^i\, \partial u^j}, x-p \right\rangle + \left\langle \frac{\partial x}{\partial u^i}, \frac{\partial x}{\partial u^j} \right\rangle = -\left\langle \frac{\partial^2 x}{\partial u^i\, \partial u^j}, tv \right\rangle + g_{ij},$$

where we have put $p = x+tv$. It is now immediate, in view of the above lemma, that the degenerate critical points of L_p occur precisely when p is a focal point for M; this is what we were seeking to prove.

The following important consequence is usually called the "theorem on the existence of a tubular neighbourhood".

11.3.3. Corollary. *Let $M \subset \mathbb{R}^q$ be a smooth, closed submanifold. Then there exists an $\varepsilon > 0$ such that the tubular neighbourhood $N_\varepsilon(M) = \{y \in \mathbb{R}^q \mid \rho(y, M) < \varepsilon\}$ is a smooth q-dimensional submanifold of $\mathbb{R}^q$, with boundary $\partial N_\varepsilon(M)$ a smooth $(q-1)$-dimensional submanifold of $\mathbb{R}^q$. The manifold $N_\varepsilon(M)$ fibres into $(q-m)$-dimensional discs $D_\varepsilon^{q-m}(x)$, with centres x (only) on M, and the manifold $\partial N_\varepsilon(M)$ fibres into spheres $S_\varepsilon^{q-m-1}(x)$, $x \in M$.*

PROOF. It suffices to take $\varepsilon < \min_{i,x} \lambda_i^{-1}(x)$, where x ranges over the closed manifold M, and $1 \le i \le m$; then in $N_\varepsilon(M)$ there will be no focal points for $M \subset \mathbb{R}^q$, and the corollary follows from the definition of focal points. □

CHAPTER 3

The Degree of a Mapping. The Intersection Index of Submanifolds. Applications

§12. The Concept of Homotopy

12.1. Definition of Homotopy. Approximation of Continuous Maps and Homotopies by Smooth Ones

Let M and N be two smooth (for simplicity of class C^∞) manifolds and let $f: M \to N$, be a smooth map between them.

12.1.1. Definition. A *smooth* (resp. *piecewise smooth, continuous*) *homotopy* (or *deformation*) of a map $f: M \to N$ of manifolds, is a smooth (resp. piecewise smooth, continuous) map

$$F: M \times I \to N \qquad (I = [0, 1])$$

of the cylinder $M \times I$ to N, with the property that $f(x, 0) = f(x)$ for all $x \in M$. Each of the maps $f_t: M \to N$ defined by $f_t(x) = F(x, t)$ is said to be *homotopic* to the initial map $f = f_0$, and the map F of the whole cylinder is called a *homotopy* (or "homotopy process").

Since clearly the relation of homotopy between maps is an equivalence relation, the set of all maps $M \to N$ homotopic to a particular map f, consists of pairwise homotopic maps; such a set is called a *homotopy class* of maps $M \to N$.

For each $l \geq 0$, one defines in the obvious way the smoothness class C^l of a smooth homotopy; in particular corresponding to $l = 0$ we have the class of all continuous homotopies. However for suitable manifolds M, N we shall now establish (on the basis of Theorem 10.1.2 above, on the approximability of continuous maps by smooth ones) the following two properties of homotopies:

(i) Any continuous map $f: M \to N$ can (for suitable M, N) be approximated arbitrarily closely by maps of class C^∞ which are (continuously) homotopic to f;
(ii) If two smooth maps $M \to N$ are continuously homotopic, then they are smoothly homotopic.

We now turn to the details.

12.1.2. Theorem. *Let M, N be smooth, compact manifolds, and let $f, g: M \to N$ be continuous maps. Then given any Riemannian metric on the manifold N, there exists a number $\varepsilon > 0$ with the property that whenever f, g satisfy $\rho(f, g) < \varepsilon$ (where ρ denotes the distance between maps—see* §10.1), *then the maps f and g are homotopic.*

PROOF. Let ρ denote also the distance between points of N, defined (as in §1.2) in terms of the given Riemannian metric on N. It follows from the compactness of N and the local uniqueness of geodesics (see §29.2 of Part I) that there exists an $\varepsilon > 0$ with the property that whenever two points $p, q \in N$ satisfy $\rho(p, q) < \varepsilon$, there is a unique geodesic arc γ joining p, q (and whose length is least among all continuous arcs from p to q so that the length of the geodesic arc is $\rho(p, q)$—see §36.2 of Part I). Now let $f, g: M \to N$, be continuous maps satisfying $\rho(f, g) < \varepsilon$ (i.e. $\max_{x \in M} \rho(f(x), g(x)) < \varepsilon$); we wish to construct a homotopy $F: M \times I \to N$ with $F(x, 0) = f(x)$, $F(x, 1) = g(x)$. Thus we seek to define $F(x, t)$ appropriately for all $(x, t) \in M \times I$. For each $x \in M$, since $\rho(f(x), g(x)) < \varepsilon$, there is, by choice of ε, a unique (shortest) geodesic arc $\gamma_{f(x), g(x)}(\tau)$ joining $f(x)$ to $g(x)$. As the image point $F(x, t)$ we choose that point on this (directed) geodesic arc which divides it in the ratio $t: (1-t)$ (see Figure 28). The continuity of F follows from the result (of the theory of ordinary differential equations) stating that the solution of a system of ordinary differential equations (in particular of the system of such equations defining the geodesics in N) depends continuously on the initial conditions. This completes the proof. □

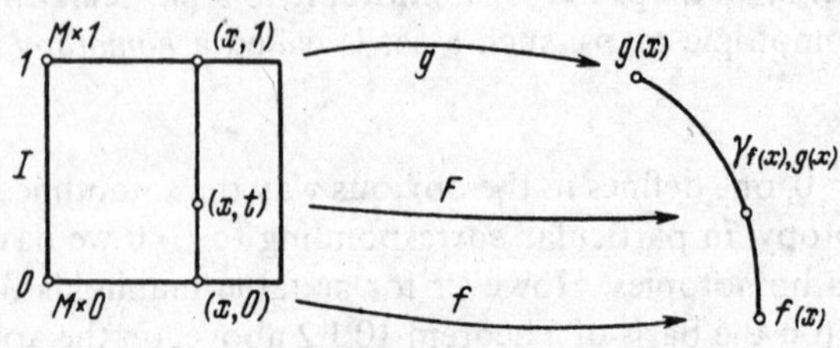

Figure 28

12.1.3. Theorem. *Any continuous map $f: M \to N$ between compact, smooth manifolds M and N, is homotopic to some smooth map $g: M \to N$.*

PROOF. Since by Theorem 10.1.2 the map f can be approximated arbitrarily closely by smooth maps, this result follows immediately from the preceding theorem. □

EXERCISE

In connexion with the latter theorem, prove that if the continuous map f is smooth on some submanifold $X \subset M$, then the smooth approximating map g homotopic to it, may be so chosen as to coincide with f on X (see the proof of Theorem 10.1.2).

From this exercise and the result (Theorem 10.1.2) on the approximability by smooth maps of any continuous map (in this case the continuous homotopy $F: M \times I \to N$ constructed in the proof of Theorem 12.1.2) we deduce immediately the following

12.1.4. Theorem. *If $f, g: M \to N$ are two smooth maps between the compact, smooth manifolds M, N, which are continuously homotopic, then they are smoothly homotopic.*

A similar line of argument leads to the following result: *Any continuous map from a compact, smooth manifold M to a compact, smooth manifold N of sufficiently large dimension, is homotopic to a smooth embedding. If the initial map is smooth (rather than just continuous) then the homotopy (with the embedding as terminal map) can be made smooth.*

12.1.5. Definition. Two smooth embeddings $f, g: M \to N$ are said to be (smoothly) *isotopic*, if there exists a smooth homotopy $F: M \times I \to N$, between the maps f and g, such that for all $t \in [0, 1]$ the map $f_t: M \to N$ defined by $f_t(x) = F(x, t)$, is a (smooth) embedding.

12.1.6. Theorem. *Any two smooth embeddings f, g of a smooth, closed manifold M into Euclidean space $\mathbb{R}^q$ are isotopic provided q is sufficiently large. (In fact $q \geq 2m + 2$ suffices, where $m = \dim M$.)*

PROOF. It can be shown that for sufficiently large q the embeddings f and g are homotopic via a continuous (and hence by a smooth) homotopy $F: M \times I \to \mathbb{R}^q$. (This can be achieved for instance by showing that for q large enough both maps f, g are continuously homotopic to a map of M onto a single point of $\mathbb{R}^q$.) In view of the second assertion of Theorem 11.1.1, the smooth homotopy F can then be smoothly deformed to a smooth embedding of $M \times I$ into $\mathbb{R}^q$ again provided q is large enough; moreover, it can be arranged that this embedding of $M \times I$ agrees with F on the "base" $M \times 0$ and the "lid" $M \times 1$ of the cylinder $M \times I$. This smooth embedding of $M \times I$ into $\mathbb{R}^q$ then serves as the desired smooth isotopy between f and g. □

The above theorems permit us largely to ignore the difference between continuous and smooth homotopies; thus in the sequel we shall employ whichever sort of homotopy—continuous, piecewise smooth, or smooth of class C^∞—is most appropriate in the particular context.

We shall in what follows denote the homotopy class of a map $f: M \to N$ by $[f]$, and the set of all homotopy classes of maps $M \to N$ by $[M; N]$.

12.2. Relative Homotopies

We shall subsequently need to consider homotopies and homotopy classes of maps satisfying certain additional conditions. We note here the most important such conditions.

(i) Let $x_0 \in M$ and $y_0 \in N$ be particular points of manifolds M and N. We shall sometimes be interested in the restricted class of maps $M \to N$ which send x_0 to y_0. The homotopies between maps from M to N where all maps are restricted in this way are termed *homotopies relative to the distinguished points* $x_0 \in M$, $y_0 \in N$.

(ii) Suppose that M and N are manifolds with boundaries ∂M and ∂N. A homotopy involving exclusively maps $M \to N$ which send ∂M to ∂N, is called a *homotopy relative to the boundaries*.

(iii) Suppose that M, N are (not necessarily compact) manifolds. If we restrict our consideration to maps $M \to N$ with the property that the complete preimage of each point of N is compact, then homotopies $F: M \times I \to N$ which also have this property, are called *proper*.

(iv) The most general kind of relative homotopy is that where the permitted maps $M \to N$ are those which send a specified subset A of M to a specified subset B of N. The totality of corresponding relative homotopy classes is in this context denoted by $[(M, A); (N, B)]$.

§13. The Degree of a Map

13.1. Definition of Degree

In this subsection our basic interest will be the theory of homotopy classes of maps between closed, oriented manifolds M and N of the same dimension n, especially in the case where N is a sphere. Let $f: M \to N$ be a smooth map, and let $y_0 \in N$ be a regular value of f (see §10.2). This means that the complete inverse image of y_0 consists of only finitely many points $x_1, \ldots, x_m$, and that if x_i^β are local co-ordinates in a neighbourhood of x_i, and y_0^α are local co-ordinates in a neighbourhood of y_0, then the Jacobian $\det(\partial y_0^\alpha/\partial x_i^\beta)$ is non-zero at x_i for each $i = 1, \ldots, m$. (The finiteness of $f^{-1}(y_0)$ follows from the second (i.e. the defining) condition for regularity.) Note that since we are also assuming our manifolds M, N to be oriented, the Jacobians of the transition functions between co-ordinates on the intersection of local co-ordinate neighbourhoods are always positive.

13.1.1. Definition. By the *degree* of smooth map $f: M \to N$ of connected, oriented, closed manifolds, with respect to a regular value $y_0 \in N$, we mean the number

$$\deg f = \sum_{f(x_i)=y_0} \operatorname{sgn} \det \left(\frac{\partial y_0^\alpha}{\partial x_i^\beta} \right), \tag{1}$$

where the x_i, $i = 1, \ldots, m$, are the preimages of y_0 under f. (Note that in view of the orientations prescribed on M and N, the signs of the Jacobians are uniquely defined on regions of overlap of charts.)

The degree has important invariance properties.

13.1.2. Theorem. *The degree of a map f (as above) is independent of the choice of the regular value y_0, and invariant under homotopies.*

PROOF. We show first that the degree of $f: M \to N$ is independent of the choice of the regular value y_0. (For regular values sufficiently close to y_0 this is immediate since the number of preimages will be the same and the signs of the Jacobians will correspond.)

Let y_0, y_1 be any distinct regular values of f, and join them by a smooth non-self-intersecting path γ in N, with non-zero tangent vector at each point. By the sort of argument used to prove Theorem 10.3.2 it can be shown that the path γ can be chosen so that f is t-regular on γ. The t-regularity of f on γ then implies that the complete inverse image $f^{-1}(\gamma)$ of γ, is a smooth one-dimensional manifold in M, with boundary made up of the two sets $f^{-1}(y_0)$ and $f^{-1}(y_1)$. (See Figure 29, where $n = 2$, the points x_{i0} are the preimages of y_0, the

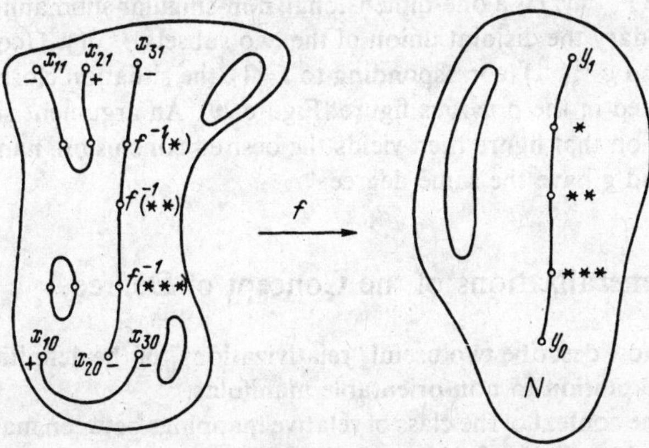

Figure 29

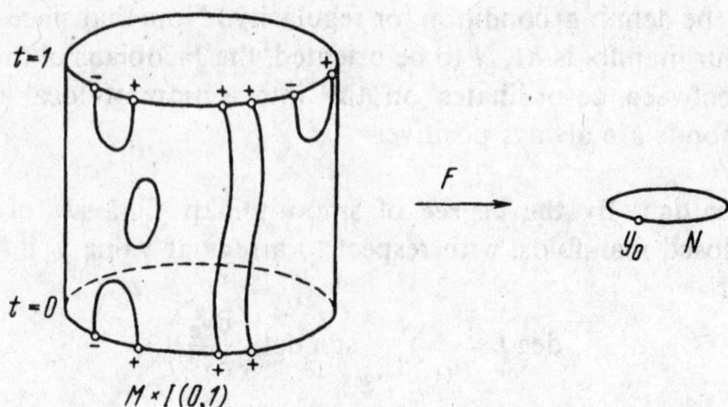

Figure 30

points x_{i1} are the preimages of y_1, and where each of these points is labelled with the sign of the corresponding Jacobian. Note that different points of the arc γ may have different numbers of preimages.) If two points of $f^{-1}(y_0)$ (or $f^{-1}(y_1)$) are the end-points of a connected component of $f^{-1}(\gamma)$ (in Figure 29 the points x_{10} and x_{30} are such), then by tracing out the image of that connected component in γ and taking into account the orientations on M and N, it is not difficult to see that these end-points must have opposite signs associated with them (as do x_{10} and x_{30} in Figure 29). If on the other hand a connected component of $f^{-1}(\gamma)$ joins a point in $f^{-1}(y_0)$ to a point in $f^{-1}(y_1)$ (as the points x_{20} and x_{31} are joined in Figure 29), then those points will have the same sign attached. The first statement of the theorem now follows.

We now turn to the second statement, namely that the degree of f is invariant under homotopies. Let $F: M\times I\to N$ be any smooth homotopy between $f(x)=F(x, 0)$ and $g(x)=F(x, 1)$; by Corollary 10.2.8 (or more precisely a slight extension of it), we may suppose that y_0 is a regular value of the whole homotopy process F, whence, as noted in §10.2, the complete inverse image $F^{-1}(y_0)$ is a smooth submanifold of M (see Figure 30, where $n=1$). Since $F^{-1}(y_0)$ is a one-dimensional, non-singular submanifold of $M\times I$, with boundary the disjoint union of the two subsets $f^{-1}(y_0)$ (corresponding to $t=0$) and $g^{-1}(y_0)$ (corresponding to $t=1$), the situation closely resembles that depicted in the previous figure (Figure 29). An argument similar to the one based on that figure then yields the desired conclusion, namely that the maps f and g have the same degree. □

13.2. Generalizations of the Concept of Degree

We shall now describe two useful "relativizations" of the definition of degree, and its adaptation to non-orientable manifolds.

(i) In the context of the class of relative mappings between manifolds-with-boundary M and N, of the same dimension n, the degree is defined in the following natural way. Consider such a map $f:(M, \partial M)\to(N, \partial N)$. Since f

maps boundary to boundary, the complete inverse image of an interior point of N must consist solely of interior points of M, and of course this is also true for any mapping relatively homotopic to f. Hence if we define the degree of the map f as in Definition 13.1.1 with respect to a regular value y_0 in the interior of N, then it follows by imitating the proof of Theorem 13.1.2, that this degree is independent of the choice of the interior regular value y_0, and is invariant under relative homotopies of f (for compact, connected, oriented M and N).

The boundaries ∂M and ∂N of M and N are closed, oriented (with the orientations induced from those on M and N) $(n-1)$-dimensional manifolds. If we suppose them to be also connected (though this is not a really crucial requirement), then we obtain the following result.

13.2.1. Theorem. *The degree of the boundary map* $f|_{\partial M}\colon \partial M \to \partial N$, *is the same as the degree of* f; *i.e.* $\deg f|_{\partial M} = \deg f$.

PROOF. We first of all by means of a small (smooth) relative homotopy deform the map $f\colon M \to N$, into one sending no interior point of M into the boundary ∂N. This is achieved in the following way. Denote by U_ε a (small) ε-neighbourhood in N of the boundary ∂N (where distance on N is defined in terms of some Riemannian metric), and let $n(y)$ be a unit vector field on U_ε, consisting of tangent vectors to N. Let $\varphi(y) \geq 0$ be a real C^∞-function on U_ε, taking the value ε at all points of ∂N, vanishing on the "opposite" boundary of U_ε, and decreasing monotonically with distance from ∂N. Extend φ to all of N by defining it to be zero outside U_ε. Write $V_\varepsilon = f^{-1}(U_\varepsilon) \subset M$. The function φ^* on M, defined for all $x \in M$ by $\varphi^*(x) = \varphi(f(x))$, clearly vanishes outside V_ε, and attains its largest value (namely ε) at the points of the complete inverse image of ∂N. Let ψ be a C^∞-function on M, vanishing on the boundary ∂M, taking the value 1 outside a (small) δ-neighbourhood of ∂M, and monotonically increasing with distance from ∂M.

We are now in a position to define the desired homotopy of f. Let the image of each point $x \in M$ move (in N) a distance $\psi(x)\ \varphi(f(x))$ along the trajectory through it (if $f(x) \in U_\varepsilon$) of the vector field $n(y)$. Clearly the points of the image of the boundary ∂M do not move (since $\psi(x) = 0$ on ∂M), and neither do the images of the points outside V_ε (since for those points $\varphi(f(x)) = 0$); all other points move through a (small) positive distance. Hence the result of this arbitrarily small deformation of f is a map g agreeing with f on ∂M such that for interior points $x \in M$ with $f(x)$ close to (or on) ∂N, we have $g(x)$ further from ∂N (assuming the vector field $n(y)$ suitably chosen).

We can now proceed to the proof of the theorem. Let y_0 be a regular value of the map f, lying in the boundary ∂N. Then by applying a homotopy like the one just described, we may suppose that the complete inverse image $f^{-1}(y_0)$ is contained in ∂M. If we imagine this regular point moved along a smooth path into the interior of N through a sufficiently small distance, then we shall observe that both the number of preimages and their associated signs are preserved, since the number of preimages or their signs can change only on passing through a critical value, and the orientation on the boundary is induced by that on N. Hence the degrees of f and of its restriction to the boundary of M coincide, as required. □

(ii) The definition of degree and the proof of its independence of the regular value and its invariance under homotopies, carry over without essential change to the class of proper maps and homotopies between connected, oriented, but not necessarily compact manifolds (see §12.2(iii)).

(iii) The degree with respect to a regular value of a map $f: M \to N$, where M is non-orientable, is not well-defined by the formula (1) since the Jacobians of the transition functions will vary in sign; however, the residue modulo 2 of the integer furnished by (1) is well defined for such maps, and so we take this residue (0 or 1) as defining the *degree of a mapping between non-orientable n-manifolds.*

13.3. Classification of Homotopy Classes of Maps from an Arbitrary Manifold to a Sphere

In the case where the image manifold N is the n-sphere S^n, the degree of a map $f: M \to S^n$, turns out to characterize the homotopy class of f completely.

13.3.1. Theorem. *A pair of smooth maps $f, g: M \to S^n$ from a closed, oriented n-dimensional manifold M to the n-sphere, are homotopic precisely if their degrees coincide.*

PROOF. We wish to show that if $\deg f = \deg g$, then f and g are homotopic. To begin with we consider the simple situation where there exists a regular value $y_0 \in S^n$ such that the number of its preimages under each of f and g is exactly $\deg f (= \deg g)$. Assuming this to be the case, we construct a homotopy between f and g as indicated in the following three steps.

Step 1. By means of a homotopy, deform f so that the complete inverse images $f^{-1}(y_0)$ and $g^{-1}(y_0)$ coincide.

Step 2. Since by assumption the signs of the respective Jacobians of f and g coincide at the points of the set $f^{-1}(y_0) = g^{-1}(y_0)$, we may further deform f to a map whose differential (i.e. induced map of tangent spaces) coincides with the differential of g at each point of that set.

Step 3. For sufficiently small $\varepsilon > 0$ we can deform both f and g in such a way that their restrictions to the ε-neighbourhood (with respect to some metric on M) of each point of $f^{-1}(y_0) = g^{-1}(y_0)$, are (coincident) linear maps (in terms of suitable local co-ordinates on that neighbourhood). By means of further deformations we can then clearly arrange for the boundaries of these ε-neighbourhoods to be mapped by both f and g to the point y^* antipodal to y_0. Since the complement in S^n of the point y_0 is diffeomorphic to Euclidean

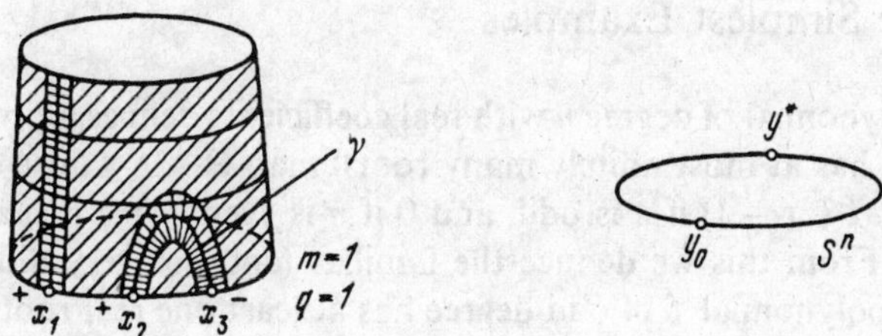

Figure 31

n-space $\mathbb{R}^n$, we may then arrange by means of yet further homotopies that the whole of the complement in M of the union of all of these ε-neighbourhoods is sent by both f and g to the point y^*.

Upon completion of Step 3, i.e. after applying all of these successive homotopies, we see that f and g coincide, as required. Hence the theorem will follow if we can show that any map $f: M \to S^n$, is homotopic to a map g having a regular point y_0 satisfying $|g^{-1}(y_0)| = \deg f$. This we shall now do. Thus let $y_0 \in S^n$ be a regular point of f such that the number of preimages of y_0 under f exceeds $|\deg f| = m$. We can still apply Step 3 above to arrive at a function (which we also denote by f even though it is merely homotopic to f) with the following canonical properties: the regular value y_0 has $m+2q$ preimages $x_1, \ldots, x_{m+2q}$, say; for some (sufficiently small) $\varepsilon > 0$, the ε-neighbourhoods of these preimages are mapped linearly by f onto the punctured sphere $S^n - \{y^*\}$, where y^* is diametrically opposite y_0, while the complement of the union of these neighbourhoods is mapped to the point y^*; and, finally, the signs of the Jacobians at the points $x_1, \ldots, x_m$ are all the same, while for each $i = 1, \ldots, q$, the Jacobians at the points x_{m+i}, x_{m+q+i} have opposite signs. We now define a homotopy $F: M \times I \to S^n$, between f and a map having exactly m preimages of the regular value y_0. To this end, for each $i = 1, \ldots, q$, let γ_i be a path in $M \times I$ joining x_{m+i} to x_{m+q+i} (as shown in Figure 31, where $m=1, q=1$), and on some sufficiently small neighbourhood of each γ_i define F so that it sends the boundary of that neighbourhood to the point y^*; this is possible in view of the difference in sign of the Jacobians at x_{m+i} and x_{m+q+i} (consider in Figure 31 the images in S^n of the segments shading the neighbourhood of γ). Define F on the ε-neighbourhoods of the points $x_1, \ldots, x_m$, as the identity homotopy (as indicated in Figure 31 for the point x_1), and let F map the rest of $M \times I$ to the point y^*. Clearly the homotopy F eliminates the preimages $x_{m+1}, \ldots, x_{m+2q}$, as required. □

Remark. In the case $n=1$ some care has to be exercised in defining the homotopy F; we invite the reader to fill in the details.

EXERCISE

As noted in §13.2(iii), the degree of a map from a non-orientable n-manifold M to the sphere S^n, is defined only as a residue modulo 2. Prove the analogue of the above theorem in this case.

13.4. The Simplest Examples

(a) Any polynomial of degree n with real coefficients defines a proper (since a polynomial has at most finitely many roots) map $\mathbb{R}\to\mathbb{R}$, whose degree (as a mapping) is 1 (or -1) if n is odd, and 0 if n is even. (This is made clear by Figure 32.) From this we deduce the familiar (and not very difficult) result that a real polynomial f of odd degree has at least one real root: for if there were a real number with no preimages under f, then we should have (by definition) $\deg f = 0$.

(b) We shall now consider maps $S^1\to S^1$, from the unit circle to the unit circle. We picture the unit circle as arising from the real line by identification of the points of the form $x+2\pi n$, for each fixed x and all integral n (and similarly the image circle is obtained by identifying the points $y+2\pi n$); then a function $y=f(x)$ from $\mathbb{R}$ to $\mathbb{R}$ defines a map S^1 to S^1 precisely if it has the property that, for all x, $f(x+2\pi)=f(x)+2\pi k$ for some fixed integer k. Examination of the graph of f then makes it clear that the degree of the map $S^1\to S^1$ defined by f is just k (see Figure 33, where $k=2$). This may be expressed in terms of an integral thus:

$$k=\deg f=\frac{1}{2\pi}\int_0^{2\pi}\left(\frac{df}{dx}\right)dx.$$

If we think of the unit circle as the curve in the complex plane defined by the

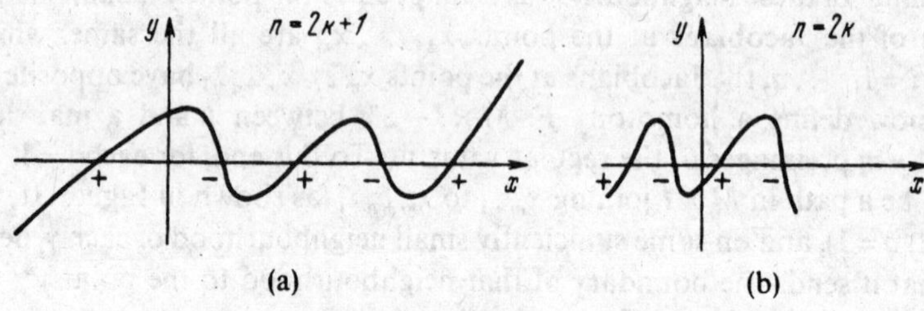

(a) (b)

Figure 32

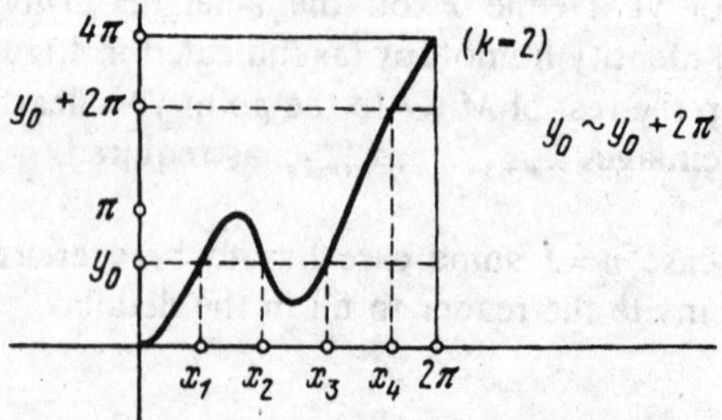

sgn x_1 = +, sgn x_2 = −, sgn x_3 = +, sgn x_4 = +, deg f = 2

Figure 33

equation $|z|=1$, then every map $S^1 \to S^1$ of degree k is homotopic to the canonical map $z \mapsto z^k$.

(c) A complex polynomial $w=f(z)$ of degree n, determines a map $f: \mathbb{R}^2 \to \mathbb{R}^2$ of the complex plane, or (if we adjoin the point at infinity) of the Riemann sphere S^2 ($\cong \mathbb{C}P^1$) to itself. We show firstly that the *degree of f as a mapping is (in absolute value) equal to the degree of f as a polynomial.* This is obvious when the polynomial is a monomial, i.e. of the form $a_0 z^n$, $a_0 \neq 0$; it then follows for any polynomial $a_0 z^n + a_1 z^{n-1} + \cdots + a_n$ of degree n ($a_0 \neq 0$), since the latter polynomial is homotopic to $a_0 z^n$ via the homotopy F defined by

$$F(z, t) = a_0 z^n + (1-t)[a_1 z^{n-1} + \cdots + a_n],$$

where t ranges from 0 to 1.

We obtain as a corollary (of this (essential) equality of the two kinds of degree of a complex polynomial) the so-called "Fundamental Theorem of Algebra".

13.4.1. Theorem (Gauss). *A complex polynomial of degree $n>0$ has at least one root.*

PROOF. If $f(z)=0$ has no solutions, then the complete inverse image $f^{-1}(0)$ is empty, whence f has degree zero (as a mapping) and hence degree zero as a polynomial. □

(As an exercise show that the degree of a rational map $S^2 \to S^2$ is (in absolute value) the larger of the (polynomial) degrees of the numerator and denominator.)

We next turn our attention to holomorphic (i.e. complex analytic) maps $f: M \to N$ between closed, complex manifolds M and N.

13.4.2. Theorem. *If the degree of f is q, then $q \geq 0$ and any regular value $y_0 \in M$ of f has exactly q preimages* (which then of course all correspond to positive Jacobians).

PROOF. It was shown (in essence) in §12.2 of Part I, that the determinant of a complex linear transformation A is never negative, since

$$\det_{\mathbb{R}} A = |\det_{\mathbb{C}} A|^2 \geq 0.$$

Hence at the preimages of the regular value y_0, the Jacobians are all positive. The theorem is now immediate from the definition of the degree of a map. □

We obtain as a corollary of this theorem (and our previous discussion) the familiar result that a complex polynomial of degree $n>0$ for which zero is a regular value, has exactly n distinct roots.

Another example of a holomorphic map is the projection of the Riemann surface of an n-valued algebraic function onto the z-plane or the Riemann sphere (i.e. onto the manifold on which the map is defined). Clearly the degree of such a map is the number n of sheets of the Riemann surface.

(d) A map f from a closed n-dimensional manifold to $\mathbb{R}^n$ is obviously proper (in fact independently of the image space). Hence the degree of such a map is defined. It is easy to see that this degree is zero, since $f(M)$ is a compact subset of $\mathbb{R}^n$, so that in view of the non-compactness of the latter space, there must exist a point y_0 sufficiently far out in $\mathbb{R}^n$ for which $f^{-1}(y_0)$ is empty.

From the vanishing of the degree of such a map f we infer immediately that any regular point of f must have an even number of preimages.

(e) Consider a map $f: (M, \partial M) \to (N, \partial N)$ between compact, connected, oriented manifolds-with-boundary M and N, whose restriction to ∂M is a diffeomorphism: $\partial M \cong \partial N$; and suppose also that this diffeomorphism respects the orientations induced on the boundaries. It follows from Theorem 13.2.1 that for such a map we must have $\deg f = 1$. Hence, in particular, if we are given a co-ordinate change $y = f(x)$ on a region U of $\mathbb{R}^n$ with smooth boundary ∂U, which is one-to-one on ∂U, then f has degree ± 1 (on the whole of U).

§14. Applications of the Degree of a Mapping

14.1. The Relationship Between Degree and Integral

We shall here investigate the behaviour under a mapping $M \to N$ (whose degree is defined) of the integral of a differential form of rank n defined on the image n-manifold N; i.e. we seek the relationship between the integral of the given form over N and the integral over M of the pullback (via the mapping $M \to N$) of the form to M. Thus let $f: M \to N$ be a smooth map of degree q between connected, closed, oriented manifolds M and N, and let Ω be a differential form on N, of rank $n = \dim M = \dim N$, expressed as $\varphi_i(y)\, dy_i^1 \wedge \cdots \wedge dy_i^n$ in terms of local co-ordinates y_i^α on the ith local co-ordinate neighbourhood of N. The integral $\int_N \Omega$ of the form Ω over N is defined as in §8.2 (in terms of a partition of unity), as is also the integral $\int_M f^*\Omega$ of the restricted form $f^*\Omega$ over M. By definition of the restriction or pullback operation (see §22.1 of Part I), we have

$$f^*\Omega = \varphi_i(f(x))\, dx_j^1 \wedge \cdots \wedge dx_j^n \det\left(\frac{\partial y_i^\alpha}{\partial x_j^\beta}\right), \qquad (1)$$

where the x_j^β are local co-ordinates on M, or, more precisely, on that region of M which is mapped under f to the chart of N with local co-ordinates y_i^α.

14.1.1. Theorem. *With $f: M \to N$, and Ω as above, we have*

$$\int_M f^*\Omega = (\deg f) \int_N \Omega.$$

PROOF. For each regular value y_0 ($\in N$) of the map f, there is a neighbourhood U say of y_0 which contains only regular values of f. If $x_1, \ldots, x_m$ are the (necessarily only finitely many) distinct preimages of y_0 under f, then, provided U is sufficiently small, its complete inverse image $f^{-1}(U)$ will have the form

$$f^{-1}(U) = U_1 \cup \cdots \cup U_m, \qquad x_j \in U_j,$$

where this union is disjoint, i.e. the U_j are pairwise disjoint. Let y_0^α be local co-ordinates on $U \subset N$, and for each $j = 1, \ldots, m$ let x_j^α be local co-ordinates on $U_j \subset M$. Since all points in the regions U_j are regular points of f, we have by the Inverse Function Theorem that the restriction of f to each U_j is one-to-one. Hence for each such restriction we may apply the formula for changing the variables of integration, obtaining

$$\int_{U_j} \varphi(y(x)) \det\left(\frac{\partial y_0^\alpha}{\partial x_j^\beta}\right) dx_j^1 \wedge \cdots \wedge dx_j^n$$
$$= \operatorname{sgn} \det\left(\frac{\partial y_0^\alpha}{\partial x_j^\beta}\right) \int_U \varphi(y)\, dy_0^1 \wedge \cdots \wedge dy_0^n.$$

By virtue of (1) (and the additivity of integrals over disjoint regions) this immediately yields

$$\int_{f^{-1}(U)} f^*\Omega = \left(\sum_j \operatorname{sgn} \det\left(\frac{\partial y_0^\alpha}{\partial x_j^\beta}\right)\right) \int_U \Omega = (\deg f) \int_U \Omega. \qquad (2)$$

Since by Sard's theorem (§10.2) the set of critical values of f has measure zero in N, their contribution to the integral $\int_N \Omega$ is zero. On the other hand, at the preimages of such points (i.e. at the critical points of f) the form $f^*\Omega$ vanishes since the Jacobian in (1) is zero; hence the contribution of the critical points to $\int_M f^*\Omega$ is also zero. The theorem now follows from (2) and the additive property of integrals. (Note that the set of regular values is an open, everywhere dense subset of N.) □

Remarks. 1. This theorem is valid in the non-compact case, provided that the map f is proper and the form Ω is "finitary" (i.e. has compact support in N), and also when M, N are manifolds-with-boundary (and f maps ∂M to ∂N).

2. If f is a boundary-preserving mapping between compact regions of $\mathbb{R}^n$, which is one-to-one between the boundaries, then (see Example (e) of §13.4) we have $|\deg f| = 1$, whence by the above theorem, the integrals of $f^*\Omega$ and Ω

over the respective regions are equal in absolute value (which is not surprising if f is one-to-one, i.e. if f is a co-ordinate change).

14.2. The Degree of a Vector Field on a Hypersurface

Let $\xi=(\xi^\alpha(x))$, $\alpha=1,\ldots,n$, denote a smooth, non-vanishing vector field defined on a region U of n-dimensional Euclidean space $\mathbb{R}^n$ with Euclidean co-ordinates $x^1,\ldots,x^n$. Since $\xi(x)\neq 0$ at all points of U, the vector field ξ gives rise to a unit vector field $n(x)=\xi(x)/|\xi(x)|$ on U, which leads in turn to the *spherical (Gauss) map* $f\colon U\to S^{n-1}$, where for each point $x\in U$, $f(x)$ is the tip of the unit vector $n(x)$ when its tail is at the origin. (Thus $f(x)$ is a point on the unit $(n-1)$-sphere, as required.) If now Q is any hypersurface entirely within U, then the degree of the restriction map $f|_Q\colon Q\to S^{n-1}$, is called the *degree of the vector field* ξ *on the hypersurface* Q.

Consider the following closed $(n-1)$-form Ω on S^{n-1} (or, if you like, on $\mathbb{R}^n\backslash\{0\}$):

$$\Omega=\frac{1}{\gamma_n}\frac{\sum_{i=1}^{n}(-1)^{i+1}x^i\,dx^1\wedge\cdots\wedge\widehat{dx^i}\wedge\cdots\wedge dx^n}{((x^1)^2+\cdots+(x^n)^2)^{n/2}}. \tag{3}$$

Here the hatted symbols are to be omitted, and γ_n is the normalizing constant chosen so that $\int_{S^{n-1}}\Omega=1$. (It is not difficult to verify that the form $\gamma_n\Omega$ is just the volume element on the unit sphere S^{n-1}, whence γ_n is the volume of S^{n-1}.) For instance in the plane $\mathbb{R}^2\backslash\{0\}$ with Euclidean co-ordinates x, y, we have

$$\Omega=\frac{1}{2\pi}\left(\frac{x\,dy-y\,dx}{x^2+y^2}\right),$$

or, in polar co-ordinates, $\Omega=d\varphi/2\pi$. In $\mathbb{R}^3\backslash\{0\}$ with Euclidean co-ordinates x, y, z, we have

$$\Omega=\frac{1}{4\pi}\left(\frac{x\,dy\wedge dz-y\,dx\wedge dz+z\,dy\wedge dx}{(x^2+y^2+z^2)^{3/2}}\right),$$

or, in spherical co-ordinates, $\Omega=(1/4\pi)|\sin\theta|\,d\theta\,d\varphi$.

From Theorem 14.1.1 we deduce the following

14.2.1. Corollary. *The degree of a non-vanishing vector field $\xi(x)$ on a closed hypersurface Q in Euclidean n-space, is equal to the integral $\int_Q f^*\Omega$, where $f\colon Q\to S^{n-1}$ is the Gauss map determined by the vector field, and Ω is the $(n-1)$-form on S^{n-1} defined in (3). It follows that if Q is given locally by a system of equations*

$$x^\alpha=x^\alpha(u^1,\ldots,u^{n-1}),\qquad \alpha=1,\ldots,n,$$

(where $x^1,\ldots,x^n$ are Euclidean co-ordinates in $\mathbb{R}^n$), then

$$\deg f = \deg_Q \xi = \frac{1}{\gamma_n}\int_Q \frac{1}{|\xi|^n}\det\begin{pmatrix} \xi^1 & \cdots & \xi^n \\ \frac{\partial \xi^1}{\partial u^1} & \cdots & \frac{\partial \xi^n}{\partial u^1} \\ \cdots & \cdots & \cdots \\ \frac{\partial \xi^1}{\partial u^{n-1}} & \cdots & \frac{\partial \xi^n}{\partial u^{n-1}} \end{pmatrix} du^1 \wedge \cdots \wedge du^{n-1}. \quad (4)$$

(The first statement is immediate from Theorem 14.1.1. The second follows easily from the definitions of Ω, of f (via $f(x)=(1/|\xi(x)|)(\xi^1(x),\ldots,\xi^n(x))$), and of $f^*\Omega$ (see §22.1 of Part I).)

In the case $n=2$, the formula (4) becomes:

$$\deg f = \frac{1}{2\pi}\oint \frac{dt}{|\xi|^2}\left(\xi^1\frac{d\xi^2}{dt} - \xi^2\frac{d\xi^1}{dt}\right),$$

where t is a parameter on the closed curve over which the integral is taken, and $\xi(t)=(\xi^1(t), \xi^2(t))$ is a non-vanishing vector field on the curve. In the case $n=3$, we have $\gamma_3=4\pi$ whence (4) becomes:

$$\deg f = \frac{1}{4\pi}\iint_Q \frac{du\,dv}{|\xi|^3}\det\begin{pmatrix} \xi^1 & \xi^2 & \xi^3 \\ \frac{\partial \xi^1}{\partial u} & \frac{\partial \xi^2}{\partial u} & \frac{\partial \xi^3}{\partial u} \\ \frac{\partial \xi^1}{\partial v} & \frac{\partial \xi^2}{\partial v} & \frac{\partial \xi^3}{\partial v} \end{pmatrix} \quad (5)$$

$$= \frac{1}{4\pi}\iint_Q \frac{du\,dv}{|\xi|^3}\left\langle \xi, \left[\frac{\partial \xi}{\partial u}, \frac{\partial \xi}{\partial v}\right]\right\rangle,$$

where [,] denotes the vector (or cross) product.

A particular situation of interest is that where the vector field $\xi(x)$ is a unit field (i.e. $|\xi|=1$), and at every point of the closed manifold Q has the direction of the exterior normal to Q. We know (from §26.2 of Part I) that in this case

$$\gamma_n(f^*\Omega) = K\,d\sigma = K\sqrt{g}\,du^1 \wedge \cdots \wedge du^{n-1}, \quad (6)$$

where K is the Gaussian curvature of the hypersurface (defined as the product of the principal curvatures), and $d\sigma = \sqrt{g}\,du^1 \wedge \cdots \wedge du^{n-1}$ is the standard volume element on the hypersurface Q determined by the metric induced on Q by the Euclidean metric on the ambient space $\mathbb{R}^n$. (Recall that when $n=2$ we have $d\sigma = dl$, the element of length on the curve Q, and $K=k$, the curvature of the curve, while for $n=3$, K is the usual Gaussian curvature of a 2-dimensional surface, and $d\sigma=\sqrt{EG-F^2}\,du \wedge dv$ is the usual element of area (see §7 of Part I).

Combining (6) with the first statement of Corollary 14.2.1, we obtain immediately the following

14.2.2. Theorem. *The integral of the Gaussian curvature over a closed hypersurface in Euclidean n-space is equal to the degree of the Gauss map of the surface, multiplied by γ_n (the Euclidean volume of the unit $(n-1)$-sphere).*

14.3. The Whitney Number. The Gauss–Bonnet Formula

Our aim in this subsection is to discover the relationship between the degree of the Gauss map of a hypersurface in Euclidean n-space and certain striking geometrical features of the hypersurface (or at least of the way it is embedded in $\mathbb{R}^n$). We shall restrict attention to the most important cases $n=2, 3$.

The case $n=2$: curves in $\mathbb{R}^2$. Let $\gamma=(x(t), y(t))$ be a closed curve "in general position" in Euclidean $\mathbb{R}^2$; by this we mean that: for all t, $x(t+2\pi)=x(t)$, $y(t+2\pi)=y(t)$; the tangent vector $(\dot{x}, \dot{y})$ is non-zero for all t; and finally every self-intersection (if there are any) is double and has the property that the two tangent vectors at the point of intersection are linearly independent (as for instance in Figure 34).

To each point of self-intersection of the closed curve γ we attach a sign (± 1) in the following way. Fix on any point t_0 of the curve which is not a point of self-intersection. Denote by $[1, 2]$ an orienting frame on $\mathbb{R}^2$ (e.g. that determined by the given Euclidean co-ordinates x, y, in that order). We now trace out the curve γ once starting from the point t_0 in the direction of increase of the parameter; when any particular point of self-intersection is first encountered we label with the numeral 1 the tangent vector at that point to the branch we are on, and when we meet that point of self-intersection again we label the tangent vector to the second branch (on which we now find ourselves) with the numeral 2 (see Figure 34). Once our tour is completed, each point of self-intersection of γ has associated with it a frame whose orientation class of course coincides either with that of the prescribed frame $[1, 2]$, or its opposite; in the former case we attach the sign $+1$ to the point, and in the latter case the sign -1. We then define the *Whitney number* $W(\gamma)$ of the plane closed curve γ (in general position) to be the sum of the signs of all its points of self-intersection.

The following result links the degree of the Gauss map to the Whitney number.

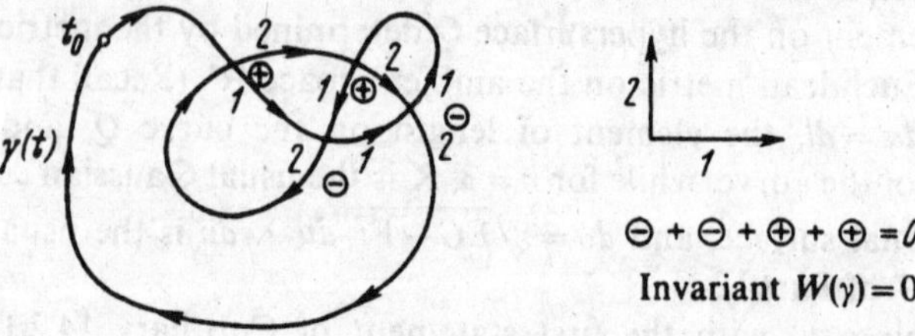

Figure 34

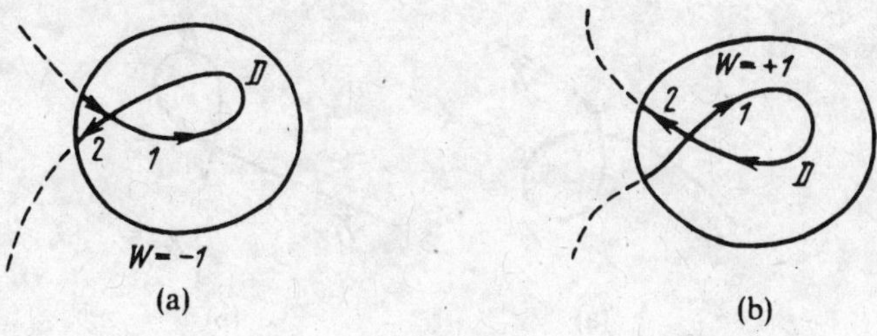

Figure 35

14.3.1. Theorem. *Let γ be a non-singular, plane closed curve in general position. Then the parity of the number of self-intersections of γ (which is the same as the parity of the Whitney number of γ) is opposite to that of the degree of the Gauss map f of γ.*

PROOF. For a simple (i.e. non-self-intersecting) plane closed curve γ it is easy to see that $\deg f = 1$ (assuming the curve γ and S^1 traced in the same sense), while clearly $W(\gamma) = 0$; hence the theorem is true in this case. With this as our point of departure we shall prove the theorem by induction on the number of self-intersections. Suppose first that our curve γ has a loop with the properties that the "starting point" t_0 is not on the loop, and no points of γ are in its interior (see Figure 35). Enclose this loop in a region D which excludes all points of γ not on the loop (as in Figure 35); subsequent alterations of γ will be carried out entirely in the region D. The (direction-preserving) alterations in question are depicted in Figure 36(a) or (b) (depending on the sense in which the loop is traced). The operation of replacing γ by $\gamma_1 \cup \gamma_2$ and then removing γ_2, decreases the degree of f by 1 in case (a), and increases the degree by 1 in case (b); clearly the Whitney number is affected in the same way. Thus the inductive step is completed in the situation where γ has such a "minimal" loop.

Suppose now that γ has no "minimal" loops. It can be shown that there will still be a loop γ_2 say (see Figure 37) which does not have the point t_0 on it, but which intersects the remainder of the curve (denoted by γ_1 in Figure 37). If we now alter the curve γ appropriately within some sufficiently small neighbourhood of the point of self-intersection of our loop we shall arrive

(a) $W=-1$ $\xrightarrow{\gamma \to \gamma_1 + \gamma_2}$ γ_1 $W=0$ γ_2 $\frac{1}{2\pi}\oint_{\gamma_2} k\,dl = -1$

(b) $W=+1$ $\xrightarrow{\gamma \to \gamma_1 + \gamma_2}$ γ_1 γ_2 $W=0$ $\frac{1}{2\pi}\oint_{\gamma_2} k\,dl = +1$

Figure 36

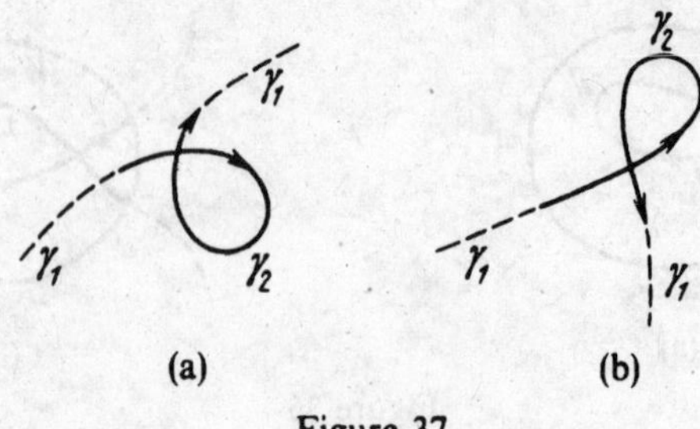

Figure 37

again at the situation depicted in Figure 36(a) or (b), except that now the simple closed curve γ_2 of that figure will intersect γ_1. However since the number of such points of intersection will of course be even, the argument concerning parities goes through as before. This completes the proof of the theorem. □

Remark. It can be shown that under the above-described alteration of γ and subsequent removal of the simple closed component γ_2, the Whitney number and the degree of f change by the same amount (and not just modulo 2). This leads to the more precise result that *the Whitney number* $W(\gamma)$ *is equal to either* $\deg f - 1$ *or* $\deg f + 1$. Whether 1 is to be added or subtracted depends on the initial choice of the starting point t_0 (as illustrated in Figure 38).

The case $n = 3$: *surfaces in* $\mathbb{R}^3$. Our aim is to express the degree of the Gauss map $f: Q \to S^2$, of a smooth, closed, oriented surface Q in $\mathbb{R}^3$, more directly in terms of the geometry of Q. Let $y_0 \in S^2$ be a regular value of the map f. By applying a suitable orthogonal transformation we may clearly arrange for the point y_0 to be at the north pole of S^2, i.e. to have co-ordinates (0, 0, 1), and, by means of a suitable, small deformation (i.e. homotopy), we may further arrange for the south pole $y_0^* = (0, 0, -1)$ to be also a regular value of f. (The simultaneous regularity of two diametrically opposite points of S^2 is equivalent to the regularity of their common image point in the real projective plane $\mathbb{R}P^2$ under the composite map

$$Q \xrightarrow{f} S^2 \to \mathbb{R}P^2,$$

from Q to $\mathbb{R}P^2$.)

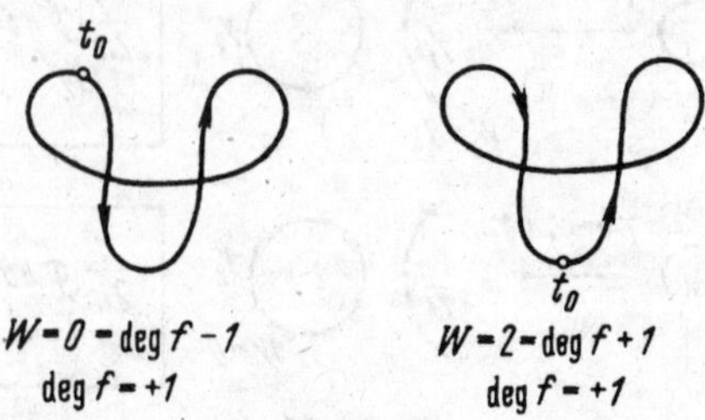

Figure 38

14.3.2. Lemma. *Let Q be a hypersurface in Euclidean 3-space, with f (the Gauss map of Q), y_0, y_0^* as above, and let φ be the height function on Q (i.e. for each $P \in Q$, $\varphi(P)$ is the z-co-ordinate of P). Then the set of critical points of the function φ coincides with the union $f^{-1}(y_0) \cup f^{-1}(y_0^*)$, and the critical points are all non-degenerate.*

PROOF. In some neighbourhood of each critical point of $\varphi: Q \to \mathbb{R}$, the surface Q is defined by an equation of the form $z = \varphi(x, y)$, and the critical points in that neighbourhood are just those satisfying grad $\varphi = 0$. Now the gradient of $\varphi(x, y)$ is zero at precisely those points of Q where the direction of the z-axis is perpendicular to Q; but by definition of the Gauss map f of Q, the latter are exactly the points sent by f to the north or south poles of S^2. Hence the set of critical points of φ is $f^{-1}(y_0) \cup f^{-1}(y_0^*)$.

It is immediate from the proof of (6) above given in §26.2 of Part I, that at each point of $f^{-1}(y_0)$ the Jacobian of the Gauss map $f: Q \to S^2$ (in terms of suitable local co-ordinates on S^2) coincides with the determinant of the Hessian of the function $z = \varphi(x, y)$ (and hence with the Gaussian curvature). Since y_0 is a regular value of the map f it follows (from the definition (10.4.1) of non-degeneracy of critical points of a function) that the points of $f^{-1}(y_0)$ are indeed non-degenerate (critical) points of φ. (Alternatively this could have been inferred immediately from Lemma 11.2.1.) By applying the same argument with the z-axis reversed, we draw the same conclusion for the points in $f^{-1}(y_0^*)$, completing the proof of the lemma. □

In connexion with this proof note the obvious equality

$$\det\left(\frac{\partial^2 \varphi}{\partial u^i \, \partial u^j}\right) = (-1)^{n-1} \det\left(\frac{\partial^2(-\varphi)}{\partial u^i \, \partial u^j}\right), \tag{7}$$

where $u^1, \ldots, u^{n-1}$ are local co-ordinates in some neighbourhood of any point of the surface Q; this equality is crucial in the distinction arising between the cases of odd and even n. Since in our case $n - 1 = 2$, it follows from (7) and the above proof that at every point of the set $f^{-1}(y_0) \cup f^{-1}(y_0^*)$ the Jacobian of the map f agrees in sign with the Gaussian curvature K (independently of the choice of the direction of the z-axis, i.e. of whether we work with φ or $-\varphi$). This, the above lemma, and the definition of the degree of a map, together yield immediately the following

14.3.3. Lemma. *With $f: Q \to S^2$ as above, we have*

$$2 \deg f = \sum_{P_j} (-1)^{\alpha(P_j)}, \tag{8}$$

where the summation is over the set of critical points P_j of the height function $z = \varphi$, and where $\alpha(P_j) = 0$ for those critical points where φ has a local maximum or minimum (where sgn $K = +1$*), and $\alpha(P_j) = 1$ for saddle points (where* sgn $K = -1$*).*

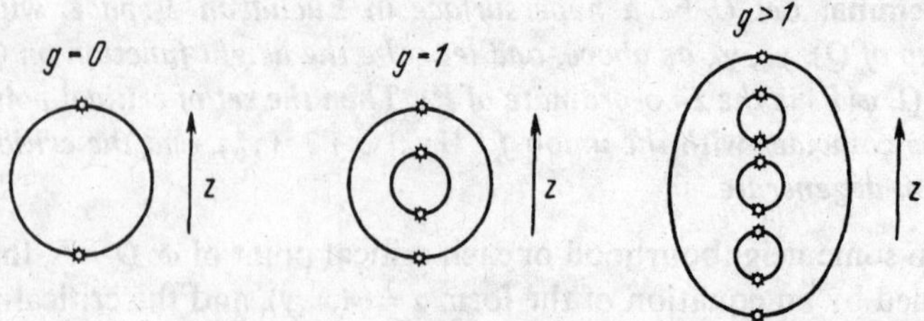

Figure 39

We shall now show that when Q is the sphere-with-g-handles, the number on the right-hand side of (8) is $2-2g$. It is easy to imagine this surface embedded in $\mathbb{R}^3$ in such a way that the height function φ has exactly one minimum point, one maximum point, and $2g$ saddle points (see Figure 39, where the critical points are indicated). For such an embedding we have by (8) together with Theorem 14.2.2 that

$$2 \deg f = \frac{1}{2\pi} \int_Q K\, d\sigma = 2-2g.$$

Of course other embeddings of the surface Q in $\mathbb{R}^3$ are possible. However at the conclusion of §37.4 of Part I we showed that the quantity $\int_Q K\, d\sigma$ is not altered by smooth variations in the metric on Q. Let $g_{ij}^{(0)}$ and $g_{ij}^{(1)}$ be any two Riemannian metrics on the surface Q, and consider the family of metrics

$$g_{ij}(t) = tg_{ij}^{(1)} + (1-t)g_{ij}^{(0)}, \qquad 0 \le t \le 1.$$

Clearly $g_{ij}(0) = g_{ij}^{(0)}$, $g_{ij}(1) = g_{ij}^{(1)}$, and $g_{ij}(t)$ is positive definite for all $t \in [0, 1]$. Hence the integral $\int_Q K\, d\sigma$ is the same for both metrics. We have thus proved (in more precise form than Corollary 37.4.3 of Part I) the following celebrated result.

14.3.4. Theorem (Gauss–Bonnet). *For the sphere-with-g-handles Q endowed with any Riemannian metric we have*

$$\frac{1}{2\pi} \int_Q K\, d\sigma = 2-2g.$$

14.4. The Index of a Singular Point of a Vector Field

In this subsection we shall study the Gauss map of a vector field in a neighbourhood of an "isolated singular point" of the vector field. We first need to define these terms. Thus let $\xi = \xi(x)$ be a vector field defined on some neighbourhood of a particular point x_0 of Euclidean space $\mathbb{R}^n$. In the conventional terminology, the point x_0 is called a *singular point of the field* ξ if $\xi(x_0) = 0$, and an *isolated* singular point of the field if it is a singular point but at all other points of some (small) neighbourhood of it $\xi(x)$ does not vanish. A singular point x_0 is said to be *non-degenerate* if

$$\det\left(\frac{\partial \xi^\alpha}{\partial x^\beta}\bigg|_{x=x_0}\right) \neq 0.$$

14.4.1. Lemma. *A non-degenerate singular point of a vector-field is necessarily isolated.*

PROOF. Consider $\xi(x)$ as a map from $\mathbb{R}^n$ to $\mathbb{R}^n$. Since the derivative $\det(\partial\xi^\alpha/\partial x^\beta) \neq 0$ at the non-degenerate singular point x_0, the Inverse Function Theorem tells us that ξ is one-to-one on some neighbourhood of x_0. Hence x_0 is an isolated singular point, as claimed. □

By the *roots* of a non-degenerate singular point x_0 of a vector field ξ we shall mean the eigenvalues $\lambda_1, \ldots, \lambda_n$ of the matrix $(\partial\xi^\alpha/\partial x^\beta)_{x=x_0}$, and by the *index* of such a point the sign

$$\operatorname{sgn}\det\left(\frac{\partial \xi^\alpha}{\partial x^\beta}\bigg|_{x=x_0}\right) = \operatorname{sgn}(\lambda_1 \ldots \lambda_n).$$

Note that if the vector field ξ happens to be the gradient field of some function f, i.e. $\xi^\alpha = \partial f/\partial x^\alpha$, then the index of a non-degenerate singular point x_0 coincides with the sign of the determinant of the Hessian:

$$\operatorname{sgn}\det\left(\frac{\partial \xi^\alpha}{\partial x^\beta}\bigg|_{x=x_0}\right) = \operatorname{sgn}\det\left(\frac{\partial^2 f}{\partial x^\alpha\,\partial x^\beta}\bigg|_{x=x_0}\right) = (-1)^{i(x_0)},$$

where $i(x_0)$ denotes the number of negative squares occurring in the quadratic form $d^2f|_{x=x_0}$ when brought into canonical form.

Let $Q_\varepsilon \cong S^{n-1}$ be a sphere with centre an isolated singular point x_0 of a vector field ξ, and with radius $\varepsilon > 0$ sufficiently small for the field ξ to be non-vanishing on Q_ε and its interior. We can then define, as in §14.2 above, the spherical Gauss map on Q_ε (relative to the field ξ):

$$f_{x_0}\colon Q_\varepsilon \to S^{n-1}.$$

14.4.2. Definition. The *index of an isolated singular point x_0 of a vector field ξ* is the degree of the Gauss map f_{x_0}:

$$\operatorname{ind}_{x_0}(\xi) = \deg f_{x_0}.$$

If the point x_0 is non-degenerate, then this definition of index is equivalent to the previous one; this is the import of the following result.

14.4.3. Theorem. *For a non-degenerate singular point x_0 of a vector field $\xi(x)$, we have*

$$\deg f_{x_0} = \operatorname{sgn}\det\left(\frac{\partial \xi^\alpha}{\partial x^\beta}\bigg|_{x=x_0}\right), \tag{9}$$

where f_{x_0} is the Gauss map defined above.

PROOF. In some neighbourhood of x_0 we have by Taylor's theorem that

$$\xi(x)=\xi^{(1)}(x)+\xi^{(2)}(x),$$

where $\xi^{(1)\beta}(x)=\partial\xi^\beta/\partial x^\gamma|_{x=x_0}(x^\gamma-x_0^\gamma)$, and $|\xi^{(2)}(x)|=o(|\xi^{(1)}(x)|)$. The homotopy $\xi(x,t)$ defined by

$$\xi(x,t)=\xi^{(1)}(x)+(1-t)\xi^{(2)}(x), \qquad 0\le t\le 1,$$

clearly has the properties that $\xi(x,0)=\xi(x)$, $\xi(x,1)=\xi^{(1)}(x)$, and in some sufficiently small neighbourhood of x_0, for all t, $0\le t\le 1$, $\xi(x,t)$ vanishes only at x_0. We have thus shown that in some neighbourhood of x_0 of the field ξ is smoothly homotopic to the linear vector field $\xi^{(1)}$ via fields for which x_0 is the only singular point (in the neighbourhood). Let ε be small enough for the sphere Q_ε with centre x_0 to be entirely contained in this neighbourhood. In the course of the above-defined deformation of ξ into $\xi^{(1)}$, the linear part of $\xi(x,t)$ remains fixed, and the map $f_{x_0}: Q_\varepsilon\to S^{n-1}$ undergoes a smooth homotopy; hence both sides of equation (9) remain constant throughout the homotopy, and it therefore suffices to verify (9) for the linear field $\xi^{(1)}$, and the corresponding map

$$f_{x_0}^{(1)}(x)=\frac{\xi^{(1)}(x)}{|\xi^{(1)}(x)|}.$$

Since $\xi^{(1)}$ is a non-singular linear transformation (from a neighbourhood of x_0 to a neighbourhood of the origin of $\mathbb{R}^n$), it follows that it is one-to-one, and that the map $f_{x_0}^{(1)}: Q_\varepsilon\to S^{n-1}$, is also one-to-one and without critical points. Hence each point y_0 on the sphere S^{n-1} is a regular value with exactly one preimage, and it is clear that the sign of the determinant of the linear transformation $\xi^{(1)}$ determines whether the orientation is preserved or reversed by $f_{x_0}^{(1)}$. This completes the proof. □

We now list the possible types of non-degenerate singular points in the simplest cases $n=2$ and 3.

(i) ($n=2$). The non-degenerate singular points of vector fields on the plane will be of one of the following types (note that in this case the index of a singular point is independent of the direction of the field):

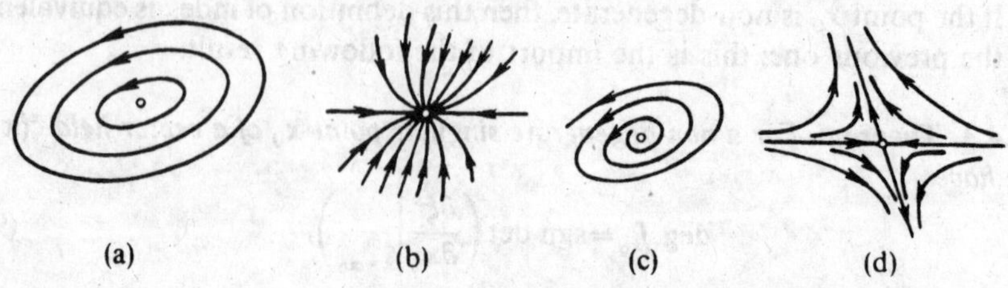

Figure 40

		Index
a centre (when both eigenvalues are purely imaginary)	Figure 40(a)	+1
a node (eigenvalues real and of the same sign)	Figure 40(b)	+1
a focal point (complex conjugate eigenvalues, not purely imaginary)	Figure 40(c)	+1
a saddle point (eigenvalues real and of opposite signs)	Figure 40(d)	−1

If the field happens to be the gradient of some two-variable function f then the possible singularities are restricted to the following types:

	Index
a local minimum point of f (Figure 40(b) with the arrows reversed—a "source")	+1
a saddle point of f (Figure 40(d))	−1
a local maximum point of f (Figure 40(b)—a "sink")	+1

(ii) ($n=3$). If the field is a gradient ($\xi^\alpha = \partial f/\partial x^\alpha$), a non-degenerate singular point is of one of the following types:

	Index
a local minimum point of f ("source")	+1
a saddle point of type 1 (when the form $d^2 f$ has only one negative square)	−1
a saddle point of type 2 (when the form $d^2 f$ has two negative squares)	+1
a local maximum of f ("sink")	−1

The classification of non-degenerate singular points of a general vector field in a region of $\mathbb{R}^3$ is as follows (note that either all three eigenvalues λ_i are real, or else one is real and the other two are complex conjugates):

	Index
a source (Re $\lambda_i \geq 0$, $i=1, 2, 3$)	+1
a saddle of type 1 (Re $\lambda_1 \geq 0$, Re $\lambda_2 \geq 0$, λ_3 real and negative)	−1
a saddle of type 2 (λ_1 real and positive, Re $\lambda_2 \leq 0$, Re $\lambda_3 \leq 0$)	+1
a sink (Re $\lambda_i \leq 0$, $i=1, 2, 3$)	−1

14.4.4. Theorem. *Let $\xi = \xi(x)$ be a vector field in $\mathbb{R}^n$ having only isolated singular points $x_1, \ldots, x_m$, and let Q be a closed oriented hypersurface in $\mathbb{R}^n$ avoiding the singular points of the vector field, and bounding a region D of $\mathbb{R}^n$. Then the degree of the vector field ξ on the surface Q, i.e. the degree of the Gauss map $Q \rightarrow S^{n-1}$ determined by the field, is equal to the sum of the indices of the singular points $x_{i_1}, \ldots, x_{i_k}$ lying in the region D.*

PROOF. For each $j=1,\ldots,m$ let $Q_{j\varepsilon}$ be the sphere of radius $\varepsilon>0$ about the point x_j, choosing ε sufficiently small for the closed balls bounded by the $Q_{j\varepsilon}$ to miss the hypersurface Q and each other. The $\tilde{D}$ obtained from D by removing from it all open balls bounded by those spheres $Q_{j\varepsilon}$ in D, clearly has boundary

$$\partial\tilde{D}=Q\cup Q_{i_1\varepsilon}\cup\cdots\cup Q_{i_k\varepsilon}.$$

Consider now the form $f^*\Omega$ on $\tilde{D}$, where $f\colon\tilde{D}\to S^{n-1}$ is the Gauss map relative to ξ, and Ω is the form defined by the formula (3). Dimensional considerations yield immediately that $d\Omega=0$ on the sphere S^{n-1}, whence also $df^*\Omega=f^*(d\Omega)=0$ on S^{n-1}. Hence by the general Stokes formula (see §8.2) (and the additivity property of integrals) we have

$$0=\int_{\tilde{D}} df^*\Omega=\int_{\partial\tilde{D}} f^*\Omega=-\int_Q f^*\Omega+\sum_{q=1}^{k}\int_{Q_{i_q\varepsilon}} f^*\Omega,$$

where the difference in sign in the final expression is accounted for by the fact that in applying the general Stokes formula, the "external" component Q of the boundary $\partial\tilde{D}$ of $\tilde{D}$ is taken with orientation opposite to that on the boundaries of the $Q_{i_q\varepsilon}$. The theorem now follows from Definition 14.4.2 and Corollary 14.2.1. □

14.5. Transverse Surfaces of a Vector Field. The Poincaré–Bendixson Theorem

A situation of particular interest is that where the hypersurface Q is itself a sphere (of large radius) with the property that at no point of it is the vector field $\xi(x)$ tangential to it (or zero on it). We call a closed surface with the latter property a *transverse surface* to the vector field ξ. In this situation the following simple result holds.

14.5.1. Lemma. *The degree of a vector field ξ on a transverse hypersurface in $\mathbb{R}^n$ is (up to sign) equal to the (normalized) integral of the Gaussian curvature over the surface. (If the transverse surface is a sphere then this (normalized) integral is (in absolute value) unity.)*

PROOF. A vector field which is transverse to a closed hypersurface $Q\subset\mathbb{R}^n$ is homotopic (in the class of fields on Q nowhere tangent to Q) to one or the other of the unit normal vector fields $\pm n(x)$ to Q, so that, degree being a homotopy invariant, we may assume our vector field on Q to be $\pm n(x)$. We know from Theorem 14.2.2 that the degree of the Gauss map (determined by $\pm n(x)$) is $\pm(1/\gamma_n)\int_Q K\,d\sigma$, where K is the Gaussian curvature on Q, $d\sigma$ is the volume element on Q, and γ_n is the volume of the unit $(n-1)$-sphere. If Q is a sphere then it is easy to calculate that $(1/\gamma_n)\int_Q K\,d\sigma=1$. This completes the proof of the lemma. □

From this lemma and Theorem 14.4.4 we deduce immediately the following result.

14.5.2. Corollary. *Any (n-1)-dimensional sphere transverse to a vector field $\xi(x)$ in $\mathbb{R}^n$ (in particular, any (appropriate) circle in $\mathbb{R}^2$) contains in its interior at least one singular point of ξ.*

Remark. It is appropriate to mention at this juncture a further situation arising in the planar case ($n=2$). It can happen that an integral curve of a planar field $\xi(x)$ is periodic, forming a simple closed curve Q. Since the degree of the field on Q is easily seen to be ± 1, it follows (again from Theorem 14.4.4) that there must be at least one singular point of ξ in the interior of Q.

Information about singular points and transverse surfaces of a vector field $\xi(x)$ is of great importance for obtaining a qualitative picture of the behaviour of the integral trajectories of the vector field, especially in the planar case. Consider for instance the situation where we have a planar field $\xi(x)$ and a simple closed curve Q transverse to it such that at each point x of Q the field $\xi(x)$ is directed towards the interior of Q, and there is exactly one singular point x_0, say, of ξ in the interior of Q, which happens to be a "source" (as shown in Figure 41). Any integral curve $\gamma=(x(t), y(t))$ of the field ξ, beginning at some point of Q (as in Figure 41) cannot reach the point x_0 since that point is a source. Denote by $\omega^+(\gamma)$ the "limit set" of this trajectory, consisting of the limit points in $\mathbb{R}^2$ of all sequences of the form $\{\gamma(t_1), \gamma(t_2), \dots\}$ where $t_i<t_{i+1}$ and $t_i \to \infty$ as $i \to \infty$. Clearly the set $\omega^+(\gamma)$ is compact, closed, and does not contain any singular points of the field ξ.

14.5.3. Theorem (Poincaré–Bendixson). *The limit set $\omega^+(\gamma)$ is a periodic integral curve of the field ξ (called a "limit cycle"), onto which the curve γ winds itself from the outside.*

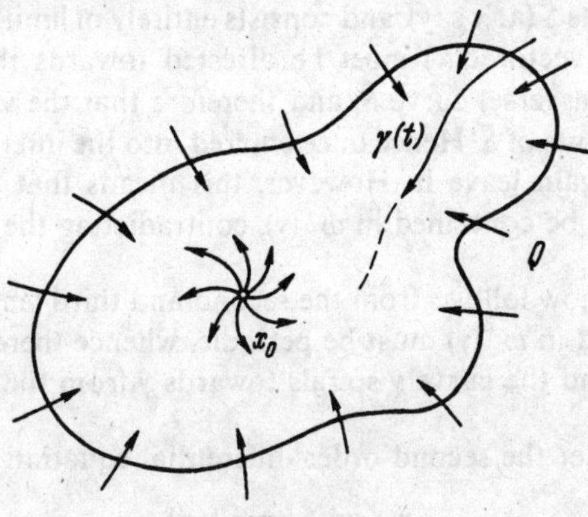

Figure 41

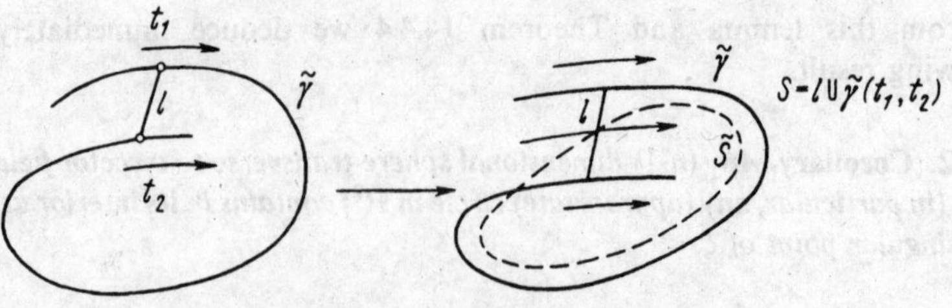

Figure 42

We split the proof up into three lemmas

14.5.4. Lemma. *For every point P in the set $\omega^+(\gamma)$, the whole integral trajectory $\tilde{\gamma}$ through P is contained in $\omega^+(\gamma)$.*

PROOF. If $P = P_0 = \lim_{i\to\infty} \gamma(t_i)$, then every other point P_t $(-\infty < t < \infty)$ on the integral curve through P is given by $P_t = \lim_{i\to\infty} \gamma(t_i + t)$. □

14.5.5. Lemma. *Let $\tilde{\gamma}$ be an integral trajectory of ξ which is not periodic, but for which the set $\omega^+(\tilde{\gamma})$ is compact and contains no singular points of ξ (as is the case for the $\tilde{\gamma}$ of the preceding lemma). Then there exists a closed curve transverse to the field ξ which crosses $\tilde{\gamma}$.*

PROOF. The hypotheses guarantee the existence on $\tilde{\gamma}$ of points $\tilde{\gamma}(t_1)$, $\tilde{\gamma}(t_2)$ arbitrarily close to one another in $\mathbb{R}^2$ but distant from one another along $\tilde{\gamma}$, i.e. such that $|t_1 - t_2| \gg 1$. Join these points by a (short) curve segment l transverse to ξ, and consider the closed curve $S = l \cup [\tilde{\gamma}(t_1, t_2)]$; it is not difficult to see (cf. Theorem 10.3.2) that the curve S can be approximated by a closed curve which is transverse to ξ, and cuts $\tilde{\gamma}$ (see Figure 42). □

14.5.6. Lemma. *Let the set $\omega^+(\gamma)$ be as in the Poincaré–Bendixson theorem. If $\tilde{\gamma}$ is an integral trajectory of ξ contained in $\omega^+(\gamma)$, then there is no closed curve transverse to ξ which crosses $\tilde{\gamma}$.*

PROOF. Suppose on the contrary that there is a transverse curve $\tilde{S}$ intersecting $\tilde{\gamma}$. Since $\tilde{\gamma}$ intersects $\tilde{S}$ (at x say), and consists entirely of limit points of $\omega^+(\gamma)$, it follows that the vector $\xi(x)$ must be directed towards the interior of the (simple, since transverse) curve $\tilde{S}$, and therefore that the vector field ξ is so directed at all points of $\tilde{S}$. Hence once entered into the interior of $\tilde{S}$, the paths γ and $\tilde{\gamma}$ never again leave it. However, this means that that portion of $\tilde{\gamma}$ outside $\tilde{S}$ cannot be contained in $\omega^+(\gamma)$, contradicting the first lemma. □

The theorem now follows from the second and third lemmas: any integral curve $\tilde{\gamma}$ contained in $\omega^+(\gamma)$ must be periodic, whence there is only one such integral curve, and the curve γ spirals towards $\tilde{\gamma}$ from the outside. □

Example. Consider the second-order differential equation

$$\ddot{x} + a\dot{x} + bx = f(\dot{x}),$$

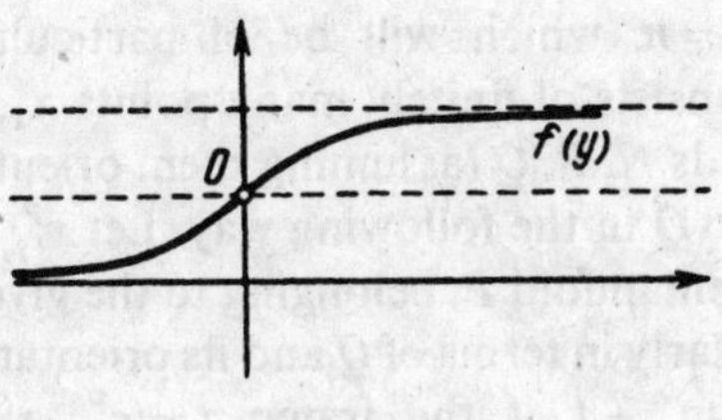

Figure 43

where f is a monotonic odd function (i.e. $f(-\dot{x})=-f(\dot{x})$) (see Figure 43), and a, b are positive. Define a vector field ξ on the "phase plane" with co-ordinates $x, y=\dot{x}$ by

$$\xi(x, y)=(\dot{x}, \dot{y})=(y, -ay-bx-f(y)).$$

It is easy to see that all sufficiently large circles with centre the origin are transverse to ξ with the field directed towards their interiors. The origin is the only singular point of ξ in the finite plane, and there the matrix $(\partial\xi^\alpha/\partial x^\beta)$ has the form

$$\begin{pmatrix} 0 & -b \\ 1 & -a+f'(0) \end{pmatrix},$$

which has eigenvalues

$$\lambda_{1,2}=\frac{p}{2}\pm\sqrt{\frac{p^2}{4}-b},$$

where $p=f'(0)-a$. The collection for the origin $(0, 0)$ to be a source is that λ_1, λ_2 be real and positive, i.e. in the present context that $f'(0)>a$ and $(f'(0)-a)^2\geq 4b$. Hence if these conditions are satisfied then by the Poincaré–Bendixson theorem the differential equation (i.e. the corresponding field ξ) has a limit cycle.

§15. The Intersection Index and Applications

15.1. Definition of the Intersection Index

Let P and Q be two (smooth) closed submanifolds of degrees p and q, respectively, of a (smooth) n-dimensional manifold N (e.g. $\mathbb{R}^n$). Recall (from §10.3) that the submanifolds P and Q *intersect transversely* (or, as we shall sometimes say, are in *general position* in N) if at each point $x\in P\cap Q$ the tangent spaces to P and Q together span the whole of the tangent space to N. As was noted in §10.3, the basic property of transversely intersecting submanifolds P, $Q\subset N$, is that their intersection $P\cap Q$ is a (smooth) $(p+q-n)$-dimensional submanifold of N.

In the case $p+q=n$, which will be of particular interest to us, the intersection $P \cap Q$ consists of finitely many points $x_1, \ldots, x_m$. Given orientations on the manifolds N, P, Q (assuming them orientable), we attach a sign to each point $x_j \in P \cap Q$ in the following way. Let $\tau^P_{(j)}$ be a tangent frame at the point x_j to the submanifold P, belonging to the given orientation class on P, and define $\tau^Q_{(j)}$ similarly in terms of Q and its orientation. Then we attach to the point x_j the sign $+1$ if the frame $\tau=(\tau^P_{(j)}, \tau^Q_{(j)})$ (which is linearly independent by virtue of the fact that P and Q intersect transversely) belongs to the given orientation class on N, and -1 if not; we denote the sign attached to x_j in this way by $\operatorname{sgn} x_j(P \circ Q)$.

15.1.1. Definition. The *intersection index* of transversely intersecting oriented submanifolds P, Q of an oriented manifold N, where $p+q=n$, is the integer

$$P \circ Q = \sum_{j=1}^{m} \operatorname{sgn} x_j(P \circ Q).$$

(If any of the manifolds P, Q, N is non-orientable we define $P \circ Q$ to be the residue modulo 2 of the number m of points in the intersection.)

15.1.2. Lemma. *With P, Q, N as above, we have $P \circ Q=(-1)^{pq} Q \circ P$.*

PROOF. This is immediate from the fact that if τ^P, τ^Q are ordered bases (i.e. frames) for p- and q-dimensional subspaces of a vector space, which together span a $(p+q)$-dimensional subspace V, then the determinant of the linear transformation of V taking the frame (τ^P, τ^Q) to the frame (τ^Q, τ^P) is just $(-1)^{pq}$. □

15.1.3. Theorem. *If two submanifolds $Q_1, Q_2 \subset N$ are homotopic (i.e. are the images of homotopic pair of embeddings of Q into N), then their intersection indices with any (appropriate) submanifold P coincide:*

$$Q_1 \circ P = Q_2 \circ P.$$

PROOF. Let $F\colon Q \times I \to N$, be a smooth homotopy such that $F(Q \times 0)=Q_1$ and $F(Q \times 1)=Q_2$. In view (essentially) of Theorem 10.3.2 and the fact that each of Q_1, Q_2 intersects P transversely, we may assume that F is t-regular on P. Hence its complete inverse image $F^{-1}(P)$ is a one-dimensional submanifold

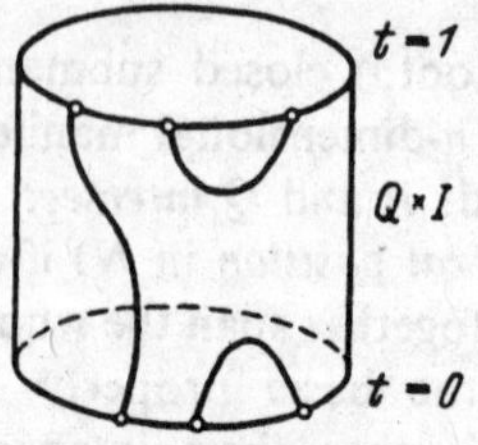

Figure 44

of the cylinder $Q \times I$, moreover with boundary the (finite) set $(Q_1 \cap P) \cup (Q_2 \cap P)$, where $Q_1 \cap P$ lies in the base $Q \times 0$, $Q_2 \cap P$ in the lid $Q \times 1$ of the cylinder, and the curve segments comprising $F^{-1}(P)$ approach the base and lid transversely (as illustrated in Figure 44). The desired conclusion now follows exactly as in the proof of the invariance of the degree of a map under homotopies (Theorem 13.1.2). (Compare Figure 44 with Figure 30.) □

15.1.4. Corollary. *The intersection index of two (transversely intersecting) closed submanifolds P, Q of Euclidean n-space is zero.*

PROOF. By translating Q along a vector $a \in \mathbb{R}^n$ whose length is sufficiently great, we obtain a (homotopic) manifold $Q_2 = Q + a$, which has empty intersection with P. (This is made possible by the compactness of P and Q.) Then obviously $Q_2 \circ P = 0$, whence by the above theorem, $Q \circ P = 0$ also. □

15.1.5. Corollary. *Any connected, closed $(n-1)$-dimensional submanifold M of $\mathbb{R}^n$ separates $\mathbb{R}^n$ into two disjoint parts (and hence is orientable).*

PROOF. Suppose the contrary. Let x be any point of M and let y_1, y_2 be two points of $\mathbb{R}^n$ close to x and on opposite sides of M on the straight line normal to M at x. (This makes sense locally since locally M is identifiable with a region of $\mathbb{R}^{n-1}$). Join the points y_1 and y_2 by a path γ which does not intersect M and close the gap in γ (between y_1 and y_2) by means of a short straight-line segment perpendicular to M and intersecting M in exactly one point. (This will be possible if y_1, y_2 are sufficiently close to x.) Denote the resulting closed curve (after suitable smoothing) by C. Then since C intersects M in exactly one point, the intersection index $C \circ M = \pm 1$, contradicting the above corollary. □

Remarks. 1. Note in connexion with this proof that we are not assuming beforehand that M is orientable; if M is non-orientable then the intersection index $C \circ M = 1$, the residue modulo 2 of the number of points in $C \cap M$, so that the argument is still valid.

2. The latter corollary ceases to hold if we weaken the hypothesis by demanding only that M be immersed in $\mathbb{R}^n$, i.e. if we allow self-intersections of M in $\mathbb{R}^n$. For instance there exist immersions (necessarily involving self-intersections) of the non-orientable 2-manifold $\mathbb{R}P^2$ in $\mathbb{R}^3$ (see [8]).

15.2. The Total Index of a Vector Field

Let ξ be a tangent-vector field on a smooth closed manifold P of dimension p, and denote by N the tangent-bundle manifold of P (of dimension $n = 2p$). (Recall from §7.1 that the points of N are the pairs (x, η) where x ranges over the points of P, and η over the tangent space to P at x.) The vector field ξ gives rise to an embedding $f_\xi: P \to N$, defined by the rule $f_\xi(x) = (x, \xi(x))$. We shall denote the image submanifold under f_ξ by $P(\xi)$, and shall in the usual way identify the submanifold $P(0) \subset N$, corresponding to the identically zero vector field, with our original manifold P.

15.2.1. Definition. A tangent vector field ξ on the manifold P is said to be *in general position* if the submanifolds $P(\xi)$ and $P = P(0)$ are in general position in the tangent bundle N.

It follows easily (from the definition of transversely intersecting submanifolds) that a vector field in general position on a manifold P can have at worst isolated singularities. Note also that if P is oriented by a frame τ^P at the (variable) point x, then the frame (τ^P, τ^P) at the point (x, η) determines a natural orientation on N.

15.2.2. Lemma. *All singular points x_j of a vector field in general position on a smooth, closed oriented manifold P, are non-degenerate. The sign ± 1 contributed by each singular point x_j (regarded as a point of the intersection $P(0) \cap P(\xi)$) to the intersection index $P(0) \circ P(\xi)$, coincides with the index of the singular point, i.e. with* sgn det $(\partial \xi^\alpha / \partial x^\beta)_{x_j}$.

PROOF. A typical point of the intersection $P(0) \cap P(\xi)$ has the form $(x_j, 0)$ where x_j is a singular point of ξ. The tangent space to $P(0)$ at this point consists of all vectors of the form $(\eta, 0)$, while the tangent space to $P(\xi)$ at the point consists of all vectors of the form $(\eta^\alpha, (\partial \xi^\alpha / \partial x_j^\beta)\eta^\beta)$, where the x_j^β are local co-ordinates on P in a neighbourhood of x_j, and in both cases η ranges over the tangent space to P at x_j. Thus if we write $J = (\partial \xi^\alpha / \partial x_j^\beta)_{x = x_j}$, and denote by τ^P an orienting frame for P at the point x_j, then the frames $\tau_1^P = \tau^P \times 0$ and $\tau_2^P = \tau^P \times J\tau^P$, will be orienting for $P(0)$ and $P(\xi)$, respectively, at their point of intersection $(x_j, 0)$. To obtain the sign to be attached to that point we need to compare the two frames (τ_1^P, τ_2^P) and (τ_1^P, τ_1^P) for the tangent space to the tangent bundle N at $(x_j, 0)$, the latter frame determining the orientation on N. Since these two frames agree in their first p vectors (namely those of τ_1^P), it is clear that the sign to be attached to $(x_j, 0)$ is that of the determinant of the $p \times p$ matrix sending the frame τ^P to the frame $J\tau^P$, i.e. of the determinant of J. This completes the proof of the lemma. □

15.2.3. Theorem. *For any closed, oriented manifold P, the sum of the indices of the singular points of any vector field ξ in general position on P coincides with the intersection index $P(0) \circ P(\xi)$ in the tangent-bundle manifold N, and is the same for all such vector fields ξ.*

PROOF. The equality between the intersection index $P(0) \circ P(\xi)$ and the sum of the indices of the singular points of ξ is immediate from the preceding lemma. To see that this number is actually independent of the field ξ, observe first that any vector field is homotopic to the zero field (via a homotopy of the form $\xi(x, t) = t\xi(x)$) so that any two vector fields ξ, η on P are homotopic to one another. Hence the embeddings $P \to P(\xi) \subset N$, $P \to P(\eta) \subset N$ determined by the fields ξ, η, are homotopic. Assuming each of these fields to be in general position on P, an appeal to Theorem 15.1.3 now gives us the desired equality $P(0) \circ P(\xi) = P(0) \circ P(\eta)$. □

15.2.4. Corollary. *If p is odd then the sum of the indices of the singular points of a vector field in general position on a closed, orientable p-manifold, is zero.*

PROOF. Let ξ be a vector field in general position on the p-manifold P. By Lemma 15.1.2 we have $P(0) \circ P(\xi) = (-1)^{p^2} P(\xi) \circ P(0) = -P(\xi) \circ P(0)$. On the other hand, in view of the fact that ξ is homotopic to the zero field it follows from Theorem 15.1.3 that $P(0) \circ P(\xi) = P(\xi) \circ P(0)$. Hence $P(0) \circ P(\xi) = 0$, as required. □

15.2.5. Corollary. *Let f be any smooth function on a closed, orientable manifold P, having only non-degenerate critical points. Then the sum*

$$\sum_{j=1}^{m} (-1)^{i(x_j)}, \tag{1}$$

(where $x_1, \ldots, x_m$ are the critical points of f and $i(x_j)$ is the number of negative squares in (the standard form of) the quadratic form $(d^2 f)_{x_j}$) is independent of f, and is zero if p is odd.

This is immediate from Theorem 15.2.3 (and the preceding corollary) with grad f in the role of ξ. (Note that the field grad f is in general position on P since by Lemma 14.4.1 non-degenerate singular points are isolated.)

The integer defined by (1) above is called the *Euler characteristic of the manifold P.* The reader may possibly be more familiar with the alternative definition of the Euler characteristic by means of generalized "triangulations" of P. We shall content ourselves with examining this alternative definition only in the case $p=2$. Thus suppose that our manifold P is a closed, orientable (2-dimensional) surface, subdivided into (curvilinear) closed triangles in such a way that any two distinct triangles intersect in the empty set, or exactly one vertex, on a single (whole) edge (i.e. have at most a single vertex or a single edge in common).

15.2.6. Definition. The *Euler characteristic of a closed orientable surface P* is the integer $\alpha_0 - \alpha_1 + \alpha_2$, where α_0 is the number of vertices, α_1 the number of edges and α_2 the number of triangles, in any triangulation.

The equivalence of this definition with our earlier one (in the case $p=2$) is shown by the following result.

15.2.7. Theorem (Hopf). *The Euler characteristic of a closed orientable surface P coincides with the sum of the indices of the singular points of any vector field ξ in general position on P.*

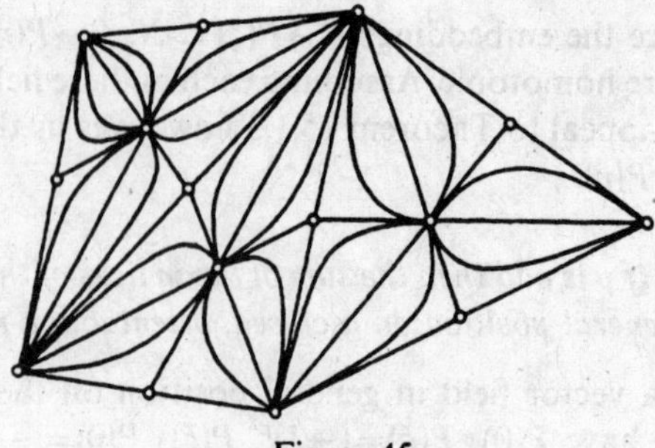

Figure 45

PROOF. In view of Theorem 15.2.3 the result will follow if we can produce any particular vector field on P for which it is valid. We construct such a particular field ξ as follows. Its singular points are to be comprised of: the vertices (of any given triangulation of P) which are all to be sinks; exactly one point in the interior of each triangle (all of which points are to be sources); and finally one point in the interior of each edge (which points are to be saddles) (see Figure 45 where some of the integral curves of the field are indicated; note also that the edges are intended to be (made up of) integral curves of the field, so that the integral curves are formed "independently" in the separate triangles of the triangulation). Since in the case $p=2$ sources and sinks have index $+1$ and saddles index -1 (see §14.4), the theorem now follows. □

For a sphere-with-g-handles the Euler characteristic is $2-2g$ (verify it!). In particular the (2-dimensional) sphere has Euler characteristic 2.

15.3. The Signed Number of Fixed Points of a Self-map (the Lefschetz Number). The Brouwer Fixed-Point Theorem

Let $f: M \to M$ be a smooth self-map of a closed, oriented n-manifold M. In this subsection we shall investigate the "fixed points" of such maps, i.e. the solutions of equations of the form $f(x)=x$. Let x_j be a fixed point of f, and let x_j^α be local co-ordinates in some neighbourhood of that fixed point; then the co-ordinates of x_j satisfy the system of equations $x_j^\alpha = f^\alpha(x_j^1, \ldots, x_j^n)$, $\alpha = 1, \ldots, n$.

15.3.1. Definitions. A fixed point x_j of a self-map $f: M \to M$, is said to be *non-degenerate* if the matrix

$$\left[\delta_{\alpha\beta} - \left(\frac{\partial f^\alpha}{\partial x^\beta}\right)_{x=x_j}\right] = (1-df)_{x=x_j}$$

is non-singular. We then define the *sign* of a non-degenerate fixed point x_j to be $\operatorname{sgn}\det(1-df)_{x_j}$, and if all of the fixed points $x_1, \ldots, x_m$ of f are non-degenerate (there are only finitely many of them since each of them is isolated and M is compact) we call the sum $\sum_j \operatorname{sgn}\det(1-df)_{x_j}$ of their signs the *signed* (or *algebraic*) *number of fixed points* of f (or the *Lefschetz number* of f), denoted by $L(f)$. (We shall say also that a self-map f is *in general position* if all of its fixed points are non-degenerate.)

Consider now the direct product $M \times M$ and two submanifolds of it of particular interest to us here:

(i) the diagonal Δ, consisting of all pairs of the form (x, x);
(ii) the graph $\Delta(f)$ of f, consisting of all pairs $(x, f(x))$.

If M is smooth, then it is easy to see that each of these is a smooth submanifold diffeomorphic to M, of the (smooth) manifold $M \times M$.

15.3.2. Theorem. *Let $f: M \to M$ be a smooth self-map of the smooth, closed, oriented n-manifold M. Then the intersection index $\Delta(f) \circ \Delta$ coincides with the Lefschetz number of f.*

PROOF. Obviously the intersection $\Delta \cap \Delta(f)$ consists precisely of the fixed points $x_1, \ldots, x_m$ of f. Let $\tau = (v_1, \ldots, v_n)$ be an orienting frame for M at the point x_j. Then the frame (τ, τ) at the point (x_j, x_j) determines an orientation of $M \times M$, the frame at (x_j, x_j) consisting of the vector-pairs $(v_1, v_1), \ldots, (v_n, v_n)$ in order, which frame we denote by $\tau \times \tau$, determines an orientation of the diagonal Δ, and, finally, we orient the graph $\Delta(f)$ by choosing at the point $(x_j, f(x_j)) = (x_j, x_j)$, the frame $\tau \times df(\tau)$ consisting of the pairs $(v_1, df(v_1)), \ldots, (v_n, df(v_n))$ in that order, where df denotes the differential of the map f at the point x_j. The $2n \times 2n$ matrix of the change from the frame (τ, τ) (consisting of the vectors $(v_1, 0), \ldots, (v_n, 0), (0, v_1), \ldots, (0, v_n)$ in that order) on $M \times M$ to the frame $(\tau \times df(\tau), \tau \times \tau)$ composed of the orienting frames on $\Delta(f)$ and Δ is easily seen to be

$$\begin{pmatrix} 1 & 1 \\ df & 1 \end{pmatrix} \sim \begin{pmatrix} 1 & 0 \\ df & 1-df \end{pmatrix},$$

which has determinant $\det(1-df)$. The theorem is now immediate from the definitions of Lefschetz number and of the intersection index. □

15.3.3. Corollary. *If the map $f: M \to M$ is homotopic to a map to a single point, then $L(f) = \pm 1$ (whence the map f has at least one fixed point).*

PROOF. The assumed homotopy of f induces a homotopy between the embeddings $M \to \Delta(f) \subset M \times M$, and $M \to M \times x_0 \subset M \times M$ (where x_0 is a particular point of M) via embeddings which are t-regular on Δ. Writing $M_0 = M \times x_0$, we therefore have by Theorem 15.1.3 that $\Delta(f) \circ \Delta = M_0 \circ \Delta$;

however since $M_0 \cap \Delta$ clearly consists of just one point (namely (x_0, x_0)), the latter intersection index is ± 1, whence the corollary. □

15.3.4. Corollary (Brouwer). *Any continuous self-map f of the closed disc D^n (i.e. closed ball) has a fixed point.*

PROOF. We may identify the disc with the lower hemisphere of $S^n \subset \mathbb{R}^{n+1}$. Let $\psi: S^n \to D^n$ be the map which leaves fixed the points of the lower hemisphere and projects the upper hemisphere onto the lower one. Then the composite map

$$S^n \xrightarrow{\psi} D^n \xrightarrow{f} D^n \subset S^n,$$

is a null-homotopic map $S^n \to S^n$ (i.e. homotopic to a map to a single point) since the image is contained in D^n which we can shrink to a point in S^n. By Theorems 10.1.2 and 12.1.2 the composite map $f \circ \psi: S^n \to S^n$ can be approximated arbitrarily closely by smooth maps homotopic to it. By the preceding corollary such smooth maps will have fixed points, and a short argument (involving nested compact subsets of S^n) then shows that $f \circ \psi$ must also have a fixed point. Since $(f \circ \psi)(S^n) \subset D^n$, any such fixed point must lie in D^n, and is therefore a fixed point of f. □

Examples. (a) As we saw in Example (b) of §13.4, the map $z \to z^n$ (or $\varphi \to n\varphi$) of the unit circle $|z| = 1$ to itself has degree n. The fixed points are the solutions of the equations $z^n = z$, $|z| = 1$, i.e. of $z^{n-1} = 1$; they are therefore just the $(n-1)$th roots of unity. It follows that (for $n > 1$) the Lefschetz number of this map is $-(n-1)$. Since every map $S^1 \to S^1$ of degree n is homotopic to the map $z \mapsto z^n$ (see §13.4, Example (b)), and the Lefschetz number is invariant under homotopies (see the proof of Corollary 15.3.3), we deduce that for any map $f: S^1 \to S^1$ of degree n, the Lefschetz number $L(f) = -(n-1)$.

(b) The map $S^2 \to S^2$ defined (for each $n > 1$) by $z \mapsto z^n$ (identifying as usual the 2-sphere with the extended complex plane or Riemann sphere) has n fixed points in the finite plane $\mathbb{C}^1 = \mathbb{R}^2$, and has the point at infinity as a further fixed point. We leave to the reader the verification that all of these fixed points are non-degenerate of sign $+1$. Thus the Lefschetz number of this map is $n+1$.

EXERCISES

1. Show that for any map $f: S^m \to S^m$ of degree n, the Lefschetz number $L(f)$ is (in absolute value) $n+1$ or $n-1$ according as m is even or odd. (In particular, the "antipodal" map $\xi \mapsto -\xi$, which has no fixed points, has degree $(-1)^{m-1}$.)
2. Calculate $L(f)$ for the linear self-map f of the m-dimensional torus T^m, defined by an integral matrix of degree m. (Recall the definition of T^m (given in §5.2) as the quotient space of $\mathbb{R}^m$ by the integer lattice.)

15.4. The Linking Coefficient

Consider a pair of smooth, closed, regular directed curves γ_1 and γ_2 in $\mathbb{R}^3$, which do not intersect one another. We may assume them to be parametrized in the form $\gamma_i(t)=r_i(t)$, $0\le t\le 2\pi$, where r denotes the radius-vector of points in $\mathbb{R}^3$.

15.4.1. Definition. The *linking coefficient* of the two curves γ_1, γ_2 parametrized as above is defined (in terms of the "Gauss integral") by

$$\{\gamma_1, \gamma_2\} = \frac{1}{4\pi} \oint_{\gamma_1} \oint_{\gamma_2} \frac{\langle [dr_1, dr_2], r_{12}\rangle}{|r_{12}|^3}, \tag{2}$$

where $r_{12}=r_2-r_1$.

Intuitively speaking the linking coefficient gives the algebraic (i.e. signed) number of loops of one contour (imagine a wire lead) around another. This interpretation is justified by the following result.

15.4.2. Theorem. (i) *The linking coefficient $\{\gamma_1, \gamma_2\}$ is an integer, and is unchanged by deformations of the closed curves γ_1, γ_2, involving no intersections of the curves with each other.*

(ii) *Let $F\colon D^2\to\mathbb{R}^3$ be a map of the disc D^2 which agrees with $\gamma_1\colon t\mapsto r_1(t)$, $0\le t\le 2\pi$, on the boundary $S^1=\partial D^2$, and is t-regular on the curve $\gamma_2\subset\mathbb{R}^3$. Then the intersection index $F(D^2)\circ\gamma_2$ is equal to the linking coefficient $\{\gamma_1, \gamma_2\}$.*

Proof. The closed curves $\gamma_i(t)=r_i(t)$, $i=1, 2$, give rise to a 2-dimensional, closed, oriented surface

$$\gamma_1\times\gamma_2\colon (t_1, t_2)\mapsto(r_1(t_1), r_2(t_2))$$

in $\mathbb{R}^6$. Since the curves are non-intersecting the map $\varphi\colon \gamma_1\times\gamma_2\to S^2$, given by

$$\varphi(t_1, t_2)=\frac{r_1(t_1)-r_2(t_2)}{|r_1(t_1)-r_2(t_2)|},$$

is well defined thereby. From §14.2 (formula (5)) we see that the degree of the map φ is equal to the integral on the right-hand side of (2) above; hence the linking coefficient is indeed an integer. Under deformations of the curves γ_1, γ_2 involving no intersections with one another, the map φ undergoes a homotopy, so that its degree, and therefore also the linking coefficient, are preserved.

We now prove (ii). If the curves are not linked (i.e. if by means of a homotopy respecting non-intersection they can be brought to opposite sides of a 2-dimensional plane in $\mathbb{R}^3$) then it can be verified directly that $\{\gamma_1, \gamma_2\}=\deg\varphi=0$. Hence by applying a homotopy as indicated in Figure 46(a) and (b), we reduce the general case of the problem of calculating the linking coefficient essentially to the simple situation shown in

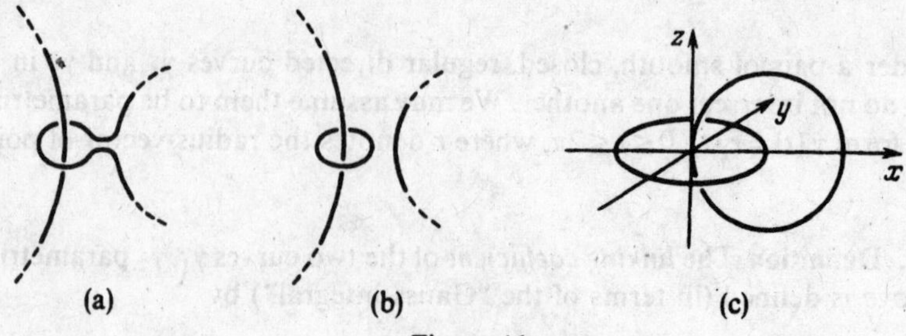

Figure 46

Figure 46(c). The calculation in this case is made especially easy by letting the radius of one of the circles go to ∞; thus we suppose γ_1 and γ_2 to be given respectively by $r_1(t_1)=(0, 0, t_1)$, $-\infty < t_1 < \infty$, and $r_2(t_2)=(\cos t_2, \sin t_2, 0)$, $0 \le t_2 \le 2\pi$. By (2) the linking coefficient for these two curves is

$$\{\gamma_1, \gamma_2\} = \frac{1}{4\pi}\int_{-\infty}^{\infty}\int_0^{2\pi} \frac{dt_1 \wedge dt_2}{(1+t_1^2)^{3/2}}$$

$$= \frac{1}{2}\int_{-\infty}^{\infty} \frac{dt_1}{(1+t_1^2)^{3/2}} = \frac{1}{2}\int_{-\infty}^{\infty} \frac{dz}{\cosh^2 z} = \frac{1}{2}\tanh z\,|_{-\infty}^{\infty} = 1,$$

where we have used the substitution $t_1 = \sinh z$. Hence for these two directed curves statement (ii) of the theorem holds. The general result now follows easily from this and the already-noted fact that linking coefficient of a pair of unlinked curves is zero. □

CHAPTER 4

Orientability of Manifolds. The Fundamental Group. Covering Spaces (Fibre Bundles with Discrete Fibre)

§16. Orientability and Homotopies of Closed Paths

16.1. Transporting an Orientation Along a Path

According to the simplest of the definitions of an orientation on a manifold given above (see Definition 1.1.3), a manifold M is oriented if the local co-ordinate systems x_j^α given on the members U_j of a covering collection of local co-ordinate neighbourhoods (or charts) for M, are such that the transition functions from one local co-ordinate system to another on the regions of overlap $U_j \cap U_k$, have positive Jacobian:

$$\det\left(\frac{\partial x_j^\alpha}{\partial x_k^\beta}\right) > 0.$$

The second definition (2.1.2) of an orientation (on an n-manifold M) was as follows: The set of tangent n-frames (for the tangent space) at each point $x \in M$ is divided into two "classes", two frames being in the same class if the linear transformation taking one into the other has positive determinant; an orientation is then said to be given on M if with each point $x \in M$ there is associated one of its classes of tangent frames in such a way that the class varies continuously with x (as x ranges over M).

We found these definitions convenient for establishing the orientability of various kinds of manifolds, for instance complex manifolds, and surfaces in $\mathbb{R}^n$ defined by non-singular systems of equations $f_1 = 0, \ldots, f_{n-k} = 0$. However our aim in the present subsection is to establish the *non-orientability* of certain manifolds, for which purpose the concept of "transporting an orientation along a path" in a manifold M, will be useful.

We now define this concept. We assume our n-manifold M to be connected, and for the sake of convenience we endow it with a metric g_{ab}. Let $\gamma=\gamma(t)$, $0 \le t \le 1$, be a piecewise smooth path (i.e. curve segment) in M, and at each point $\gamma(t)$ of the path, let $\tau^n(t)$ be a linearly independent tangent n-frame to M, depending continuously on t.

16.1.1. Definition. The orientation class of the frame $\tau^n(t)|_{t=1}$ at the point $\gamma(1)$, will be called the *result of transporting the orientation class of the frame $\tau^n(0)$ at the point $\gamma(0)$ along the path γ.*

The operation of transporting an orientation along a path has the following three basic properties.

(i) For each point $x \in M$, *there is a (sufficiently small) neighbourhood of x with the property that for any point y in that neighbourhood, the result of transporting an orientation from x to y along any path entirely contained in the neighbourhood, is uniquely defined.*

This property is immediate from the fact that in any manifold each point has a neighbourhood contained in a chart of the manifold (and therefore identifiable with a region of $\mathbb{R}^n$).

(ii) *For any given piecewise smooth path γ, the result of transporting an orientation class along γ exists, and is independent of the frame $\tau^n(t)$ along the curve* (subject of course to the conditions that the frame depend continuously on t, and that the initial frame $\tau^n(0)$ be in a specific orientation class).

The existence follows from the fact that it is always possible to parallel transport a frame along a smooth or piecewise smooth curve segment on a manifold endowed with a Riemannian metric (see §1.2 above, and §29.1 of Part I). That the result of transporting an orientation class along a path γ is independent of the choice of frame $\tau^n(t)$ along γ, is proved as follows. Let $\tau_1^n(t)$ and $\tau_2^n(t)$ be two fields of frames on γ which at $t=0$ belong to the same orientation class. Denote by $A(t)$ the matrix transforming the frame $\tau_1^n(t)$ to the frame $\tau_2^n(t)$ at time t. Then $\det A(t) \neq 0$ for all t, and $\operatorname{sgn} \det A(0) = +1$. In view of the continuity of the dependence of the orientation classes of τ_1^n and τ_2^n on t, it follows that $\operatorname{sgn} \det A(t) = +1$ for all t.

(iii) *If two piecewise smooth paths $\gamma_1(t)$ and $\gamma_2(t)$ have the same initial and terminal points, and one can be deformed into the other by means of a piecewise smooth homotopy holding fixed the end-points $x_0=\gamma_1(0)=\gamma_2(0)$ and $x_1=\gamma_1(1)=\gamma_2(1)$, then the results of transporting an orientation class along the two paths, coincide.*

This can be seen as follows. Let $F(t, s)$, $0 \le t \le 1$, $0 \le s \le 1$, be a homotopy of the kind assumed in the above statement; thus $F(t, 0)=\gamma_1(t)$, $F(t, 1)=\gamma_2(t)$, and for each fixed t, the path $F(t, s)$, $0 \le s \le 1$, is piecewise smooth. Let $\tau^n(t)$ be a (continuously varying) field of frames on the curve $\gamma_1(t)=F(t, 0)$, and for each fixed t parallel transport the frame $\tau^n(t)$ along the path $F(t, s)$, $0 \le s \le 1$. (The metric on $M \times I$ in terms of which this parallel transport is carried out, is the natural one defined by the inner product $\langle(\xi, \eta), (\xi, \eta)\rangle = g_{ab}\xi^a\xi^b + |\eta|^2$, where ξ is a tangent vector to M at x, and η is a tangent vector to $t \in I = [0, 1]$.) It follows from the definition of parallel transport (see §29.1 of Part I) that the result of such parallel transport (carried out for every t) will be a continuously varying field of frames on the curve $\gamma_2(t) = F(t, 1)$, and that moreover since the (common) end-points of the two paths are held fixed during the homotopy, the orientation classes of the initial and final frames of this field are respectively the same as the initial and final frames of the field of frames $\tau^n(t)$ on γ_1.

From these properties we derive the following

16.1.2. Theorem. *A connected manifold M is orientable if and only if transport around any closed path (i.e. one having coincident initial and terminal points, also called a "loop") preserves the orientation class.*

PROOF. If there is a closed path beginning and ending at a point x_0 which reverses the orientation (i.e. such that the result of parallel transporting a frame τ^n around the path from x_0 back to x_0 is a frame in the opposite orientation class), then the manifold is non-orientable since clearly it is not possible to associate with each point of the manifold (or, more particularly, with each point of this closed path) a continuously varying orientation class.

For the converse, suppose that all closed paths beginning and ending at x_0 preserve the orientation. We now define an orientation of the manifold M by choosing arbitrarily an orientation class at the point x_0, and assigning to every other point x_1 the orientation class resulting from transporting the orientation class at x_0 along a path from x_0 to x_1. This procedure will yield an orientation of M provided the assignment of orientation class to the (arbitrary) point x_1 is independent of the path γ. Now if γ_1, γ_2 are two paths from x_0 to x_1, then transport of the initially chosen orientation class at x_0 along them will necessarily result in the same orientation class at x_1, since otherwise the closed path $\gamma_2^{-1} \circ \gamma_1$ would reverse the orientation. (Here γ_2^{-1} denotes the path γ_2 traced in the opposite direction, and the product path $\gamma_2^{-1} \circ \gamma_1$ is (by definition) the path $q(s)$, $0 \le s \le 2$, where $q(s) = \gamma_1(s)$ for $0 \le s \le 1$, and $q(s) = \gamma_2(2 - s)$ for $1 \le s \le 2$.) □

16.2. Examples of Non-orientable Manifolds

(a) The *Möbius strip* (or band) is the manifold with co-ordinates (φ, t), $0 \le \varphi \le 2\pi$, $-1 \le t \le 1$, with the identification $(0, t) \sim (2\pi, -t)$, $-1 \le t \le 1$ (see Figure 47).

It is easy to see that the closed curve $\gamma = \{(\varphi, 0) | 0 \le \varphi \le 2\pi\}$ interchanges the orientation classes. (Check it!)

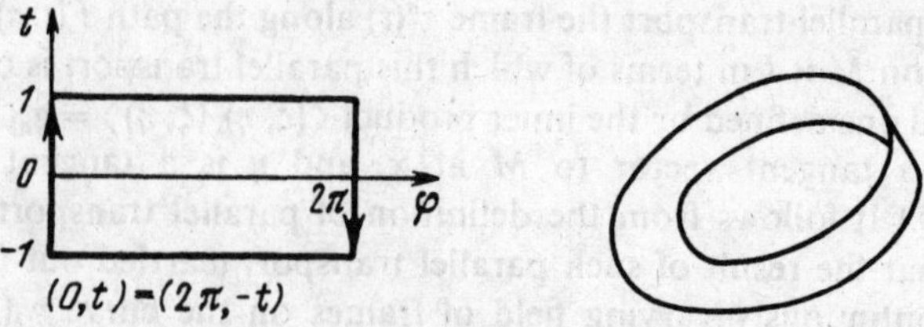

Figure 47. The Möbius strip in $\mathbb{R}^3$ (a "one-sided surface").

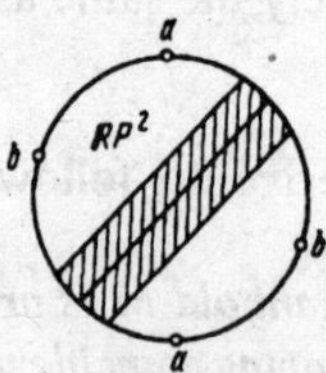

Figure 48

(b) As we saw in §2.2, the projective plane $\mathbb{R}P^2$ can be realized as a disc D^2 with each pair of diametrically opposite points of the boundary $\partial D^2 = S^1$ identified. It is intuitively clear that a small neighbourhood of the projective line $\mathbb{R}P^1 \subset \mathbb{R}P^2$ (shown in Figure 48, with the neighbourhood shaded) through the origin of co-ordinates, will then be a Möbius strip. Hence the closed curve $\mathbb{R}P^1$ interchanges the orientation classes, whence by the above theorem the surface $\mathbb{R}P^2$ must be non-orientable.

EXERCISE

Prove that the manifold $\mathbb{R}P^n$ is orientable for odd n, and non-orientable for even n.

(c) The *Klein bottle* is the 2-manifold with co-ordinates (t, τ), $0 \leq t \leq 1$, $0 \leq \tau \leq 1$, where for each t the points $(t, 0)$ and $(1 - t, 1)$ are identified, and for each τ the points $(0, \tau)$ and $(1, \tau)$ are identified (see Figure 49 where the sides of the square to be identified are labelled with the same letter, and the manner of identification is indicated by the arrows). We leave it to the reader to show that the Klein bottle is non-orientable.

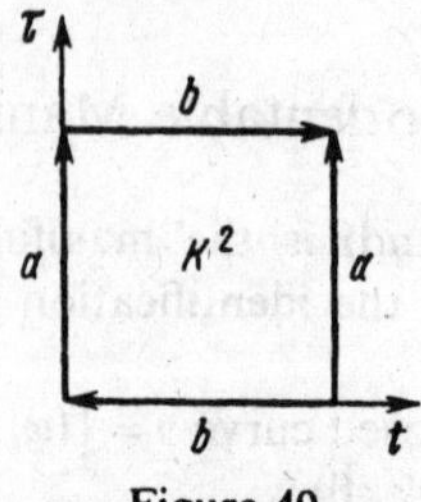

Figure 49

§17. The Fundamental Group

17.1. Definition of the Fundamental Group

Let M be an arbitrary connected manifold (or more generally a path-wise connected topological space). Given any two continuous (or piecewise smooth) paths $\gamma_1(t)$, $0 \le t \le 1$, and $\gamma_2(t)$, $1 \le t \le 2$, such that the terminal point of γ_1 coincides with the initial point of γ_2, we can (as we have already seen) "multiply" them as follows.

17.1.1. Definition. The *product* of the paths γ_1 and γ_2 is the path $\gamma_2 \circ \gamma_1 = q(t)$, $0 \le t \le 2$, defined by

$$q(t) = \gamma_1(t), \qquad 0 \le t \le 1,$$

$$q(t) = \gamma_2(t), \qquad 1 \le t \le 2.$$

17.1.2. Definition. The *inverse* $\gamma^{-1}(t)$ of a path $\gamma(t)$ in M is the path $\gamma(t)$ traversed in the reverse direction; thus $\gamma^{-1}(t) = \gamma(1-t)$ if $0 \le t \le 1$.

17.1.3. Definition. We say that two paths $\gamma_1(t)$, $\gamma_2(\tau)$ are *equivalent* if they differ only by a strictly monotonic change of parameter: $t = t(\tau)$, $dt/d\tau > 0$, $\gamma_1(t(\tau)) = \gamma_2(\tau)$.

Henceforth by a (directed) path we shall mean, strictly speaking, a class of equivalent paths; in practice we shall choose whichever parametrization is most convenient for the purpose at hand.

Consider the totality of all closed paths beginning and ending at a particular point x_0 of M. This set is usually denoted by $\Omega(x_0, M)$, while more generally the set of all paths in M from x_0 to some other point x_1 is denoted by $\Omega(x_0, x_1, M)$. The loops in $\Omega(x_0, M)$ can be multiplied in accordance with Definition 17.1.1, with the product path remaining in $\Omega(x_0, M)$; furthermore, $\Omega(x_0, M)$ contains an identity element e (or 1) under this product operation, namely the constant (or null) path, defined by $e(t) = x_0$ for all t.

Observe that if in a product $\gamma_2 \circ \gamma_1$ of two directed paths, we replace the component paths γ_1 and γ_2 by paths $\tilde{\gamma}_1$ and $\tilde{\gamma}_2$, respectively, homotopic to them via homotopies fixing the end of γ_1 and the beginning of γ_2 (which coincide), then the new product $\tilde{\gamma}_2 \circ \tilde{\gamma}_1$, is homotopic to the original product $\gamma_2 \circ \gamma_1$. Hence in particular it makes sense to speak of the *product of (relative) homotopy classes of (directed) loops beginning and ending at a particular point* x_0 of M.

17.1.4. Theorem. *Let M be a connected manifold (or more generally a pathwise connected topological space), and let x_0 be a point of M. Then the (relative) homotopy classes of (directed) paths in $\Omega(x_0, M)$ form a group under the*

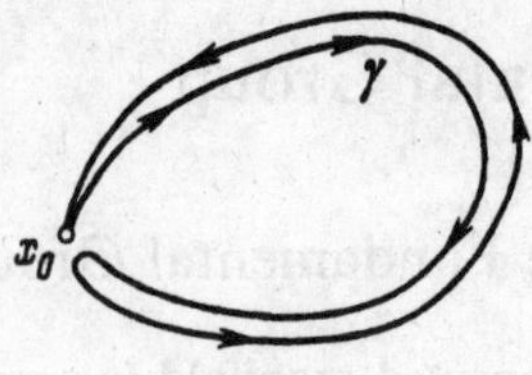

Figure 50

operation of multiplication of these classes. The group-theoretical inverse of a homotopy class is the homotopy class of the inverse of any path from that class, and the identity element is the homotopy class of the null path.

Anticipating the validity of this result we stop to bestow on this group a name and a notation. It is called the *fundamental group with respect to the base point* x_0 of the manifold M, and is denoted by $\pi_1(M, x_0)$. (We emphasize again that during all homotopies the beginnings and ends of paths are assumed fixed at x_0.)

PROOF OF THEOREM 17.1.4. We first show that given any path $\gamma(t)$, $0 \leq t \leq 1$, beginning and ending at x_0, the path $\gamma^{-1} \circ \gamma$ is homotopic to e (see Figure 50). The deformation of the path $\gamma^{-1} \circ \gamma$ to e is performed essentially by shrinking it along itself all the way back to the point x_0, in the following precise manner. Let $q = q(\tau)$ be the map of the interval $[0, 2]$ onto the interval $[0, 1]$ which folds the former interval in two at the point $\tau = 1$:

$$q(\tau) = \tau, \qquad \tau \leq 1,$$

$$q(\tau) = 2 - \tau, \qquad \tau \geq 1.$$

The map q can by means of an obvious homotopy be deformed into the constant map $\tilde{q}(\tau) \equiv 0$, moreover in such a way that the end-points (corresponding to $\tau = 0$, $\tau = 2$) are throughout the homotopy process fixed at 0 (note that $\tau = 1$ does not, of course, correspond to an end-point). Now clearly $\gamma^{-1} \circ \gamma$ is by definition just $\gamma(q(\tau))$; hence by applying the above homotopy from q to the constant map $\tilde{q}$, we induce a homotopy of $\gamma^{-1} \circ \gamma$ to the null path in M.

It remains to show only that the multiplication on $\Omega(x_0, M)$ is associative. However, this is easy since if we define the "triple" product $\gamma_1 \circ \gamma_2 \circ \gamma_3$ of three given (suitably parametrized) paths $\gamma_1, \gamma_2, \gamma_3$, to be the path $q(\tau)$, $0 \leq \tau \leq 3$, with $q = \gamma_3$ for $\tau \leq 1$, $q = \gamma_2$ for $1 \leq \tau \leq 2$, and $q = \gamma_1$ for $\tau \geq 2$, then clearly this path coincides (as usual to within monotonically increasing changes of parameter) with each of the paths $(\gamma_1 \circ \gamma_2) \circ \gamma_3$ and $\gamma_1 \circ (\gamma_2 \circ \gamma_3)$. This completes the proof of the theorem. □

Finally in this subsection, we consider the effect on the fundamental group (with base point x_0) of our manifold M, of a continuous map $f: M \to N$ from M to another manifold N. Under such a map each path $\gamma = \gamma(t)$ in M is sent to the path $f(\gamma(t))$ in N, and under this induced mapping of the set of paths in M, a product of paths goes to the product of the image paths, and homotopic paths in M go into homotopic paths in N. Furthermore any homotopy $F(x, t) = f_t$, of the map f (i.e. $f_0 = f$) with fixed image of the distinguished point x_0 (i.e. such that $F(x_0, t) \equiv f(x_0)$) induces a homotopy of the image $f(\gamma) \subset N$ of any loop γ in M beginning and ending at x_0. From this the following facts are readily inferred.

17.1.5. Theorem. *Any continuous map $f: M \to N$ between manifolds M and N (or more generally between pathwise connected topological spaces) induces a homomorphism between fundamental groups:*

$$f_*: \pi_1(M, x_0) \to \pi_1(N, y_0), \qquad \text{where } f(x_0) = y_0,$$

which is unaltered by homotopies of f throughout which x_0 is sent to y_0. Hence in particular if $M = N$ and the map f is homotopic to the constant map $M \to x_0$, then the homomorphism f_* is trivial (i.e. sends every element to 1). If on the other hand the map f is homotopic to the identity map 1_M, then the homomorphism f_* is an isomorphism.

17.2. The Dependence on the Base Point

Our next result elucidates the dependence of the fundamental group $\pi(M, x_0)$ on the base point x_0. The essential ingredient in this theorem is an operation (to be described below) yielding the group $\pi_1(M, x_1)$ from the group $\pi_1(M, x_0)$ via a path γ joining x_0 to x_1; we shall refer to this operation as that of *transporting the fundamental group of M along γ.*

17.2.1. Theorem. *Any path γ from a point x_0 to a point x_1 of M, determines an isomorphism*

$$\gamma^*: \pi_1(M, x_1) \to \pi_1(M, x_0),$$

which depends only on the (relative) homotopy class of γ (relative to homotopies leaving fixed the initial and terminal points x_0 and x_1, i.e. to homotopies which only involve paths from $\Omega(x_0, x_1, M)$). If the path γ is closed, i.e. if $x_0 = x_1$, (so that γ itself represents an element of $\pi_1(M, x_0)$, which element we denote also by γ), then the isomorphism γ^* is an inner automorphism:

$$\gamma^*(\alpha) = \gamma^{-1}\alpha\gamma.$$

PROOF. The map γ^* is defined as follows. If γ_1 is a path representing an element of the group $\pi_1(M, x_1)$, then $\gamma^*(\gamma_1)$ is to be the homotopy class containing the path $\gamma^{-1} \circ \gamma_1 \circ \gamma$ (see Figure 51). Since under this map the class of a product $\gamma_1 \circ \gamma_2$ is sent to the class containing the path $\gamma^*(\gamma_1 \circ \gamma_2) = \gamma^{-1} \circ \gamma_1 \circ \gamma \circ \gamma^{-1} \circ \gamma_2 \circ \gamma$, which is clearly in the homotopy class containing $\gamma^*(\gamma_1) \circ \gamma^*(\gamma_2)$, we have that the mapping γ^* is a homomorphism.

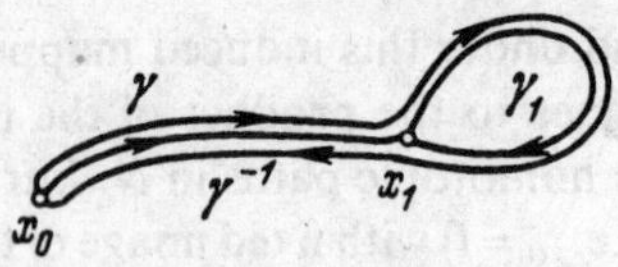

Figure 51

To see that γ^* is one-to-one and onto, consider in place of the path γ, its inverse γ^{-1}, joining x_1 to x_0, and in place of $\pi_1(M, x_1)$ the group $\pi_1(M, x_0)$. We are led as before to a homomorphism

$$(\gamma^{-1})^*: \pi_1(M, x_0) \to \pi_1(M, x_1).$$

It is then easily verified that the composite maps $\gamma^* \circ (\gamma^{-1})^*$ and $(\gamma^{-1})^* \circ \gamma^*$ are the identity maps (of $\pi_1(M, x_0)$ and $\pi_1(M, x_1)$ respectively).

The proof is completed by noting that the second statement of the theorem, relating to the situation where $x_0 = x_1$, is now immediate from the definition of γ^*. □

17.3. Free Homotopy Classes of Maps of the Circle

We consider next the problem of classifying the "free" homotopy classes of maps of the circle S^1 to a connected manifold M (or pathwise connected topological space). Here by "free" we mean that no "initial" point of the circle is distinguished.

17.3.1. Theorem. *The set $[S^1, M]$ of homotopy classes of maps $S^1 \to M$, is in one-to-one correspondence with the classes of conjugate elements (or "conjugacy classes") of the group $\pi_1(M, x_0)$ (for any base point x_0, in view of the preceding theorem).*

Proof. On the circle S^1 (locally) co-ordinatized by φ, $0 \le \varphi \le 2\pi$, denote by φ_0 the point corresponding to $\varphi = 0$, and let x_0 be any particular point of M. As a first step we show that every map $\gamma: S^1 \to M$ is homotopic to a map sending φ_0 to x_0. Let γ_1 be a path in M joining x_0 to the point $\gamma(\varphi_0)$, and consider the closed path $q(\tau) = \gamma_1^{-1}\gamma\gamma_1$, $0 \le \tau \le 3$, beginning and ending at x_0:

$$q = \gamma_1, \qquad 0 \le \tau \le 1,$$

$$q = \gamma, \qquad 1 \le \tau \le 2,$$

$$q = \gamma_1^{-1}, \qquad 2 \le \tau \le 3.$$

As a map of the circle the closed path q is clearly homotopic to γ, and moreover has the property that it sends the point φ_0 to x_0.

Thus each homotopy class of maps $S^1 \to M$, includes at least one element of $\pi_1(M, x_0)$. Suppose now that α_1, α_2 are two (representative paths of) elements of $\pi_1(M, x_0)$, which are homotopic as maps $S^1 \to M$; in the course of

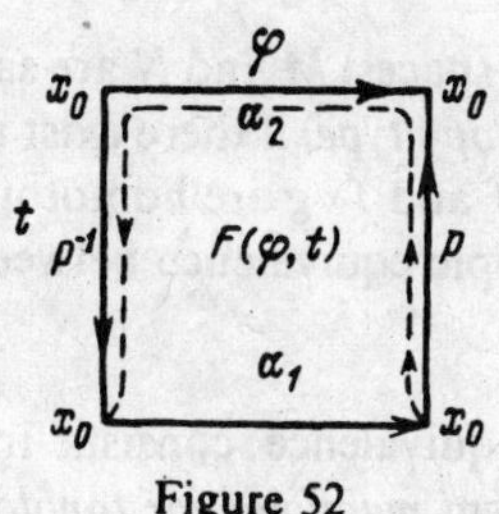

Figure 52

Figure 53

such a homotopy $F(\varphi, t)$ say, $0 \leq \varphi \leq 2\pi$, $0 \leq t \leq 1$, between them, the image of the initial point φ_0 moves along a closed path p say, beginning and ending at x_0:

$$F(\varphi, 0) = \alpha_1, \qquad F(\varphi, 1) = \alpha_2, \qquad F(\varphi_0, t) = p(t).$$

It is not difficult to see that the paths α_1 and $p^{-1}\alpha_2 p$ are homotopic via a homotopy fixing x_0, i.e. that they define the same element of $\pi_1(M, x_0)$ (see Figure 52).

For the converse, suppose we are given loops α_1, α_2 and p all beginning and ending at x_0, such that $\alpha_2 = p^{-1}\alpha_1 p$. Then the paths α_1 and $\alpha_2 = p^{-1}\alpha_1 p$ are freely homotopic (i.e. as maps from S^1 to M); this can be seen from Figure 53 where there is depicted a region of the plane with two points a, b removed, which points are encircled by the paths p and α_1, respectively. It is intuitively clear that the path $\alpha_2 = p^{-1}\alpha_1 p$, indicated by the dotted curve, can be deformed around the puncture a, into the path α_1, by pushing the point where it both begins and ends around the path p. (Imagine the path α_2 to be a contractible thread or elastic band.) □

17.4. Homotopic Equivalence

An "open" manifold M of dimension n (e.g. a region of Euclidean space $\mathbb{R}^n$) can often be deformed within itself into a subset of smaller dimension (which need not in general be a submanifold), whose fundamental group (as well as other invariants) is significantly easier to calculate. The precise formulation of this idea involves the concept of "homotopic equivalence" of manifolds, which we now define. Let M and N be two manifolds (or, more generally, topological spaces), and let f, g be two continuous maps (smooth or piecewise smooth in the case of manifolds) between them (in opposite directions):

$$f: M \to N, \qquad g: N \to M.$$

The composite maps $g \circ f: M \to M$, and $f \circ g: N \to N$, are then self-maps of M and N. As usual we denote by 1_M and 1_N the identity maps on M and N respectively.

17.4.1. Definition. Two manifolds (or topological spaces) M and N are said to be *homotopically equivalent* (or of the *same homotopy type*) if there exist maps f, g (as above) such that the composite maps $g \circ f$ and $f \circ g$ are homotopic to 1_M and 1_N, respectively. The relation of homotopic equivalence between M and N is denoted by $M \sim N$.

The fundamental property of homotopic equivalence consists in the following: *If M and N are homotopically equivalent manifolds (or topological spaces), then given any manifold (or topological space) K, there is a natural one-to-one correspondence between the sets* $[K, M]$ *and* $[K, N]$ *of homotopy classes of maps.*

PROOF. The maps f and g (as in the definition) determine in a natural way (via composition) maps between the sets of homotopy classes:

$$f_*: [K, M] \to [K, N], \qquad g_*: [K, N] \to [K, M].$$

It is immediate from the fact that $g \circ f$ and $f \circ g$ are homotopic to the respective identity maps, that $(g \circ f)_*$ and $(f \circ g)_*$ are the appropriate identity maps. Since, as is easily shown, $(g \circ f)_* = g_* \circ f_*$ and $(f \circ g)_* = f_* \circ g_*$, it follows that f_* and g_* are mutual inverses, whence the desired conclusion. □

The concept of homotopic equivalence is amenable to the mild modification (or "relativization") wherein distinguished points x_0, y_0 of the manifolds M, N, respectively, are specified. It is demanded that the maps f and g send x_0 to y_0 and y_0 to x_0, and also that all maps involved in the homotopies from $g \circ f$ to 1_M, and $f \circ g$ to 1_N, fix x_0 or y_0, as the case may be. It can be shown that for sufficiently "nice" topological spaces, such as for instance connected manifolds, the condition that M, N be homotopically equivalent in this relative sense, is no more stringent than the condition that they be "freely" homotopically equivalent (i.e. in the former sense).

For this relativized version of homotopic equivalence, the above fundamental property becomes: *If the spaces M and N with distinguished points* x_0 *and* y_0 *are homotopically equivalent (relative to those points), then given any space K with distinguished point* k_0*, there is a natural one-to-one correspondence between the sets of relative homotopy classes of maps* $K \to M$ *and* $K \to N$*, sending* k_0 *to* x_0 *and* y_0 *respectively.*

The proof of this is as in the case of free homotopic equivalence. By taking K to be the circle S^1 it readily follows that in this situation M and N have (naturally) isomorphic fundamental groups.

17.5. Examples

(a) Euclidean n-space $\mathbb{R}^n$, and any null-homotopic region of it (i.e. continuously contractible over itself to a point), is homotopically equivalent to a point $x_0 \in \mathbb{R}^n$; in symbols $\mathbb{R}^n \sim x_0$. This is easily proved as follows.

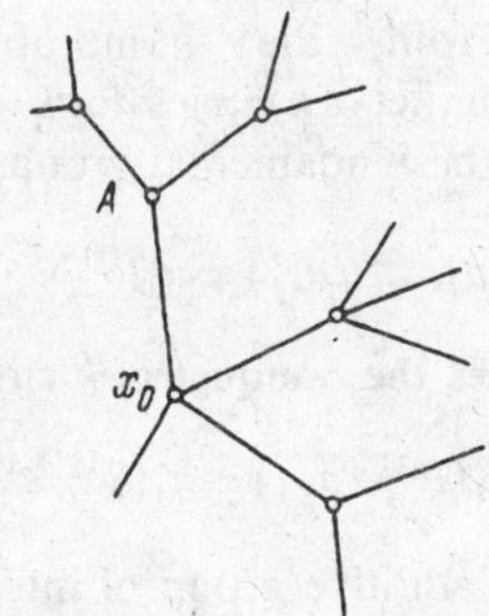

Figure 54. (Part of) a tree $A \sim x_0$.

Consider the embedding $f: \{x_0\} \to \mathbb{R}^n$, and the constant map $g: \mathbb{R}^n \to \{x_0\}$. On the one hand we obviously have $g \circ f = 1_{x_0}$, while on the other hand the composite map $f \circ g: \mathbb{R}^n \to \mathbb{R}^n$, which sends every point of $\mathbb{R}^n$ to x_0, is homotopic to the identity map $1_{\mathbb{R}^n}$ via the homotopy $F(x, t) = tx + (1 - t)x_0$. This proves the above statement for $\mathbb{R}^n$. We leave to the reader the details of the (analogous) proof for an arbitrary subset A of $\mathbb{R}^n$ contractible within itself to a point $x_0 \in A$.

Examples of such contractible subsets are furnished by the (open or closed) disc D^n or any region homeomorphic to it, and by any tree A (i.e. one-dimensional complex (or "graph") without cycles—see Figure 54). (Note that a tree is not in general a manifold.) All of these spaces are homotopically equivalent to a point, and therefore have trivial fundamental group:

$$\pi_1(\mathbb{R}^n, x_0) = 1, \qquad \pi_1(D^n, x_0) = 1, \qquad \pi_1(A, x_0) = 1.$$

(b) Consider the plane $\mathbb{R}^2$ with k points $a_1, \ldots, a_n$ removed. The resulting region $\mathbb{R}^2 \setminus \{a_1, \ldots, a_k\}$ is homotopically equivalent to a "bouquet" of k circles, all joined at a single point (which space is again not, generally speaking, a manifold). (See Figure 55, where it is intuitively clear how $\mathbb{R}^2 \setminus \{a_1, a_2\}$ can be deformed to the bouquet A.) In particular the region $\mathbb{R}^2 \setminus \{a\}$, obtained by removing from $\mathbb{R}^2$ a single point a, is homotopically equivalent to the circle S^1. (As an exercise write down a formula for the

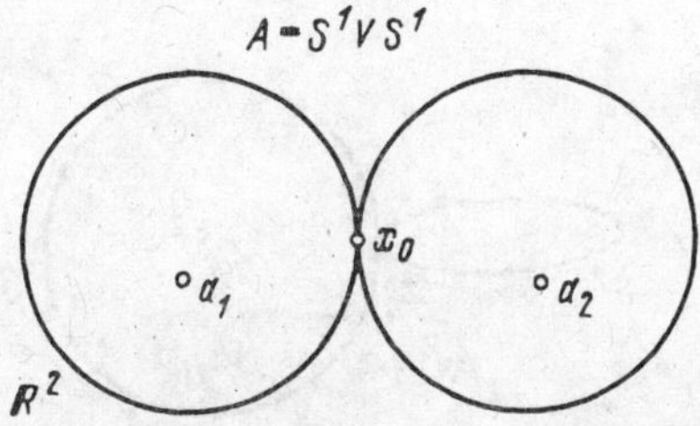

Figure 55. A bouquet of two circles.

relevant homotopy.) From the homotopy equivalence between $\mathbb{R}^2 \backslash \{a_1, \ldots, a_k\}$ and the bouquet of k circles (for $k = 2$ a "figure eight") we infer an isomorphism between the fundamental groups:

$$\pi_1(\mathbb{R}^2 \backslash \{a_1, \ldots, a_k\}) \simeq \pi_1(S_1^1 \vee \cdots \vee S_k^1),$$

where $S_1^1 \vee \cdots \vee S_k^1$ denotes the bouquet of k circles. In particular,

$$\pi_1(\mathbb{R}^2 \backslash \{a_1\}) \simeq \pi_1(S^1) \simeq \mathbb{Z},$$

where $\mathbb{Z}$ here denotes the additive group of integers (i.e. the infinite cyclic group); the second isomorphism ($\pi_1(S^1) \simeq \mathbb{Z}$) follows from §13.4, Example (b).

(c) The region obtained by removing a single point a from $\mathbb{R}^3$, is homotopically equivalent to the 2-sphere S^2. More generally if we remove k points $a_1, \ldots, a_k$ from $\mathbb{R}^3$ we obtain a region homotopically equivalent to a bouquet of k 2-spheres. The region $\mathbb{R}^3 \backslash \mathbb{R}^1$ is homotopically equivalent to the circle S^1. (Verify these statements!) These homotopic equivalences yield the following isomorphisms between fundamental groups:

$$\pi_1(\mathbb{R}^3 \backslash \{a_1, \ldots, a_k\}) \simeq \pi_1(S_1^2 \vee \cdots \vee S_k^2) = 1, \qquad \pi_1(\mathbb{R}^3 \backslash \mathbb{R}^1) \simeq \pi_1(S^1) \simeq \mathbb{Z}.$$

(See Example (e) below for the triviality of $\pi_1(S_1^2 \vee \cdots \vee S_k^2)$.)

(d) Consider the 3-sphere $S^3 \cong \mathbb{R}^3 \cup \{\infty\}$, and the region U of S^3 obtained by removing some (non-self-intersecting, i.e. embedded) circle S^1; thus we may write $U = S^3 \backslash S^1$. As an exercise show that if S^1 is unknotted in S^3 (for instance, if it is embedded as the circle $x^2 + y^2 = 1$ in the plane $\mathbb{R}^2 \subset \mathbb{R}^3 \subset S^3$), then the region $U = S^3 \backslash S^1$ is homotopically equivalent to the circle S^1, and the region $V = U \backslash \{\text{point}\} \cong \mathbb{R}^3 \backslash S^1$, is homotopically equivalent to the bouquet (or "wedge product") formed from the circle S^1 and the sphere S^2 (which incidentally is not of course a manifold) (see Figure 56). Show also that (provided S^1 is unknotted in $\mathbb{R}^3$)

$$\pi_1(\mathbb{R}^3 \backslash S^1) \simeq \mathbb{Z}.$$

(e) We conclude by proving that *for all $n > 1$ the fundamental group of the n-sphere S^n is trivial*:

$$\pi_1(S^n, x_0) = 1 \quad \text{for } n > 1.$$

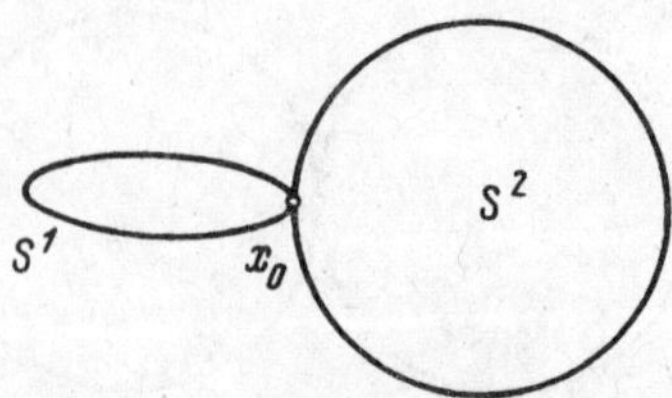

Figure 56. The bouquet $S^1 \vee S^2$.

PROOF. Let $f: S^1 \to S^n$ be any piecewise smooth map. Since $n > 1$ every regular value $y_0 \in S^n$ of f, must be without preimages. Since by Sard's theorem (see §10.2) f must have regular values, it follows that the image of S^1 under f lies in $S^n \backslash \{y_0\}$ for some point y_0. Since $S^n \backslash \{y_0\} \cong \mathbb{R}^n$, it follows that $f(S^1)$ is contractible to a point in S^n. □

Remark. A similar argument shows that for any manifold K of dimension $< n$, the set $[K, S^n]$ of homotopy classes is trivial (i.e. contains just one element).

In the sequel we shall calculate the fundamental groups of various other concrete manifolds (and topological spaces), in particular of bouquets of circles (or, equivalently, the punctured plane), of the closed 2-dimensional surfaces, and of regions of the form $\mathbb{R}^3 \backslash S^1$ where S^1 may be knotted in $\mathbb{R}^3$.

17.6. The Fundamental Group and Orientability

From the results of §16.1 it is immediate that given any (relative) homotopy class of closed paths beginning and ending at the base point x_0 of a connected manifold M, either every path in that class preserves the orientation of a tangent frame transported around the path, or every path in the class reverses the orientation, i.e. the homotopy class as a whole preserves or reverses the orientations of tangent frames. If we assign the number $+1$ to those homotopy classes preserving orientation and -1 to those reversing it, we obtain a homomorphism

$$\sigma: \pi_1(M, x_0) \to \{\pm 1\} \simeq \mathbb{Z}_2,$$

where $\{\pm 1\}$ denotes the obvious multiplicative group, and $\mathbb{Z}_2$ is the group of additive integers modulo 2 (i.e. the cyclic group of order 2): $\sigma(\gamma) = \operatorname{sgn} \gamma = \pm 1$. By Theorem 16.1.2 this homomorphism is trivial if M is orientable, and non-trivial if M is non-orientable, since in the latter case there exist closed paths reversing the orientation of tangent frames. Hence that theorem leads to the following

17.6.1. Corollary. *The fundamental group of a non-orientable manifold is non-trivial, and in fact can be mapped homomorphically onto the cyclic group of order* 2.

For the Möbius strip we have $\pi_1(M) \simeq \mathbb{Z}$, since it can be contracted within itself to the circle S^1 along its middle (see Figure 47). For the projective plane $\mathbb{R}P^2$ we deduce that $\pi_1(\mathbb{R}P^2) \neq 1$; we shall subsequently show (in §19.2) that in fact $\pi_1(\mathbb{R}P^2) \simeq \mathbb{Z}_2$.

§18. Covering Maps and Covering Homotopies

18.1. The Definition and Basic Properties of Covering Spaces

The concept of a covering map derives from the study of the graphs of multivalued functions having a fixed number of values (i.e. m-valued for fixed m), and whose branches cannot be separated.

Let M, N be manifolds of the same dimension and consider a map $f: M \to N$ with the following two properties:

(i) The map f is to be non-singular; i.e. its Jacobian is non-zero at every point x of the manifold M:

$$\det\left(\frac{\partial y^\alpha}{\partial x^\beta}\right) \neq 0,$$

where the x^β are co-ordinates in a neighbourhood of x, and the y^α are co-ordinates in a neighbourhood of the point $y = f(x) \in N$.

(ii) There should exist around each point $y \in N$, a neighbourhood $U_j \subset N$ whose complete inverse image $f^{-1}(U_j)$ is the union $V_{1j} \cup V_{2j} \cup \cdots$ of a set of pairwise disjoint regions V_{kj}, $k = 1, 2, \ldots$, with the property that the restriction $f|_{V_{kj}}: V_{kj} \to U_j$, of f to each region V_{kj} is a diffeomorphism between V_{kj} and U_j. It is further required that N be covered by finitely or countably many such regions U_j and in such a way that each point $y \in N$ lies in only finitely many of the U_j. (Sometimes the more stringent requirement is imposed that each compact subset of N intersect only finitely many of the U_j.) The analogous property is required to hold also for the covering of the manifold M by the regions V_{kj}.

18.1.1. Definition. A map $f: M \to N$ (M and N of the same dimension) with properties (i) and (ii) above, is called a *covering map* (or simply *covering*). (In fact property (ii) suffices for the definition since clearly (i) follows from (ii).) The manifold N is called the *base* of the covering map, and M its *covering space*. The complete inverse image $f^{-1}(y)$ of any point $y \in N$ is called a *fibre* of the covering. The number of regions V_{kj} in the complete inverse image $f^{-1}(U_j)$ (or, what amounts to the same thing, the number of points in $f^{-1}(y)$ for $y \in U_j$) is referred to as the *number of sheets* (corresponding to U_j) of the covering; if this number is finite and constant for all U_j, say equal to m, then the covering is said to be *m-sheeted*.

We shall suppose from now on that the manifold N is connected. We say that a given covering is *indecomposable* if the manifold M is also connected. The covering with $M \cong N \times F$ where the fibre F is (necessarily) a finite or countable discrete topological space (i.e. consists of finitely or countably many isolated points) and f is (essentially) the projection map, is called the *trivial covering*.

18.1.2. Lemma. *If the base space N is connected then the number of sheets of a covering is independent of the choice of the point $y \in N$.*

PROOF. Let y_0 and y_1 be any two distinct points of N and join them by a piecewise smooth, non-self-intersecting path $\gamma(t)$, $0 \le t \le 1$. Subdivide the interval $[0, 1]$ into K equal subintervals, denoting by δ_k the subinterval $k/K \le t \le (k+1)/K$, for $k = 0, 1, \ldots, K-1$. Choose K sufficiently large for each of the corresponding segments $\gamma(\delta_k)$ of the path $\gamma(t)$ to lie entirely within one of the regions U_{jk} of N specified in defining the given covering. By definition of a covering, for each k the complete preimage $f^{-1}(\gamma(\delta_k))$ is a union

$$f^{-1}(\gamma(\delta_k)) = \delta_{jk,\,1} \cup \delta_{jk,\,2} \cup \ldots,$$

of pairwise disjoint path segments, where the segment $\delta_{jk,\,q}$ is contained in the region $V_{jk,\,q}$, and is mapped by the covering map f homeomorphically onto the path segment $\gamma(\delta_k)$. Hence as a point moves continuously along any particular segment $\gamma(\delta_k)$, each point of its complete preimage moves continuously along whichever of the path segments $\delta_{jk,\,1}$, $\delta_{jk,\,2}, \ldots,$ it lies on, without merging with any other point in the complete preimage. It follows that the points of the (closed) segment $\gamma(\delta_k)$ all have the same number of preimages under f. Now having reached the end of the segment $\gamma(\delta_k)$, we apply the same argument to the next segment $\gamma(\delta_{k+1})$ lying within the region $U_{j(k+1)}$. After K repetitions of the argument, starting from y_0, we reach y_1. Hence y_0 and y_1 have the same number of preimages and the lemma is proved. □

From this proof we deduce the following

18.1.3. Lemma. *Let $f\colon M \to N$ be a covering map with connected base space N. The complete inverse image of a piecewise smooth, non-self-intersecting path γ (with distinct end-points y_0 and y_1) is (diffeomorphic to) the direct product of the path γ with any fibre F, i.e. is the union of as many pairwise disjoint copies of the path γ as there are sheets in the covering: $f^{-1}(\gamma) \cong \gamma \times F$. Each of these component paths is projected by f diffeomorphically onto the path γ in the base space N.*

PROOF. If we label the points of the fibre corresponding to the point $y_0 = \gamma(0)$ $(t = 0)$ with the positive integers $1, 2, \ldots,$ then we can introduce on the set $f^{-1}(\gamma)$ co-ordinates t, n, $0 \le t \le 1$, $n = 1, 2, \ldots,$ in the following way: for $t = 0$ the points of the fibre $f^{-1}(y_0) = f^{-1}(\gamma(0))$ are assigned co-ordinates $(0, n)$ according to the labelling. As we move continuously along γ towards y_1, i.e. as t varies continuously from 0 to 1, for each t we label the points of the fibre $f^{-1}(\gamma(t))$ with the integers $1, 2, \ldots,$ in a continuous manner made clear by the preceding proof. Then if a point of $f^{-1}(\gamma(t))$ is labelled with n, we assign it the co-ordinates (t, n). This completes the proof. □

18.1.4. Definition. With $f: M \to N$ as above, we say that a path $\mu(t)$ in M *covers* a path $\gamma(t)$ in N if $f(\mu(t)) = \gamma(t)$.

From Lemma 18.1.3 we deduce the following

18.1.5. Corollary. *Given any piecewise smooth path $\gamma(t)$ in the base space N, there exists a covering path $\mu(t)$ in M, which is uniquely determined by specifying a single point $\mu(t_0)$ say, in M such that $f(\mu(t_0))$ lies on γ (i.e. by specifying a point in the fibre over any point of γ).*

(For the proof it essentially suffices to subdivide $\gamma(t)$ into non-self-intersecting segments, and apply Lemma 18.1.3 to each of these segments.)

Let K be a manifold (or more generally a topological space), let $q: K \to N$ be a (piecewise smooth, say) map from K to the base space N of a covering map $f: M \to N$, and let $F: K \times I \to N$ be a (piecewise smooth) homotopy of the map q (so that $F(x, 0) = q(x)$ for all $x \in K$).

18.1.6. Theorem (On the Covering Homotopy). *Let q, f and F be as above, and let $\tilde{q}: K \to M$ be a map which "covers" q, in the sense that $f \circ \tilde{q} = q$. Then there is a unique homotopy $\tilde{F}: K \times I \to M$ of $\tilde{q}$ which covers the given homotopy $F: K \times I \to N$ of q, i.e. such that $f \circ \tilde{F} = F$, and $\tilde{F}(x, 0) = \tilde{q}(x)$ for all $x \in K$.*

PROOF. Under the homotopy F of the map q, each point $q(x)$ is moved in N along the path defined by $\gamma_x(t) = F(x, t)$. The initial point $\gamma_x(0) = q(x)$ of each such path is covered by the point $\tilde{q}(x) \in M$ (i.e. $\tilde{q}(x)$ is in the fibre above $q(x)$). The theorem now follows from Corollary 18.1.5 together with the observation that the (recipe for obtaining the) covering path depends continuously (even smoothly) on the initial point $\gamma_x(0) = q(x)$ of the path in the base N. □

18.2. The Simplest Examples. The Universal Covering

The examples are as follows.

(a) Let $M = \mathbb{R}^1$ (the real line) and $N = S^1$, and define the covering f by $f(t) = e^{2\pi i t}$, where t co-ordinatizes $\mathbb{R}^1$. Here the number of sheets is infinite.

(b) Let $M = S^1$ and $N = S^1$, and define the covering map by $f(z) = z^n$ (for all z such that $|z| = 1$); this covering clearly has n sheets. The same formula $z \mapsto z^n$ defines a covering with the roles of base N and covering space M both filled by the space $\mathbb{R}^2 \setminus \{0\} \cong \mathbb{C}^*$.

(c) Let $M = S^n$, $N = \mathbb{R}P^n$, and define the covering map $f: S^n \to \mathbb{R}P^n$ to be the obvious one identifying each pair of diametrically opposite points of S^n. Here the number of sheets is 2. The group epimorphism $SU(2) \to SO(3)$ (noted in §§13.2, 14.3 of Part I) represents a particular case of this covering map:

$$SU(2) \cong S^3 \to \mathbb{R}P^3 \cong SO(3).$$

Another example of a 2-sheeted covering is furnished by the group epimorphism $S^3 \times S^3 \cong SU(2) \times SU(2) \to SO(4)$, with kernel $\{(1, 1), (-1, -1)\}$ (see §14.3 of Part I).

(d) Let $M = \mathbb{R}^n$ and consider the (additive) subgroup $\mathbb{Z}^n$ consisting of those vectors in $\mathbb{R}^n$ with integer components. As defined in §4.1, the torus T^n is the quotient group $\mathbb{R}^n/\mathbb{Z}^n$ with the natural manifold structure. (When $n = 1$ we obtain just the circle S^1.) The natural map $\mathbb{R}^n \to T^n$ is a covering map (verify it!).

(e) Consider the Euclidean plane $\mathbb{R}^2$ with Euclidean co-ordinates x, y, and the subgroup G of the isometry group of $\mathbb{R}^2$, generated by the two transformations T_1, T_2 defined by:

$$T_1(x, y) = (x, y + 1), \qquad T_2(x, y) = (x + \tfrac{1}{2}, -y).$$

By identifying all points mapped to one another by elements of the group G (i.e. the points in each orbit under the action of G on $\mathbb{R}^2$), it is intuitively clear that we obtain the Klein bottle K^2, since the group G identifies opposite sides of the $\frac{1}{2} \times 1$ rectangle depicted in Figure 57, in the manner indicated by the arrows. The projection map $f\colon \mathbb{R}^2 \to K^2$ is then an infinite-sheeted covering map. It is not difficult to see that the generators T_1, T_2 are linked by the single defining relation $T_2^{-1}T_1T_2T_1 = 1$, whence (or alternatively by direct calculation) it follows that the subgroup G_1 of index 2 in G, generated by the transformations T_1, T_2^2, is isomorphic to $\mathbb{Z}^2$. Identification under the action of the group G_1 on $\mathbb{R}^2$ yields the torus T^2 (see Example (d)) since $T_2^2(x, y) = (x + 1, y)$, i.e. the orbit space $\mathbb{R}^2/G_1$ (or quotient group from another point of view) is the torus T^2, which furnishes a 2-sheeted covering space for the Klein bottle since each orbit of $\mathbb{R}^2$ under the action of G splits into two orbits under the action of $G_1 \simeq \mathbb{Z}^2$.

(f) We indicate by visual means covering spaces for the figure eight (or bouquet of two circles), and the bouquet $S^1 \vee S^2$ formed from the circle and 2-sphere. (Note incidentally that neither of these spaces is a manifold.) The covering spaces in question are shown in Figure 58; the covering maps, both denoted by f, are to be understood as the projections downwards. In particular, it is clear from that diagram that the covering space of $N = S^1 \vee S^2$ depicted there is topologically just a collection of

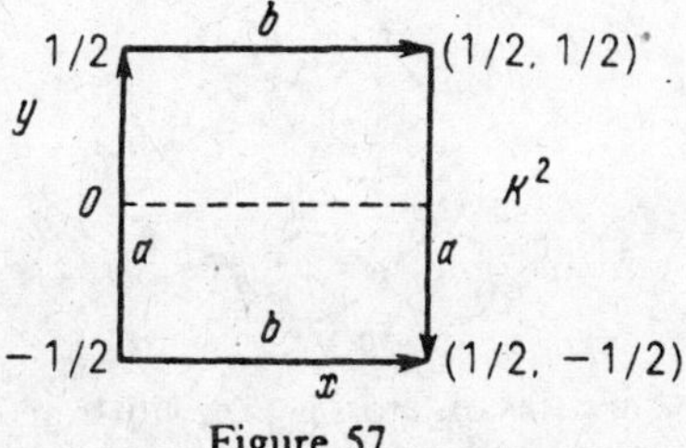

Figure 57

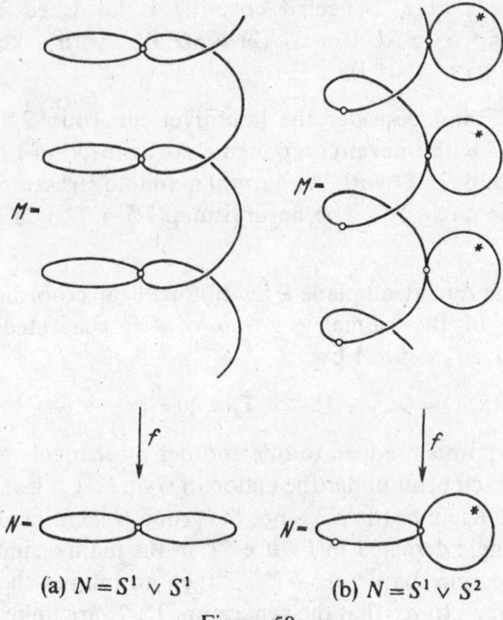

(a) $N = S^1 \vee S^1$ (b) $N = S^1 \vee S^2$

Figure 58

2-spheres attached to the points with integer co-ordinate of the real line $\mathbb{R}^1$ (depicted as a helix).

(g) We conclude our examples in this subsection (they will continue in the next) with a further example of a covering space for the figure eight $S^1 \vee S^1$ (see Figure 59). (This covering is "universal" in the sense of the definition below.) Here M is an infinite tree of "crosses", i.e. having exactly 4 edges

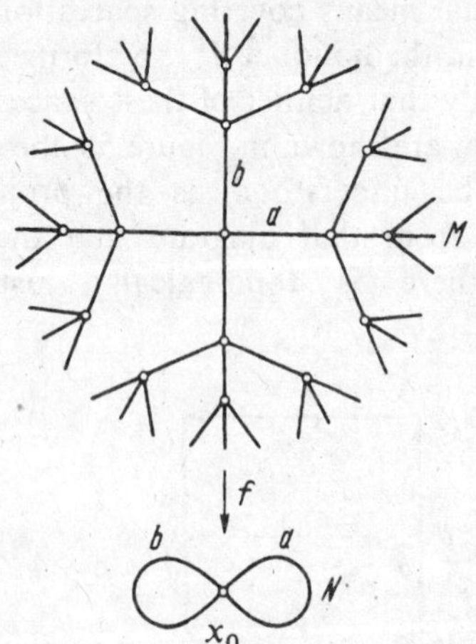

Figure 59. Here M is an infinite tree of "crosses", i.e. with exactly 4 edges incident with each vertex. Being a tree, M has no "cycles", and is therefore contractible to a point.

emanating from each vertex. Being a tree, M has no "cycles", and is therefore contractible (over itself) to a point. The centres of the crosses (i.e. the vertices of the tree) are all preimages of the single point x_0 of $S^1 \vee S^1$. Each of the edges of the tree projects onto either the circle a or the circle b of $S^1 \vee S^1$, and of the 4 edges at each vertex, two projects onto a and two onto b.

We conclude this subsection by introducing the following important concept, which we shall exploit in calculating certain fundamental groups.

18.2.1. Definition. A covering $f: M \to N$ is said to be *universal* if $\pi_1(M) = 1$ (i.e. if the space M is simply-connected).

Of the coverings considered in the above examples, the following are universal:

(a) $\mathbb{R}^1 \to S^1$;

(c) $\begin{cases} S^n \to \mathbb{R}P^n \text{ for } n \geq 2; \\ SU(2) \times SU(2) \to SO(4); \end{cases}$

(d) $\mathbb{R}^n \to T^n$;

(e) $\mathbb{R}^2 \to K^2$;

(f) $M \to S^1 \vee S^2$ (see Figure 58(b));

(g) the tree $M \to S^1 \vee S^1$.

The remaining coverings fail to be universal in view of the fact that the respective covering spaces are not simply-connected.

18.3. Branched Coverings. Riemann Surfaces

In this subsection we continue our list of examples. Before doing so however we prove two theorems showing how to obtain covering maps from general maps of closed manifolds.

Thus suppose that M and N are closed smooth manifolds of the same dimension n. We first show that under these conditions any regular map $f: M \to N$ (i.e. a map with nowhere vanishing Jacobian) is automatically a covering map.

18.3.1. Theorem. *A regular map $f: M \to N$ of closed smooth manifolds (of the same dimension) is a finite-sheeted covering map.*

PROOF. By the Inverse Function Theorem, around each point x of the manifold M there is a neighbourhood V_x such that the restriction of f to V_x is a diffeomorphism. Hence in view of the compactness of the manifold M, each point $y \in N$ can have only finitely many preimages. Let $x_1, \ldots, x_m$ be the distinct preimages under f of any particular $y \in N$, and let $V_1, \ldots, V_m$ be respectively pairwise non-intersecting neighbourhoods of these preimages, on each of which f is a diffeomorphism. Then for some sufficiently small neighbourhood U_y of y we shall have $f^{-1}(U_y) \subset V_1 \cup \cdots \cup V_m$, whence $f^{-1}(U_y)$ is the disjoint union of the regions contained in the V_i, the restriction of f to each of which is a diffeomorphism between that region and U_y. This completes the proof. □

Now suppose that while M and N are as before closed smooth manifolds (of the same dimension n), the map $f: M \to N$ is no longer everywhere regular, i.e. its Jacobian vanishes on some non-empty set $A \subset M$. Generally speaking the set A will have dimension $n-1$; however, it can happen that its dimension is less than $n-1$. An important situation where this occurs is that where n is even, the manifolds M and N are both complex analytic, and the map $f: M \to N$ is complex analytic (i.e. holomorphic). In this case the vanishing of the Jacobian is expressible by means of a single complex (analytic) equation in terms of complex local co-ordinates, whence the (real) dimension of the set A of singular points cannot exceed $n-2$ (and therefore A cannot separate M into disjoint parts).

This condition on the dimension of A leads to the following conclusion.

18.3.2. Theorem. *Suppose that the dimension of the set A of zeros of the Jacobian of a map $f: M \to N$ between smooth, closed, connected n-dimensional manifolds does not exceed $n-2$ (so that the set $f(A)$ does not separate N into disjoint parts). Then setting $N' = N \setminus f(A)$ and $M' = M \setminus f^{-1}(f(A))$, it follows that M' is connected, and that the map $f: M' \to N'$ (obtained by restricting $f: M \to N$) is a finite-sheeted covering map.*

Before giving the proof we introduce the concomitant terminology: The initial map $f: M \to N$ is called a *covering map branched along* $f(A)$, and the points of the set $f(A)$ are called the *branch points* of the (branched) covering f.

PROOF OF THE THEOREM. For (small) $\varepsilon > 0$ denote by U_ε the (open) ε-neighbourhood of the subset $f(A)$ of N, and write $N_\varepsilon = N \setminus U_\varepsilon$, $M_\varepsilon = M \setminus f^{-1}(U_\varepsilon)$. Then for sufficiently small ε, M_ε and N_ε are compact, connected manifolds-with-boundary, and f induces a regular map $f_\varepsilon: M_\varepsilon \to N_\varepsilon$. By arguing exactly as in the proof of Theorem 18.3.1 it follows that f_ε is a finite-sheeted covering map, and then the desired conclusion is obtained by letting $\varepsilon \to 0$ (which does not alter the number of sheets). □

Note in connexion with this proof that the assumption that A has dimension at most $n-2$ was used solely to ensure the connectedness of M' and N'. (Recall that our underlying assumption is that the base space, at least, is connected.)

These theorems are relevant to the important class of examples provided by non-singular Riemann surfaces Γ defined by non-singular, complex analytic, in particular algebraic, equations in the (z, w)-plane $\mathbb{C}^2$. Thus for instance the equation

$$\Phi(z, w) = w^n + a_1(z)w^{n-1} + \cdots + a_n(z) = 0,$$

where $a_1, \ldots, a_n$ are polynomials in z, determines the Riemann surface of the n-valued function $w(z)$ (see §4.2).

Let Γ be a Riemann surface defined by such a function Φ. The projection $f: \Gamma \to \mathbb{C}$ extends to a projection (which we also denote by f) of the closed Riemann surface $\hat{\Gamma}$ (which may include "points at infinity" in $\mathbb{C}P^2$—see §4.2) onto the Riemann sphere $\mathbb{C}P^1 \cong S^2$. Put $M = \hat{\Gamma}$ and $N = S^2$. In this situation the set $f(A)$ of branch points of the Riemann surface $\hat{\Gamma}$, consists of points in the plane $\mathbb{C}$ possibly together with the point ∞. Denote by N' the plane $\mathbb{C} \cong \mathbb{R}^2 \cong S^2 \backslash \{\infty\}$ with the branch points z_α removed. The complete preimage $f^{-1}(z_\alpha)$ of each branch point consists of points $(z_\alpha, w_{\alpha j}) = P_{\alpha j}$ of the surface Γ such that (see the remark below)

$$\left.\frac{\partial \Phi}{\partial w}\right|_{z = z_\alpha,\ w = w_{\alpha j}} = 0.$$

Denote by M' the manifold obtained by removing from $\hat{\Gamma}$ all preimages of the z_α (for all α), and all the points of $f^{-1}(\infty)$. Then by Theorem 18.3.2 the map $f: M' \to N'$ is an n-sheeted covering map.

Remark. It is not difficult to see (essentially from the complex analogue of the Implicit Function Theorem—see §12.3 of Part I) that the preimages on the surface Γ under the map $f: (z, w) \mapsto z$, of the branch points z_α, satisfy the pair of equations

$$\Phi(z, w) = 0, \qquad \frac{\partial \Phi(z, w)}{\partial w} = 0. \tag{1}$$

We now consider more specific cases (resuming our list of examples of covering maps from where it left off).

Examples (continued). (h) Consider the case $\Phi(z, w) = w^2 - P_n(z) = 0$ (of a hyperelliptic Riemann surface). It was shown in §12.3 of Part I that in this case the Riemann surface is non-singular if and only if $P_n(z)$ has no multiple roots. Here

$$N' = \mathbb{C} \backslash \{z_\alpha\}, \qquad M' = \Gamma \backslash \left(\bigcup_\alpha f^{-1}(z_\alpha) \right),$$

where the z_α are the branch points (and by (1) satisfy $P_n(z_\alpha) = 0$), and the covering map $f: M' \to N'$ is 2-sheeted.

(i) In the slightly more general case $\Phi(z, w) = w^k - P_n(z) = 0$, we obtain in a similar manner a k-sheeted covering map $f: M' \to N'$.

(j) Consider next any polynomial of degree n in the two variables z, w; assuming the coefficient of w^n is non-zero, such a polynomial may be brought into the form

$$\Phi(z, w)=w^n+\sum_{i>1} a_i(z)w^{n-i},$$

where for each i the degree of the polynomial $a_i(z)$ does not exceed i. For the "general" such polynomial the corresponding Riemann surface will have $n(n-1)$ branch points z_α, obtained by solving the system $\Phi=0$, $\partial\Phi/\partial w=0$ in $\mathbb{C}^2$ (see (1)). Then with M' and N' as before we obtain an n-sheeted covering map $f\colon M'\to N'$.

(k) Suppose finally that the function $\Phi(z, w)$ is not algebraic (though complex analytic), and, as always, that the surface $\Phi=0$ in $\mathbb{C}^2$ is non-singular. It can be shown that the branch points z_α in the z-plane are at most countable in number; we need to impose the requirement that they form a discrete subset of $\mathbb{C}$. The resulting covering with base $N'=\mathbb{C}\setminus\{z_\alpha\}$ will generally speaking have infinitely many sheets. The simplest example of this kind is provided by the equation $z-e^w=0$. Here $w=\ln z$, and $z_\alpha=0$ is the only branch point. Hence we obtain the "logarithmic" covering map

$$f\colon M'\to N'=\mathbb{C}\setminus\{0\}.$$

We leave it to the reader to show that M' is diffeomorphic to the plane $\mathbb{C}$.

18.4. Covering Maps and Discrete Groups of Transformations

Our aim in this subsection is to introduce an important class of coverings related to so-called "discrete groups of transformations" of a manifold. We begin with the definition of the latter term.

18.4.1. Definition. Let M be a smooth manifold (or more general topological space) and let G be a group of diffeomorphisms of M (homeomorphisms if M is a more general topological space). The group G is called a *discrete group of transformations* of M if the points of each orbit of M under the action of G are "discretely distributed" (meaning that around each point y of M there exists a neighbourhood U containing no other point of the orbit $G(y)$, with the property that each of its images $g(U)$ under the elements g of G either coincides with or is disjoint from U), and if furthermore the number of elements of G fixing any point y is finite (i.e. the stabilizer in G of each point of M is finite). We shall further say that a discrete group acts *freely* on M if for every $y\in M$ the stabilizer of y in G is the identity subgroup (i.e. $g(y)=y$ implies $g=1$). (Thus a freely acting discrete group G of transformations of M is characterized by the property that around each point y of M there is a neighbourhood U such that $g(U)\cap U$ is empty for all $g\neq 1$ in G.)

In the case where M is a manifold, the transformations comprising the discrete group G will in practice often (though not always) be isometries of M, endowed with some Riemannian metric.

18.4.2. Definition. We say that a *covering map* $f: M \to N$ *is determined by a freely acting discrete group* G *of transformations* $M \to M$, if the fibre $F = f^{-1}(y)$ above each point $y \in N$, is an orbit of G. The base space N can then be identified with the orbit space of M under (the action of) G, and we write $N = M/G$. Such a covering is said to be *regular*, or a *principal fibre bundle with respect to the discrete group* G. (In Chapter 6 below we shall consider principal fibre bundles also with respect to non-discrete groups of transformations.)

The first nine (from (a) to (i)) of the above examples of covering maps are indeed determined by various discrete groups of transformations. On the other hand the covering spaces of Example (j) (general algebraic Riemann surfaces) and of Example (k) (excepting the simple logarithmic branched covering considered there) are generally speaking not determined by freely acting discrete groups of transformations.

§19. Covering Maps and the Fundamental Group. Computation of the Fundamental Group of Certain Manifolds

19.1. Monodromy

We now introduce the important concept of the "monodromy representation" σ arising from a covering space, and the associated "monodromy group" (also called the "group of holonomies"). Thus let y_0 be a point in the base N of a given covering map $f: M \to N$, and let $\{x_1, x_2, \ldots\}$ be the fibre $F = f^{-1}(y_0)$ above y_0, with its elements enumerated arbitrarily. Let γ be a closed path in N beginning and ending at y_0, which with permissible looseness of language we may regard as an element of $\pi_1(N, y_0)$. By Corollary 18.1.5, for each point $x_j \in F$ there is a unique path μ in M beginning at x_j and covering the path γ; we can assume γ and μ parametrized with the same parameter t so that $f(\mu(t)) = \gamma(t)$ for all t, i.e. as γ is traced out in the base N, μ is traced out above in M, beginning at x_j. Once γ is traced out, we shall have come back to the point $y_0 = \gamma(1)$, so that on the covering path μ the corresponding end-point will be again a point of the fibre F; we denote this point by $x_{\sigma(j)} = \mu(1)$. Thus corresponding to our path γ we have obtained a permutation $\sigma(\gamma)$ of (the subscripts of) the points of the fibre F:

$$\sigma(\gamma): x_j \mapsto x_{\sigma(j)}.$$

By Theorem 18.1.6 (appropriately relativized) the permutation $\sigma(\gamma)$ depends only on the homotopy class in $\pi_1(N, y_0)$ to which γ belongs. Since, as is readily seen, we have also $\sigma(\gamma^{-1})=\sigma(\gamma)^{-1}$ and $\sigma(\gamma_1\circ\gamma_2)=\sigma(\gamma_1)\circ\sigma(\gamma_2)$, it follows that σ defines a homomorphism from the fundamental group $\pi_1(N, y_0)$ to the group of permutations of the fibre F (or in other words a representation of $\pi_1(N, y_0)$ by permutations on F). The representation σ is called the *monodromy* or *discrete holonomy* of the covering, and the image group $\sigma(\pi_1(N, y_0))$ is the *monodromy group.*

We shall now investigate the monodromies of the covering spaces considered in the examples in §§18.2, 18.3.

(a) Here we have $f: \mathbb{R}^1\to S^1, t\mapsto e^{2\pi it}$. As we saw in §17.5, the group $\pi_1(S^1)$ is isomorphic to $\mathbb{Z}$. The complete inverse image of the point $\varphi_0=0$ of the circle, consists of the integer points $0, \pm1, \pm2, \ldots$ of the real line, so that the points of the fibre $f^{-1}(0)$ come ready-labelled with the integers. It is then easy to see that the monodromy σ represents the obvious generator a of $\pi_1(S^1, 0)$ by the translation

$$\sigma(a): n\mapsto n+1, \qquad n\in\mathbb{Z}.$$

(b) Here $f: S^1\to S^1$ was defined by $z\mapsto z^n$ (where $|z|=1$). The preimages of the point 1 are the points $z_k=\exp(2\pi ki/n)$, $k=0, 1, \ldots, n-1$. The element of the monodromy group corresponding to the obvious generator a of $\pi_1(S^1, 1)$ is readily seen to be the cyclic permutation

$$\sigma(a)=\begin{pmatrix}0 & 1 & \cdots & n-1\\ 1 & 2 & \cdots & 0\end{pmatrix}.$$

(c) Under the map $f: S^n\to\mathbb{R}P^n$, each point $y_0\in\mathbb{R}P^n$ has as its preimages antipodal points x_1 and x_2 of S^n. Let $a\in\pi_1(\mathbb{R}P^n)$ denote the homotopy class containing the projection in $\mathbb{R}P^n$ of a path $\tilde{\gamma}$ in S^n joining the points x_1 and x_2. Then the transformation $\sigma(a)$ interchanges these points: $x_1\mapsto x_2$, $x_2\mapsto x_1$, so that σ maps $\pi_1(\mathbb{R}P^n)$ onto $\mathbb{Z}_2$, the cyclic group of order 2. (In the next subsection we shall show that in fact $\pi_1(\mathbb{R}P^n)\simeq\mathbb{Z}_2$.)

In a similar manner it can be shown that under the monodromy representation afforded by the covering $f: SU(2)\times SU(2)\to SO(4)$, the fundamental group $\pi_1(SO(4))$ is mapped onto $\mathbb{Z}_2$. (In this case also the fundamental group is in fact isomorphic to $\mathbb{Z}_2$.)

(d) Here the covering map is $f: \mathbb{R}^n\to T^n$. For each $j=1, \ldots, n$, we denote by $\tilde{\gamma}_j$ the straight-line path in $\mathbb{R}^n$ joining the origin O to the point $(0, \ldots, 1, \ldots, 0)$ with jth co-ordinate 1 and the remaining co-ordinates all zero. The projections $a_j=f(\tilde{\gamma}_j)$, which are clearly closed paths in T^n, have their homotopy classes represented under the monodromy by the (infinite-degree) permutation

$$\sigma(a_j): (m_1, \ldots, m_j, \ldots, m_n)\mapsto(m_1, \ldots, m_j+1, \ldots, m_n),$$

where here $(m_1, \ldots, m_n)$ denotes a typical integral lattice point of $\mathbb{R}^n$ (i.e. a typical point of the fibre above O). Given that (the homotopy classes of) the a_j actually generate $\pi_1(T^n, O)$, it follows that σ maps $\pi_1(T^n, O)$ onto $\mathbb{Z}^n$, the direct product of n copies of the additive integers. (We shall see in the next subsection that in fact $\pi_1(T^n) \simeq \mathbb{Z}^n$.)

Of the remaining examples we examine explicitly the monodromy representation in (g) and (j) only. (The reader may like to investigate for himself the monodromy in the rather simple examples (e) and (f).)

(g) Here $f: M \to N$ is the universal covering of the figure eight $N = S^1 \vee S^1$. Write $a_1 = a \in \pi_1(S^1 \vee S^1)$, $a_2 = b \in \pi_1(S^1 \vee S^1)$, where the paths a and b are as in Figure 59. Then in the monodromy group the element $\sigma(a_1)$ is the graph-isomorphism of the tree M (shown in Figure 59) which sends every vertex to the position of the adjacent vertex to the right, and $\sigma(a_2)$ that which moves each vertex into the position of the vertex immediately above it. It follows easily that the image in the monodromy group of any non-empty word of the form

$$a_{i_1}^{n_1} a_{i_2}^{n_2} \cdots a_{i_k}^{n_k}$$

(where $k \geq 1$, the integers n_q are non-zero, the i_q are either 1 or 2, and $i_q \neq i_{q+1}$) leaves no vertex of the tree fixed, and so in particular must be non-trivial. Since a_1 and a_2 generate $\pi_1(S^1 \vee S^1)$, we infer that the monodromy group (and hence, in this case, $\pi_1(S^1 \vee S^1)$ also) is isomorphic to the free group on two free generators.

(j) It was noted previously that for a general type of 2-variable polynomial (monic, of degree n in w and of total degree n) there are $n(n-1)$ non-degenerate branch points of the associated covering map f of the Riemann surface, which we shall denote by z_{jk}, where $j = 1, \ldots, n$; $k = 1, \ldots, n$, and $j \neq k$. As before we take $N' = \mathbb{C} \setminus \{z_{jk}\}$, the complex plane with the branch points removed. Choosing a base point y_0 arbitrarily in N', let $x_1, \ldots, x_n$ denote the points in $f^{-1}(y_0)$, the fibre above y_0, and for each pair j, k ($j \neq k$) let a_{jk} denote a simple closed path enclosing the branch point z_{jk} (see Figure 60). It can be shown (cf. §4.2) that the branch points z_{jk} can be paired off in such a way that, after suitable re-indexing (of the a_{jk} and the x_q), both a_{jk} and a_{kj} are sent under the monodromy representation σ to the permutation which transposes x_j and x_k and leaves the remaining points of the fibre fixed.

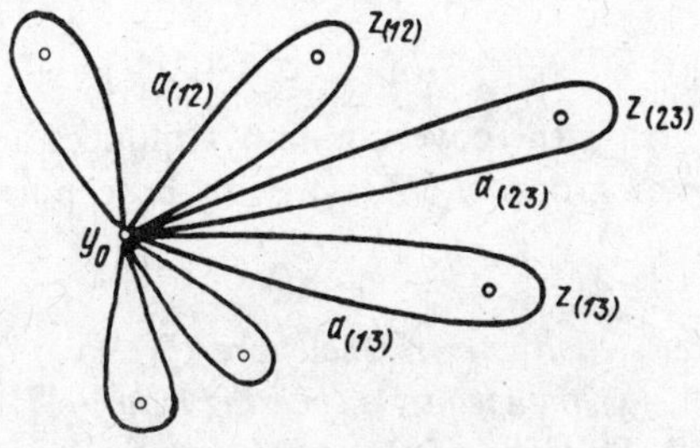

Figure 60

Hence the monodromy group of the covering is the group of all $n!$ permutations on n symbols, i.e. the symmetric group of degree n (since that group is generated by its transpositions and the a_{jk} generate $\pi_1(N', y_0)$). (That each of the permutations $\sigma(a_{jk})$ moves only two of the points of the fibre is a consequence of the fact that for a Riemann surface of this general type the degenerate points of the projection $\Gamma \to \mathbb{C}$ have low "order of degeneracy"; we leave to the reader the precise investigation of these details.)

The conclusion that the monodromy group of the general algebraic Riemann surface coincides with the full symmetric group (of degree equal to the number of sheets) turns out to have an important consequence. The reader might like to attempt the proof of the following result: *If a multi-valued function $w = w(z)$ is given by an expression in z involving only the usual arithmetic operations in combination with arbitrary radicals $\sqrt[k]{\ }$, then the corresponding Riemann surface has soluble monodromy group.* (Recall that a group G is *soluble* if it has a finite chain of normal subgroups $\{1\} < G_1 < \cdots < G_r = G$, beginning with the identity subgroup and ending with G, all of whose factors G_{i+1}/G_i are abelian.)

Since, as is well known, no symmetric group of degree ≥ 5 is soluble (having as its only proper normal subgroup the "alternating group", consisting of the even permutations, which is non-abelian), we deduce the following celebrated result.

19.1.1. Theorem (Abel). *For the roots of the general polynomial of degree ≥ 5 in one indeterminate, there is no formula in the coefficients involving only the usual arithmetic operations ($+$, $-$, $\times$, $\div$) in combination with (arbitrary) radicals $\sqrt[k]{\ }$.*

19.2. Covering Maps as an Aid in the Calculation of Fundamental Groups

Let $f: M \to N$ be a covering map, choose $y_0 \in N$ arbitrarily, and let $x_1, x_2, \ldots$ denote the points of the fibre $f^{-1}(y_0)$ above y_0. As usual we denote by σ the monodromy representation afforded by the covering map f; thus via σ, the fundamental group $\pi_1(N, y_0)$ acts on the fibre $F = f^{-1}(y_0)$:

$$\sigma(\alpha): x_j \mapsto x_{\sigma(j)} \qquad (\alpha \in \pi_1(N, y_0)).$$

It is clear that for each j, the covering map f induces a homomorphism f_* from the fundamental group $\pi_1(M, x_j)$ (with base point x_j) to the group $\pi_1(N, y_0)$.

19.2.1. Theorem. *For each x_j in the fibre $f^{-1}(y_0)$, the homomorphism $f_*: \pi_1(M, x_j) \to \pi_1(N, y_0)$, induced by a covering map $f: M \to N$, is an embedding (i.e. a monomorphism). The subgroup $f_*\pi_1(M, x_j)$ of $\pi_1(N, y_0)$ consists precisely* *of those elements* α *of* $\pi_1(N, y_0)$ *whose images* $\sigma(\alpha)$ *under the monodromy representation, leave the point* x_j *fixed (i.e. which act trivially on* x_j*). The subgroups* $f_*\pi_1(M, x_j)$ *and* $f_*\pi_1(M, x_k)$ *corresponding to distinct points* x_j *and* x_k *of the fibre, are conjugate to one another by means of any element* γ *of* $\pi_1(N, y_0)$ *which sends* x_j *to* x_k*, i.e.*

$$\gamma^{-1} f_*\pi_1(M, x_j)\gamma = f_*\pi_1(M, x_k),$$

for any $\gamma \in \pi_1(N, y_0)$ *such that* $\sigma(\gamma)$ *maps* x_j *to* x_k.

PROOF. We first show that f_* is one-to-one. To this end let $\alpha = \alpha(t) \in \pi_1(M, x_j)$ be such that $f_*(\alpha) = 1$; we wish to show that then $\alpha = 1$. Write $\gamma(t) = f(\alpha(t))$; then by assumption $\gamma(t)$ is contractible to the point y_0 in N by means of a homotopy keeping the end-points fixed at y_0 (for all τ); let $F(t, \tau)$ denote such a homotopy. Since $f(\alpha(t)) = \gamma(t) = F(t, 0)$, we are in a situation where the "theorem on the covering homotopy" (18.1.6) can be applied (with γ and α in the roles of q, $\tilde{q}$, respectively); we can therefore conclude from that theorem that in particular there is a homotopy which contracts the path $\alpha(t)$ in M to the point x_j. Hence f_* is a monomorphism, as claimed.

Leaving to the reader the easy second statement in the theorem, we proceed to the third and last. Let α be any path in $\pi_1(M, x_j)$, and let $\tilde{\gamma}(t)$ be any path, also in M, joining the point $x_k = \tilde{\gamma}(0)$ to the point $x_j = \tilde{\gamma}(1)$. Then by Theorem 17.2.1 the map $\alpha \mapsto \tilde{\gamma}^{-1}\alpha\tilde{\gamma}$ defines an isomorphism between the groups $\pi_1(M, x_j)$ and $\pi_1(M, x_k)$. Write $\gamma(t) = f(\tilde{\gamma}(t))$; then γ, being a closed path in N, represents an element of $\pi_1(N, y_0)$, and further, by construction of $\tilde{\gamma}$, has the property that $\sigma(\gamma)$ sends x_j to x_k. It is clear that the above isomorphism $\pi_1(M, x_j) \to \pi_1(M, x_k)$ induces (via the covering map) an isomorphism $f_*\pi_1(M, x_j) \to f_*\pi_1(M, x_k)$, given by

$$f_*(\alpha) \mapsto \gamma^{-1} f_*(\alpha)\gamma \in f_*\pi_1(M, x_k),$$

Since, essentially by reversing the order of construction of $\tilde{\gamma}$ and γ, this conclusion holds for any $\gamma \in \pi_1(N, y_0)$ for which $\sigma(\gamma)$ maps x_j to x_k, the proof is complete. □

EXERCISES

1. Show that corresponding to any subgroup H of the fundamental group $\pi_1(N)$ of an arbitrary manifold N, there is a covering map $f: M \to N$ such that $f_*\pi_1(M) = H$; hence in particular every manifold has a universal covering.

2. Prove that if two covering maps $f: M \to N$ and $f': M' \to N$, with the same base manifold N, are such that the subgroups $f_*\pi_1(M)$ and $f'_*\pi_1(M')$ coincide, then the coverings are "equivalent", i.e. there exists a homeomorphism $\varphi: M \to M'$ such that $f' \circ \varphi = f$. (This explains the use of the definite article in the phrase "the universal covering".)

Remark. In both of these exercises the requirement that N be a manifold can be significantly relaxed; for instance, as we have seen, such spaces as the figure eight, and the bouquet of the circle and the 2-sphere, possess universal coverings (see Figures 58 and 59).

19.2.2. Theorem. *If a covering map $f: M \to N$ is determined by a freely acting discrete group G of transformations $M \to M$, and the manifold (or topological space more generally) M is simply-connected (i.e. $\pi_1(M) = 1$), then*

$$\pi_1(N) \simeq G.$$

PROOF. Let y_0 be any point of N, and fix on any point x_0 in the fibre $f^{-1}(y_0)$ above y_0. We shall establish a one-to-one correspondence between the points of the fibre (each of which has the form $g(x_0)$ for some $g \in G$) and the elements of the group $\pi_1(N, y_0)$. With this our aim, take any path $\gamma_1 \in \pi_1(N, y_0)$, and consider the (unique) path in M covering γ_1 and beginning at the point x_0; this covering path will terminate at that point x_1 which is the image of x_0 under the permutation $\sigma(\gamma_1)$. Let g_1 be such that $g_1(x_0) = x_1$; the correspondence $\gamma_1 \leftrightarrow g_1$ is then the one we are seeking. We now show that this does indeed define a one-to-one correspondence between $f^{-1}(y_0)$ and $\pi_1(N, y_0)$. Suppose first that $\gamma_1, \gamma_2 \in \pi_1(N, y_0)$ both correspond to g_1; then the path $\gamma_1^{-1}\gamma_2$ has the property that $\sigma(\gamma_1^{-1}\gamma_2)$ fixes x_0, whence by the simple-connectedness of M and the second statement in Theorem 19.2.1, the paths γ_1 and γ_2 are homotopic to one another in M. The one-to-one-ness in the other direction is an almost immediate consequence of the freeness of the action of G. That every $g_1 \in G$ corresponds to some $\gamma_1 \in \pi_1(N, y_0)$ follows easily from the connectedness of M. Thus we have our desired one-to-one correspondence; that it is an isomorphism follows from the facts that σ is a homomorphism of $\pi_1(N, y_0)$, and that the action of the permutation $\sigma(\gamma_1)$ on the fibre is identical with the action of the corresponding $g_1 \in G$. □

The following result generalizes the preceding one to the situation where the covering space M is not necessarily simply-connected, i.e. is not the universal covering space.

19.2.3. Theorem. *Given (as in the preceding theorem) a covering map $f: M \to N$ which is a principal fibre bundle with respect to a discrete group G, i.e. is determined by a freely acting discrete group G of transformations $M \to M$, then the group G, regarded as a group of permutations of any fibre $F = f^{-1}(y_0)$, coincides with the monodromy group $\sigma(\pi_1(N, y_0))$. Furthermore for any $x_j \in F$, the subgroup $f_*\pi_1(M, x_j)$ is independent of x_j, and is a normal subgroup of the fundamental group $\pi_1(N, y_0)$, and the quotient group $\pi_1(N, y_0)/f_*\pi(M, x_j)$ is isomorphic to the monodromy group.*

PROOF. Exactly as in the proof of the preceding theorem we associate with each $\gamma_1 \in \pi_1(N, y_0)$ a unique element $g_1 \in G$, in such a way that the map $\varphi: \pi_1(N, y_0) \to G$ defined by $\gamma_1 \mapsto g_1$ is an epimorphism, and the map $\sigma(\gamma_1) \mapsto g_1$ is an identification of the monodromy group $\sigma\pi_1(N, y_0)$ with G regarded as a group of permutations of the fibre $f^{-1}(y_0)$. By the second statement in Theorem 19.2.1, the subgroup $f_*\pi_1(M, x_j)$ (for x_j in the fibre) consists of those elements α of $\pi_1(N, y_0)$ such that $\sigma(\alpha)$ fixes x_j; now since the actions of the monodromy group and the group G on the fibre $f^{-1}(y_0)$ are the same, and since G acts freely, it follows that $\sigma(\alpha)$ is the identity element, i.e. that $f_*\pi_1(M, x_j)$ is precisely the kernel of the epimorphism φ. Hence the subgroup $f_*\pi_1(M, x_j)$ is the same subgroup for all x_j in the fibre, is normal in $\pi_1(N, y_0)$, and the quotient $\pi_1(N, y_0)/f_*\pi_1(M, x_j)$ is isomorphic to G. This completes the proof of the theorem. □

EXERCISE

Prove that for a general (not necessarily regular as in the preceding two theorems) covering map, the monodromy group is isomorphic to the quotient group $\pi_1(N, y_0)/P$, where P is the normal subgroup $\bigcap_j f_*\pi_1(M, x_j)$.

We now examine anew some of our examples, in the light of the above results.

Examples. (a) $\mathbb{R}^1 \to S^1$. Here the covering is determined by the action of the discrete group $\mathbb{Z}$ $(\simeq \pi_1(S^1))$ on $\mathbb{R}^1$ by means of translations through integer distances. (See Example (a) of §§18.2 and 19.1.)

(b) The covering map $S^n \to \mathbb{R}P^n$ is determined by the action of the group $G \simeq \mathbb{Z}_2$, whose non-trivial element is the reflection $x \mapsto -x$ of the sphere $S^n \subset \mathbb{R}^{n+1}$. Since the sphere is simply connected for $n > 1$, Theorem 19.2.2 applies to yield $\pi_1(\mathbb{R}P^n) \simeq \mathbb{Z}_2$. (See Example (c) of §§18.2 and 19.1.)

(c) The covering $\mathbb{R}^n \to T^n$ is determined by the discrete group $G \simeq \mathbb{Z}^n$, which acts on the n-tuples of $\mathbb{R}^n$ via translations. Since the covering is universal, Theorem 19.2.2 again applies to give $\pi_1(T^n) \simeq \mathbb{Z}^n$. (See Example (d) of §§18.2 and 19.1.)

(d) The covering $\mathbb{R}^2 \to K^2$ is determined by the group G generated by the transformations T_1, T_2 of $\mathbb{R}^2$ given by $T_1(x, y) = (x, y+1)$, $T_2(x, y) = (x + \frac{1}{2}, -y)$ (see Example (e) of §18.2). (The group G, which has a simple structure, is defined abstractly by these two generators T_1, T_2 together with the single relation $T_2^{-1} T_1 T_2 T_1 = 1$.) Since $\mathbb{R}^2$ is simply connected, it follows as before that $\pi_1(K^2) \simeq G$.

(e) The universal covering of $S^1 \vee S^1$ is determined by the discrete free group on two free generators acting in the manner described in Example (g) of §19.1. Hence $\pi_1(S^1 \vee S^1)$ is isomorphic to the free group of rank 2.

It can be shown similarly that the fundamental group $\pi_1(S^1 \vee \cdots \vee S^1)$ of a bouquet of k circles is the free group on k free generators. Consequently the fundamental group of a region of the form $\mathbb{R}^2 \setminus \{x_1, \ldots, x_k\}$, i.e. the plane with k distinct points removed, is also free of rank k (since such a region is contractible to (i.e. homotopically equivalent to) a bouquet of k circles).

(f) The fact that $\pi_1(S^1 \vee S^2) \simeq \mathbb{Z}$ can be established using the universal covering of Example (e) of §18.2 (see Figure 58(b)). The covering space has the form of a line $\mathbb{R}^1$ at the integer points of which 2-spheres are attached. The discrete group $G \simeq \mathbb{Z}$ moves this line through integer distances, thereby sending the attached spheres into one another. (Recall also (from §17.5) that the space $S^1 \vee S^2$ is homotopically equivalent to the region $\mathbb{R}^3 \setminus S^1$, provided the circle is unknotted in $\mathbb{R}^3$; and also to $\mathbb{R}^3 \setminus (\mathbb{R}^1 \cup \{x_0\})$.)

19.3. The Simplest of the Homology Groups

19.3.1. Definition. The *one-dimensional* (or *first*) *homology group* $H_1(M)$ of a (connected) manifold (or more general topological space) M is the quotient group of the fundamental group of M by its commutator subgroup:

$$H_1(M) = \pi_1(M)/[\pi_1, \pi_1],$$

where $[\pi_1, \pi_1]$ denotes the subgroup of $\pi_1(M)$ generated by all "commutators" $aba^{-1}b^{-1}$, $a, b \in \pi_1(M)$. The group operation in $H_1(M)$ is usually written additively; thus $a \mapsto [a]$, $ab \mapsto [a]+[b]$ under the natural homomorphism $\pi_1 \to H_1$.

Now let ω be any closed 1-form on a manifold N (so that $d\omega \equiv 0$ on N), and suppose that $\gamma_1(t), \gamma_2(t)$ are two closed paths in N both beginning and ending at a point $y_0 \in N$. Then if γ_1 and γ_2 are homotopic (as usual via a homotopy throughout which y_0 is held fixed), we shall have

$$\oint_{\gamma_1} \omega = \oint_{\gamma_2} \omega;$$

this follows from an application of the general Stokes formula (see the exercise at the end of §8.2) in essence to the integral of $d\omega$ over the image set in N under the map $F: [0, 1] \times [0, 1] \to N$, which has (oriented) boundary $\gamma_1 \cup \gamma_2^{-1}$, taking into account the assumption $d\omega \equiv 0$. Hence for each closed form ω on N, the map

$$\gamma \mapsto \oint_{\gamma} \omega \tag{1}$$

is well defined as a map of the group π_1. It follows from the familiar properties of integrals that

$$\oint_{\gamma_1\gamma_2} \omega = \oint_{\gamma_1} \omega + \oint_{\gamma_2} \omega = \oint_{\gamma_2\gamma_1} \omega,$$

$$\oint_{\gamma^{-1}} \omega = -\oint_{\gamma} \omega,$$

whence the map (1) defines a linear function on the first homology group

$H_1(N)=\pi_1(N)/[\pi_1, \pi_1]$, with real or complex values. Thus the standard procedure whereby a circuit integral is calculated by deforming the path γ, is seen here to amount to replacing the closed path γ by one equivalent to it in the homology group (i.e. in the same "homology class").

Suppose now that $[\gamma] \in H_1(N)$ is a torsion element ("homology class"), i.e. has finite order m, say, in $H_1(N)$. Then $m[\gamma]=0$ in $H_1(N)$, whence

$$0=\oint_{m[\gamma]} \omega = m \oint_{[\gamma]} \omega,$$

and therefore $\oint_{[\gamma]} \omega = 0$. It follows that the map (1) defining a linear function on the homology group, is actually perhaps more appropriately regarded as defining a linear function on the group $\tilde{H}_1(N)$ obtained as the quotient group of $H_1(N)$ by its torsion subgroup (i.e. by the subgroup consisting of all finite-order elements). The group $\tilde{H}_1(N)$ is called the *reduced homology group*.

Examples. (a) As noted in Example (f) of the preceding subsection, the fundamental group $\pi_1(N)$ of the planar region $N=\mathbb{R}^2\setminus\{x_1, \ldots, x_k\}$ is the free group of rank k. Hence the first homology group $H_1(N)$ is isomorphic to the free abelian group $\mathbb{Z}^k$ (of integer lattice points), which is torsion-free.

(b) The fundamental group $\pi_1(\mathbb{R}P^n)$ of real projective n-space is isomorphic to $\mathbb{Z}_2$ (by Example (b) of the preceding subsection). Hence $H_1(\mathbb{R}P^n)$ is also $\mathbb{Z}_2$, and the reduced homology group $\tilde{H}_1(\mathbb{R}P^n)$ is trivial.

(c) We saw in Example (d) of the preceding subsection that the fundamental group $\pi_1(K^2)$ of the Klein bottle is isomorphic to the abstract group on two generators T_1, T_2 defined by the single relation $T_2^{-1}T_1T_2T_1=1$. In the abelianized group $H_1(K^2)$ this relation takes the form $2[T_1]=0$. Hence $H_1(K^2)\simeq\mathbb{Z}_2\oplus\mathbb{Z}$, and $\tilde{H}_1(K^2)\simeq\mathbb{Z}$.

We now resume our discussion, the upshot of which to this point was that each closed 1-form on a manifold N determines a linear function on the reduced first homology group with values in $\mathbb{R}$ or $\mathbb{C}$.

Sometimes it is useful to consider linear functions taking values other than real or complex ones (for instance the so-called "characters" whose values lie in the set of reals modulo 1, i.e. on the circle S^1). Generally speaking we are compelled in such cases to work with the full homology group $H_1(N)$ as domain, the group $\tilde{H}_1(N)$ not sufficing. As an example we may take the homomorphism defined in §17.6:

$$\sigma: \pi_1(N)\to(\pm 1)\simeq\mathbb{Z}_2,$$

where for each path (class) $\gamma \in \pi_1(N)$, the image $\sigma(\gamma)$ is $+1$ or -1 according as transport around γ preserves or reverses an orientation. As noted in §17.6, for every non-orientable manifold this homomorphism is non-trivial. In particular for $\mathbb{R}P^2$, for which $\pi_1\simeq\mathbb{Z}_2$, we have $\sigma(\gamma)=-1$ for all $\gamma\neq 1$. For the Klein bottle K^2, whose fundamental group is as we have seen generated by two

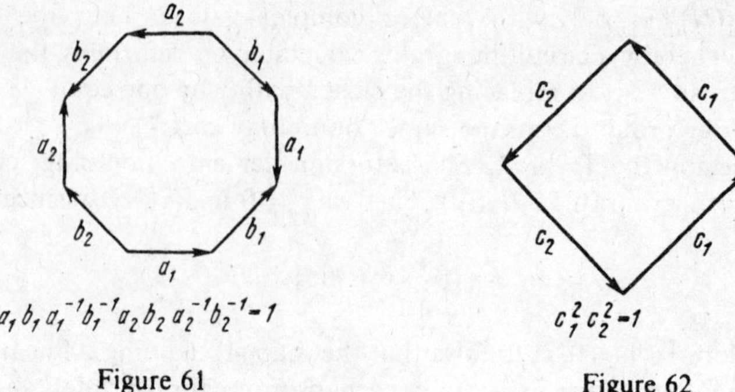

Figure 61 Figure 62

generators T_1, T_2 such that $2[T_1]=0$ in the group $H_1(K^2)$, the above "orientation" function σ is given by

$$\sigma(T_1)=+1, \qquad \sigma(T_2)=-1.$$

We note in conclusion that a technique for calculating the fundamental group of the complement of a knotted circle embedded in $\mathbb{R}^3$, will be described below (in §26).

19.4. Exercises

1. The orientable closed surface M_g^2, known as the "sphere-with-g-handles", can be obtained by identifying appropriate pairs of edges of a $4g$-gon (as indicated in Figure 61 where $g=2$). Prove that the group $\pi_1(M_g^2)$ is defined abstractly by the presentation in terms of generators $a_1, b_1, \ldots, a_g, b_g$, and the single relation

$$\prod_{i=1}^{g} a_ib_ia_i^{-1}b_i^{-1}=1.$$

2. The non-orientable surface N_μ^2 is obtained by glueing edges as indicated in Figure 62 (for the case $\mu=2$). Prove that $\pi_1(N_\mu^2)$ is presented by generators $c_1, \ldots, c_\mu$ with the single relation $c_1^2c_2^2\cdots c_\mu^2=1$.
3. Calculate the fundamental group of the unit tangent bundle of the manifold M_g^2 (see §7.1).

§20. The Discrete Groups of Motions of the Lobachevskian Plane

A description of all possible discrete groups of motions of the Euclidean plane and of Euclidean 3-space can easily be gleaned from §20 of Part I. (For instance in order to obtain all the orientation-preserving discrete subgroups of the isometry group of the Euclidean plane, we need merely add to the list of such groups which preserve "translation-invariant" lattices of points, given in §20 of Part I, the finite cyclic subgroups of arbitrary order.) There the close connexion between many of these groups and "crystal" lattices of points in the plane and in space was made evident enough. A classification of the discrete subgroups of the isometry group of the Lobachevskian plane (see §10.1 of Part I) can be carried out along similar lines; it is our aim in the present section to describe this classification (for the finitely generated discrete groups at least), for the most part suppressing proofs in view of the greater complexity of the arguments than in the Euclidean case. (To describe the discrete groups of motions of 3-dimensional Lobachevskian space is a much more difficult problem, which we shall eschew altogether.)

Our interest in the discrete groups of motions of the Lobachevskian plane stems from the close connexion they have with the 2-dimensional closed manifolds and their fundamental groups. We have already seen examples of the analogous such connexion in the Euclidean case: Recall (from among the various 2-dimensional surfaces hitherto described) the realization of the 2-dimensional torus T^2 as the quotient space $\mathbb{R}^2/\mathbb{Z} \oplus \mathbb{Z}$ of the Euclidean plane, obtained by identifying the points mapped to one another under the action of the discrete group $\mathbb{Z}(a) \oplus \mathbb{Z}(b)$, where a, b denote the translations of $\mathbb{R}^2$ along the vectors (1, 0), (0, 1), respectively (see §§4.1, 18.2). We saw in §19.2 that since this discrete group acts freely on the (simply-connected) space $\mathbb{R}^2$, it is isomorphic to the fundamental group of the torus: $\pi_1(T^2) \simeq \mathbb{Z} \oplus \mathbb{Z}$. On the other hand in view of the simple-connectedness of the sphere S^2, and perhaps the fact that its Gaussian curvature is positive, it is not surprising that the 2-sphere can not be realized as an orbit space of the plane (as unbranched covering) under the action of any discrete group. It turns out that the remaining closed orientable surfaces (namely those of genus > 1, the spheres with more than one handle) can however be obtained as quotient spaces of the Lobachevskian plane under the action of suitable finitely generated, freely acting, discrete subgroups of the isometry group, isomorphic to the respective fundamental groups. Since in each case the discrete group consists of isometries of the Lebachevskian plane, it follows that the resulting quotient manifold will be automatically endowed with a metric, "induced" in this sense by the Lobachevskian metric and so defining a constant negative curvature on that manifold. (Note that in the case of the torus the fundamental group also acts as a group of (Euclidean) isometries of $\mathbb{R}^2$.)

Before embarking on the promised classification we make the further remark that groups acting discretely on the Lobachevskian plane arise also in connexion with the problem of classifying the one-dimensional complex analytic manifolds. Every connected complex analytic manifold X can be obtained as a quotient $\tilde{X}/G$ where $\tilde{X}$ is a simply-connected complex manifold (the universal covering space of X), and the group G acts discretely and freely on $\tilde{X}$ as a group of self-biholomorphisms (see §4.1), whence by Theorem 19.2.2 it is isomorphic to the fundamental group $\pi_1(X)$ of the manifold X. (It can be shown that all such groups G yielding a given complex manifold X as the quotient space $\tilde{X}/G$ of its universal covering space $\tilde{X}$, are conjugate to one another in the group of all self-biholomorphisms of the manifold $\tilde{X}$.) Now it turns out that up to biholomorphic equivalence there are in all only three connected, simply-connected, one-dimensional complex manifolds, namely:

(i) the complex projective line $\mathbb{C}P^1$ (see §2.2);
(ii) the affine complex line $\mathbb{C}^1$, i.e. the complex plane; and
(iii) the open unit disc

$$\{z \mid z \in \mathbb{C}, |z| < 1\}$$

in the complex plane.

Hence the aforementioned problem (of classifying the one-dimensional complex manifolds) reduces to that of describing the groups of self-biholomorphisms which act freely and discretely on each of these three complex manifolds. The connexion which we have been leading up to, between this and discrete groups of isometries of the Lobachevskian plane, is revealed in the third statement of the following proposition (whose proof we omit).

20.1. Proposition. (i) *Every self-biholomorphism of the manifold $\mathbb{C}P^1$ has a fixed point.*
(ii) *Every discretely and freely acting group G of self-biholomorphisms of the complex plane $\mathbb{C}^1$ such that the quotient space $\mathbb{C}^1/G$ is compact, consists of translations $z \mapsto z+a$, where a ranges over the vectors of some 2-dimensional lattice of points in $\mathbb{C}^1$.*
(iii) *Every self-biholomorphism of the unit disc has the form*

$$z \mapsto \theta\frac{z-\alpha}{1-\bar{\alpha}z},$$

where $|\theta|=1$ and $|\alpha|<1$; it follows that the group of all self-biholomorphisms of the unit disc coincides with the group of orientation-preserving isometries of the Poincaré model of the Lobachevskian plane (see §13.2 of Part I).

We now return to our main concern, namely the description of the (finitely generated) discrete groups of isometries of the Lobachevskian plane L^2. We begin by associating with each such discrete group a (Lobachevskian) convex polygon, called the "fundamental region" for the group. We shall throughout this section use one or the other of the following two models of the Lobachevskian plane: the upper half-plane (of the complex plane), endowed with the metric $dl^2=(dx^2+dy^2)/y^2$, and the unit disc with metric $dl^2=(dr^2+r^2\,d\varphi^2)/(1-r^2)^2$. (See §10.1 of Part I for the derivation of these metrics.) A "discrete group of transformations" (i.e. diffeomorphisms) of a manifold (in the present context the Lobachevskian plane), which we defined in §18.4 above, is readily seen to be characterized also by the following property: Each pair of (not necessarily distinct) points x, y of L^2 (the Lobachevskian plane) have (open) neighbourhoods U_x, U_y and that the intersection $g(U_x) \cap U_y$ is non-empty for only finitely many $g \in G$. Hence in particular, as in the definition given in §18.4, for each $x \in L^2$, the stabilizer G_x of x in the group G is a finite subset (in fact subgroup).

20.2. Definition. Let G be a discrete group of transformations of the Lobachevskian plane L^2, consisting of Lobachevskian isometries. A subset D of L^2 is called a *fundamental region* for the group G if:

(i) D is a closed set;
(ii) the images $G(D)$ of the set D together cover the entire plane L^2;
(iii) some (sufficiently small) neighbourhood of each point of L_2 intersects only finitely many of the image sets $g(D)$, $g \in G$;
(iv) the image of the set of interior points of D under any non-identity element of G, intersects D trivially, i.e. $g(\text{Int } D) \cap \text{Int } D = \varnothing$ for all $1 \neq g \in G$, where $\text{Int } D = D \setminus \partial D$.

It can be shown that for any finitely generated discrete group of isometries of L^2, there exists a fundamental region which is a (Lobachevskian) convex polygon with a finite number of sides (together, of course, with the polygon's interior).

Now let G be any finitely generated discrete group of motions of L^2, and let D be a convex "polygon" serving as fundamental region for G. By definition of the fundamental region the images $g(D)$ for distinct $g \in G$, do not overlap, and together cover the whole of L^2; thus these regions form a "tessellation" of the Lobachevskian plane, with "cells" the $g(D)$. We shall say that two cells are *adjacent* if their intersection is a one-manifold, i.e. a curve-segment. By suitably adding (as indicated schematically in Figure 63) further vertices to the polygonal boundary of the fundamental region D (the angle at all such vertices being then of course π), we can arrange that the intersection $D_1 \cap D_2$ of any pair D_1, D_2 of adjacent cells is exactly a common side of these two polygons. This done, we shall have for each side a of the cell D, a unique

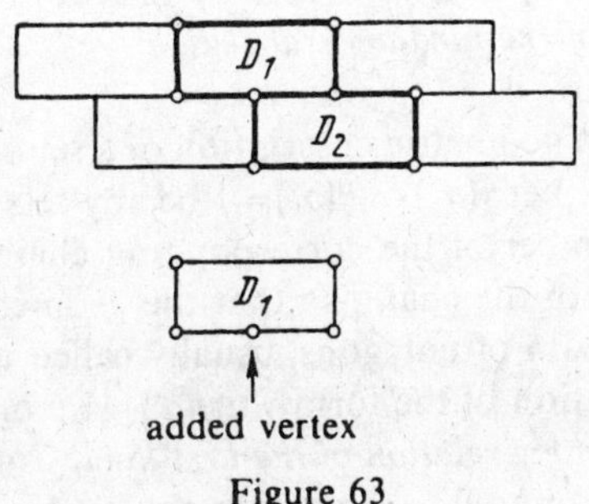

Figure 63

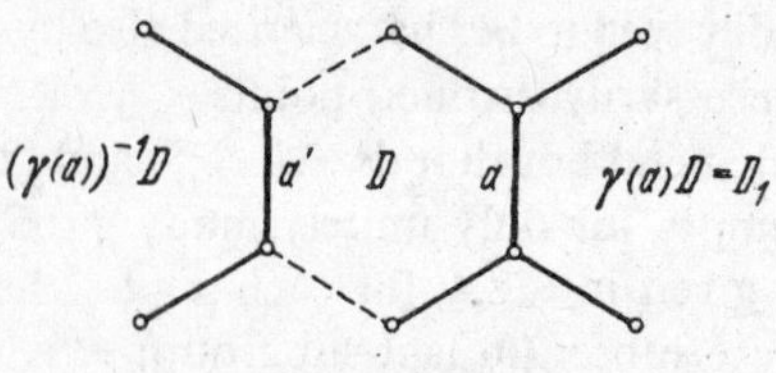

Figure 64

cell D_1 adjacent to D with the side a in common; we denote by $\gamma(a)$ the unique group element which sends D onto D_1. Under this transformation there will be a side a' say, of D which is sent to the side a of D_1, i.e. $\gamma(a)a' = a$. It follows that $\gamma(a') = \gamma(a)^{-1}$, and $a'' = (a')' = a$ (see Figure 64); thus the map $\gamma(a)$ gives rise to an involuntary (i.e. of order two) permutation $a \mapsto a'$ of the set of sides of the fundamental region D, unless $a = a'$ for all a. If for any particular a we have $a = a'$ then $\gamma(a)$ fixes the side a (as a whole), and hence must either reflect D in the side a, or else rotate D through the angle π about the mid-point of the side a; in either case it follows that $(\gamma(a))^2 = 1$.

The following lemma is an immediate consequence of the fact that two cells $\gamma_1(D)$ and $\gamma_2(D)$ are adjacent precisely if $\gamma_2^{-1}\gamma_1(D)$ and D are adjacent.

20.3. Lemma. *Two cells $\gamma_1(D)$ and $\gamma_2(D)$ are adjacent if and only if for some side a of D we have $\gamma_2 = \gamma_1\gamma(a)$.*

We shall call a finite sequence

$$D = D_0, D_1, \ldots, D_k$$

of cells a *chain*, if D_{i-1} is adjacent to D_i for $i = 1, \ldots, k$. Take any such chain of cells, and for each $i = 1, \ldots, k$, define $\gamma_i \in G$ by $D_i = \gamma_i D_0$. Since D_{i-1} and D_i are adjacent we have by the above lemma that $\gamma_i = \gamma_{i-1}\gamma(a_i)$ for some side a_i of D; similar successive decompositions of $\gamma_{i-1}, \gamma_{i-2}, \ldots, \gamma_1$ yield finally $\gamma_k = \gamma(a_1)\gamma(a_2)\cdots\gamma(a_k)$ for some finite sequence $a_1, a_2, \ldots, a_k$ of sides of the cell D. Since there is for every cell $\hat{D}$ a chain of cells beginning with D and terminating with $\hat{D}$, we deduce immediately the following result.

20.4. Theorem. *The group G is generated by the transformations $\gamma(a)$ where a ranges over the sides of the fundamental region.*

We shall now give a geometric description of a set of defining relations for G on these generators. Let $\gamma(a_1)\cdots\gamma(a_k) = 1$ be any relation on the generators $\gamma(a)$; then the last member of the corresponding chain of cells will coincide with the initial cell D of the chain, so that the geometric counterpart of the relation is a closed chain of polygons, usually called a *cycle* (see Figure 65). We shall call any relation of the form $\gamma(a)\gamma(a') = 1$ (corresponding to a cycle D_0, D_1, D_0) an *elementary relation of the first kind*. Consider next any vertex of D, and the (finitely many) cells sharing this vertex. These cells, taken in

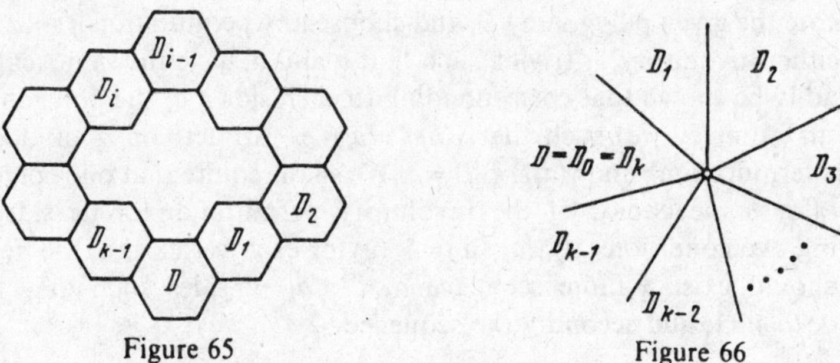

Figure 65

Figure 66

order around the vertex, form a cycle (see Figure 66); the relation corresponding to such a cycle will be called an *elementary relation of the second kind*. It can be shown that these two kinds of "elementary" relations actually suffice to define G:

20.5. Theorem. *The elementary relations of the first and second kind together form a set of (abstract) defining relations for the group G on the generators $\gamma(a)$; i.e. every relation between the $\gamma(a)$ is a group-theoretical consequence of them.*

Thus this result (whose proof we omit) gives a geometric description, of a sort, of the finitely generated groups which can occur as discrete groups of isometries of the Lobachevskian plane.

We now turn our attention to the converse problem, namely that of constructing a discrete group G from a given fundamental polygon. Thus suppose that we are given in L^2 a convex Lobachevskian polygon with a finite number of sides. We shall assume to begin with that our polygon has no vertices "at infinity", meaning that, while the polygon may be unbounded (as for instance in Figure 67), it should not have two sides with a common "vertex at infinity" (as in Figure 68). (Since the points of the boundary circle (in the Poincaré model of L^2) do not actually belong to L^2, we regard an unbounded Lobachevskian straight line segment (forming a side of the polygon, like AB in Figure 67), as not having on it a vertex at infinity; however, for polygons like that shown in Figure 68, having two sides which meet "at infinity", we do regard the polygon as having a vertex at infinity.) Note that, as before, we allow our polygon to have vertices at which the angle is π.

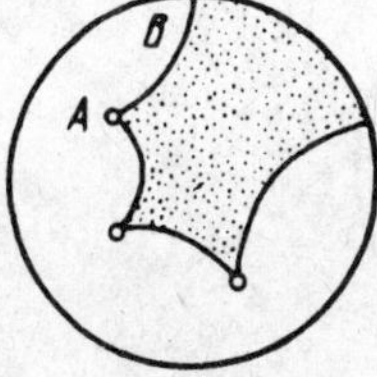

Figure 67

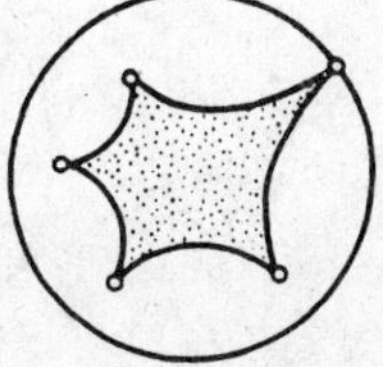

Figure 68

Denote the given polygon by D, and choose any permutation $a \mapsto a'$, of its sides, either involutory or trivial, such that a and a' have the same length. It can readily be shown that corresponding to each side a of the polygon there exists an isometry $\gamma(a)$ such that $\gamma(a)a' = a$ (i.e. $\gamma(a)$ acts on a' as does the chosen permutation), and $\gamma(a)D \cap D = a$. We shall require that our polygon D satisfy (for some choice of the involutory permutation of its sides) the following two conditions: (i) $\gamma(a)\gamma(a') = 1$; (ii) for each vertex A of the polygon there should exist a finite sequence $a_1, \ldots, a_k$ of sides such that firstly, $\gamma(a_1)\cdots\gamma(a_k) = 1$, and secondly the sequence

$$D, \gamma(a_1)D, \gamma(a_1)\gamma(a_2)D, \ldots, \gamma(a_1)\cdots\gamma(a_k)D$$

of polygons is a "cycle of the second kind" about the vertex A in the sense that all the polygons in the sequence share the vertex A, each is adjacent to its successor, they do not overlap, and together they cover some neighbourhood of A. The result we have in mind (whose proof we omit) states that these conditions, which are clearly necessary, are also sufficient for there to exist an appropriate discrete group:

20.6. Theorem. *If the preceding conditions* (i) *and* (ii) *are satisfied by the polygon D (and the chosen involutory permutation of its sides), then the $\gamma(a)$ generate a discrete group of isometries of the Lobachevskian plane, for which the given polygon D is a fundamental region.*

By way of an example, we now consider the discrete groups arising in this way from the simplest sort of such polygon.

Example. Let D be a polygon without any vertices at all (like that shown in Figure 69). We pair off the edges arbitrarily (allowing pairing of some edges with themselves), and for each pair a, a' we select isometries $\gamma(a), \gamma(a')$ satisfying $\gamma(a)a' = a$, $\gamma(a)\gamma(a') = 1$, $\gamma(a)D \cap D = a$: such isometries always exist, since given any two straight lines l, l' there is an isometry interchanging them, and sending either of the half-planes determined by l onto an arbitrarily prescribed one of the half-planes determined by l'. If no side of our given polygon-without-vertices is paired with itself, then it is not difficult to see that the discrete group generated by the $\gamma(a)$ (where a ranges over all sides) is a free group. On the other hand if some sides are paired with themselves (as

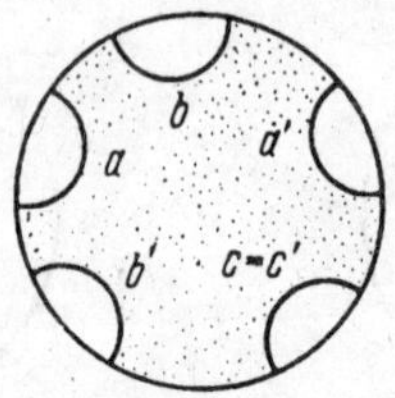

Figure 69

indicated in Figure 69), then we shall obtain some non-trivial relations (of the form $(\gamma(c))^2 = 1$). If every side is paired with itself, i.e. $a' = a$ for all a, then we obtain a free product of 2-cycles (choosing for instance $\gamma(a)$ for each a to be the reflection in a).

We now turn to the case where the given polygon D (with, as before, finitely many sides) has a vertex at infinity (as in Figure 68). We choose some involutory (or trivial) permutation $a \mapsto a'$ of the sides D, and corresponding to each side a a Lobachevskian isometry $\gamma(a)$ satisfying $\gamma(a)a' = a$, $\gamma(a)D \cap D = a$, $\gamma(a)\gamma(a') = 1$ (cf. the previous case). For any vertex A of the polygon D (including the vertex at infinity) these conditions imply the existence of a sequence $a_1, \ldots, a_q, \ldots$ of sides of D such that the polygonal regions

$$D, \gamma(a_1)D, \gamma(a_1)\gamma(a_2)D, \ldots$$

are as shown in Figure 70. (Note that in that diagram the symbol a_i, which denotes a certain side of D, is used also to denote its image under $\gamma(a_1) \cdots \gamma(a_{i-1})$.) Let

$$A, A_1, A_2, \ldots$$

be the corresponding sequence of vertices, obtained by taking successive images of A under the same transformations; we shall say that the two sequences (of sides and of vertices) are *generated by the vertex A*. Since our polygon has only finitely many sides (and each side joins at most two vertices), it follows that both of these sequences will be periodic; we call the smallest common period p of the two sequences the *period of the vertex A*. (Note that we might instead have proceeded in the counterclockwise direction around A, starting with the other side of D incident with A.) We shall say that the sequence $A_1, A_2, \ldots, A_{p-1}$ constitutes a *cycle generated by the vertex A*. (Note that this discussion all makes sense even if A is the vertex at infinity.)

We now suppose that the isometries $\gamma(a)$ generate a discrete group having D as a fundamental region. Then if A is an ordinary vertex of the polygon D (i.e. not the vertex at infinity) the sequence of cells around A depicted in Figure 70, must close up, whence (see below) there exists a positive integer m such that

$$[\gamma(a_1)\gamma(a_2) \cdots \gamma(a_p)]^m = 1. \tag{1}$$

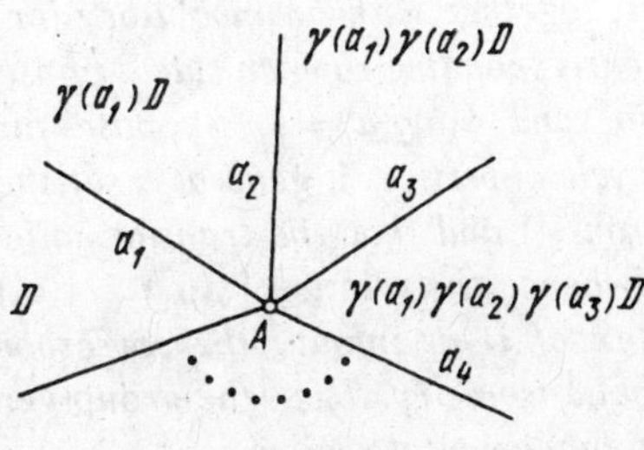

Figure 70

(Note that even though the vertex A will (by definition of p) be back in its original position after p steps of the sort indicated in Figure 70, it does not necessarily follow that all cells having A as a vertex will then have been exhausted.) The number m is called the *multiplicity of the vertex A* (not to be confused with the period of A). Clearly in order for the word $[\gamma(a_1)\gamma(a_2)\cdots\gamma(a_p)]^m$ to correspond precisely to the full cycle of all cells with A as a vertex, we must have

$$\sum_{i=1}^{p} (\angle A_i) = \frac{2\pi}{m}, \tag{2}$$

where $\angle A_i$ denotes the size of the angle of D at the vertex A_i. If we impose the condition that the transformation $[\gamma(a_1)\cdots\gamma(a_p)]^m$ be orientation-preserving, then the relation (1) will follow from (2). It is easy to verify that the relations resulting in this way from other vertices of the cycle generated by the vertex A (as well as that obtained by proceeding around A in the counterclockwise direction) are all equivalent to the relation (1).

Since it is, obviously, impossible for a cycle of cells sharing the vertex at infinity to close up, no relation is yielded in the above-described manner by that vertex. There is however the following result, given without proof. (Recall from §13.2 of Part I that the orientation-preserving isometries of the Lobachevskian plane are, in the Klein model, just the linear fractional transformations $z \mapsto (az+b)/(cz+d)$ where a, b, c, d are real and $ad-bc=1$.)

20.7. Lemma. *For the vertex at infinity the transformation* $\gamma(a_1)\cdots\gamma(a_p)$ *(defined as above) is a parabolic motion, i.e. the real matrix* $\begin{pmatrix} a & b \\ c & d \end{pmatrix}$ *(or its negative) corresponding (in the Klein model of the Lobachevskian plane) to the transformation* $\gamma(a_1)\cdots\gamma(a_p)$, *is similar (i.e. conjugate) to the matrix* $\begin{pmatrix} 1 & 1 \\ 0 & 1 \end{pmatrix}$.

We are now in a position to state (without proof) the result we have been aiming at, which tells us essentially that the above necessary conditions for the given polygon D to be a fundamental domain for the discrete group generated by the $\gamma(a)$, are in fact also sufficient.

20.8. Theorem. *Suppose we are given in the Lobachevskian plane a polygon D with finitely many sides, together with an involutory (or trivial) permutation* $a \mapsto a'$ *of its sides, and corresponding to each side a an isometry* $\gamma(a)$ *satisfying* $\gamma(a)a' = a$, $\gamma(a)D \cap D = a$, *and* $\gamma(a)\gamma(a') = 1$. *Suppose further that for each ordinary vertex A, the cycle of vertices it generates satisfies condition* (2) *above for some m (depending on A) and that the transformation* $[\gamma(a_1)\cdots\gamma(a_p)]^m$ *is orientation-preserving (whence the relation* $[\gamma(a_1)\cdots\gamma(a_p)]^m = 1$). *Suppose finally that for each vertex of D at infinity the corresponding transformation* $\gamma(a_1)\cdots\gamma(a_p)$ *is a parabolic isometry. Then the group generated by the* $\gamma(a)$ *is discrete and has D as a fundamental region.*

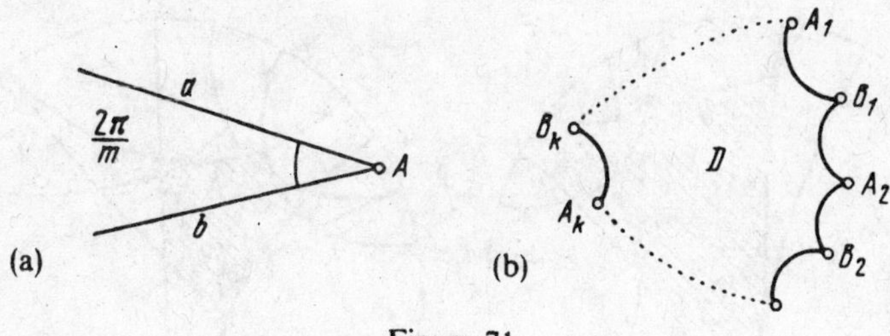

Figure 71

Examples. (a) Consider the (2-sided) polygon of Figure 71(a), and the group generated by the reflections in each of the sides. Here the vertex A has period 2 and multiplicity m.

(b) With the polygon D as in Figure 71(b), suppose that the angle at each A_s is $2\pi/m_s$ where m_s is a positive integer, and that the two edges meeting at each A_s are of equal length. Suppose further that

$$\sum_{i=1}^{k} \angle B_i = \frac{2\pi}{m},$$

for some positive integer m. For each s the clockwise rotation γ_s of the plane through the angle $2\pi/m_s$ about the vertex A_s sends one of the two edges incident with A_s, to the other. A direct verification shows that the γ_s satisfy the hypotheses of Theorem 20.8, whence we conclude that they generate a discrete group with D as fundamental region. Corresponding to each vertex A_s we obtain the relation $\gamma_s^{m_s}=1$, while the vertices B_s yield relations equivalent to $(\gamma_1 \cdots \gamma_k)^m = 1$ (and these constitute a full set of defining relations for the discrete group).

(Note that there are very few analogous such polygons in the Euclidean plane, since (as is readily verified) the equation

$$\sum_{i=1}^{k} \frac{1}{m_i} + \frac{1}{m} = k-1$$

holds for such (Euclidean) polygons, whence $k \leq 4$. On the other hand there are infinitely many essentially different such polygons (and consequently such discrete groups) in the Lobachevskian plane.)

(c) Consider the $4k$-gon in the Lobachevskian plane shown in Figure 72(a), where it is assumed that the angle-sum is 2π, and that for each i the sides a_i and a_i' have the same length, as also do b_i and b_i'. We then have the

20.9. Proposition. *The orientation-preserving isometries* α_i, β_i *of the Lobachevskian plane, uniquely defined by the conditions* $\alpha_i\colon a_i \to a_i'$, $\beta_i\colon b_i \to b_i'$,

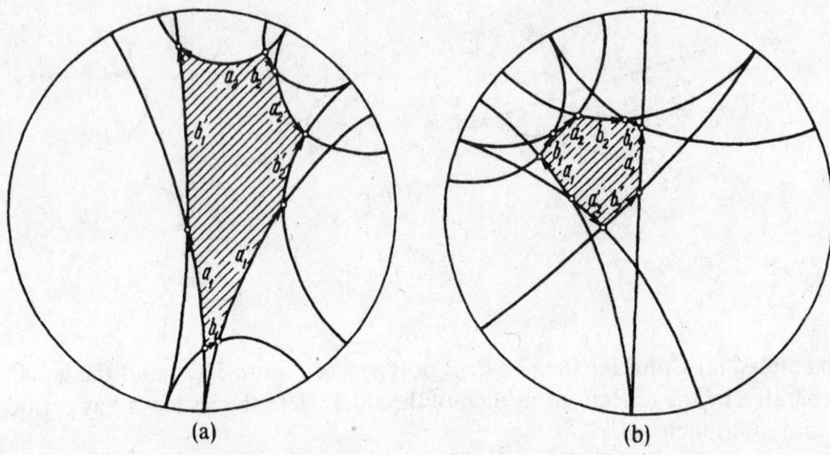

Figure 72

generate a discrete group which acts fixed-point freely, is defined abstractly by the single relation

$$\alpha_1\beta_1\alpha_1^{-1}\beta_1^{-1}\cdots\alpha_k\beta_k\alpha_k^{-1}\beta_k^{-1}=1,$$

and has the given 4k-gon as fundamental region.

This discrete group is isomorphic to the fundamental group of a Riemann surface of genus k, i.e. of the sphere-with-k-handles, and the given $4k$-gon affords a canonical diagrammatic representation of that surface—see Exercise 1 of §19.4. From that exercise and the definition of a covering defined by a freely acting discrete group (18.4.2), we deduce the

20.10. Corollary. *The universal covering space of the closed, orientable surface of genus $g>1$ (i.e. the sphere-with-g-handles M_g^2) is the Lobachevskian plane.*

EXERCISE

Show that if in Example (c) above we consider instead the polygon in Figure 72(b) (rather than that of Figure 72(a)), the resulting orientable surface is the same.

We conclude this particular discussion with the statement (without proof) of a "finiteness" result for discrete groups.

20.11. Theorem. *If a convex fundamental polygon of a discrete group of motions of the Lobachevskian plane has finite area, then it has finitely many sides (and if there are excursions to infinity, then there are only finitely many of them).*

We now turn to the consideration of the so-called "Möbius group" and the classification of linear-fractional transformations. (The relevance of this to the Lobachevskian plane will be recalled below.)

The set of all linear-fractional transformations of the Riemann sphere, or extended complex plane $\mathbb{C}P^1 \cong \mathbb{C} \cup \{\infty\} \cong S^2$, forms a group which is sometimes called the *Möbius group*, denoted by "Möb". There is an obvious isomorphism

$$\text{Möb} \simeq SL(2, \mathbb{C})/\{\pm 1\}.$$

(Recall that the centre of $SL(2, \mathbb{C})$ is precisely $\{\pm 1\}$.) From the theory of the Jordan normal form of matrices, we know that every matrix σ from $SL(2, \mathbb{C})$ is conjugate to a matrix either of the form $\begin{pmatrix} \lambda & 1 \\ 0 & \lambda \end{pmatrix}$, corresponding to the linear-fractional transformation $z \mapsto z + 1/\lambda$ $(\lambda = \pm 1)$, or of the form $\begin{pmatrix} \lambda & 0 \\ 0 & \mu \end{pmatrix}$ (where of course $\mu = 1/\lambda$), which corresponds to a linear-fractional transformation of the form $z \mapsto cz$. A non-identity transformation σ of the former conjugacy type is called *parabolic*, while if of the latter type it is called *elliptic* if $|c| = 1$ and *hyperbolic* if $c \in \mathbb{R}$ and $c > 0$; the remaining possible non-identity linear-fractional transformations are lumped together under the term *loxodromic transformations*. Note that this taxonomy (from which incidentally the identity transformation and its negative are excluded) is to be applied indifferently to matrices in $SL(2, \mathbb{R})$ and to the corresponding linear-fractional transformations (i.e. elements of the Möbius group).

20.12. Lemma. *A matrix σ from $SL(2, \mathbb{C})$, not equal to ± 1, is:*

(i) *parabolic if and only if* $\operatorname{tr}\sigma = \pm 2$;
(ii) *elliptic if and only if* $\operatorname{tr}\sigma$ *is real and* $|\operatorname{tr}\sigma| < 2$;
(iii) *hyperbolic if and only if* $\operatorname{tr}\sigma$ *is real and* $|\operatorname{tr}\sigma| > 2$.

(*Hence the matrix $\sigma \neq \pm 1$ will be loxodromic if and only if* $\operatorname{tr}\sigma$ *is not real.*)

It is clear from this lemma that the group $SL(2, \mathbb{R})$ does not contain loxodromic elements. We now restrict our attention to this group, giving a characterization of its elements in terms of the fixed points of the corresponding linear-fractional transformations. (Every non-identity Möbius transformation has two fixed points (sometimes "merged" into a single fixed point).)

Before describing this characterization we recall the connexion between $SL(2, \mathbb{R})$ and the Lobachevskian plane: The orientation-preserving isometries of that plane are, in the Klein model, precisely the linear-fractional transformations $z \mapsto (az+b)/(cz+d)$ with a, b, c, d real (and $ad - bc = 1$) (see §13.2 of Part I). Hence there is a homomorphism from $SL(2, \mathbb{R})$ to the full isometry group of the Lobachevskian plane (in fact onto the connected component of the identity of that group), with kernel the centre $\{\pm 1\} \simeq \mathbb{Z}_2$ of $SL(2, \mathbb{R})$.

We begin our promised geometric characterization of the elements of $SL(2, \mathbb{R})$ with the following result (again omitting the proof):

20.13. Lemma. *A matrix σ from $SL(2, \mathbb{R})$, not equal to ± 1, is:*

(i) *parabolic if and only if it has only one fixed point (which must then be on the extended real line $\mathbb{R} \cup \{\infty\}$);*

(ii) *elliptic if and only if it has one fixed point in the open upper half-plane $H = \{z \in \mathbb{C} \mid \text{Im } z > 0\}$, and one in the open lower half-plane; and*

(iii) *hyperbolic if and only if it has two distinct fixed points on the extended real line $\mathbb{R} \cup \{\infty\}$.*

Let G be a discrete subgroup of $SL(2, \mathbb{R})$. A point $z \in H$ is called an *elliptic point of the group G*, if z is a fixed point of some elliptic transformation in G; similarly a point of $\mathbb{R} \cup \{\infty\}$ is a *parabolic point* of G if it is the fixed point of some parabolic transformation in G. We now list the simplest geometric properties of the linear-fractional transformations in $SL(2, \mathbb{R})$ (with the usual permitted imprecision of statement) of the three types.

1. Hyperbolic Transformations. (i) Every circle passing through both fixed points of a hyperbolic transformation is sent onto itself under that transformation, and each of the two connected segments of such a circle having the fixed points as end-points, is also sent onto itself by the transformation.

(ii) Each of the two connected regions into which the extended complex plane is divided by a circle through the fixed points, is sent onto itself by the hyperbolic transformation.

(iii) Every circle orthogonal to a circle through the fixed points is sent to another such circle.

(iv) The fixed points of a hyperbolic transformation are inverse points with respect to any such circle, i.e. to any circle orthogonal to some circle through the fixed points. (Points A, B are *inverse points* with respect to a circle with centre O and radius R if the points O, A, B lie on a (Euclidean) half-line emanating from O, and $|OA| \cdot |OB| = R^2$.)

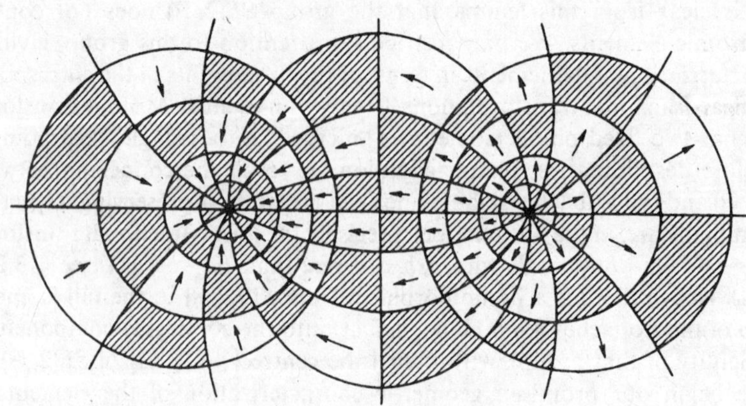

Figure 73

Two such orthogonal systems (or "pencils") of circles are depicted in Figure 73, where it is also indicated how the regions into which the circles subdivide the plane move under the hyperbolic transformation: each shaded region is sent onto the neighbouring (unshaded) region in the direction of the arrows.

2. Parabolic Transformations. (i) Each circle C passing through the fixed point of a parabolic transformation, is sent by that transformation onto a circle tangent to it (i.e. to C) at the fixed point.

(ii) There is a family of circles all tangent to one another at the fixed point (i.e. a "tangent pencil" of circles) and defined by a single parameter, each of which is sent onto itself by the parabolic transformation.

(iii) The interior of every circle preserved (i.e. sent onto itself) by a parabolic transformation is mapped onto itself (i.e. is also preserved).

The effect of a parabolic transformation is conveyed by Figure 74; again each shaded (unshaded) region is to be imagined as moved over onto the neighbouring unshaded (shaded) region in the direction indicated by the arrows.

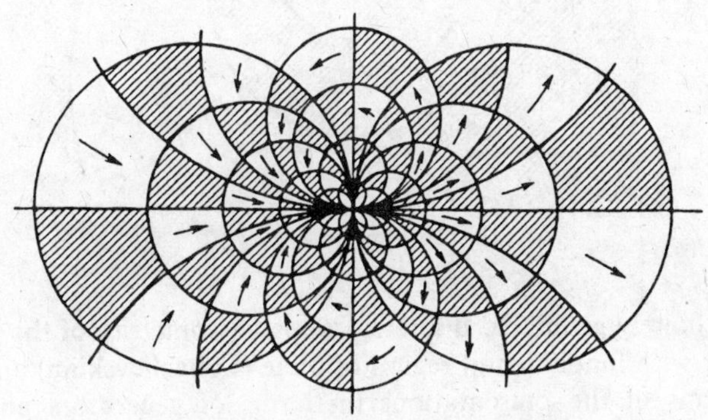

Figure 74

3. Elliptic Transformations. (i) Each circular arc joining the two fixed points of an elliptic transformation, is sent by that transformation to another such circular arc.

(ii) Every circle orthogonal to the circles through the fixed points is preserved.

(iii) The interior of each such circle is preserved.

(iv) The fixed points are inverse points with respect to every such circle (orthogonal to the circles through the fixed points).

The effect of an elliptic transformation of the plane, may be gathered from Figure 75.

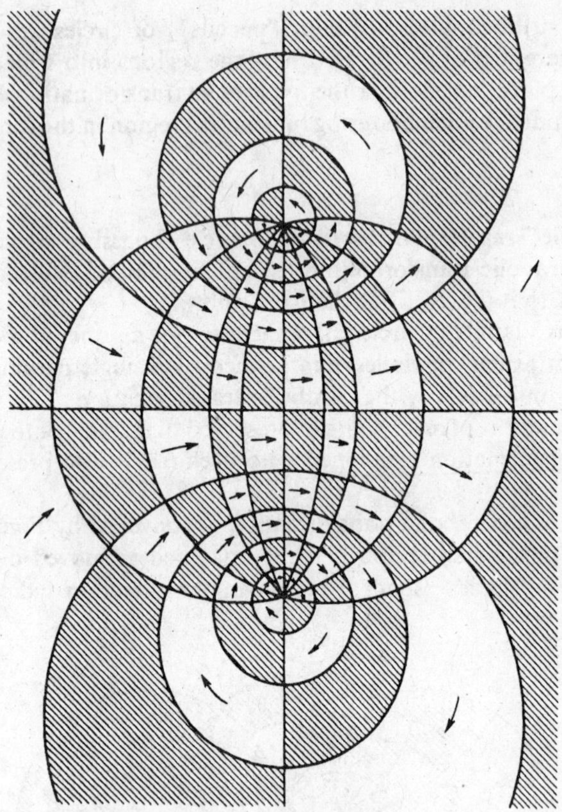

Figure 75

For the finitely generated discrete groups of isometries of the Lobachevskian plane with fundamental region of finite (Lobachevskian) area, explicit general forms of the presentations (in terms of generators and defining relations) are known; we now give these presentations and look at some of their properties.

Among such groups those which are orientation-preserving are of particular importance; they are generally termed *Fuchsian groups*. We begin by giving the possible presentations of these groups. It turns out that for any given Fuchsian group G we can find generators $x_1, \ldots, x_r; a_1, b_1, \ldots, a_g, b_g$ (for some $r \geq 0, g \geq 0$) in terms of which a set of defining relations for the group is given by

$$x_1^{m_1} = 1, \ldots, x_r^{m_r} = 1, \qquad x_1 x_2 \cdots x_r a_1 b_1 a_1^{-1} b_1^{-1} \cdots a_g b_g a_g^{-1} b_g^{-1} = 1,$$

where $m_1, \ldots, m_r \geq 2$, and some or all of the m_i may be ∞ (in which case the corresponding relations $x_i^\infty = 1$ may be deleted). The integer g is called the *genus*, the m_i the *periods*, the number of occurrences of any particular period in the sequence $m_1, \ldots, m_r$, the *multiplicity* of that period, and the $(r+1)$-tuple $(g; m_1, \ldots, m_r)$ the *F-signature* of the Fuchsian group. As the introduction of this terminology might lead the reader to suspect, these are, essentially, invariants of the group; this is the substance of the following result. (Again we content ourselves with the statement alone; the invariance of g is easily seen, however, since it is the rank of the commutator quotient group $G/[G, G]$.)

Proposition. *If a Fuchsian group with F-signature $(g; m_1, \ldots, m_r)$ is isomorphic to another Fuchsian group with F-signature $(g'; m'_1, \ldots, m'_{r'})$, then $g=g'$, $r=r'$, and there is a permutation φ of the set $\{1, \ldots, r\}$ such that $m'_i = m_{\varphi(i)}$ for $i=1, \ldots, r$.*

(The converse statement is also valid: If φ is any permutation of the set $\{1, \ldots, r\}$, then any Fuchsian group with F-signature $(g; m_1, \ldots, m_r)$ is isomorphic to any group with F-signature $(g; m_{\varphi(1)}, \ldots, m_{\varphi(r)})$.)

It is natural to ask for the conditions a sequence $(g; m_1, \ldots, m_r)$ with $g, r \geq 0$, $m_i \geq 2$, must satisfy for it to be the F-signature of some Fuchsian group. The precise condition turns out to be

$$\mu(g; m_1, \ldots, m_r) = 2g - 2 + \sum_{i=1}^{r} \left(1 - \frac{1}{m_i}\right) > 0.$$

It can be shown that in fact the (Lobachevskian) area of the fundamental region of a Fuchsian group of F-signature $(g; m_1, \ldots, m_r)$ is just $\pi\mu(g; m_1, \ldots, m_r)$.

It can also be shown that every element of finite order of a Fuchsian group G of F-signature $(g; m_1, \ldots, m_r)$ with generators as above, is conjugate in G to a power of one of the elements $x_1, \ldots, x_r$ (which are of orders $m_1, \ldots, m_r$, respectively), and furthermore that no two non-trivial powers of distinct x_i's are conjugate in G.

We now consider the wider class consisting of those (finitely generated, discrete) groups whose fundamental region has finite area; these are called the *non-Euclidean crystallographic groups.*

As a preliminary to giving the possible presentations of such groups, we define an NEC-*signature* to be an ordered quadruple of the form

$$(g, \varepsilon, [m_1, \ldots, m_r], \{c_1, \ldots, c_k\}),$$

where: g, called as before the *genus*, is a non-negative integer; $\varepsilon = \pm 1$; $m_1, \ldots, m_r$, the *periods*, take their values from among the positive integers ≥ 2 or ∞; and the c_i, called *cycles*, are finite sequences of integers ≥ 2:

$$c_1 = (n_{11}, \ldots, n_{1s_1}), \ldots, c_k = (n_{k1}, \ldots, n_{ks_k}), \qquad n_{ij} \geq 2.$$

(Note that r (the number of periods) or k (the number of cycles) may be zero.) Thus the generic NEC-signature may be written in more detailed fashion as

$$(g, \pm, [m_1, \ldots, m_r]; \{(n_{11}, \ldots, n_{1s_1}), \ldots, (n_{k1}, \ldots, n_{ks_k})\}).$$

Given any NEC-signature, we define a corresponding NEC-*group* in terms of generators and relations, as follows.

(i) In the case of an "orientable" signature ($\varepsilon = +1$) the generators and relations of the corresponding NEC-group are as in the following table.

Element of the signature	Generators	Defining relations
the period m_i	x_i	$x_i^{m_i} = 1$
the cycle $c_i = (n_{i1}, \ldots, n_{is_i})$	e_i $c_{i0}, c_{i1}, \ldots, c_{is_i}$	$c_{is_i} = e_i^{-1} c_{i0} e_i$, $c_{i,j-1}^2 = c_{ij}^2 = (c_{i,j-1} c_{ij})^{n_{ij}} = 1$
$g, +1$	$a_1, b_1, \ldots, a_g, b_g$	$x_1 \cdots x_r e_1 \cdots e_k a_1 b_1 a_1^{-1} b_1^{-1} \cdots$ $\cdots a_g b_g a_g^{-1} b_g^{-1} = 1$

(ii) In the case $\varepsilon = -1$, the generators and defining relations of the corresponding NEC-group are obtained by replacing the last row of the above table by the following one:

$g, -1$	$a_1, \ldots, a_g$	$x_1 \cdots x_r e_1 \cdots e_k a_1^2 \cdots a_g^2 = 1$

The significance of these presentations lies in the fact that every non-Euclidean crystallographic group has a presentation as an NEC-group.

We conclude by giving a specific example (of a discrete group of motions of the Lobachevskian plane) of the type considered in Example (c) above. We take as fundamental region a regular $4g$-gon with each of its angles equal to $\pi/2g$, and with centre at, say, the centre of the unit circle (assuming that we are working in Poincaré's model), as shown in Figure 76 in the case $g = 2$. We pair off each side of this $4g$-gon with the side diametrically opposite to it, and denote by $A_1, \ldots, A_{2g}$ (Lobachevskian) translations each of which shifts one side (of each pair of diametrically opposed sides) into the position of the other, and where for each $k = 1, \ldots, 2g-1$, the direction along which the translation A_{k+1} is performed, is obtained from the direction of A_k by rotating the latter through the angle $\pi - \pi/2g$ (or in other words by conjugating A_k by the transformation B_g which rotates the plane

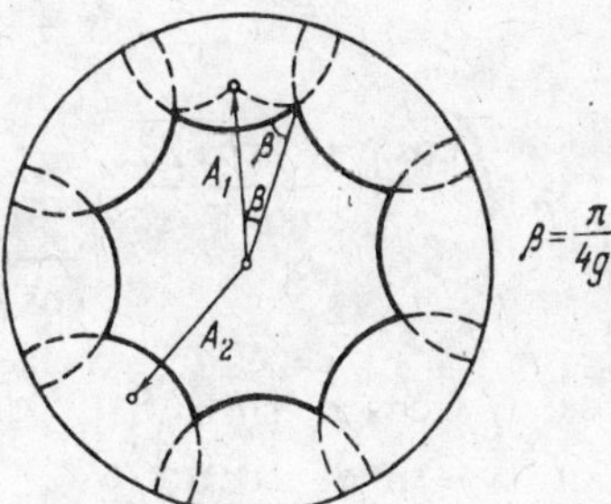

Figure 76. The transformations A_1 and A_2 translate the centre of the octagon to the positions indicated by the arrows.

through the angle $\pi-\pi/2g$ about the centre of the circle). (The relationship between A_1 and A_2 (typical of that between A_k and A_{k+1}) can be seen from Figure 76.) The transformations $A_1, \ldots, A_{2g}$ satisfy the relation $A_1 \cdots A_{2g} A_1^{-1} \cdots A_{2g}^{-1} = 1$ (verify!)

It is not difficult to obtain explicit matrices in $SL(2, \mathbb{R})$ corresponding to the transformations $A_1, \ldots, A_{2g}$ (now of course considered as acting on the upper half-plane model). Under the transformation $z=(1+iw)/(1-iw)$ between the Poincaré model (in the z-plane) and the Klein model (the upper half of the w-plane), the centre of the unit circle corresponds to the point i; hence we may assume, by choosing our polygon so that in the upper half-plane model it has a side perpendicular to the imaginary axis, that A_1 preserves the imaginary axis, whence it needs must have the form $w \mapsto \lambda w$, where $\lambda = e^l$, with l the double of the length of a leg of the triangle with angles $\pi/2$, $\pi/4g$, $\pi/4g$ (this is clear from Figure 76). A straightforward calculation of the length of such a leg yields (verify it!)

$$l = 2 \ln \frac{\cos\beta + \sqrt{\cos 2\beta}}{\sin\beta}, \qquad \beta = \frac{\pi}{4g}.$$

As already noted, the matrices $A_2, \ldots, A_{2g}$ are obtained from A_1 by successive conjugation by B_g, the matrix of the rotation of the plane through the angle $\pi[(2g-1)/2g]$ about the point i; thus

$$A_k = B_g^{-k+1} A_1 B_g^{k-1}, \qquad k = 2, \ldots, 2g,$$

where

$$B_g = \begin{pmatrix} \cos\pi\dfrac{2g-1}{4g} & \sin\pi\dfrac{2g-1}{4g} \\ -\sin\pi\dfrac{2g-1}{4g} & \cos\pi\dfrac{2g-1}{4g} \end{pmatrix}$$

Hence finally

$$A_k = \begin{pmatrix} \cos\alpha & \sin\alpha \\ -\sin\alpha & \cos\alpha \end{pmatrix}^{-k+1} \begin{pmatrix} \dfrac{\cos\beta+\sqrt{\cos 2\beta}}{\sin\beta} & 0 \\ 0 & \dfrac{\sin\beta}{\cos\beta+\sqrt{\cos 2\beta}} \end{pmatrix}$$

$$\times \begin{pmatrix} \cos\alpha & \sin\alpha \\ -\sin\alpha & \cos\alpha \end{pmatrix}^{k-1},$$

$$\alpha = \pi\frac{2g-1}{4g}, \qquad \beta = \frac{\pi}{4g}, \qquad k = 1, \ldots, 2g.$$

EXERCISE

Show that the group with generators $A_1, \ldots, A_{2g}$ and defining relation $A_1 \cdots A_{2g} A_1^{-1} \cdots A_{2g}^{-1} = 1$, is isomorphic to the group on generators $a_1, b_1, \ldots, a_g, b_g$ with defining relation $a_1 b_1 a_1^{-1} b_1^{-1} \cdots a_g b_g a_g^{-1} b_g^{-1} = 1$.

CHAPTER 5

Homotopy Groups

§21. Definition of the Absolute and Relative Homotopy Groups. Examples

21.1. Basic Definitions

The homotopy groups of a manifold or more general topological space M, which we shall shortly define, represent (as will become evident) the most important of the invariants (under homeomorphisms) of the space M. The one-dimensional homotopy group of M is, by definition, just the fundamental group $\pi_1(M, x_0)$. The zero-dimensional homotopy group $\pi_0(M, x_0)$ does not, generally speaking, exist: its elements are, by somewhat loose analogy with the general definition of the homotopy groups given below, the pathwise connected components of the space M, from amongst which there is distinguished a "trivial" element, namely the component containing the base point x_0; however only in certain cases does this set come endowed with a natural group structure. The two most important such instances are as follows:

(a) If M is a Lie group (see §2.1 for the definition) then the component M_0 of the identity element $x_0 = 1$, is a normal subgroup, so that the set $\pi_0(M, x_0) = M/M_0$ does have a natural group structure, namely that of the quotient group.

For example if $M = O(n)$, then $\pi_0(M, x_0) \simeq \mathbb{Z}_2$, since two matrices from $O(n)$ are connected precisely if their determinants are equal (see §4.4 of Part I). In the case $M = O(1, n)$, we have $\pi_0(M, x_0) \simeq \mathbb{Z}_2 \oplus \mathbb{Z}_2$ since two elements of $O(1, n)$ are connected precisely if their determinants are equal and either they both preserve or both reverse the direction of flow of the time (cf. §6.2 of Part I).

(b) The *loop space* $M=\Omega(x_0, N)$ of a space N (relative to a point x_0) has as its points the paths γ in N beginning and ending at the point x_0, with the natural topology on them (defined for instance by the maximum distance between such paths with respect to a metric on N (cf. §10.1)). From §12.1 (see in particular Theorem 12.1.2) it follows that two points (i.e. loops in N with base point x_0) in the loop space are connected if and only if they are homotopic, so that the set $\pi_0(M, e)$ (where e denotes the constant path $\gamma(t)\equiv x_0$) coincides with the group $\pi_1(N, x_0)$ (by definition of the fundamental group).

We are now ready to embark on the definition of the "higher" homotopy groups $\pi_i(M, x_0)$; the reader will see that it is simply a generalization of the definition of $\pi_1(M, x_0)$, with the i-dimensional closed disc (or closed ball) D^i in the role played by the interval $[0, 1]\cong D^1$ in the definition of the fundamental group.

21.1.1. Definition. An *element of the homotopy group* $\pi_i(M, x_0)$ is a homotopy class of maps $D^i\to M$, which send the boundary S^{i-1} of the disc, to the point x_0 (and where the homotopies permitted are all relative to the point x_0, i.e. S^{i-1} is to be sent to x_0 throughout such homotopies).

It follows that each element of $\pi_i(M, x_0)$ is determined by a homotopy class of maps $S^i\to M$ which send a prescribed point s_0 of the sphere S^i to x_0; clearly this provides an equivalent characterization of the elements of $\pi_i(M, x_0)$. (By analogy with the case of the loop space introduced above, we may therefore say that the elements of the group $\pi_i(M, x_0)$ are just the connected components of the "mapping space" of maps $S^i\to M$ for which $s_0\mapsto x_0$.)

Having defined the "carrier" of the group $\pi_i(M, x_0)$, the next step is to introduce the operation justifying our calling it a "group". Consider the i-dimensional sphere S^i and identify S^{i-1} in the normal way with the equator of S^i. Choose the point s_0 to be on the equator S^{i-1}, and consider the obvious map ψ from S^i to the bouquet of two spheres $S_1^i \vee S_2^i$, sending the equator to the single point (also denoted by s_0) at which the bouquet is gathered, i.e. to the point common to S_1^i and S_2^i (see Figure 77). It is clear that at all points of S^i except those on the equator, ψ is one-to-one and orientation-preserving. If now we are given two maps α: $S_1^i\to M$, $\alpha(s_0)=x_0$, and β: $S_2^i\to M$, $\beta(s_0)=x_0$, then we define the *product map* $\alpha\beta$ to be the map $S^i\to M$ coinciding with $\alpha\circ\psi$ on the upper hemisphere $D+$, and with $\beta\circ\psi$ on the hemisphere D^-:

$$\alpha\beta(x)=\begin{cases}\alpha\psi(x) & \text{for } x\in D^+,\\ \beta\psi(x) & \text{for } x\in D^-.\end{cases} \tag{1}$$

It is obvious that $\alpha\beta(s_0)=x_0$, and easy to see that the product of two maps homotopic to α and β respectively, is homotopic to $\alpha\beta$ (where the permitted

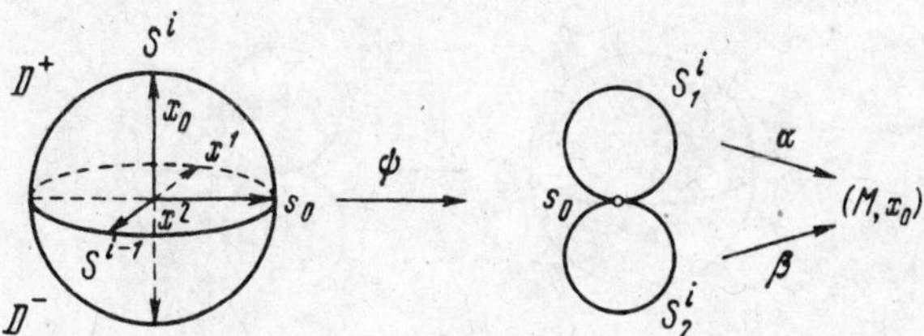

Figure 77

homotopies are, as usual, those throughout which $s_0 \mapsto x_0$). Hence we may define the *product of two homotopy classes of maps* $S^i \to M$, $s_0 \mapsto x_0$, as the homotopy class containing the product of any two representative maps from the respective classes.

21.1.2. Theorem. *Under the operation of multiplication of homotopy classes of maps $S^i \to M$, $s_0 \mapsto x_0$, the set $\pi_i(M, x_0)$ of such classes is a group, which is, moreover, commutative for $i > 1$.*

PROOF. Since we have already proven this result for $i = 1$ (Theorem 17.1.4), we may assume that $i > 1$.

(i) We first prove that the operation is commutative, i.e. that $\alpha\beta$ is homotopic to $\beta\alpha$. As usual we consider the i-sphere S^i to be the hypersurface $\sum_{j=0}^{i} (x^j)^2 = 1$ in $\mathbb{R}^{i+1}$ $(x^0, x^1, \ldots, x^i)$. We take the equator to be the set of points on this sphere with $x^0 = 0$, and take $s_0 = (0, 1, 0, \ldots, 0)$. Consider the rotation f_φ of the sphere through the angle φ, $0 \leq \varphi \leq \pi$, which rotates the (x^0, x^2)-plane (through φ) and leaves the orthogonal complement of that plane pointwise fixed. For $\varphi = 0$ this is of course the identity map, while the rotation corresponding to $\varphi = \pi$ interchanges D^+ and D^-. Hence the family of maps f_φ, $0 \leq \varphi \leq \pi$, determines a homotopy $F: I \times S^i \to M$ (where $I = [0, \pi]$) given by $F(\varphi, x) = \alpha\beta(f_\varphi(x))$ which in essence interchanges the two maps α and β. Hence $\alpha\beta$ and $\beta\alpha$ are homotopic.

(ii) We next establish associativity, i.e. that $(\alpha\beta)\gamma$ is homotopic to $\alpha(\beta\gamma)$. With the sphere S^i as in part (i) above, divide the lower hemisphere $D^-(x^0 \leq 0)$ into two halves: $D^- = D_1^- \cup D_2^-$, where $x^1 \leq 0$ on D_1^- and $x^1 \geq 0$ on D_2^-. Define in the intuitively obvious way a map ψ from S^i to a bouquet of three i-spheres, and then map these in turn to M via the given maps α, β, γ respectively (see Figure 78). It is then readily seen that the resulting composite map $S^i \to M$ is

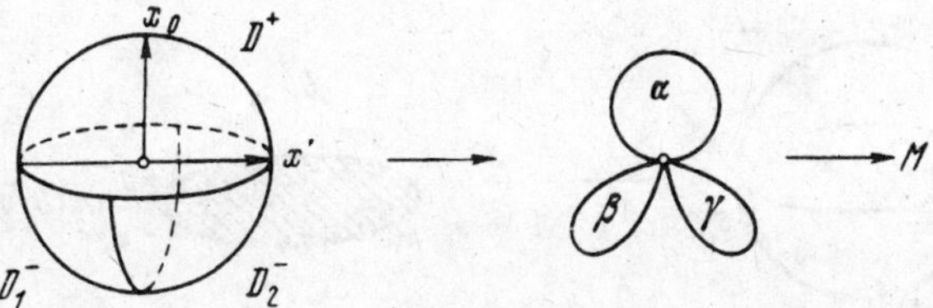

Figure 78

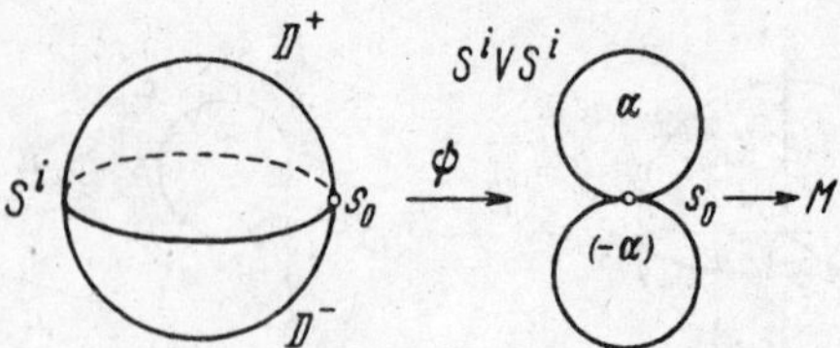

Figure 79

in the homotopy class of both $(\alpha\beta)\gamma$ and $\alpha(\beta\gamma)$. (We leave the precise details to the reader.)

(iii) Finally we prove that inverses exist. Given any map $\alpha: (S^i, s_0) \to (M, x_0)$, we shall show that the homotopy class of the map $\bar{\alpha}: (S^i, s_0) \to (M, x_0)$, defined by

$$\bar{\alpha}: (x^0, x^1, \ldots, x^i) \mapsto \alpha(-x^0, x^1, \ldots, x^i),$$

is inverse (with respect to the above-defined operation on $\pi_i(M, x_0)$) to the class containing α. To do this we compute the product $\alpha\bar{\alpha}$ using, as before, the map ψ: $S^i \to S_1^i \vee S_2^i$. The map $\alpha\psi$ is considered on the upper hemisphere $D^+(x^0 \geq 0)$, while the map $\bar{\alpha}\psi$ is considered on the lower hemisphere D^- where it is given by the formula $\bar{\alpha}\psi(x^0, x^1, \ldots, x^i) = \alpha\psi(-x^0, x^1, \ldots, x^i)$ (see Figure 79). The maps $\alpha\psi$ and $\bar{\alpha}\psi$ combine (as in (1)) to yield the product $\alpha\bar{\alpha} = f: S^i \to M$, which sends the points $y = (x^0, \ldots, x^i)$ and $y^* = (-x^0, \ldots, x^i)$ to the same point of M, i.e. $f(y) = f(y^*)$. It follows that f can be expressed as a composite $f = g \circ \pi$, where π: $S^i \to D^i$ is the projection map, so that $\pi(y) = \pi(y^*)$ (see (Figure 80), and $g = \alpha\psi$: $D^i \to M$ (identifying D^+ with D^i). Hence f is homotopic to the constant map (by means of a homotopy throughout which s_0 is mapped to x_0). This completes the proof of the theorem. □

At this stage there will, naturally enough, be only a few spaces whose higher homotopy groups $\pi_i(M, x_0)$, $i > 1$, we can calculate. For instance:

(a) $\pi_i(M, x_0) = 0$ (in additive notation since the groups are abelian for $i > 1$) for any contractible topological space (e.g. for $M = \mathbb{R}^n$, D^n, a tree, etc.);
(b) $\pi_i(S^n) = 0$ for $i < n$ (see the remark in §17.5), and $\pi_n(S^n) \simeq \mathbb{Z}$ (see §13.3).

We may however easily enlarge our supply of examples by means of the following simple result.

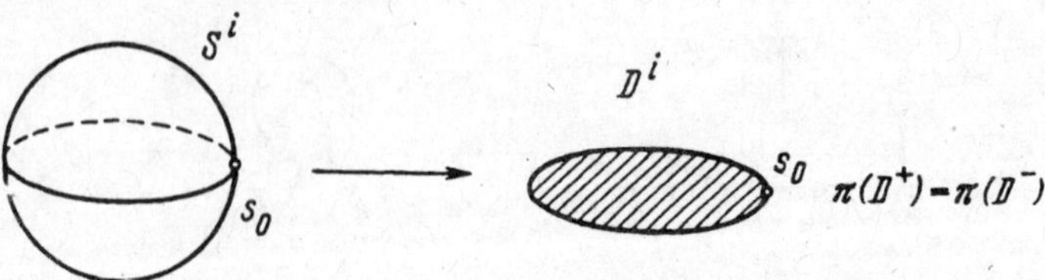

Figure 80

21.1.3. Proposition. *For a direct product $M \times N$ of topological spaces we have*

$$\pi_i(M \times N) \simeq \pi_i(M) \times \pi_i(N).$$

PROOF. Any map $f: S^i \to M \times N$ is determined by its component maps $f_1: S^i \to M$ and $f_2: S^i \to N$, obtained by projecting; thus $f = (f_1, f_2)$. The desired conclusion is in essence a consequence of the fact that under a homotopy of the map f, the components f_1 and f_2 undergo deformation independently. □

21.2. Relative Homotopy Groups. The Exact Sequence of a Pair

Given a topological space, a non-empty subset A of M, and a point x_0 of A, the ith *relative homotopy group* $\pi_i(M, A, x_0)$, $i \geq 1$, is defined as follows. Its elements are the (relative) homotopy classes of maps $\alpha: D^i \to M$, which send the boundary S^{i-1} of D^i to A, and a prescribed point s_0 of that boundary to x_0; thus we might denote such maps more explicitly by

$$\alpha: (D^i, S^{i-1}, s_0) \to (M, A, x_0).$$

We shall now define for $i \geq 2$ a binary operation on the set $\pi_i(M, A, x_0)$ under which it is a group, in fact for $i \geq 3$ an abelian group. (In the case $i = 1$, i.e. for $\pi_1(M, A, x_0)$, there is, generally speaking, no natural group operation.) The definition of the operation in question is completely analogous to that of the group multiplication in the "absolute" homotopy groups $\pi_i(M, x_0)$, given in the preceding subsection. Thus if $\alpha, \beta \in \pi_i(M, A, x_0)$, then the product $\alpha\beta$ is the map of the disc D^i (given in $\mathbb{R}^i$ by the inequality $\sum_{j=1}^{i} (x^j)^2 \leq 1$) defined as follows. As before we denote by ψ the map $D^i \to D_1^i \vee D_2^i$ which pinches the "waist" $D^{i-1}(x^1 = 0)$ of the disc D^i to the single point s_0 (see Figure 81). If, as indicated in that diagram, we realize α and β as maps of the discs D_1^i and D_2^i respectively, then the composite map

$$D^i \xrightarrow{\psi} D_1^i \vee D_2^i \to M$$

(cf. (1) above) is taken as defining the product $\alpha\beta$ (which clearly does represent

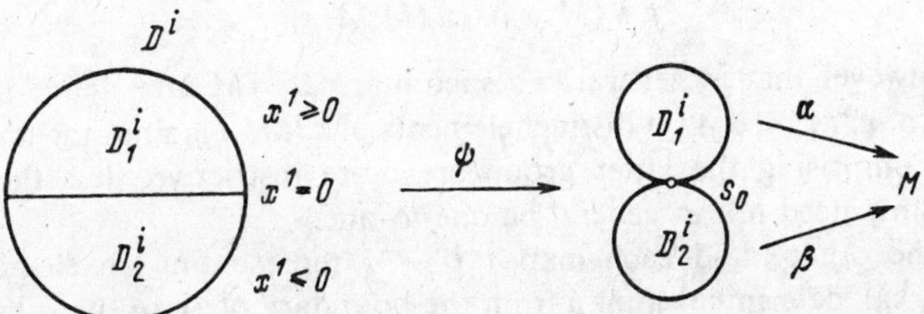

Figure 81

an element of $\pi_i(M, A, x_0)$). (Note that this definition does not work in the case $i=1$.)

In the case $i=2$ the boundary $\partial D^2 = S^1$ is one-dimensional. Hence for appropriate A, essentially since S^1 is always to be mapped to A (i.e. essentially for the same reason that $\pi_1(M, x_0)$ need not be commutative), the above-defined operation on $\pi_2(M, A, x_0)$ may also fail to be commutative. However by imitating the proof of Theorem 21.1.2 it can without difficulty be shown that under the operation, for each $i \geq 2$ the set $\pi_i(M, A, x_0)$ is a group, which is commutative for $i \geq 3$; we leave the precise details to the reader. (Note incidentally that if $A = \{x_0\}$ then the relative groups $\pi_i(M, A, x_0)$ become just the "absolute" homotopy groups $\pi_i(M, x_0)$.)

Analogously to the case of the fundamental group (see Theorem 17.1.5), any continuous map

$$f\colon M \to N, \quad A \to B, \quad x_0 \mapsto y_0,$$

between manifolds (or more general topological spaces) M and N induces in the natural way a homomorphism

$$f_*\colon \pi_i(M, A, x_0) \to \pi_i(N, B, y_0), \tag{2}$$

associating with (the homotopy class of) each map $D^i \to M$, (the homotopy class of) the composite map

$$D^i \to M \xrightarrow{f} N.$$

It is easy to see (as it was for the fundamental group) that this homomorphism is unaffected by homotopies of $f\colon M \to N$ throughout which A continues to be mapped to B and x_0 to y_0.

Each map $D^i \to M$ under which the boundary $\partial D^i = S^{i-1}$ is sent to the point x_0, determines, of course, an element of $\pi_i(M, x_0)$, and also determines an element of $\pi_i(M, A, x_0)$. Since two such maps defining the same element of $\pi_i(M, x_0)$ (i.e. homotopic via a homotopy throughout which S^{i-1} is sent to x_0) certainly define the same element of $\pi_i(M, A, x_0)$, we obtain a homomorphism

$$j\colon \pi_i(M, x_0) \to \pi_i(M, A, x_0).$$

(Note however that in general two such maps $D^i \to M$ may define the same element of $\pi_i(M, A, x_0)$ yet distinct elements of $\pi_i(M, x_0)$, since the homotopy classes comprising the latter group are more restrictive; thus the homomorphism j need not in general be one-to-one.)

On the other hand each map $f\colon D^i \to M$ representing an element α of $\pi_i(M, A, x_0)$, determines a map from the boundary of D^i to the subspace A:

$$f|_{\partial D^i}\colon S^{i-1} \to A, \qquad s_0 \mapsto x_0.$$

It is clear that homotopies of f throughout which S^{i-1} continues to be sent to A and s_0 to x_0 (i.e. which keep f in the homotopy class α), induce homotopies of the restricted map $f|_{\partial D^i}$ throughout which s_0 continues to be sent to x_0. Hence restriction of the maps $f\colon D^i \to M$ representing elements of $\pi_i(M, A, x_0)$ yields a mapping

$$\partial\colon \pi_i(M, A, x_0) \to \pi_{i-1}(A, x_0),$$

which is in fact a homomorphism (called the the "boundary homomorphism") since, as is easily seen, the restriction of a product of maps $f, g\colon D^i \to M$ (sending S^{i-1} to A and s_0 to x_0) is the product of the restrictions $f|_{\partial D^i}$ and $g|_{\partial D^i}$.

Finally, the inclusion map $i\colon A \to M$, which maps A identically, gives rise in the usual way (see (2)) to a corresponding "inclusion homomorphism"

$$i_*\colon \pi_i(A, x_0) \to \pi_i(M, x_0).$$

Before formulating our result concerning the homomorphisms j, i_*, ∂, we remind the reader that the *kernel*, denoted by Ker φ, of a group homomorphism $\varphi\colon G \to H$ is the normal subgroup of G consisting of those elements α of G such that $\varphi(\alpha) = 1$, while Im φ denotes the image group $\varphi(G)$ (a subgroup of H).

21.2.1. Theorem. *The homomorphisms j, i_*, ∂ satisfy the following "exactness" conditions:*

$$\operatorname{Ker} j = \operatorname{Im} i_*;$$

$$\operatorname{Ker} i_* = \operatorname{Im} \partial;$$

$$\operatorname{Ker} \partial = \operatorname{Im} j;$$

or, in other words, the sequence

$$\cdots \xrightarrow{\partial} \pi_i(A, x_0) \xrightarrow{i_*} \pi_i(M, x_0) \xrightarrow{j} \pi_i(M, A, x_0) \xrightarrow{\partial} \pi_{i-1}(A, x_0) \to \cdots$$

of groups and homomorphisms, is "exact".

PROOF. (i) Ker j = Im i_*. We show first that Ker $j \subset$ Im i_*. Each element of $\pi_i(M, x_0)$ is represented by a map $\alpha\colon D^i \to M$ satisfying $\alpha(\partial D^i) = x_0$. The condition for such an element to lie in Ker j is (by definition) that there exist a homotopy $\alpha_t (0 \le t \le 1)$ with $\alpha_0 = \alpha$ and $\alpha_1(D^i) \subset A$ (and throughout which of course $\partial D^i \to A$ and $s_0 \mapsto x_0$). By restricting this homotopy, we get a homotopy $\partial D^i \times [0, 1] \to A$, which, since $\alpha_0(\partial D^i) = x_0$, yields a map $D^i \to A$ (by pinching the base of the cylinder $\partial D^i \times [0, 1]$ to the single point s_0). By combining this map with the map $\alpha_1\colon D^i \to A$ (i.e. by re-capping the (pinched) cylinder with D^i and applying α_1 to this cap), we obtain a map $S^i \to A$ which sends s_0 to x_0, and also represents an element of Im i_*. Since this map also (as is not difficult to see) represents the same element of $\pi_i(M, x_0)$ as does α, the desired inclusion follows.

The reverse inclusion is a consequence of the fact that any map $f\colon D^i \to A$

satisfying $f(\partial D^i)=x_0$, represents the identity element in $\pi_i(M, A, x_0)$, since clearly f is homotopic to the map sending D^i to the point x_0, via a homotopy throughout which ∂D^i (in fact D^i) continues to be mapped to A (and s_0 to x_0) (i.e. since the disc can be contracted within A to the point x_0).

(ii) Ker $\partial=$ Im j. To see that Ker $\partial \subset$ Im j, consider any element of Ker ∂. A representative map $\alpha: D^i \to M$ (for which of course $S^{i-1} \to A$, $s_0 \mapsto x_0$) of such an element will have the property that its restriction $\alpha|_{S^{i-1}}: S^{i-1} \to A$ to the boundary of D^i is null-homotopic, i.e. there is a homotopy $\alpha_t: S^{i-1} \to A$ $(0 \leq t \leq 1)$ with $\alpha_0=\alpha|_{S^{i-1}}$, $\alpha_t(s_0)=x_0$, $\alpha_1(S^{i-1})=x_0$. Since $\alpha_1(S^{i-1})=x_0$, this homotopy yields a map from a disc D^i (obtained by pinching to the point $s_0=(s_0, 1)$ the top rim $S^{i-1}\times 1$ of the cylinder $S^{i-1}\times[0,1]$) to A. By combining this map with the map $\alpha: D^i \to M$ (i.e. by filling in the base of the (pinched) cylinder with D^i and applying α to it), we obtain a map $S^i \to M$ sending the distinguished point s_0 to x_0, and therefore representing an element of Im j. Since this map clearly represents the same element of $\pi_i(M, A, x_0)$ as does α, the desired inclusion follows.

The reverse inclusion is easy since under any map $\alpha: D^i \to M$ representing an element of Im j, the boundary of D^i is mapped to a point, so that the boundary map applied to such an element yields the identity element of $\pi_{i-1}(A, x_0)$.

(iii) Im $\partial=$ Ker i_*. Consider an element of Ker $i_* \subset \pi_i(A, x_0)$, represented by a map $\alpha: S^i \to A$, with $s_0 \mapsto x_0$. By definition of Ker i_* there must exist a homotopy $\alpha_t: S^i \to M$ $(0 \leq t \leq 1)$ such that $\alpha_0=\alpha$, $\alpha_t(s_0)=x_0$, and $\alpha_1(S^i)=x_0$. Since $\alpha_1(S^i, 1)=x_0$ (i.e. α_1 pinches the top of the cylinder $S^i\times[0,1]$ to a point), this homotopy yields a map $F: D^{i+1} \to M$ of the $(i+1)$-dimensional disc, under which the boundary $\partial D^{i+1}=S^i=(S^i, 0)$ goes (via $\alpha=\alpha_0$) to A, and s_0 to x_0. Hence F represents an element of $\pi_{i+1}(M, A, x_0)$, and since ∂F is just α, we conclude that Ker $i_* \subset$ Im ∂.

For the reverse inclusion observe that if $F \in \pi_{i+1}(M, A, x_0)$, then $\partial F: S^i \to A$ is homotopic (in M) to the constant map to the point x_0, via a homotopy throughout which $s_0 \mapsto x_0$. This concludes the proof of the theorem. □

Example. Taking $M=D^n$ and $A=\partial D^n=S^{n-1}$, we have $\pi_n(D^n, S^{n-1}, x_0)\simeq \mathbb{Z}$, and for $i<n$, $\pi_i(D^n, S^{n-1}, x_0)=0$.

To see this consider the exact sequence

$$\pi_n(D^n) \xrightarrow{j} \pi_n(D^n, S^{n-1}) \xrightarrow{\partial} \pi_{n-1}(S^{n-1}) \xrightarrow{i_*} \pi_{n-1}(D^{n-1}).$$

Since for all $i\geq 0$ the ball D^i is contractible, we have for $n>1$

$$\pi_n(D^n)=\pi_{n-1}(D^{n-1})=0,$$

which forces Im $j=0$ and Im $i_*=0$. Since by the above theorem Im $j=$ Ker ∂, we deduce that also Ker $\partial=0$, i.e. the homomorphism $\partial: \pi_n(D^n, S^{n-1}) \to \pi_{n-1}(S^{n-1})$ is actually a monomorphism (i.e. one-to-one). Since, again invoking the above theorem, Im $\partial=$ Ker $i_*=\pi_{n-1}(S^{n-1})$, the map ∂ is onto, and is therefore an isomorphism between $\pi_n(D^n, S^{n-1})$ and $\pi_{n-1}(S^{n-1})$. Since $\pi_{n-1}(S^{n-1}) \simeq \mathbb{Z}$ (see §13.3) it follows that $\pi_n(D^n, S^{n-1}) \simeq \mathbb{Z}$ as required. We leave the proof of the (easy) second statement to the reader.

A similar argument shows that, more generally, if M is contractible then since $\pi_j(M)=0$ for all $j \geq 0$, we have $\pi_n(M, A) \simeq \pi_{n-1}(A)$ for $n \geq 1$. (Verify!)

§22. Covering Homotopies. The Homotopy Groups of Covering Spaces and Loop Spaces

22.1. The Concept of a Fibre Space

Let X and Y be topological spaces and $f: X \to Y$ a continuous map. (In some of the examples considered below X will be an infinite-dimensional function space of a certain kind, called a "path space".) For any smooth manifold (or more general topological space) K and mappings

$$\varphi: K \to Y, \qquad \tilde{\varphi}: K \to X,$$

we say that $\tilde{\varphi}$ *covers* φ if $f\tilde{\varphi} = \varphi$.

22.1.1. Definition. Given a continuous map $f: X \to Y$ between topological spaces X and Y, we call the triple (X, f, Y) a *fibre space* (or *Serre fibration*) *with respect to a topological space* K, if for any homotopy $\Phi = \{\varphi_t\}: K \times I \to Y$ $(0 \leq t \leq 1)$ of any map $\varphi: K \to Y$ covered by a prescribed map $\tilde{\varphi}: K \to X$, there exists a homotopy $\tilde{\Phi} = \{\tilde{\varphi}_t\}: K \times I \to X$, of $\tilde{\varphi}$ (i.e. $\tilde{\varphi}_0 = \tilde{\varphi}$), which covers Φ (i.e. $f\tilde{\varphi}_t = \varphi_t$, $0 \leq t \leq 1$), and which has the further property that it is "stationary" whenever φ_t is; i.e. any $k \in K$ which is held constant by φ_t for some segment δ of $[0, 1]$ (in symbols $\varphi_t(k) = \text{const.}$ for $t \in \delta$), is also held constant by $\tilde{\varphi}_t$ for t in that segment (i.e. $\tilde{\varphi}_t(k) = \text{const.}$ for all $t \in \delta$). Such a homotopy $\tilde{\Phi}$ is called a *covering homotopy* for Φ, and we say also that the map f has the *covering homotopy property* with respect to K. The space Y is called the *base space*, the space X the *total space*, and f the *projection* of the fibre space. For each point $y \in Y$ the complete inverse image $F_y = f^{-1}(y)$ is called the *fibre* above y.

In those cases of a map with the covering homotopy property which actually arise in practice, there is usually given a precise prescription for obtaining the points of X covering the positions of a point moving in the base space Y; this prescription is required to depend continuously and multiplicatively (i.e. must respect multiplication of paths) on the path traced out by the point moving in Y (under the influence of some homotopy), and on the initial position of the point. The conditions are more precisely as follows.

(i) With each continuous path $\gamma(t)$: $[0, 1]\to Y$ in the base space Y, and initial point $x_0 \in X$ satisfying $f(x_0)=y_0=\gamma(0)$, there is associated a unique continuous path $\tilde{\gamma}(t, x_0)$: $[0, 1]\to X$, satisfying $\tilde{\gamma}(0, x_0)=x_0$ and $f\tilde{\gamma}(t, x_0)=\gamma(t)$. The path $\tilde{\gamma}(t, x_0)$ (obtained by "lifting" $\gamma(t)$) is required to depend continuously on the path $\gamma(t)$ and the choice of the initial point x_0.

(ii) The prescription should respect multiplication of paths; i.e. the product of two paths γ_1, γ_2 in the base space Y (with initial point $x_0 \in X$, say) should lift under the prescription to the product $\tilde{\gamma}_1(t, x_0)\circ\tilde{\gamma}_2(\tau, x_1)$ (where $0\le t, \tau\le 1$), provided the initial point x_1 given along with γ_2 is $x_1=\tilde{\gamma}_1(1, x_0)$. (Note that we have now changed over to writing a product of paths in the order left-to-right.)

(iii) The lift $\tilde{\gamma}$ of the constant path $\gamma(t)=y_0$, $0\le t\le 1$, should be the constant path $\tilde{\gamma}(t)=x_0$, $0\le t\le 1$.

Observe that (given a prescription satisfying these three conditions) for each path $\gamma(t)$ joining y_0 to y_1 in Y, the totality of paths $\tilde{\gamma}(t, x)$ where x ranges over $f^{-1}(y_0)$, affords us in the obvious way a map (denoted by $\tilde{\gamma}$: $f^{-1}(y_0)\to f^{-1}(y_1)$) which "transports" the points of the fibre $f^{-1}(y_0)$ to those of $f^{-1}(y_1)$, and which has the properties that $(\widetilde{\gamma_1\circ\gamma_2})=\tilde{\gamma}\circ\tilde{\gamma}_2$, and the self-map $\tilde{\gamma}\circ\tilde{\gamma}^{-1}$ of the fibre $F=f^{-1}(y_0)$ is homotopic to the identity map 1_F.

Exercise

Prove that this "transport" map between fibres is a homotopy equivalence (so that if X and Y are path-connected spaces the fibres of the fibre space are all homotopically equivalent).

22.1.2. Definition. A prescription satisfying the above three conditions is called a *homotopy connexion* for the fibre space.

Examples (a) A covering map (see §18.1) defines a fibre space having all of its fibres discrete. The covering homotopy property and transport between points of fibres were discussed for this case in §§18.1 and 19.1 respectively.

(b) Given a (connected) smooth manifold M (or more general topological space) and a point x_0 of M, we denote by $X=E(x_0)$ the space of all paths $\gamma(t)$, $0\le t\le 1$, beginning at x_0 and ending at any point of M (i.e. $\gamma(1)$ is to vary arbitrarily with γ). (We leave it to the reader to furnish the precise definition of the natural topology on $E(x_0)$ in terms of that on M (see §10.1).) Taking $M=Y$ we obtain (as shall be proved below) a fibre space (X, f, Y) where f: $E(x_0)\to M$ (i.e. $X\to Y$) is defined by $f(\gamma)=\gamma(1)$. For each point y of M the fibre $f^{-1}(y)$ above y is called the *path space* (between x_0 and y), denoted by $\Omega(x_0, y, M)$. (Recall that this notation for the set of paths from x_0 to y in M was introduced in §17.1.) As promised we now prove the

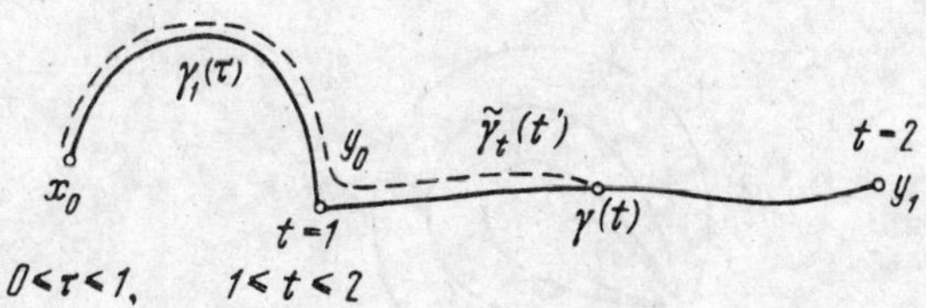

Figure 82

22.1.3. Lemma. *The triple* $(E(x_0), f, M)$ *is a fibre space.*

PROOF. It suffices to construct a homotopy connexion for the map f. Thus let $\gamma(t)$ be any path in M from y_0 to y_1, where we may suppose γ is parametrized so that $1 \leq t \leq 2$ (with $\gamma(1) = y_0$ and $\gamma(2) = y_1$). Take as prescribed initial point of $X = E(x_0)$ covering y_0, any path $\gamma_1(\tau)$, $0 \leq \tau \leq 1$, beginning at x_0 (since it belongs to $E(x_0)$) and ending at y_0 (since it belongs to $f^{-1}(y_0)$). Our prescription for obtaining the covering of the path $\gamma(t)$ by a path in the space $X = E(x_0)$ is as follows: for each t, $1 \leq t \leq 2$, we take as the point in X covering the point $\gamma(t)$, the path (i.e. point of $E(x_0)$)$\tilde{\gamma}_t(t') \in \Omega(x_0, \gamma(t), M)$, $0 \leq t' \leq t$, defined by the formulae

$$\tilde{\gamma}_t(t') = \gamma_1(t'), \qquad 0 \leq t' \leq 1,$$

$$\tilde{\gamma}_t(t') = \gamma(t'), \qquad 1 \leq t' \leq t.$$

(See Figure 82 where the path $\tilde{\gamma}_t(t')$ is indicated by the dotted line.) For each particular value of t $(1 \leq t \leq 2)$, we can re-parametrize the path $\tilde{\gamma}_t(t')$ by the parameter $t'' = t'/t$; then as t' varies from 0 to t, the parameter t'' varies from 0 to 1.

Since this covering of the path $\gamma(t)$ in the base space $Y = M$ by the path $\tilde{\gamma}_t$ in the space $X = E(x_0)$ clearly depends continuously on the initial point (i.e. on the path $\gamma_1(\tau)$ above y_0) and on the path $\gamma(t)$ in the base space, and since the other two properties of a homotopy connexion are easily verified, the lemma follows. □

22.2. The Homotopy Exact Sequence of a Fibre Space

Let (X, f, Y) be a fibre space, let $y_0 \in Y$, and choose any point f_0 in the fibre $F = f^{-1}(y_0)$ above y_0. Then (see §21) the homotopy groups $\pi_i(X, f_0)$, $\pi_i(F, f_0)$, $\pi_i(X, F, f_0)$ can be arranged naturally in the exact sequence (corresponding to the pair X, F)

$$\cdots \to \pi_i(F) \xrightarrow{i_*} \pi_i(X) \xrightarrow{j} \pi_i(X, F) \xrightarrow{\partial} \pi_{i-1}(F) \to \cdots. \tag{1}$$

Since under the projection $f\colon X \to Y$, the fibre $F = f^{-1}(y_0)$ is of course sent to the single point y_0, we obtain a homomorphism

$$f_*\colon \pi_i(X, F, f_0) \to \pi_i(Y, y_0).$$

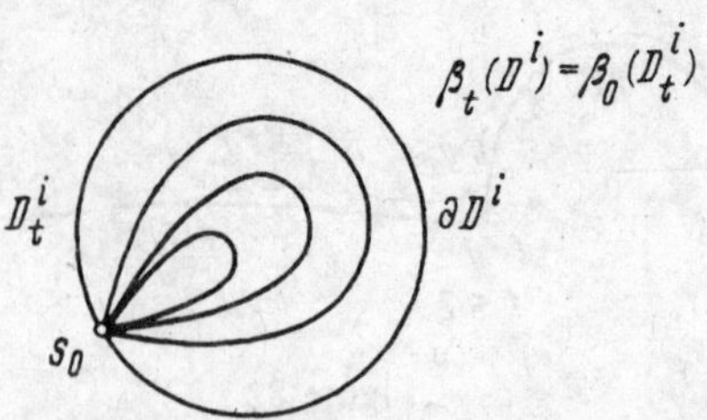

Figure 83

22.2.1. Theorem. *The homomorphism* f_* *is an isomorphism:* $\pi_i(X, F, f_0) \simeq \pi_i(Y, y_0)$. *(Hence the exact sequence* (1) *above yields the following "homotopy exact sequence of the fibration"* (X, f, Y)*:*

$$\cdots \to \pi_i(F) \xrightarrow{i_*} \pi_i(X) \xrightarrow{f_* \circ j} \pi_i(Y) \xrightarrow{\partial} \pi_{i-1}(F) \to \cdots, \tag{2}$$

where $\pi_i(X, F)$ *has been identified with* $\pi_i(Y)$.)

PROOF. We first show that the kernel of f_* is trivial (whence f_* is one-to-one). Thus consider any element $\hat{\alpha} \in \pi_i(X, F, f_0)$ such that $f_*(\hat{\alpha})$ is the identity element of $\pi_i(Y, y_0)$, and let $\alpha: D^i \to X$ (with $\partial D^i \to F$, $s_0 \mapsto f_0$) be a map representing the relative homotopy class $\hat{\alpha}$. If we write $\beta = f\alpha$, then since β represents $f_*(\hat{\alpha})$ which is the identity element of $\pi_i(Y, y_0)$, there must exist a homotopy $\beta_t: D^i \to Y$ satisfying $\beta_t(\partial D^i) = y_0$, $\beta_0 = \beta$, and $\beta_1(D^i) = y_0$. By definition of a fibre space, we can cover the homotopy β_t with a homotopy $\alpha_t: D^i \to X$, satisfying $\alpha_0 = \alpha$, $\alpha_t(s_0) = f_0$. Since $\beta_t(\partial D^i) = y_0$ and $\beta_1(D^i) = y_0$, we must then have $\alpha_t(\partial D^i) \subset F$ (for all t) and $\alpha_1(D^i) \subset F$, so that α represents the identity element of the group $\pi_i(X, F)$, as required.

To show that f_* is onto we construct for each $\beta: D^i \to Y$ with $\beta(\partial D^i) = y_0$, a map $\alpha: D^i \to X$ (with $\alpha(\partial D^i) \subset F$, $\alpha(s_0) = f_0$) such that $\beta = f\alpha$. Now since the disc is contractible, there exists a homotopy $\beta_t: D^i \to Y$ $(0 \le t \le 1)$ satisfying $\beta_0 = \beta$, $\beta_t(s_0) = y_0$, and $\beta_1(D^i) = y_0$ (see Figure 83). (Note that we certainly cannot have $\beta_t(\partial D^i) = y_0$ throughout this homotopy unless β represents the identity element of $\pi_i(Y, y_0)$.) We now cover the map $\beta_1: D^i \to y_0$, by the map $\alpha_1: D^i \to f_0$, and with this as starting point cover the whole homotopy β_t (in reverse) by a homotopy α_t, $1 \ge t \ge 0$, at the conclusion of which we shall have a map $\alpha_0: D^i \to X$ satisfying $\alpha(s_0) = f_0$, and $f\alpha_0 = \beta_0 = \beta$. (Here we are invoking the properties of a fibre space.) Since $\beta(\partial D^i) = y_0$, we must have $\alpha_0(\partial D^i) \subset F$, whence we see that we have found an element of $\pi_i(X, F)$ (namely that represented by α_0) which is a preimage of β under f_*. This completes the proof of the theorem. □

Remark. In view of this theorem we can define a convenient group structure on the set $\pi_1(X, F, f_0)$ by means of the bijection $f_*: \pi_1(X, F, f_0) \to \pi_1(Y, y_0)$. Note also that in the final few terms of the exact sequence (2), namely

$$\cdots \to \pi_1(F, f_0) \xrightarrow{i_*} \pi_1(X, f_0) \xrightarrow{f_* \circ j} \pi_1(Y, y_0) \xrightarrow{\partial} \pi_0(F),$$

(where X and Y are assumed connected), it can happen that the image of the homomorphism $f_* \circ j\colon \pi_1(X, f_0) \to \pi_1(Y, y_0)$, not necessarily being the kernel of any succeeding group homomorphism, fails to be a normal subgroup of $\pi_1(Y, y_0)$, so that on the set of (right, say) cosets of that image group in $\pi_1(Y, y_0)$, no natural group structure is defined.

In the case where $f\colon X \to Y$ is a covering map, the fibre $F = f^{-1}(y_0)$ is discrete, so that for $i \geq 2$ any (continuous) map $D^i \to X$, with $\partial D^i \to F$, $s_0 \to f_0$, must in fact send ∂D^i to f_0, whence $\pi_i(X, F, f_0) = \pi_i(X, f_0)$. This and the above theorem together yield the

22.2.2. Corollary. *If $f\colon X \to Y$ is a covering map, then the discreteness of the fibre $F = f^{-1}(y_0)$ implies that*

$$\pi_i(X, F, f_0) = \pi_i(X, f_0) \quad \text{for } i \geq 2,$$

whence

$$\pi_i(X, f_0) \simeq \pi_i(Y, y_0) \quad \text{for } i \geq 2.$$

(The relationship between the fundamental groups $\pi_1(X, f_0)$ and $\pi_1(Y, y_0)$ was examined in §19.2.)

22.2.3. Corollary. *For the fibre space $(E(x_0), f, M)$ where $E(x_0)$ is the space of paths in M beginning at $x_0 \in M$, and the fibre $f^{-1}(y)$ above each $y \in M$ is the path space $\Omega(x_0, y, M)$, we have $\pi_i(E(x_0)) = 0$, whence*

$$\pi_i(\Omega(x_0, y, M)) \simeq \pi_{i+1}(M).$$

PROOF. The space $E(x_0)$ can be contracted over itself to a point (i.e. is contractible) by means of the homotopy $\varphi_t\colon E(x_0) \to E(x_0)(1 \geq t \geq 0)$ defined by

$$\varphi_t(\gamma(\tau)) = \gamma_t(\tau),$$

where γ_t is the segment of the path $\gamma(\tau)$ $(0 \leq \tau \leq 1)$ from 0 to t (with parameter $\tau' = \tau/t$, $0 \leq \tau' \leq 1$). At the conclusion of this homotopy we obtain the constant path $\gamma_0(\tau) \equiv \gamma(0) = x_0$, whence $\varphi_0(E(x_0))$ is indeed a point (of $E(x_0)$).

Hence $\pi_i(E(x_0), x_0) = 0$ for all $i \geq 0$, and the relevant segment of the homotopy exact sequence of the fibre space $(E(x_0), f, M)$ then looks like

$$\begin{array}{ccccccc} \pi_i(E(x_0)) & \xrightarrow{f_*} & \pi_i(M, y) & \xrightarrow{\partial} & \pi_{i-1}(\Omega(x_0, y, M)) & \xrightarrow{i_*} & \pi_{i-1}(E(x_0)). \\ \| & & & & & & \| \\ 0 & & & & & & 0 \end{array}$$

The exactness of this sequence implies that ∂ is an isomorphism, since $\operatorname{Im} f_* = \operatorname{Ker} \partial = 0$, and $\operatorname{Im} \partial = \operatorname{Ker} i_* = \pi_{i-1}(\Omega(x_0, y, M))$. This completes the proof of the corollary. □

Examples. By applying Corollary 22.2.2 to the examples of covering maps considered in §18.2, we obtain the following isomorphisms of homotopy groups for $i \geq 2$:

(a) $\pi_i(S^1, s_0) \simeq \pi_i(\mathbb{R}^1, x_0) = 0$ (Example (a) of §18.2).

(b) $\pi_i(\mathbb{R}P^2, x_0) \simeq \pi_i(S^2, s_0)$ (Example (c) of §18.2). In particular, therefore, $\pi_2(\mathbb{R}P^2, x_0) \simeq \mathbb{Z}$ (see the conclusion of §21.1).

(c) $\pi_i(T^n, x_0) = 0$ (Example (d) of §18.2).

(d) $\pi_i(K^2, x_0) = 0$ (Example (e) of §18.2).

(e) $\pi_i(S^1 \vee S^1) = 0$ (Example (g) of §18.2). Since in fact the universal covering space of a bouquet of any number of circles is a tree, we have more generally $\pi_i(S^1 \vee \cdots \vee S^1) = 0$ (for $i \geq 2$). If $U = \mathbb{R}^2 \setminus \{a_1, \ldots, a_k\}$, the plane with k punctures, then since U is homotopically equivalent to a bouquet of k circles, we have (for $i \geq 2$) $\pi_i(U) = 0$ also.

(f) From Example (f) of §18.2 it follows that for $i \geq 2$

$$\pi_i(S^2 \vee S^1) \simeq \pi_i(\cdots \vee S^2 \vee S^2 \vee \cdots),$$

since the universal covering space (described in that example) is the real line with 2-spheres attached at the integer points. Since the space $V = \mathbb{R}^3 \setminus S^1$ (i.e. $\mathbb{R}^3$ with an unknotted circle removed) is homotopically equivalent to $S^2 \vee S^1$ (see §17.5), its homotopy groups will be isomorphic to those of $S^2 \vee S^1$.

(g) For all 2-dimensional surfaces (both closed and non-closed) except $\mathbb{R}P^2$ and S^2, the universal covering space is (to within a homeomorphism) $\mathbb{R}^2$. (The proof of this for the sphere-with-g-handles ($g \geq 1$) was adumbrated in §20; see also Exercise 1 of §19.4.) Hence for all of these surfaces $\pi_i = 0$ for all $i \geq 2$.

22.3. The Dependence of the Homotopy Groups on the Base Point

We shall now elucidate the manner in which the higher homotopy groups $\pi_i(M, x_0)$ ($i \geq 2$) depend on the choice of the base point $x_0 \in M$. Let x_0 and x_1 be two points of M, and $\gamma(t)$, $1 \leq t \leq 2$, a path from x_1 to x_0 (i.e $x_0 = \gamma(2)$, $x_1 = \gamma(1)$). Let $\hat{\alpha}$ be any element of $\pi_i(M, x_1)$, and let $\alpha: D_1^i \to M$, $s_0 \mapsto x_1$ be any map of the unit disc D_1^i, representing $\hat{\alpha}$. We shall now define a corresponding map $\gamma^*(\alpha): D_2^i \to M$, of the disc of radius 2 (see Figure 84). On the region $1 \leq \sum (x_j)^2 \leq 4$ between the discs (which region can clearly be identified

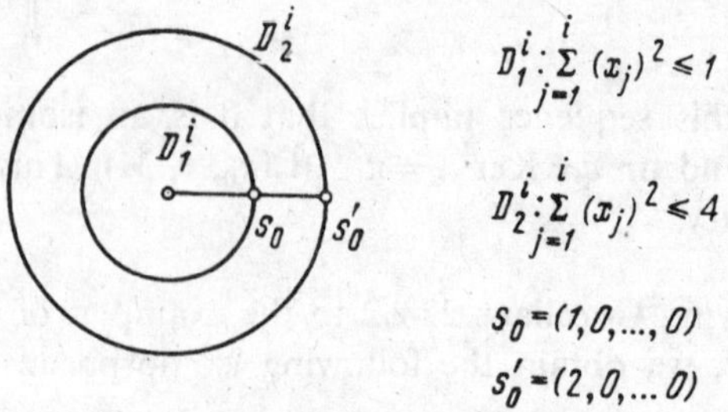

Figure 84

with $S^{i-1}\times[1,2]$) we define a map $\tilde{\gamma}: S^{i-1}\times[1,2]\to M$, by setting $\tilde{\gamma}(y,t)=\gamma(t)$ for all $y\in S^{i-1}$, $1\le t\le 2$. We then define $\gamma^*(\alpha): D_2^i\to M$ by setting

$$\gamma^*(\alpha)=\alpha \quad \text{on } D_1^i\subset D_2^i,$$

$$\gamma^*(\alpha)=\tilde{\gamma} \quad \text{on the region } \{1\le\sum(x_j)^2\le 4\}=S^{i-1}\times[1,2].$$

(Thus $\gamma^*(\alpha)$ maps ∂D_2^i (see Figure 84) to x_0.) We are now in a position to state the result we have been aiming at.

22.3.1. Theorem. (i) *The mapping $\alpha\mapsto\gamma^*(\alpha)$ (just defined) depends only on the homotopy class of the path γ joining x_0 and x_1, and defines an isomorphism*

$$\gamma^*: \pi_i(M,x_1)\to\pi_i(M,x_0).$$

Thus in particular for simply-connected spaces this isomorphism is independent of the path γ.

(ii) *If the path γ is closed, representing therefore an element of $\pi_1(M,x_0)$, then the correspondence $\alpha\mapsto\gamma^*(\alpha)$ defines an action of the group $\pi_1(M,x_0)$ on the group $\pi_i(M,x_0)$ (by means of automorphisms of the latter). If $f: X\to M$ is the universal covering map, determined by a freely acting, discrete group Γ of homeomorphisms of X (see §18.4), then the action of Γ on X induces an action of Γ on the group $\pi_i(X)\simeq\pi_i(M)$, which coincides (under the natural isomorphism $\Gamma\simeq\pi_1(M,x_0)$) with the action of $\pi_1(M,x_0)$ on $\pi_i(M,x_0)$. (Note also that in view of* (i) *above and the simple-connectedness of X (so that $\pi_1(X)=1$), the group $\pi_i(X,x)$ is independent of the choice of the base point x in the sense that for any two points x' and x'' the isomorphism $\pi_i(X,x')\to\pi_i(X,x'')$ of part* (i) *is independent of the path γ joining x' and x''.)*

(iii) *There is a natural one-to-one correspondence between the set of free homotopy classes of maps $S^i\to M$, and the set of orbits of the action of the group $\pi_1(M,x_0)$ on $\pi_i(M,x_0)$. (Thus in particular if $\pi_1(M,x_0)=1$, there is a one-to-one correspondence between the free homotopy classes of maps $S^i\to M$ and the elements of $\pi_i(M,x_0)$.)*

PROOF. Part (i) and the first statement of (ii) are established by an argument similar to that used in proving Theorem 17.2.1, while the argument for (iii) is similar to that of Theorem 17.3.1. (These two theorems are the analogues (for the fundamental groups) of the corresponding parts of the present theorem; the proofs used for the case $i=1$ here carry over to the cases $i>1$.) We shall therefore give the proof only of what is essentially new in the theorem, namely that part of (ii) concerned with the universal covering map $f: X\to M$ and the action of the discrete group Γ on the groups $\pi_i(X)\simeq\pi_i(M,x_0)$.

The isomorphism between Γ and $\pi_1(M,x_0)$ was established in Theorem 19.2.2. The action of Γ on $\pi_i(X,x')$ is defined as follows: The action of each element $g\in\Gamma$ on X gives rise in the obvious way to an isomorphism $g^*: \pi_i(X,x')\to\pi_i(X,x'')$, where $x''=g(x')$; and then the desired action is obtained by mapping $\pi_i(X,x'')$ back to $\pi_i(X,x')$ by means of the canonical isomorphism $\gamma_X^*: \pi_i(X, x'') \to \pi_i(X, x')$, which is independent of the path γ_X joining x'' and x' since X is simply-connected. (Here we are using (i) with γ_X, X, x'', x' in the roles of γ, M, x_1, x_0.) This action of Γ is then transferred to $\pi_i(M, x_0)$ via the isomorphism $f'_*: \pi_i(X, x') \to \pi_i(M, x_0)$ (where $f(x') = x_0$) of Corollary 22.2.2. Thus the action of $g \in \Gamma$ on $\pi_i(X, x')$ is given by the automorphism $\gamma_X^* g^*$, and the action of g on $\pi_i(M, x_0)$ by the automorphism

$$f'_* \gamma_X^* g^* (f'_*)^{-1} = f'_* \gamma_X^* (f''_*)^{-1} f''_* g^* (f'_*)^{-1},$$

where $f''_*: \pi_i(X, x'') \to \pi_i(M, x_0)$ is the canonical isomorphism of Corollary 22.2.2. Now $f''_* g^* (f'_*)^{-1}$ is just the identity automorphism of $\pi_i(M, x_0)$ (essentially since the points of M can be identified with the orbits of X under the action of Γ), and $f'_* \gamma_X^* (f''_*)^{-1}$ is easily seen to be just γ^*, where γ is the closed path in M (beginning and ending at x_0) obtained from γ_X by projecting. Hence the action of $g \in \Gamma$ on $\pi_i(M, x_0)$ coincides with the action of the element of $\pi_1(M, x_0)$ represented by γ. This completes the proof. □

Examples. (a) For the covering $f: S^2 \to \mathbb{R}P^2$ we have $\Gamma \simeq \pi_1(\mathbb{R}P^2) \simeq \mathbb{Z}_2$, where the generator of Γ is the map $g: S^2 \to S^2$ defined by $g(x) = -x$, which reverses orientation. Hence the action of the element g on the group $\pi_2(\mathbb{R}P^2) \simeq \pi_2(S^2) \simeq \mathbb{Z}$ (see Example (b) of the preceding subsection) with generator $1 \in \mathbb{Z}$ is given by $g^*(1) = -1$.

Note that for $\mathbb{R}P^3 \cong SO(3)$ (see Example (b) of §2.2), we have again $\Gamma \simeq \pi_1 \simeq \mathbb{Z}_2$, with generator g say $(g^2 = 1)$, and $\pi_3(\mathbb{R}P^3) \simeq \pi_3(S^3) \simeq \mathbb{Z}$ (see the conclusion of §21.1) with generator $1 \in \mathbb{Z}$. However here the action of Γ on $\pi_3(\mathbb{R}P^3)$ is trivial: $g(1) = 1$. (Verify this!)

(b) For the universal covering $X \to S^2 \vee S^1$, with the space X realized as the real line $\mathbb{R}^1$, $-\infty < t < \infty$, with a copy S_n^2 of the 2-sphere attached at each integer point $t = n$, $n = 0, \pm 1, \pm 2, \ldots$ (see Figure 58(b)), the group $\Gamma \simeq \mathbb{Z}$ is generated by the transformation g defined by

$$g: t \mapsto t + 1,$$

$$g: S_n^2 \to S_{n+1}^2, \qquad n = 0, \pm 1, \pm 2, \ldots .$$

It is intuitive that the group $\pi_2(X)$ is the direct sum of infinitely many copies of $\mathbb{Z}$ (indexed by the integers $0, \pm 1, \pm 2, \ldots$), with generators $a_n: S^2 \to S_n^2$, where each of the maps a_n has degree 1. By definition of the action of Γ on π_2, we have

$$g(a_n) = a_{n+1}, \qquad n = 0, \pm 1, \pm 2, \ldots .$$

Since $\Gamma \simeq \pi_1(S^2 \vee S^1)$ and $\pi_2(X) \simeq \pi_2(S^2 \vee S^1)$, we may write a typical element a of the group $\pi_2(S^2 \vee S^1)$ in the form

$$a = \sum_{i=-\infty}^{\infty} \lambda_i g^i(a_0),$$

where the λ_i are integers all but finitely many of which are zero.

Note finally that since the bouquet $S^2 \vee S^1$ is homotopically equivalent to $\mathbb{R}^3 \backslash S^1$, where the circle removed was unknotted in $\mathbb{R}^3$, the same conclusions hold for it.

EXERCISE

Let $U \subset \mathbb{R}^3$ be a solid torus with a single interior point removed. Calculate the groups $\pi_1(U)$, $\pi_2(U)$, and the action of π_1 on π_2.

22.4. The Case of Lie Groups

We now investigate the homotopy groups $\pi_i(M)$ in the case that the manifold M is a Lie group (see §§2.1, 2.3 for the definition and basic properties).

22.4.1. Theorem. *If M is a Lie group then $\pi_1(M)$ is abelian, and the action of $\pi_1(M)$ on every group $\pi_i(M)$ is trivial.*

PROOF. Any two maps $f, g: K \to M$ (where K is any manifold) may be multiplied by using the group operation on M: $fg(k)=f(k)g(k)$. If $f(k_0)=g(k_0)=1$, then of course we also have $fg(k_0)=1$; furthermore the continuity of the multiplication in M implies that if f and g are homotopic to f' and g' respectively, then fg is homotopic to $f'g'$; hence the above multiplication of maps defines a group structure on the set $[K, M]$ of homotopy classes of maps (and also on the set of (relative) homotopy classes of maps sending k_0 to 1).

If we take $K=S^1$, so that now $f, g: S^1 \to M$ represent elements of $\pi_1(M, 1)$, then, as we shall now show, the product fg defined by $fg(x)=f(x)g(x)$ represents the product in π_1 of (the classes of) f and g, i.e. the product determined by the Lie multiplication coincides (to within a homotopy) with the product in the fundamental group $\pi_1(M, 1)$.

To see this observe first that by means of suitable homotopies we can deform the given maps f and g so that

$$f(x)=1 \quad \text{for } x \in D^-,$$

$$g(x)=1 \quad \text{for } x \in D^+,$$

where D^+ is the upper semicircle $y \geq 0$ of the circle $S^1 \subset \mathbb{R}^2$ defined by $x^2+y^2=1$, and D^- the lower semicircle $y \leq 0$. (We may also suppose that the distinguished point s_0 of S^1 is the point $(1, 0)$.) This done, it is almost immediate that the product of these two paths f and g in the fundamental group $\pi_1(M, 1)$ (i.e. in the usual sense according to which one path is followed by the other), coincides with the Lie product of the maps f and g, namely

$$fg(x)=f(x)g(x)=\begin{cases} g(x), & x \in D^-, \\ f(x), & x \in D^+. \end{cases}$$

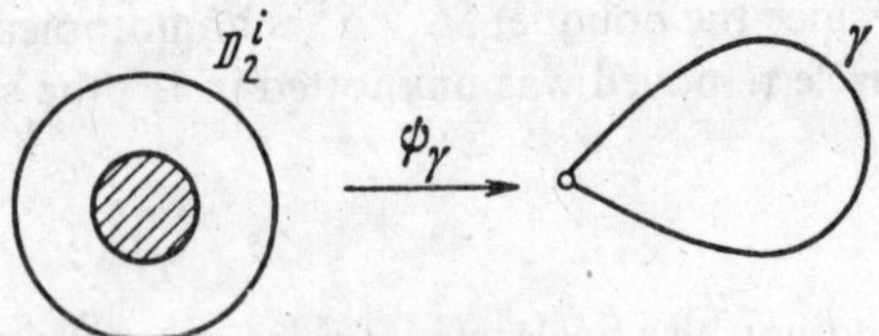

Figure 85

Since for these particular representatives f and g we obviously have $f(x)g(x)=g(x)f(x)$, the commutativity of $\pi_1(M, 1)$ now follows.

We now turn to the second claim, that the action of $\pi_1(M)$ on the $\pi_i(M)$ $(i>1)$ is trivial. As a means to proving this we shall express this action in terms of the Lie multiplication on M. As before we write D_2^i for the disc of radius 2, defined by the inequality $\sum_{j=1}^{i}(x^j)^2 \leq 4$, D_1^i for the disc of radius 1 concentric with it, and denote by $S^{i-1}\times I$ the region between the discs, defined by $1 \leq \sum (x^j)^2 \leq 4$. The proof is now carried out in two steps, each of which involves the definition of a map $D_2^i \to M$.

(i) Each path $\gamma(t)$, $1 \leq t \leq 2$, representing an element of $\pi_1(M, 1)$ determines a map $\gamma\varphi\colon S^{i-1}\times I \to M$:

$$S^{i-1}\times I \xrightarrow{\varphi} I \xrightarrow{\gamma} M,$$

where φ is defined by $\varphi(s, t)=t$. Since $\gamma\varphi(s, 1)$ is the identity element 1 of M, we can extend this map to a map $\psi_\gamma\colon D_2^i \to M$ of the whole disc D_2^i, by setting $\psi_\gamma(D_1^i)=1$; thus the map $\psi_\gamma\colon D_2^i \to M$ maps D_1^i (including of course its boundary) to the point $1 \in M$, and so is clearly homotopic to the map $D_2^i \to 1 \in M$ via a homotopy throughout which ∂D_1^i (in fact D_1^i) continues to be sent to 1. (Note that the image under ψ_γ is a one-dimensional subspace of M; see Figure 85.)

(ii) Next let $\alpha\colon D_1^i \to M$, $\alpha(\partial D_1^i)=1$, be a representing map for any element of $\pi_i(M, 1)$, and extend α to a map $\tilde{\alpha}\colon D_2^i \to M$, by setting $\tilde{\alpha}(D_2^i \backslash D_1^i)=1 \in M$. Consider the (Lie) product map $\tilde{\alpha}\psi_\gamma(x)=\tilde{\alpha}(x)\psi_\gamma(x)$, $x \in D_2^i$. (Since $\tilde{\alpha}(D_2^i \backslash D_1^i)=1=\psi_\gamma(D_1^i)$, it follows that $\tilde{\alpha}(x)\psi_\gamma(x)=\psi_\gamma(x)\tilde{\alpha}(x)$.) It is clear from the definitions of these two maps that this product $\tilde{\alpha}(x)\psi_\gamma(x)$ is just the map $\gamma^*\alpha\colon D_2^i \to M$ (defined in the preceding subsection) since $\tilde{\alpha}\psi_\gamma$ restricts to α on D_1^i and projects the annular region $S^{i-1}\times I$ outside D_1^i onto the path γ. Now as noted in (i), ψ_γ is homotopic to the map $D_2^i \to 1 \in M$, via a homotopy ψ_τ say, where $\psi_0=\psi_\gamma$, $\psi_1(x)=1$, and $\psi_\tau(\partial D_2^i)=1$, $0 \leq \tau \leq 1$. The (Lie) product $\tilde{\alpha}\psi_\tau(x)=\tilde{\alpha}(x)\psi_\tau(x)$ therefore defines a homotopy between the maps $\gamma^*\alpha$ and α (throughout which $\partial D_2^i \to 1$). Hence the action of $\pi_1(M, 1)$ on $\pi_i(M, 1)$ is trivial, as claimed. □

Exercises

1. Prove that if M is a Lie group then as for $i=1$ so also for $i>1$ is the group operation in $\pi_i(M)$ given by the Lie multiplication of maps $fg(x)=f(x)g(x)$.

2. Extend Theorem 22.4.1 to the more general situation of "H-spaces". (An *H-space* is a topological space H on which there is defined a continuous multiplication (i.e. binary operation) $\psi: H \times H \to H$, with respect to which there is an identity element $1 \in H$, i.e. an element satisfying $\psi(x, 1) = \psi(1, x) = x$ for all $x \in H$.)

In connexion with the latter exercise we note that in fact the conclusions of Theorem 22.4.1 hold under the even weaker hypothesis that the multiplication in the space H has a "homotopic identity element", i.e. an element 1 such that the maps $H \to H$ defined by $x \mapsto \psi(x, 1)$ and $x \mapsto \psi(1, x)$ are each homotopic to the identity map. The loop space $H = \Omega(x_0, M)$ (see the beginning of §21.1) provides an example of such a generalized H-space (as the reader will readily be able to verify). In fact the loop space $H = \Omega(x_0, M)$ is a "generalized H-group" in the sense that it has the following two properties:

(i) under the operation $\psi: H \times H \to H$, $\psi(h, g) = hg$ (the path-product of h and g) there is corresponding to each element $h \in H$ a "homotopic inverse", i.e. a path $h^{-1} = \varphi(h)$ such that the map $H \to H$ defined by $h \mapsto hh^{-1}$, is homotopic to the constant map of H to the homotopic identity element of H, i.e. to the constant path $H \to x_0 \in M$;

(ii) the multiplication $\psi: H \times H \to H$ is "homotopically associative", i.e. the maps $H \times H \times H \to H$ defined by

$$(h_1, h_2, h_3) \mapsto \psi(\psi(h_1, h_2), h_3) = (h_1 h_2) h_3,$$

$$(h_1, h_2, h_3) \mapsto \psi(h_1, \psi(h_2, h_3)) = h_1(h_2 h_3),$$

are homotopic. It is clear that in the loop space $\Omega(x_0, M)$ we may take as homotopic inverse to $h = \gamma(t)$ the path $\gamma^{-1}(t)$. The homotopic associativity of path-multiplication in the loop space is a consequence of the following fact.

EXERCISE

The space (it is also a group) of all self-homomorphisms of the interval $[0, 1]$ which fix the end-points (i.e. of all monotonic continuous changes of parameter $t \in [0, 1]$) is contractible.

By Corollary 22.2.3 (with $y = x_0$) we have

$$\pi_{i+1}(M, x_0) \simeq \pi_i(\Omega(x_0, M), e), \qquad i \geq 1, \tag{3}$$

where e denotes the constant path $\gamma(t) \equiv x_0$. The group $\pi_0(\Omega(x_0, M))$ of connected components of $\Omega(x_0, M)$ (which as we saw at the very beginning of this chapter is the same as $\pi_1(M, x_0)$), acts on $\pi_i(\Omega, e)$, (where $\Omega = \Omega(x_0, M)$) via the map determined by

$$\alpha \mapsto \gamma^{-1} \alpha \gamma,$$

where α represents an element of $\pi_i(\Omega, e)$, γ an element of $\pi_0(\Omega) = \pi_1(M, x_0)$, and $\gamma^{-1}\alpha\gamma: D^i \to \Omega$ is defined at each point x of D^i to be the path product of the three paths γ^{-1}, $\alpha(x)$, γ. (Cf. Figure 51.)

EXERCISE

Show that this action $\alpha \mapsto \gamma^{-1}\alpha\gamma$ of $\pi_0(\Omega)$ on $\pi_i(\Omega, e)$ coincides with the standard action (defined in Theorem 22.3.1(ii)) of $\pi_1(M)$ on $\pi_{i+1}(M)$ (under the canonical isomorphism (3)).

For Lie groups, or, more generally, H-spaces with a homotopic identity, the dependence of the homotopy groups on the choice of base point is of little importance, since, essentially, in view of the triviality of the action of π_1 on the π_i, transfer to a new base point is independent of the path from the old to the new one. (This is of course the case also—for simpler reasons—for any simply-connected topological space.) It follows that in such cases the set $[S^i, M]$ of free homotopy classes of maps is in one-to-one correspondence with (the underlying set of) $\pi_i(M)$, or in other words that the base point need not be specified, being in the above sense inessential.

22.5. Whitehead Multiplication

We shall now describe an interesting multiplication between the elements of the various homotopy groups of a space X, called "Whitehead multiplication".

Consider the direct product $M = S^i \times S^j$ of spheres and the bouquet (or "co-ordinate cross") $A = S^i \vee S^j = (S^i \times s_0'') \cup (s_0' \times S^j)$, where $s_0' \in S^i$, $s_0'' \in S^j$ are distinguished points of S^i, S^j respectively. (As distinguished point of $M = S^i \times S^j$ we shall take the point $s_0 = (s_0', s_0'')$.) Consider further the natural map f

$$D^{i+j} = D^i \times D^j \xrightarrow{f} S^i \times S^j,$$

defined to be the direct product of the standard maps (of degree $+1$) $\alpha\colon D^i \to S^i$, $\partial D^i \to s_0'$, and $\beta\colon D^j \to S^j$, $\partial D^j \to s_0''$. (Note that these maps represent generating elements of the groups $\pi_i(S^i, s_0') \simeq \mathbb{Z}$, and $\pi_j(S^j, s_0'') \simeq \mathbb{Z}$.) Since $\alpha(\partial D^i) = s_0'$ and $\beta(\partial D^j) = s_0''$, the map f sends the boundary $\partial D^{i+j} = \partial(D^i \times D^j) = [(\partial D^i) \times D^j] \cup [D^i \times (\partial D^j)]$ onto the co-ordinate cross $A = S^i \vee S^j \subset M = S^i \times S^j$. Hence the map f represents an element of the relative homotopy group $\pi_{i+j}(S^i \times S^j, S^i \vee S^j, s_0)$. The image ∂f of f under the boundary map (see §21.2)

$$\partial\colon \pi_{i+j}(S^i \times S^j, S^i \vee S^j, s_0) \to \pi_{i+j-1}(S^i \vee S^j, s_0),$$

will be crucial in our definition of "Whitehead multiplication", which definition we are now in a position to embark on. Given a space X, and base point $x_0 \in X$, the *Whitehead product* of an element a of $\pi_i(X, x_0)$ with an element b of $\pi_j(X, x_0)$ is a certain element $[a, b]$ of $\pi_{i+j-1}(X, x_0)$, defined as follows. Let $a\colon (S^i, s_0') \to (X, x_0)$ and $b\colon (S^j, s_0'') \to (X, x_0)$ be maps representing their namesakes in the homotopy groups. The maps a, b combine to yield a map $a \vee b\colon S^i \vee S^j \to X$, sending $s_0 (= (s_0', s_0''))$, the point at which the bouquet

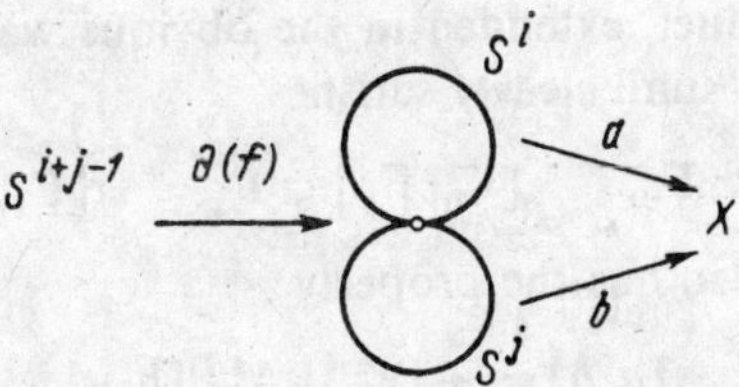

Figure 86

$S^i \vee S^j$ is gathered) to the base point x_0 of X (see Figure 86). The Whitehead product $[a, b]: S^{i+j-1} \to X$, is now defined as (the representative of) the composite of $a \vee b$ and the previously defined map ∂f (as indicated in Figure 86):

$$[a, b]: S^{i+j-1} \xrightarrow{\partial f} S^i \vee S^j \xrightarrow{a \vee b} X.$$

This completes the definition of $[a, b]$ as an element of $\pi_{i+j-1}(X, x_0)$.

If we define orientations on D^i and D^j by means of (fields of) frames τ^i and τ^j respectively, then the orientation on $D^i \times D^j = D^{i+j}$ determined by the frame (τ^i, τ^j) differs by the sign $(-1)^{ij}$ from that defined on $D^j \times D^i$ by the frame (τ^j, τ^i). From this can be inferred the following property of Whitehead multiplication:

$$[a, b] = (-1)^{ij}[b, a]. \tag{4}$$

(Note that here we are using additive notation for the group operation in $\pi_{i+j-1}(X, x_0)$.)

We shall be chiefly concerned with Whitehead multiplication when $i, j \geq 2$. In such cases the groups π_i, π_j are abelian, and we shall therefore write them additively. (It can be shown that, in additive notation, the Whitehead product $[a, b]$ is bilinear in a and b.) To begin with however we examine the situation when $i = 1$.

In the case $i = 1, j = 1$, the product $[a, b]$ coincides with the ordinary group commutator $[a, b] = aba^{-1}b^{-1}$ in the group $\pi_1(X, x_0)$. (Prove it!)

When $i = 1, j \geq 2$, the product $[a, b]$ is expressible in terms of the action of $\pi_1(X, x_0)$ on $\pi_j(X, x_0)$:

$$[a, b] = a^*(b) - b, \tag{5}$$

where $a \in \pi_1$, $b \in \pi_j$. (Prove this also!)

Proceeding to the case $i, j \geq 2$, we form the direct sum of the (additively written) odd-indexed groups π_k, $k \geq 2$, and the direct sum of the even-indexed ones:

$$\Gamma_0 = \pi_3(X) + \pi_5(X) + \pi_7(X) + \cdots + \pi_{2q+1}(X) + \cdots;$$

$$\Gamma_1 = \pi_2(X) + \pi_4(X) + \pi_6(X) + \cdots + \pi_{2q}(X) + \cdots.$$

The Whitehead product, extended in the obvious way to these groups (or rather to their direct sum) clearly satisfies

$$[\Gamma_0, \Gamma_0] \subset \Gamma_0, \quad [\Gamma_0, \Gamma_1] \subset \Gamma_1, \quad [\Gamma_1, \Gamma_1] \subset \Gamma_0,$$

and, in view of (4), also has the property

$$[a, b] = (-1)^{(m+1)(n+1)}[b, a],$$

where $a \in \Gamma_m$, $b \in \Gamma_n$, $m, n = 0, 1$.

Exercise

Show that any three elements $a \in \Gamma_m$, $b \in \Gamma_n$, $c \in \Gamma_p$, $m, n, p = 0, 1$, satisfy the "generalized Jacobi identity"

$$(-1)^{(p+1)(m+1)}[[a, b], c] + (-1)^{(m+1)(n+1)}[[c, a], b] + (-1)^{(n+1)(p+1)}[[b, c], a] = 0.$$

(In the current literature, the "$\mathbb{Z}_2$-graded" algebra $\Gamma_0 \oplus \Gamma_1$ (with the operations of addition and Whitehead multiplication), and algebras with similar properties, have, since their appearance in quantum physics, come to be called "Lie hyperalgebras".)

Computation of Whitehead products in specific cases can be difficult. In the case of the sphere S^{2n} ($n \geq 1$), the Whitehead square $[a, a]$ of a generator a of the group $\pi_{2n}(S^{2n}) \simeq \mathbb{Z}$, turns out to be an element of infinite order in, of course, $\pi_{4n-1}(S^{2n})$; we shall compute this element explicitly for $n = 1$ in the next section. On the other hand for the odd-dimensional spheres S^{2n+1}, it follows from (4) that a generator a of $\pi_{2n+1}(S^{2n+1}) \simeq \mathbb{Z}$ has Whitehead square of order 0 or 2, since

$$[a, a] = (-1)^{(2n+1)(2n+1)}[a, a] = -[a, a],$$

whence $2[a, a] = 0$ in the group $\pi_{4n+1}(S^{2n+1})$.

Proposition. *Whitehead multiplication on H-spaces (and so, in particular, on Lie groups) is trivial:* $[a, b] = 0$ *for all* $a \in \pi_i$, $b \in \pi_j$.

Proof. If $i = j = 1$ then the theorem follows from the commutativity of $\pi_1(M)$ (where M is any H-space), since as noted above, the Whitehead product coincides with the ordinary group commutator. If $i = 1$, $j \geq 2$, the theorem follows from (5) in view of the triviality of the action of $\pi_1(M)$ on $\pi_j(M)$ (see Theorem 22.4.1).

Thus we may suppose $i, j \geq 2$. Represent a and b by maps $\alpha: D^i \to M$, $\partial D^i \to 1$, and $\beta: D^j \to M$, $\partial D^j \to 1$, and consider their "H-space" product

$$\alpha\beta: D^i \times D^j \to M,$$

defined by $\alpha\beta(x, y) = \alpha(x)\beta(y)$. A direct verification shows that the restriction of $\alpha\beta$ to the boundary $\partial(D^i \times D^j) = S^{i+j-1}$ is just the map $[\alpha, \beta]: S^{i+j-1} \to M$ used above to define the Whitehead product. Since $\alpha\beta$ is homotopic to the map $D^{i+j} \to 1$, it follows that the map $[\alpha, \beta]$ is homotopic to the map $S^{i+j-1} \to 1$, i.e. $[a, b] = 0$ in $\pi_{i+j-1}(M)$. □

§23. Facts Concerning the Homotopy Groups of Spheres. Framed Normal Bundles. The Hopf Invariant

23.1. Framed Normal Bundles and the Homotopy Groups of Spheres

In our investigation (in §14.4) of non-degenerate singular points of vector fields and the invariants associated with them, a substantial part was played by the degrees of maps $S^n \to S^n$, or, what amounts to the same thing (in view of Theorem 13.3.1), the fundamental groups $\pi_n(S^n)$, which, as noted earlier (in §21.1), are all isomorphic to $\mathbb{Z}$. (It was also noted in §21.1 that for $i<n$, $\pi_i(S^n)$ is trivial.) Our aim in the present section is to identify some of the groups $\pi_{n+k}(S^n)$ for $k \geq 1$. These groups arise in connexion with the problem of classifying up to homotopies the non-singular (i.e. non-vanishing) vector fields $n(x)$ in Euclidean space $\mathbb{R}^n$, satisfying the condition $n(x) \to n_0$ as $|x| \to \infty$ (see §§25.5, 32 below). If we impose the further requirement that $|n|=1$, then such a vector field determines (and is determined by) a map (the Gauss map—see §14.2)

$$\mathbb{R}^n \cup \{\infty\} \to S^{n-1},$$

where $\infty \mapsto n_0 \in S^{n-1}$. Since $\mathbb{R}^n \cup \{\infty\} \cong S^n$, the classification problem naturally reduces to that of computing $\pi_{n+1}(S^n)$. More generally if $n(x)$ is any vector-function satisfying the same condition at infinity but where now $n(x) = (\xi^1(x), \ldots, \xi^m(x))$ with m permitted to be different from n, then assuming $|n|=1$, we again obtain a map $S^n \to S^{m-1}$, so that the homotopy classification problem for such vector-functions comes down to that of computing $\pi_n(S^{m-1})$.

Thus so far (in this book) our knowledge of the homotopy groups of spheres stands as follows: $\pi_i(S^1)=0$, $i>1$ (Example (a) of §22.2); $\pi_n(S^n) \simeq \mathbb{Z}$; $\pi_i(S^n)=0$, $i<n$. We shall now sketch a geometric method of identifying the homotopy groups of spheres in cases other than these, by considering the complete inverse images of regular points of the maps of interest.

Let $f: S^{n+k} \to S^n (k \geq 1)$ be any map, which by Theorem 12.1.3 we may assume to be smooth, and let $s_0 \in S^n$ be any regular value of f. Choose a system of local co-ordinates $\varphi_1, \ldots, \varphi_n$ for a neighbourhood of $s_0 \in S^n$, such that s_0 is the origin with respect to these co-ordinates and the gradients grad φ_i, $i=1, \ldots, n$ (with respect to whatever co-ordinates are initially given on some chart containing s_0) are linearly independent at s_0. The regularity of f with respect to the point s_0 implies (via the Implicit Function Theorem and the fact that S^{n+k} is closed) that the complete preimage $f^{-1}(s_0)$ is a smooth k-dimensional closed manifold:

$$f^{-1}(s_0)=W^k\subset S^{n+k}.$$

To be more explicit, the manifold W^k is given equationally by

$$\tilde{\varphi}_1=0,\ldots,\tilde{\varphi}_n=0,$$

where the functions $\tilde{\varphi}_i$ are defined on a certain neighbourhood of W^k (namely the complete preimage of the neighbourhood of s_0 co-ordinatized by the φ_i) by $\tilde{\varphi}_i(x)=f^*\varphi_i(x)=\varphi_i f(x)$, and, in view of the regularity of f with respect to s_0, have the property that at every point of $f^{-1}(s_0)=W^k$, their gradients grad $\tilde{\varphi}_i$ are linearly independent. It is then intuitively clear that with respect to a suitably chosen Euclidean metric on $S^{n+k}\backslash\{\infty\}\cong\mathbb{R}^{n+k}$ (where the point ∞ is taken outside W^k), these gradients will be orthogonal to W^k. We have thus associated with our given map f the pair (W^k, τ^n) where $W^k\,(=f^{-1}(s_0))$ is a closed submanifold of S^{n+k}, and τ^n (=(grad $\tilde{\varphi}_1,\ldots,$ grad $\tilde{\varphi}_n$), where $\tilde{\varphi}_i=f^*\varphi_i=\varphi_i f(x)$ with φ_i as above) is a field of frames defined on $W^k\subset S^{n+k}\backslash\{\infty\}\cong\mathbb{R}^{n+k}$, and normal to W^k with respect to some Euclidean metric on $\mathbb{R}^{n+k}$.

23.1.1. Definition. A pair (W^k, τ^n) consisting of a closed k-dimensional manifold $W^k\subset\mathbb{R}^{n+k}$, on which there is defined a non-degenerate field τ^n of normal n-frames, is called a *framed normal bundle* (without boundary), or *manifold equipped with a field of normal frames.*

Note that the field τ^n in combination with the standard orientation of $\mathbb{R}^{n+k}$, determines an orientation of W^k (namely that corresponding to a field τ^k of tangent frames such that (τ^k, τ^n) defines the standard orientation of $\mathbb{R}^{n+k}$).

Consider next a smooth homotopy $F\colon S^{n+k}\times I\to S^n$, regular with respect to the point $s_0\in S^n$, between two maps $f_0, f_1\colon S^{n+k}\to S^n$. The complete preimage $V^{k+1}=F^{-1}(s_0)$ of s_0 in $S^{n+k}\times I$, is, analogously to the above, given equationally in $S^{n+k}\times I$ by

$$\Phi_1=0,\ldots,\Phi_n=0,$$

where $\Phi_j=F^*\varphi_j=\varphi_j\circ F$, and, by virtue of the regularity of F with respect to s_0, the gradients grad Φ_j are linearly independent at every point of V^{k+1}. Thus corresponding to the given homotopy F, we have constructed the pair $(V^{k+1}\subset S^{n+k}\times I, \tau^n)$, where the field of frames $\tau^n=$(grad $\Phi_1,\ldots,$ grad Φ_n) is normal to V^{k+1} (relative to some Euclidean metric on $\mathbb{R}^{n+k}\times I$). The intersections $W_0^k=V^{k+1}\cap(S^{n+k}\times 0)$ and $W_1^k=V^{k+1}\cap(S^{n+k}\times 1)$, of V^{k+1} with the respective boundary components of $S^{n+k}\times I$, are manifolds equipped with (induced) fields of normal frames, namely those obtained by restricting τ^n; in fact these are just the framed normal bundles obtained, as above, from the maps $f_0, f_1\colon S^{n+k}\to S^n$. The regularity of F (with respect to $s_0\in S^n$), in particular on the boundary of $S^{n+k}\times I$, implies that at no point of W_0^k or W_1^k is the tangent space to V^{k+1} contained in the tangent space to the boundary of $S^{n+k}\times I$ (i.e. to whichever of the components $S^{n+k}\times 0$ or $S^{n+k}\times 1$ the point happens to lie in); or, in other words, V^{k+1} is not tangential to the boundary of $S^{n+k}\times I$ (at $t=0, 1$). It follows, again by choosing Euclidean co-ordinates suitably in $\mathbb{R}^{n+k}$, that we may suppose that V^{k+1} "approaches the boundary of $S^{n+k}\times I$ at right angles", meaning that at each point of W_0^k and W_1^k there is a tangent vector to V^{k+1} which is normal to the boundary of $S^{n+k}\times I$ at that point. (We met with a similar, though simpler, situation in §13.1; there we had $k=0$, so that the manifolds W_0^k, W_1^k were finite sets of points, the orientation-determining field of frames τ^n at those points reduced to an assignment of signs ± 1, and this orientation was continued from those points onto the one-dimensional manifold V^{k+1}.)

23.1.2. Definition. A pair $(V^{k+1}\subset S^{n+k}\times I, \tau^n)$ consisting of a manifold-with-boundary V^{k+1} embedded in such a way that (with respect to the Euclidean metric on $S^{n+k}\backslash\{\infty\}\cong\mathbb{R}^{n+k}$) it "approaches the boundary of $S^{n+k}\times I$ at right angles" (see above), together with a normal, non-degenerate field τ^n of n-frames on V^{k+1}, is called a *framed normal bundle with boundary.*

This definition allows us to introduce a natural "equivalence" between framed normal bundles.

23.1.3. Definition. Two framed normal bundles (W_1^k, τ_1^n), (W_2^k, τ_2^n) (where $W_j^k\subset\mathbb{R}^{n+k}$, $j=1, 2$) are said to be *equivalent* if there exists a framed normal bundle (V^{k+1}, τ^n), $V^{k+1}\subset S^{n+k}\times I$, whose respective boundary components (equipped with the restrictions of the field of frames τ^n) are precisely the given framed normal bundles:

$$W_1^k=V^{k+1}|_{t=0},\qquad \tau_1^n=\tau^n|_{t=0};$$

$$W_2^k=V^{k+1}|_{t=1},\qquad \tau_2^n=\tau^n|_{t=1}.$$

Alternatively we may define (W_1^k, τ_1^n) and (W_2^k, τ_2^n) to be equivalent if the framed normal bundle (W^k, τ^n) where $W^k=W_1^k\cup W_2^k$ (W_1^k and W_2^k being assumed disjoint), and τ^n coincides with τ_1^n on W_1^k, and on W_2^k is obtained from τ_2^n by replacing the first vector of τ_2^n by its negative, is equivalent (in the sense of the definition) to the empty framed normal bundle (so that in particular the V^{k+1} of the above definition avoids $S^{n+k}\times 1$ and meets $S^{n+k}\times 0$ in $W_1^k\cup W_2^k$).

The relevance of these definitions to the problem of finding the homotopy groups of spheres is made evident by the following theorem.

23.1.4. Theorem. *There is a natural one-to-one correspondence between the equivalence classes of (closed) framed normal bundles (W^k, τ^n), $W^k\subset\mathbb{R}^{n+k}$, and the elements of the group $\pi_{n+k}(S^n)$.*

Proof. Any map $f\colon S^{n+k}\to S^n$, regular with respect to the point s_0 of S^n, gives rise, in the manner described above, to a closed framed normal bundle

(W^k, τ^n), and any homotopy $F: S^{n+k} \times I \to S^n$ between two such maps determines an equivalence between their corresponding framed normal bundles.

As the first step in establishing the converse we show that any (closed) framed normal bundle (W^k, τ^n), $W^k \subset \mathbb{R}^{n+k}$, determines, in an appropriate fashion, a map $S^{n+k} \to S^n$. With this aim in mind, we observe first that in view of the presence on W^k of the normal field of frames τ^n, for sufficiently small $\varepsilon > 0$ the ε-neighbourhood W_ε of the manifold $W^k \subset \mathbb{R}^{n+k}$ is diffeomorphic to $W^k \times D_\varepsilon^n$, where D_ε^n denotes the n-dimensional disc of radius ε and (variable) centre $x \in W^k$, contained in the n-dimensional plane orthogonal to W^k at x, determined by τ^n. We now "extend" the projection $W_\varepsilon \cong W^k \times D_\varepsilon^n \to D_\varepsilon^n$ to a map $f: S^{n+k} \to S^n$ by identifying the boundary of D_ε^n with a point (thus turning it into an S^n), and mapping the whole of the complement of $W_\varepsilon \cong W^k \times D_\varepsilon^n$ in S^{n+k} to that point. It is then not difficult to see that the framed normal bundle constructed from the map f in the prescribed manner is essentially just the initially given one (W^k, τ^n).

We leave to the reader the (analogous) construction of an appropriate map $F: S^{n+k} \times I \to S^n$ from a given framed normal bundle-with-boundary (V^{k+1}, τ^n), $V^{k+1} \subset S^{n+k} \times I$. This concludes the proof of the theorem. □

Note that in §13.3, in the course of showing (in effect) that $\pi_n(S^n) \simeq \mathbb{Z}$ (see Theorem 13.3.1) we made tacit use of framed normal bundles in the particular case $k=0$.

Remark. The addition of any two elements of the homotopy group $\pi_{n+k}(S^n)$ can be interpreted in a natural way as taking the union of two non-intersecting (in fact bounded away from each other) framed normal bundles in $\mathbb{R}^{n+k}$. (This follows from the respective definitions; verify it!)

In the next result we establish a general property of equivalence classes of closed framed normal bundles.

23.1.5. Theorem. *For each pair $n, k \geq 1$, every equivalence class of closed framed normal bundles (W^k, τ^n), contains a connected framed normal bundle. (This is false for $k=0$.)*

PROOF. In view of the assumed compactness, it suffices to show that a (non-connected) closed framed normal bundle (W^k, τ^n) of the form $(W_1^k, \tau_1^n) \cup (W_2^k, \tau_2^n)$ where $W_1^k \cap W_2^k$ is empty, and W_1^k, W_2^k are connected (and where, as usual, $W^k \subset \mathbb{R}^{n+k} \subset S^{n+k}$), is equivalent to a connected one. To this end consider a smooth non-self-intersecting path $\gamma(\tau)$, $0 \leq \tau \leq 1$ (i.e. a 1-dimensional submanifold-with-boundary) from some point x_0 of W_1^k to some point x_1 of W_2^k; we shall require of this path that it both leave W_1^k and approach W_2^k at right angles, in fact in the directions of the first vectors m_{11}, m_{12} of the respective frames $\tau_j^n = (m_{1j}, \ldots, m_{nj})$, $j = 1, 2$. For each τ, $0 \leq \tau \leq 1$, let $\mathbb{R}^k(\tau)$ be a k-dimensional plane (varying continuously with τ)

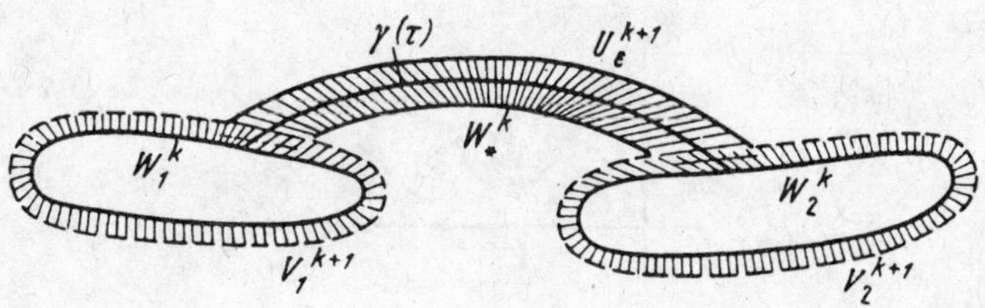

Figure 87

orthogonal to the path γ at the point $\gamma(\tau)$, and at $\tau=0, 1$ coinciding with the tangent planes to $W_1^k(\tau=0)$ and $W_2^k(\tau=1)$ (see Figure 87). Writing $\tau^n=(m_1,\ldots,m_n)$ $(=\tau_j^n=(m_{1j},\ldots,m_{nj})$ on W_j^k, $j=1,2)$, we extend the field of frames $\tau^{n-1}=(m_2,\ldots,m_n)$ from the end-points $\gamma(0)$, $\gamma(1)$ along the path γ in such a way that at each point $\gamma(\tau)$ it is normal to the plane $\mathbb{R}^k(\tau)$. We next let U_ε^{k+1} denote the "k-dimensional thickening" of the path γ in the directions determined at each of its points $\gamma(\tau)$ by the plane $\mathbb{R}^k(\tau)$; i.e. U_ε^{k+1} is the union of the k-dimensional ε-neighbourhoods in $\mathbb{R}^k(\tau)$ of the points $\gamma(\tau)$, $0\le\tau\le1$ (see Figure 87). (Since $\mathbb{R}^k(0)$ and $\mathbb{R}^k(1)$ are the tangent planes at the points $\gamma(0)$, $\gamma(1)$ of W_1^k, W_2^k respectively, it follows that at these points the thickening of γ occurs in those tangent planes.) As indicated in Figure 87, we also thicken the manifolds W_1^k and W_2^k in the direction of the first vector m_1 (which is m_{11} on W_1^k and m_{12} on W_2^k) of the field $\tau^n(=\tau_1^n$ on W_1^k and τ_2^n on $W_2^k)$, obtaining manifolds V_1^{k+1} and V_2^{k+1}, say. The union

$$V^{k+1}=V_1^{k+1}\cup U_\varepsilon^{k+1}\cup V_2^{k+1}$$

is then a $(k+1)$-dimensional manifold, with boundary

$$\partial V^{k+1}=W_1^k\cup W_2^k\cup W_*^k,$$

where since W_1^k and W_2^k are connected, the component W_*^k is also connected. By means of small adjustments of the above thickenings (i.e. by means of a modest homotopy) we can arrange that the manifold V^{k+1} is smooth, i.e. without sharp corners (or, equivalently, that its boundary component W_*^k is smooth (assuming W_1^k, W_2^k smooth to begin with)).

The field of frames $\tau^{n-1}=(m_2,\ldots,m_n)$, normal to W^k and to the path γ, obviously extends to a field of frames normal to the whole of V^{k+1}, spanning at each point of V^{k+1} the orthogonal complement in $\mathbb{R}^{n+k}$ of the tangent space to V^{k+1} at that point. Our situation is thus at present as follows: We have a manifold-with-boundary $V^{k+1}\subset\mathbb{R}^{n+k}$, and defined on it a field of frames τ^{n-1} normal to V^{k+1} in $\mathbb{R}^{n+k}$; on the boundary of V^{k+1} we have a field of frames $\tau^n=(m_1,\tau^{n-1})$, where m_1 is an internal normal to the boundary ∂V^{k+1}, tangent to V^{k+1}. Thus the boundary ∂V^{k+1} together with the field of frames τ^n constitutes a framed normal bundle $(\partial V^{k+1},\tau^n)$ in $\mathbb{R}^{n+k}$. We now split off part of the proof as a lemma concerning this particular framed normal bundle.

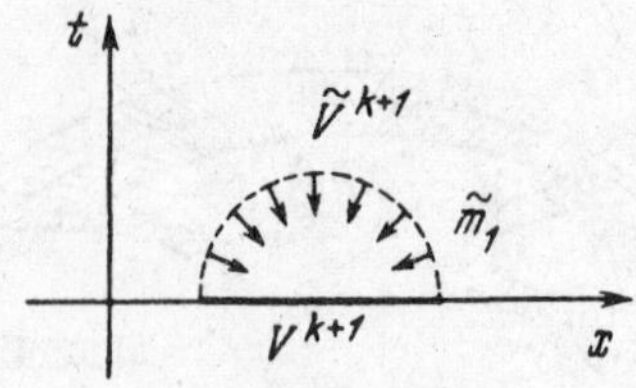

Figure 88

23.1.6. Lemma. *The framed normal bundle $(\partial V^{k+1}, \tau^n)$ constructed above is equivalent to the empty bundle.*

PROOF. Let $t(x)$ be a (numerical) function on V^{k+1} such that $t(x)=0$ on the boundary ∂V^{k+1}, $1>t(x)>0$ in the interior of V^{k+1}, and the graph $(x, t(x))$ is a smooth submanifold $\tilde{V}^{k+1} \subset V^{k+1} \times I \subset S^{n+k} \times I$, approaching the boundary $\partial \tilde{V}^{k+1} = \partial V^{k+1} \times 0 = \partial V^{k+1}$ normally (see Figure 88). We now define on $\tilde{V}^{k+1}$ a field of frames $\tilde{\tau}^n$ normal to $\tilde{V}^{k+1}$ in $\mathbb{R}^{n+k} \times I$. With this aim in view, we first lift the field of frames $\tau^{n-1} = (m_2, \ldots, m_n)$ from V^{k+1} to $\tilde{V}^{k+1}$ "trivially", i.e. we parallel transport the vectors $m_2, \ldots, m_n$ from each point $x \in V^{k+1} \subset \mathbb{R}^{n+k} \times 0$ to the corresponding point $\tilde{x} = (x, t(x)) \in \tilde{V}^{k+1} \subset \mathbb{R}^{n+k} \times I$, vertically above it. This furnishes us with a field of frames $\tilde{\tau}^{n-1} = (\tilde{m}_2, \ldots, \tilde{m}_n)$ on $\tilde{V}^{k+1}$, and it only remains to define a first vector field $\tilde{m}_1$ on $\tilde{V}^{k+1}$. Now the manifolds $V^{k+1} = V^{k+1} \times 0$ and $\tilde{V}^{k+1}$ together bound a region $U \subset V^{k+1} \times I$, so we take the field $\tilde{m}_1$ to be the unit vector field normal to $\tilde{V}^{k+1}$ in $V^{k+1} \times I$, directed into the interior of U, and furthermore such that at $t=0$, i.e. at the points of $\partial \tilde{V}^{k+1} = \partial V^{k+1} \times 0 = \partial V^{k+1}$ this field coincides with the unit field m_1 normal to ∂V^{k+1} and internal to V^{k+1} (i.e. with the (normalized) first vector of the field of frames τ^n on ∂V^{k+1}—see Figure 88). The resulting framed normal bundle-with-boundary $(\tilde{V}^{k+1}, \tilde{\tau}^n)$ then has as its boundary (equipped with the restriction of $\tilde{\tau}^n$) the framed bundle $(\partial V^{k+1}, \tau^n)$. Since $\tilde{V}^{k+1} \subset \mathbb{R}^{n+k} \times I$ avoids the opposite boundary $\mathbb{R}^{n+1} \times 1$ of $\mathbb{R}^{n+k} \times I$, it follows that $(\partial V^{k+1}, \tau^n)$ is equivalent to the empty bundle, as required. □

The theorem is almost immediate from this lemma, since the equivalence with zero of the framed normal bundle $\partial V^{k+1} = (W_1^k \cup W_2^k) \cup W_*^k$ (framed as above), amounts to the equivalence of the framed normal bundles $(W^k = W_1^k \cup W_2^k, \tau^n)$ and W_*^k (framed appropriately), and then the connectedness of W_*^k finally yields the desired conclusion. □

23.2. The Suspension Map

In what follows we shall need the following facts about the homotopy groups of the special orthogonal groups:

(i) $\pi_1(SO(2)) \simeq \mathbb{Z}$, since $SO(2) \cong S^1$;

(ii) $\pi_1(SO(3)) \simeq \mathbb{Z}_2$, since $SO(3) \cong \mathbb{R}P^3$ (see §2.2);
(iii) $\pi_1(SO(n)) \simeq \mathbb{Z}_2$ for $n \geq 3$ (see §24.4 in Chapter 6);
(iv) The homomorphism $\pi_i(SO(n)) \to \pi_i(SO(n+1))$ induced by the obvious embedding $SO(n) \to SO(n+1)$ is for $i < n-1$ an isomorphism, while for $i = n-1$, it is merely an epimorphism (i.e. homomorphism onto) which for even n has infinite cyclic kernel. (This is easily seen in the case $n = 2$; the general case will be established in Chapter 6, §24.3. In fact it will be shown there that, more generally, for any smooth manifold (or complex) M of dimension i, the map $[M, SO(n)] \to [M, SO(n+1)]$ induced by the natural embedding $SO(n) \to SO(n+1)$, is bijective if $i < n-1$, and remains surjective in the case $i = n-1$.)

By Theorem 12.1.6 any embedding of a smooth closed k-dimensional manifold M into $\mathbb{R}^{2k+q}$ where $q \geq 2$, is "unknotted", by which we mean that any two smooth embeddings $f, g: M \to \mathbb{R}^{2k+q}$ $(q \geq 2)$ are isotopic. Such an isotopy $f_t: M \to \mathbb{R}^{2k+q}$ (where $0 \leq t \leq 1$, $f_0 = f$, $f_1 = g$, and by definition of isotopy each f_t is an embedding) can of course be regarded as a smooth embedding $F: M \times I \to \mathbb{R}^{2k+q} \times I$, defined by $F(x, t) = (f_t(x), t)$. If, as is natural, we take paths in $\mathbb{R}^{2k+q} \times I$ of the form $x \times I$, $x \in \mathbb{R}^{2k+q}$, to be perpendicular to the cross-section $\mathbb{R}^{2k+q} \times t$ for every t (i.e. if we take the sides of the cylinder $\mathbb{R}^{2k+q} \times I$ to be perpendicular to these cross-sections) then we can clearly extend any field of frames τ^{k+q} normal to $M \times O = M$ (regarded as a submanifold of $\mathbb{R}^{2k+q}$) to a field of frames $\hat{\tau}^{k+q}$ normal to $M \times I$ in $\mathbb{R}^{2k+q} \times I$. Hence in considering framed normal bundles (M, τ^{k+q}), $M \subset \mathbb{R}^{2k+q}$ $(q \geq 2)$, the particular manner in which M is embedded will in the present context be immaterial since any such framed normal bundle is equivalent to one with M embedded in an arbitrarily prescribed manner.

Thus the question of classifying (up to equivalence) the framed normal bundles (M, τ^n), $M \subset \mathbb{R}^{n+k}$, reduces, at least for large enough n, to that of classifying for any fixed embedding $M \subset \mathbb{R}^{n+k}$, the possible fields of normal frames τ^n on M in $\mathbb{R}^{n+k}$. Now given any particular field of normal frames τ^n, any other defining the same orientation on M can be obtained from τ^n (to within an equivalence) by means of a rotation at each point. Thus with each map $M \to SO(n)$ there is associated in this way an equivalence class of "equipages" of fields of normal frames for $M \subset \mathbb{R}^{n+k}$; this correspondence is not however one-to-one, since homotopic maps $M \to SO(n)$ will clearly yield equivalent framed normal bundles over $M \subset \mathbb{R}^{n+k}$.

The upshot of the preceding discussion, together with fact (iv) (or rather the more general parenthetical statement following it) is, therefore, that provided $n \geq k+2$ neither the embedding $M \subset \mathbb{R}^{k+n}$, nor the homotopy classes of maps $M \to SO(n)$ (to which correspond classes of normal frames on $M \subset \mathbb{R}^{n+k}$), depends on n. In order to make this somewhat vague statement precise we now introduce the *suspension homomorphism* $E: \pi_{n+k}(S^n) \to \pi_{n+k+1}(S^{n+1})$. Let (M, τ^n) be a k-dimensional framed normal bundle in

$\mathbb{R}^{n+k}$, and consider the embedding $M \subset \mathbb{R}^{n+k} \subset \mathbb{R}^{n+k+1}$. We adjoin to the given frame τ^n an additional vector m_0 normal to $\mathbb{R}^{n+k}$ in $\mathbb{R}^{n+k+1}$, and then define the *suspension map* E by $E(M, \tau^n)=(M,(m_0, \tau^n))$. In view of the natural one-to-one correspondence guaranteed by Theorem 23.1.4, this does indeed define a map $E: \pi_{n+k}(S^n) \to \pi_{n+k+1}(S^{n+1})$; that this map is a homomorphism is a consequence of the remark following the proof of that theorem. (The suspension map for framed normal bundles-with-boundary is defined analogously.) From the statement preceding this definition it now plausibly follows that for $n \geq k+2$ the suspension homomorphism is actually an isomorphism:

$$E: \pi_{n+k}(S^n) \cong \pi_{n+k+1}(S^{n+1}), \qquad n \geq k+2.$$

These are the so-called "stable" homotopy groups of spheres.

In the case $n=k+1$ we have:

(i) Any closed k-dimensional manifold can be embedded in $\mathbb{R}^{2k+1}$ (Whitney's theorem (11.1.1));

(ii) given any particular embedding $M \subset \mathbb{R}^{2k+1}$ there is a natural surjection from the set of homotopy classes of maps $M \to SO(k+1)$ (or, equivalently, the set of equivalence classes of normal frames on $M \subset \mathbb{R}^{2k+1}$) onto the set of homotopy classes of maps $M \to SO(k+1+q)$, $q>1$ (or, equivalently, the set of classes of normal frames on $M \subset \mathbb{R}^{2k+1+q}$). (This follows from the more general parenthetical statement following (iv) above.)

These facts imply that in this case the suspension homomorphism

$$E: \pi_{2k+1}(S^{k+1}) \to \pi_{2k+2}(S^{k+2})$$

is in fact an epimorphism.

We conclude this subsection by redefining the suspension map E more directly as follows. Let S^{n+k} be the equator of S^{n+k+1}, and S^n the equator of S^{n+1}. Then given any map $f: S^{n+k} \to S^n$, we define $Ef: S^{n+k+1} \to S^{n+1}$ to be the most natural extension map of f; i.e. for each point x of S^{n+k} we simply map the points of the "line of longitude" on S^{n+k+1} passing through x to the corresponding points of the line of longitude on S^{n+1} passing through $f(x)$; thus in particular the north pole of S^{n+k+1} goes under Ef to the north pole of S^{n+1}, and the south pole to the south pole.

23.3. Calculation of the Groups $\pi_{n+1}(S^n)$

We are now in a position to calculate the groups $\pi_{n+1}(S^n)$ using Theorems 23.1.4 and 23.1.5. By those theorems each $\alpha \in \pi_{n+1}(S^n)$ can be represented by a one-dimensional, connected, closed framed normal bundle in $\mathbb{R}^{n+1}$, i.e. by a circle $S^1 \subset \mathbb{R}^{n+1}$ equipped with a field of normal frames τ^n.

(i) The case $n>2$. In this case all embeddings $S^1 \subset \mathbb{R}^{n+1}$ are isotopic, so that we may assume S^1 is the unit circle $(x^1)^2+(x^2)^2=1$ in the plane $\mathbb{R}^2$ defined by $x^3=x^4=\cdots=x^{n+1}=0$. This circle $S^1 \subset \mathbb{R}^2 \subset \mathbb{R}^{n+1}$ can in particular be equipped "trivially" with the field of normal frames $\tau_0^n=(m, e_3, e_4, \ldots, e_{n+1})$ where m is the external unit normal vector to S^1 in $\mathbb{R}^2$, and the e_j are the standard basis vectors in $\mathbb{R}^{n+1}$. From the resulting framed normal bundle (S^1, τ_0^n), which clearly corresponds to the zero element of the group $\pi_{n+1}(S^n)$, every framed normal bundle over S^1 can, up to equivalence, be obtained by means of a rotation $\tau_0^n \to A(x)\tau_0^n$ of the frame τ_0^n at each point $x \in S^1$, so that here for each x, $A(x)$ denotes an element of $SO(n)$ (and the map $x \mapsto A(x)$ may be assumed smooth). Since the map $A: S^1 \to SO(n)$ represents an element $[A]$ of $\pi_1(SO(n))$, this yields (see the preceding subsection and the exercise below) a one-to-one correspondence between the equivalence classes of framed normal bundles over S^1 and the elements of $\pi_1(SO(n))$. Since by (iii) of the preceding subsection $\pi_1(SO(n)) \simeq \mathbb{Z}_2$ for $n>2$, it follows that there are precisely two framed normal bundles over S^1, whence (by Theorem 23.1.4) $\pi_{n+1}(S^n) \simeq \mathbb{Z}_2$ for $n>2$.

Exercise

Verify that the above-described correspondence is indeed one-to-one; i.e. that the pair (S^1, τ_1^n) where τ_1^n is a field of normal frames afforded by the non-trivial element of $\pi_1(SO(n))$, is not equivalent to the trivial framed normal bundle over the circle.

(ii) The case $n=2$. Here the manner in which S^1 is embedded in $\mathbb{R}^3$ might be expected on the face of it (since knots exist in $\mathbb{R}^3$) to have some influence on our problem. However as it turns out, it does not: It can be shown (we omit the proof) that every framed normal bundle over $S^1 \subset \mathbb{R}^3$ is equivalent to some framed normal bundle over an unknotted circle $S^1 \subset \mathbb{R}^3$. Thus in this case also we may take our circle in $\mathbb{R}^3$ to the unit circle $(x^1)^2+(x^2)^2=1$ in $\mathbb{R}^2 \subset \mathbb{R}^3$. Then, as in case (i), up to equivalence every field of normal frames τ^2 on $S^1 \subset \mathbb{R}^2 \subset \mathbb{R}^3$ is obtained from the trivial one τ_0^2 by means of a rotation $A(x)$ of τ_0^2 at each point $x \in S^1$, so that once again we can associate with each homotopy class of maps

$$A: S^1 \to SO(2) \cong S^1$$

an equivalence class of framed normal bundles over $S^1 \subset \mathbb{R}^2 \subset \mathbb{R}^3$ in a well-defined manner which moreover ensures that every equivalence class of bundles is associated with some class of maps. Since $\pi_1(SO(2)) \simeq \mathbb{Z}$, we index in the natural way particular fields of normal frames on $S^1 \subset \mathbb{R}^2 \subset \mathbb{R}^3$ arising as above from the elements of $\pi_1(SO(2))$, denoting them by $\tau_{(m)}^2$, $m \in \mathbb{Z}$. It can be shown (cf. the exercise above) that the framed normal bundles $(S^1, \tau_{(m)}^2)$, $S^1 \subset \mathbb{R}^2 \subset \mathbb{R}^3$, are pairwise inequivalent, whence certainly $\pi_3(S^2)$ is infinite. It is not difficult to show further (using the remark following the proof of Theorem 23.1.4) that the union of $(S^1, \tau_{(m)}^2)$ and $(S^1, \tau_{(q)}^2)$ is equivalent to $(S^1, \tau_{(m+q)}^2)$, so that in fact $\pi_3(S^2) \simeq \mathbb{Z}$. (This will be given an alternative proof in Chapter 6.)

We conclude this subsection by introducing an important invariant of a smooth map $f: S^3 \to S^2$. If $y_0, y_1 \in S^2$ are regular values of f, and we write

$M_0 = f^{-1}(y_0)$, $M_1 = f^{-1}(y_1)$, then the *Hopf invariant* $H(f)$ of the map f is defined to be the linking coefficient of the preimages of y_0 and y_1, i.e. $H(f) = \{M_0, M_1\}$ (see §15.4 for the definition). We leave it as an exercise for the reader to show that $H(S^1, \tau^2_{(m)}) = m$.

EXERCISES

1. Show that $H(f)$ is a homotopy invariant. Find $H(f)$ for a Serre fibration $f: S^3 \to S^2$. Prove that $H([a, a])$ is always even (see §22.5).

2. Find the appropriate generalization of the Hopf invariant to the elements of $\pi_{4n-1}(S^{2n})$, and construct elements of these groups for which it is non-trivial.

3. Let ω be a volume 2-form on the 2-sphere S^2 satisfying $\int_{S^2} \omega = 1$, and let $f: S^3 \to S^2$ be any smooth map. Verify that the form $f^*(\omega)$ on the 3-sphere S^3 is "exact", i.e. there exists a 1-form ω_1 such that $f^*(\omega) = d\omega_1$. Show that the number $\int_{S^3} f^*(\omega) \wedge \omega_1$ is an integer which is in fact equal to the Hopf invariant $H(f)$.

23.4. The Groups $\pi_{n+2}(S^n)$

We shall restrict our attention to the "stable" case $n > 3$. (For $n \leq 3$ it turns out that both $\pi_4(S^2)$ and $\pi_5(S^3)$ are isomorphic to $\mathbb{Z}_2$; we shall prove this for $\pi_4(S^2)$ only, in Chapter 6.)

By Theorem 24.1.5 every element of $\pi_{n+2}(S^n)$ is represented by a connected, 2-dimensional, closed framed normal bundle (M, τ^n), $M \subset \mathbb{R}^{n+2}$. Since by our assumption $n+2 \geq 6$, the embedding $M \subset \mathbb{R}^{n+2}$ is "unknotted". Now it is a well-known theorem (which we have hinted at earlier) that every orientable closed surface is homeomorphic to a sphere-with-g-handles for some $g \geq 0$ (see §17 of Part III). Since for our purposes the manner in which M is embedded in $\mathbb{R}^{n+2}$ is immaterial, we may therefore suppose that $M \subset \mathbb{R}^3 \subset \mathbb{R}^{n+2}$, where $x^4 = \cdots = x^{n+2} = 0$ on $\mathbb{R}^3$. Proceeding somewhat as earlier we first equip M "trivially" with the field of frames $\tau_0^n = (m_1, e_4, \ldots, e_{n+2})$ where m_1 is the unit internal normal to M in $\mathbb{R}^3$ and the e_j are the standard basis vectors in $\mathbb{R}^{n+2}$. It follows as in the proof of Lemma 22.1.6 that the pair (M, τ_0^n) represents the zero element of $\pi_{n+2}(S^n)$, since the field $\tau_0^{n-1} = (e_4, \ldots, e_{n+2})$ is defined also on the region V of $\mathbb{R}^3$ bounded by the surface M. Hence, as before, any field of normal frames τ^n on $M \subset \mathbb{R}^3 \subset \mathbb{R}^{n+2}$ (determining the same orientation of M) is, up to equivalence, obtained by rotating τ_0^n at each point of M:

$$\tau^n = A(x)\tau_0^n.$$

In this way each element of $[M, SO(n)]$ is made to correspond to an equivalence class of framed normal bundles over M, or, in view of Theorem 22.1.4, to an element of $\pi_{n+2}(S^n)$ (and the correspondence is "onto").

Now let α be an element of $\pi_1(M)$ representable by a smoothly embedded (i.e. non-self-intersecting) directed circle $S^1 \subset M$, and denote by n_1 a normal vector field on S^1 in M such that the frame (τ, n_1), where τ is the tangent vector to S^1 in M, determines the same orientation as that initially prescribed on M. In terms of such representatives for (appropriate) elements of $\pi_1(M)$ we now define the *Arf function* Φ (on the set of such elements α of $\pi_1(M)$ and taking its values in $\mathbb{Z}_2$) corresponding to a given framed normal bundle (M, τ^n) in $\mathbb{R}^{n+2}$, as follows: For each $\alpha \in \pi_1(M)$ representable by $S^1 \subset M \subset \mathbb{R}^{n+2}$ as above, restrict the field of frames τ^n to this S^1. The framed normal bundle (S^1, τ^{n+1}) in $\mathbb{R}^{n+2}$, where $\tau^{n+1} = (n_1, \tau^n)$ with n_1 as above, corresponds (by Theorem 23.1.4) to a unique element of $\pi_{n+2}(S^{n+1})$; we then take $\Phi(\alpha)$ to be the image of that element under the isomorphism $\pi_{n+2}(S^{n+1}) \simeq \mathbb{Z}_2$.

Exercises

1. For $\alpha, \beta \in \pi_1(M^2)$ such that α, β and $\alpha\beta$ are all representable by embedded circles $S^1 \subset M^2$, we have
$$\Phi(\alpha\beta) = \Phi(\alpha) + \Phi(\beta) + \alpha \circ \beta \pmod 2,$$
where $\alpha \circ \beta$ is the intersection index (see §15.1) of the circles representing (as described above) α and β.

2. Let $\alpha_1, \ldots, \alpha_g, \beta_1, \ldots, \beta_g$ be $2g$ embedded circles on the sphere-with-g-handles M_g^2, representing generators of the group $\pi_1(M_g^2)$, and satisfying the (defining) relation $\alpha_1\beta_1\alpha_1^{-1}\beta_1^{-1} \cdots \alpha_g\beta_g\alpha_g^{-1}\beta_g^{-1} = 1$. (We note that it follows that the various intersection indices are then given by $\alpha_i \circ \beta_j = \delta_{ij}$, $\alpha_i \circ \alpha_j = \beta_i \circ \beta_j = 0$.) Prove that the sum
$$\Phi(M_g^2, \tau) = \sum_{i=1}^{g} \Phi(\alpha_i)\Phi(\beta_i)$$
is independent of the choice of such circles α_i, β_i on M_g^2. Prove then that the condition $\Phi(M_g^2, \tau) = 0$ is both necessary and sufficient for there to exist such circles $\alpha_1, \ldots, \alpha_g, \beta_1, \ldots, \beta_g$ for which also $\Phi(\alpha_1) = \Phi(\alpha_2) = \cdots = \Phi(\alpha_g) = 0$.

If α is a non-trivial element of $\pi_1(M)$ for which $\Phi(\alpha) = 0$, then by performing suitable "surgery" on $M = M_g^2$, we can change it to a surface with a smaller number of handles; i.e. we can find an equivalent framed normal bundle over a surface of smaller genus. This process involves in essence sewing in a thickened disc (of thickness 2ε) across one of the g "holes" of M (namely that determined by α), and then taking as our new surface M_* the surface M together with the top and bottom of the disc, but with the vertical boundary of the disc (now identified with a band in M) removed from M. (It can be shown that the condition $\Phi(\alpha) = 0$ is both necessary and sufficient for the resulting bundle over M_* to be equivalent to the original one (see Exercise 3 below).

In more precise terms the procedure is as follows. Let α denote also the unknotted, embedded, directed circle $S^1 \subset M$ used to represent $\alpha \in \pi_1(M)$ in the course of defining the Arf function Φ above. We define a map $\varphi: S^1 \times (-\varepsilon, \varepsilon) \to M$, of the "thickened circle" or "band" $S^1 \times (-\varepsilon, \varepsilon) = \partial D^2 \times (-\varepsilon, \varepsilon)$, by taking in the obvious way $\varphi(\partial D^2 \times 0) = \alpha \subset M$, and as images of the vertical segments $x \times (-\varepsilon, \varepsilon)$, segments of length 2ε of geodesics in M normal to α. We now form the 3-dimensional smooth manifold-with-boundary

$$W=(M\times I)\bigcup_{\varphi}(D^2\times(-\varepsilon,\varepsilon)),$$

where the two pieces are glued together using the function

$$\varphi\colon \partial D^2\times(-\varepsilon,\varepsilon)\to M\times 1.$$

The boundary ∂W is then the (disjoint) union of $M\times 0$ and the manifold M_* we were seeking to construct. It is not difficult to see that M_* does indeed have genus $g-1$.

We now embed W in $\mathbb{R}^{n+2}\times[0,1]$ in such a way that it approaches the boundaries $\mathbb{R}^{n+2}\times 0$ and $\mathbb{R}^{n+2}\times 1$ orthogonally, and intersects them in M and M_*:

$$W\cap(\mathbb{R}^{n+2}\times 0)=M,\qquad W\cap(\mathbb{R}^{n+2}\times 1)=M_*.$$

The desired equivalence of the initial framed normal bundle (M,τ^n) with some framed normal bundle over M_*, we leave as an

Exercise

The given field of normal frames τ^n on $M\subset\mathbb{R}^{n+2}$ can be extended to $W\subset\mathbb{R}^{n+2}\times[0,1]$ if and only if $\Phi(\alpha)=0$.

It follows from Exercise 1 that provided $g\geq 2$ there does exist a non-trivial element α of $\pi_1(M)$ (representable by an embedded circle) for which $\Phi(\alpha)=0$. Hence after performing a succession of surgeries of the above kind, we shall arrive finally either at a framed normal bundle over the sphere S^2, or a framed normal bundle over the torus $T^2\subset\mathbb{R}^3\subset\mathbb{R}^{n+2}$. (That there exists no non-trivial element of $\pi_1(T^2)\simeq\mathbb{Z}\oplus\mathbb{Z}$ at which the Arf function takes the value 0, follows from Exercise 1 and the facts: (1) that if α,β denote the obvious generators of $\pi_1(T^2)$, then $\Phi(\alpha)=\Phi(\beta)=\alpha\circ\beta=1$; and (2) that the elements representable by embedded circles in T^2 are precisely those of the form $\alpha^k\beta^l$ with $(k,l)=1$.) Now since $\pi_2(SO(n))=0$ (see Chapter 6, §24.4) it follows that every framed normal bundle over $S^2\subset\mathbb{R}^{n+2}$ is equivalent to the empty bundle, i.e. corresponds to the zero element of $\pi_{n+2}(S^n)$; hence assuming our original framed normal bundle (M,τ^n), $M\subset\mathbb{R}^{n+2}$, non-trivial, the above-described reduction process must end with an equivalent framed normal bundle over T^2. It can be shown (again we omit the details) that over $T^2\subset\mathbb{R}^{n+2}$ there is up to equivalence just one non-trivial framed normal bundle, whence we infer that $\pi_{n+2}(S^n)\simeq\mathbb{Z}_2$ for $n>3$. (We note that by bringing to bear the highly-developed techniques of the theory of 3- and 4-dimensional manifolds, the method outlined here can be made to yield also the groups $\pi_{n+3}(S^n)$.)

In conclusion we mention a few further facts, obtainable by means of complicated algebraic methods:

(i) All of the groups $\pi_{n+k}(S^n)$ are finitely generated (abelian); except for the cases $k=0$ and $n=2l$, $k=2l-1$, they are all finite, and the exceptional groups $\pi_{4l-1}(S^{2l})$ are each isomorphic to the direct sum of $\mathbb{Z}$ and some finite group.

(ii) We list most of the known "stable" groups $\pi_{n+k}(S^n)$, $n>k+1$, in the following table ($k=0,\dots,15$):

0	$\mathbb{Z}$	8	$\mathbb{Z}_2 \oplus \mathbb{Z}_2$
1	$\mathbb{Z}_2$	9	$\mathbb{Z}_2 \oplus \mathbb{Z}_2 \oplus \mathbb{Z}_2$
2	$\mathbb{Z}_2$	10	$\mathbb{Z}_2$
3	$\mathbb{Z}_{24}$	11	$\mathbb{Z}_{504}$
4	0	12	0
5	0	13	$\mathbb{Z}_3$
6	$\mathbb{Z}_2$	14	$\mathbb{Z}_2 \oplus \mathbb{Z}_2$
7	$\mathbb{Z}_{240}$	15	$\mathbb{Z}_{480} \oplus \mathbb{Z}_2$

CHAPTER 6
Smooth Fibre Bundles

§24. The Homotopy Theory of Fibre Bundles (Skew Products)

24.1. The Concept of a Smooth Fibre Bundle

A *smooth fibre bundle* is a composite object, made up of:

(i) a smooth manifold E, called the *total* (or *bundle*) *space*;
(ii) a smooth manifold M, called the *base space*;
(iii) a smooth surjective map $p: E \to M$, the *projection*, whose Jacobian is required to have maximal rank $n = \dim M$ at every point;
(iv) a smooth manifold F, called the *fibre*;
(v) a Lie group G of smooth transformations (self-diffeomorphisms) of the fibre F (where by definition of "Lie transformation group" the action $G \times F \to F$ is smooth on $G \times F$); this group is called the *structure* or *bundle group* of the fibre space;
(vi) a "fibre bundle structure" linking the above entities, and defined as follows. The base space M comes with a particular system of local co-ordinate neighbourhoods U_α (called the *co-ordinate neighbourhoods* or *charts* of the fibre bundle), above each of which the co-ordinates of the direct product are introduced via a diffeomorphism $\varphi_\alpha: F \times U_\alpha \to p^{-1}(U_\alpha)$ satisfying $p\varphi_\alpha(y, x) = x$; each map φ_α will thus induce from the natural co-ordinatization of the direct product $F \times U_\alpha$, a co-ordinatization of the complete inverse image $p^{-1}(U_\alpha)$ of U_α. The transformations $\lambda_{\alpha\beta} = \varphi_\beta^{-1}\varphi_\alpha: F \times U_{\alpha\beta} \to F \times U_{\alpha\beta}$, where $U_{\alpha\beta} = U_\alpha \cap U_\beta$, are called the *co-ordinate transformations* (or *transition functions*) of the fibre bundle. In view of the above condition on the φ_α, every transformation $\lambda_{\alpha\beta}$ has

the form $\lambda_{\alpha\beta}(y, x)=(T^{\alpha\beta}(x)y, x)$; it is further required that for all α, β, x, the transformation $T^{\alpha\beta}(x)\colon F\to F$ is an element of G. Consequently each transition function $\lambda_{\alpha\beta}$ gives rise to a smooth map from the region $U_{\alpha\beta}$ to G:

$$T^{\alpha\beta}\colon U_{\alpha\beta}\to G, \qquad x\mapsto T^{\alpha\beta}(x).$$

Observe also that from the definition of the functions $T^{\alpha\beta}(x)$ it follows that

$$T^{\alpha\beta}(x)=(T^{\beta\alpha}(x))^{-1} \quad \text{and} \quad T^{\alpha\beta}(x)T^{\beta\gamma}(x)T^{\gamma\alpha}(x)=1, \tag{1}$$

where the second equation is understood as holding on the region of intersection $U_\alpha\cap U_\beta\cap U_\gamma$. This noted, the definition of fibre bundle is complete. (We remark that for a general fibre bundle, it is required only that E, M, F be topological spaces and G a topological transformation group.)

The simplest example of a fibre bundle is furnished by the projection of the direct product of two manifolds onto the first factor, with trivial structure group; we call such fibre bundles *trivial*.

Among all fibre bundles, the "principal fibre bundles" and "vector bundles" are of particular importance. A *principal fibre bundle* is defined to be a fibre bundle whose fibre F coincides with the bundle group G, which acts on the fibre $F=G$ by multiplication on the right, i.e. by means of the right translations $R_g\colon G\to G$, $R_g(x)=xg$.

24.1.1. Theorem. *For any principal fibre bundle there is a natural smooth, free, left action of the bundle group G on the total space E with the property that the orbits of E under this action coincide with the fibres $p^{-1}(x)$, $x\in M$ (and so are in natural one-to-one correspondence with the points of the base space M). It follows that every principal fibre bundle is obtained from the free left action of a Lie group of transformations on a manifold E (with "free left action" as defined below).*

Before giving the proof we make the following remark. We first remind the reader yet again that a smooth, left action of a Lie group G on a manifold E,

$$(g, y)\mapsto g(y), \qquad y\in E, \quad g\in G,$$

is by definition one which is smooth in both the arguments g and y (see §5.1), and that a group action is free if for each y we have $g(y)=y$ only if $g=1$. In the present context all smooth actions which occur will tacitly be assumed to satisfy the following two additional conditions:

(i) the orbits should be "uniformly distant" from one another relative to some metric on E (and so not as in Figure 89(c));
(ii) around each point $y_0\in E$ there should be an n-dimensional disc D_ε^n (where $n=\dim M$) not tangential to the orbit containing y_0, which intersects each orbit sufficiently close to the orbit of y_0 in exactly one point (cf. Figure 89(b) where this is not the case).

(In Chapter 4 (see especially §19) we were concerned with the particular

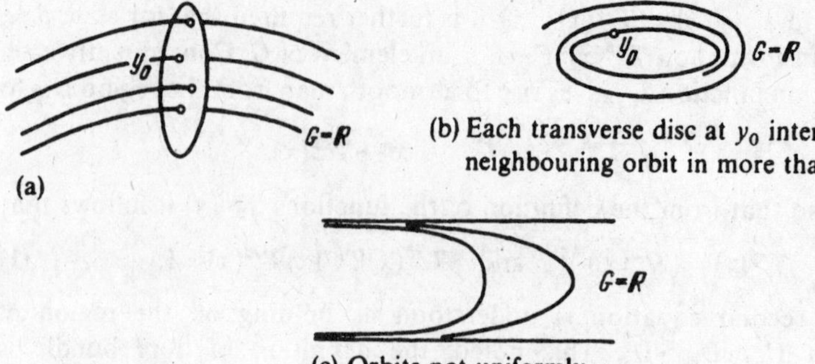

(a)

(b) Each transverse disc at y_0 intersects every neighbouring orbit in more than one point.

(c) Orbits not uniformly distant from one another.

Figure 89

case (of covering spaces) where the manifolds E and M have the same dimensions n, the group G is discrete, and the orbits of the action of G on E are discrete sets of points. The second of the above two conditions is in that situation clearly satisfied since each point $y_0 \in E$ has a neighbourhood D_ε^n containing no other points of the orbit of y_0, and exactly one point of every orbit sufficiently near to it.)

Proof of Theorem 24.1.1. Crucial to the argument will be the fact that left and right translations commute:

$$R_{g_1} L_{g_2}(y) = L_{g_2} R_{g_1}(y) = g_2 y g_1.$$

We define a left action of G on the set $F \times U_\alpha = G \times U_\alpha$ above each co-ordinate neighbourhood U_α, by the formula

$$L_{g_1}(g, x) = (g_1 g, x);$$

this action (which is clearly free) then transfers to an action on the region $p^{-1}(U_\alpha) \subset E$ via the diffeomorphism $\varphi_\alpha: G \times U_\alpha \to p^{-1}(U_\alpha)$ (see part (vi) of the definition of a fibre bundle). For these actions (of G on each of the regions $p^{-1}(U_\alpha)$) to extend to a single action of G on the whole of E, they will have to coincide on the regions of overlap $p^{-1}(U_{\alpha\beta})$ (where $U_{\alpha\beta} = U_\alpha \cap U_\beta$); we now verify this. We have two left actions of G on $p^{-1}(U_{\alpha\beta})$, given by

$$p^{-1}(U_{\alpha\beta}) \xrightarrow{\varphi_\alpha^{-1}} G \times U_{\alpha\beta} \xrightarrow{L_{g_1}} G \times U_{\alpha\beta} \xrightarrow{\varphi_\alpha} p^{-1}(U_{\alpha\beta}),$$

$$p^{-1}(U_{\alpha\beta}) \xrightarrow{\varphi_\beta^{-1}} G \times U_{\alpha\beta} \xrightarrow{L_{g_1}} G \times U_{\alpha\beta} \xrightarrow{\varphi_\beta} p^{-1}(U_{\alpha\beta}).$$

Thus we need to show that for any point $y \in p^{-1}(U_{\alpha\beta})$,

$$\varphi_\alpha L_{g_1} \varphi_\alpha^{-1}(y) = \varphi_\beta L_{g_1} \varphi_\beta^{-1}(y),$$

or, equivalently, applying φ_β^{-1} to both sides,

$$\lambda_{\alpha\beta} L_{g_1} \varphi_\alpha^{-1}(y) = L_{g_1} \varphi_\beta^{-1}(y).$$

Since for each fixed $x \in U_{\alpha\beta}$, $\lambda_{\alpha\beta}$ is a right translation of G (see (vi) in the definition of fibre bundle), and since as noted above left and right translations of G commute, it follows that $\lambda_{\alpha\beta}L_{g_1}=L_{g_1}\lambda_{\alpha\beta}$, and then the above equation follows from $\lambda_{\alpha\beta}\varphi_\alpha^{-1}=\varphi_\beta^{-1}$. Hence the left actions of G on the regions $p^{-1}(U_\alpha)$ agree on the regions of overlap, and therefore combine to yield a left action of G on E, as required. (It is easy to see that this action is free.) □

The structure of a fibre bundle with base space M, bundle group G and fibre F is determined essentially by the transition functions $\lambda_{\alpha\beta}$, or, equivalently, by the prescription of maps $T^{\alpha\beta}\colon U_{\alpha\beta}\to G$ satisfying the conditions (1). It follows that the fibre F is a feature of the fibre bundle which is readily variable in the sense that if we are given a representation of G as a group of transformations of a different fibre F', then the same maps $T^{\alpha\beta}\colon U_{\alpha\beta}\to G$, will define a new fibre bundle with fibre F' (and the same base M), which we shall say is *associated* with the original fibre bundle. In particular the right regular representation of G acting on itself as fibre yields a principal fibre bundle associated with the original fibre bundle; we conclude from this that:

Every fibre bundle can be obtained as a fibre bundle associated with some (unique) principal fibre bundle; hence the problem of classifying fibre bundles in general, reduces to that of classifying the principal fibre bundles.

Important classes of fibre bundles. (a) *Coverings.* Here the fibre is discrete (a set of isolated points), and the group G is the monodromy group of the covering. As noted earlier (in §18.4) the principal fibre bundles are in this case called also "regular coverings"; they were defined in §18.4 in terms of the free action of a discrete group $G=\sigma(\pi_1(M))$ on the covering space. For examples of coverings see §§18, 19.

(b) *Vector bundles* (see §7.1). Here the fibre is $\mathbb{R}^n$, and the group G acts on F as a subgroup of $GL(n,\mathbb{R})$. Naturally distinguished from among these are the *orthogonal vector bundles* ($G\subset O(n)$), and the *complex bundles* ($F=\mathbb{C}^n$, $G\subset GL(n,\mathbb{C})$), including the *unitary vector bundles* ($G\subset U(n)$).

(c) *Fibre bundles related to the tangent bundle.* In §7.1 we introduced in passing the concept of the *tangent n-frame bundle E* over a given manifold M, consisting of pairs (x,τ^n) where x ranges over the points of M, and for each x, τ^n ranges over the ordered bases for the tangent space to M at the point x. This clearly defines a fibre bundle with projection $p\colon E\to M$ and bundle group taken to be $GL(n,\mathbb{R})$. Since the fibre can be naturally identified with G, with right multiplication as the action, the tangent n-frame bundle with base space M is actually a principal fibre bundle. The corresponding action of the group $G=GL(n,\mathbb{R})$ on E (see Theorem 24.1.1) is then easily seen to be given by:

$$A(x,\tau^n)=(x,A(\tau^n)),$$

for $A\in GL(n,\mathbb{R})$, $(x,\tau^n)\in E$.

If the manifold M is endowed with a Riemannian metric (g_{ij}), then it is natural for us to restrict the tangent frames τ^n at each point to be orthonormal with respect to this metric, thereby obtaining a principal fibre bundle $E_O \to M$ with group $O(n)$. If in addition the manifold M is orientable, then by further restricting the permitted frames at each point to belong to the orientation class determined by a particular orientation of M, we obtain a principal fibre bundle $E_{SO} \to M$ with group $SO(n)$. If on the other hand M is a complex manifold of complex dimension n, then it is natural to restrict the frames at each point to complex ones, thence obtaining a principal fibre bundle $E_{\mathbb{C}} \to M$ with group $GL(n, \mathbb{C})$, and the presence on the manifold M of an Hermitian metric gives rise in the analogous way to a principal fibre bundle $E_U \to M$ with group $U(n)$.

Other fibre bundles related to tangent bundles, including those mentioned in §7.1, are obtained via association with the above principal ones. The best-known of these have fibres as follows:

(i) $F = \mathbb{R}^n$ (the tangent bundle);
(ii) $F = S^{n-1}$ (where each point of M is paired with unit tangent vectors, or, equivalently, rays);
(iii) $F = \mathbb{R}P^{n-1}$ (where each point is paired with straight lines or "directions");
(iv) $F = V_{n,k}$ (where each point is paired with orthonormal k-frames in $\mathbb{R}^n$);
(v) $F = \Lambda^k(\mathbb{R}^n)^*$ (skew-symmetric k-forms at each point);
(vi) $F = \mathbb{R}^n \otimes \cdots \otimes \mathbb{R}^n \otimes (\mathbb{R}^n)^* \otimes \cdots \otimes (\mathbb{R}^n)^*$ (tensors of type (p, q)).

(d) *Homogeneous spaces.* For any closed subgroup H of a Lie group G we defined (in §5) the corresponding *homogeneous space* $M = G/H$ of right cosets of H in G, as the quotient space of G determined by the projection $p: G \to G/H = M$. This yields a principal fibre bundle with group H acting on the bundle space G by left multiplication:

$$g \mapsto hg \qquad (g \in G, h \in H).$$

The orbits of this action are then the right cosets of H, i.e. the points of the base space M. Thus a homogeneous space is the base space of a principal fibre bundle. Various examples of homogeneous spaces were considered in §5. (See also some of the examples in §§18, 19.)

(e) *Normal bundles over a submanifold.* Let M be an n-manifold smoothly embedded in an $(n+k)$-manifold endowed with a Riemannian metric. We define (as in §7.2) the *normal bundle* over M to consist of pairs (x, τ) where x ranges over M, and τ ranges over the vectors normal to M at the point x. Clearly this yields a fibre bundle with group $O(k)$ and fibre $\mathbb{R}^k$.

Occasionally the structure group of a fibre bundle will play no essential role, or we shall be led to the consideration of objects which, though like fibre bundles in every other aspect, lack a structure group G, so that the transition functions $\lambda_{\alpha\beta}$ are determined by arbitrary self-diffeomorphisms of the fibre F.

In such cases we shall simply assume as structure group the group of all self-diffeomorphisms of F, denoted by diff F. (If F is orientable, then we can distinguish in diff F the subgroup diff ^+F of orientation-preserving self-diffeomorphisms of F.)

We now turn to the natural question of the most appropriate definition of a "map (or 'morphism') between fibre bundles".

24.1.2. Definition. If $(E, M, p\colon E \to M, F, G)$ and $(E', M', p'\colon E' \to M', F, G)$ are two fibre bundles with common fibre F and common group G, then we say that a map $\tilde{f}\colon E \to E'$ between the total spaces is a *fibre bundle map* (or "morphism") if it preserves the fibre bundle structure, i.e. if:

(i) the map $\tilde{f}$ "respects fibres"; in other words there is a map $f\colon M \to M'$ such that $p'\tilde{f} = fp$ (and then f will be uniquely determined by this condition);

(ii) on the fibre above each point x of E, the map $\tilde{f}_F\colon F \to F$ (obtained essentially by restricting $\tilde{f}$) is a transformation belonging to the structure group G. The precise definition of the maps $\tilde{f}_F$ is as follows: according to part (vi) of the definition of a fibre bundle, above each co-ordinate neighbourhood $U_\alpha \subset M$ we are given a diffeomorphism $\varphi_\alpha\colon F \times U_\alpha \to p^{-1}(U_\alpha) \subset E$, and above each co-ordinate neighbourhood $U'_{\beta'} \subset M'$ a diffeomorphism $\varphi'_{\beta'}\colon F \times U'_{\beta'} \to p'^{-1}(U'_{\beta'})$; hence above each region of the form $W_{\alpha\beta'} = U_\alpha \cap f^{-1}(U'_{\beta'})$ we have the composite map

$$F \times W_{\alpha\beta'} \xrightarrow{\varphi_\alpha} p^{-1}(U_\alpha) \xrightarrow{\tilde{f}} p'^{-1}(U'_{\beta'}) \xrightarrow{(\varphi'_{\beta'})^{-1}} F \times U'_{\beta'},$$

which (again in view of part (vi) of the definition of a fibre bundle) for each point $x \in W_{\alpha\beta'}$ has the form $(y, x) \mapsto (Ty, f(x))$, where $T = T_x$ is some transformation of the fibre F. It is this transformation T (denoted above by $\tilde{f}_F$) which for each $x \in M$ is required to belong to the structure group G.

24.1.3. Definition. A map between two fibre bundles with common base $M' = M$ is called a *fibre bundle equivalence* if the induced map $f\colon M \to M$ of the base is the identity map.

Below (see §24.4) we shall investigate the problem of classifying fibre bundles up to equivalence, particularly in some cases of special interest (e.g. when the base space is a sphere). We shall in particular show that every fibre bundle over the n-dimensional disc D^n (or over $\mathbb{R}^n$) is equivalent to the direct product, i.e. to the trivial fibre bundle over D^n (or $\mathbb{R}^n$).

24.2. Connexions

We now introduce the concept of a "connexion" on a fibre bundle with total space E, base M, projection $p\colon E \to M$, fibre F, and structure group G. To begin with we shall disregard the given structure group G, i.e. as our structure group we shall take instead, as noted above, the group diff F consisting of all self-diffeomorphisms of the fibre.

A fibre bundle equipped with a connexion can be thought of intuitively as follows: we have a family $\{F_x\}$ of spaces (fibres) (where the parameter x ranges throughout the base space M), whose union $\bigcup_x F_x$ is the total space E. Given any path $\gamma(t)$, $a \le t \le b$, in the base M, the "connexion" provides us with a rule for "parallel transporting" the fibre F along the path γ from one end to the other, i.e. a map (in fact a diffeomorphism)

$$\varphi_\gamma\colon F_{x_0} \to F_{x_1}, \qquad x_0 = \gamma(a), \quad x_1 = \gamma(b),$$

satisfying the following natural requirements:

(i) φ_γ should depend continuously on the path γ;
(ii) $\varphi_{\gamma_1\gamma_2} = \varphi_{\gamma_1}\varphi_{\gamma_2}$; $\varphi_{\gamma^{-1}} = (\varphi_\gamma)^{-1}$; if the path γ is constant, then φ_γ should be the identity map;
(iii) φ_γ should be independent of the parametrization of the path γ.

We now give the precise definition of a connexion on a fibre space, and then show how, starting from this definition, one arrives at the maps φ_γ.

24.2.1. Definition. Given a fibre bundle with base space M of dimension n, a *connexion of general type* on the fibre bundle (with structure group disregarded) is a "distribution on E" which associates with each point y of the total space E an n-dimensional tangential "direction" (i.e. an n-dimensional subspace of the tangent space to E at y) which varies smoothly with y, and is transverse to the fibre through y (i.e. is mapped one-to-one under the map of tangent spaces induced by the projection $p\colon E \to M$). These n-dimensional "directions" at each point of E are called the *horizontal directions of the connexion*. A smooth curve $\tilde{\gamma} = \tilde{\gamma}(t)$ in E is said to be *horizontal* if for every t its tangent vector belongs to the horizontal direction at the point $\tilde{\gamma}(t)$.

24.2.2. Lemma. *Any smooth fibre bundle can be endowed with a connexion of general type.*

PROOF. Let (g_{ij}) be any Riemannian metric on the total space E of our (arbitrary) fibre bundle. (That such a metric exists was shown, at least for compact manifolds, in §8.2.) A connexion on the fibre bundle is then obtained by taking the n-dimensional horizontal direction at each point $y \in E$ to be the n-dimensional subspace of the tangent space to E at y, orthogonal to the fibre containing y. □

24.2.3. Lemma. *Given a general connexion on a fibre bundle (E, M, p, F) with compact fibre F, then corresponding to each piecewise smooth path $\gamma(t)$, $0 \le t \le 1$, in the base M, and each point $y_0 \in E$ in the fibre above $\gamma(0)$, i.e. such that $p(y_0) = \gamma(0)$, there is precisely one horizontal path $\tilde{\gamma}(t)$ in E covering $\gamma(t)$ and beginning at y_0, i.e. with the properties $p\tilde{\gamma}(t) = \gamma(t)$, $0 \le t \le 1$, and $\tilde{\gamma}(0) = y_0$.*

PROOF. For any sufficiently small, non-self-intersecting segment δ of the path γ, the complete inverse image $p^{-1}(\delta)$ will be the total space E_δ of a fibre bundle with fibre F and base space δ. The horizontal directions of the given connexion, in the total space E of the original fibre bundle, will determine a unique one-dimensional horizontal direction at each point of E_δ. The integral curves on E_δ of the vector field obtained by taking for example a unit vector in the horizontal direction at each point of E_δ, will then be horizontal curves. The conclusion of the lemma as it applies path segment δ now follows from the existence and uniqueness (given sufficient smoothness) of an integral curve with a prescribed initial point, and then the full conclusion is obtained by repeating the above argument for a (finite) succession of segments exhausting the path γ. (Note that the compactness of the fibre was used (implicitly) to ensure that the integral curves above the segments δ do not make "excursions to infinity".) □

Remark. Without the assumption of compactness of the fibre, the conclusion of the lemma does not in fact necessarily hold; to ensure that it does one needs to place restrictions on the general connexion to ensure that the horizontal curves above path segments in the base space, obtained as in the above proof, do not make excursions towards infinity. For "differential-geometric" connexions (see §25.1), which do take into account the structure group G, this condition is fulfilled automatically.

Thus given a smooth fibre bundle on which there is defined a general connexion, and which has compact fibre (or for which at least the conclusion of Lemma 24.2.3 holds) then in view of that lemma there is, corresponding to any piecewise smooth path $\gamma(t)$, $a \le t \le b$, in the base, a map

$$\varphi_\gamma: F_{x_0} \to F_{x_1}$$

from the fibre above $x_0 = \gamma(a)$ to the fibre above $x_1 = \gamma(b)$, defined by $\varphi_\gamma(y_0) = \tilde{\gamma}(b)$ for each $y_0 \in F_{x_0}$. (That φ_γ is smooth is a consequence of the smoothness of the dependence of the horizontal curve $\tilde{\gamma}$ on its initial point y_0; see the proof of Lemma 24.2.3.) It is obvious that the map φ_γ is independent of the parametrization of the path γ, and that

$$\varphi_{\gamma_1\gamma_2} = \varphi_{\gamma_1}\varphi_{\gamma_2}, \qquad \varphi_{\gamma^{-1}} = (\varphi_\gamma)^{-1}.$$

Thus we have constructed from the definition of a general connexion the promised maps φ_γ between fibres; as was noted earlier they are called *parallel transporting maps of the fibre, determined by the connexion.*

An assignment to each path γ in M, of a map φ_γ between the fibres above the end-points of γ, satisfying the above three conditions, is termed an *abstract connexion* on the fibre bundle. It follows from those conditions that, given any point x_0 of M, the correspondence φ under which each closed path γ in M beginning and ending at x_0, is associated with the map φ_γ, is a homomorphism from the "H-group" $\Omega(x_0, M)$ (see §22.4) to the group $G = \text{diff } F$:

$$\varphi\colon \Omega(x_0, M) \to G = \text{diff } F,$$

$$\gamma \mapsto \varphi_\gamma\colon F \to F.$$

The image $\varphi(\Omega)$, which is a subgroup of G, is called the *group of holonomies* of the given connexion; it generalizes the concept of the monodromy group of a covering, introduced in §19.1.

24.2.4. Definition. A *G-connexion* (or *connexion compatible with the action of G*) on a fibre bundle with structure group G, is a family (or "field") of horizontal directions in the total space E (i.e. a general connexion) with the property that the corresponding parallel transporting maps φ_γ all belong to the group G.

(The existence of G-connexions on fibre bundles will be established in §25; see especially Lemma 25.1.4.)

In practice the structure group of a fibre bundle usually coincides with the group of holonomies. In the next section (§25) we shall give a global definition in differential-geometric terms of the concept of a G-connexion.

24.3. Computation of Homotopy Groups by Means of Fibre Bundles

Like coverings, fibre bundles (with compact fibre) have the "covering homotopy property" (see §18.1).

24.3.1. Theorem. *A smooth fibre bundle with compact fibre has the covering homotopy property with respect to piecewise smooth maps of any manifold K into the base space M and the total space E, and piecewise smooth homotopies of such maps.*

PROOF. Let $F\colon K \times I \to M$ be a piecewise smooth homotopy of the map $f_0\colon K \to M$, and suppose that f_0 is covered by the map $\tilde{f}_0\colon K \times 0 \to E$, i.e. $p\tilde{f}_0 = F|_{K \times 0} = f_0$. By Lemma 24.2.2 we may assume that our fibre bundle has defined on it a general connexion. By Lemma 24.2.3 such a connexion affords a well-defined procedure for covering a piecewise smooth path γ (traced out by a point moving in the base M) by a path in E depending continuously on the path γ and the initial point of the covering path. By covering in this way the paths traced out in M by the image points of K during the homotopy F, we obtain the desired homotopy of $\tilde{f}_0$ covering the given homotopy of f_0. □

It follows that a fibre bundle has the defining property of a "Serre fibration" (see Definition 22.1.1), so that 21.2.1 and 22.2.1 apply to yield immediately the

24.3.2. Corollary. *The homotopy groups* $\pi_j(E, F, y_0)$ *and* $\pi_j(M, x_0)$, *where* $x_0 = p(y_0)$, *are isomorphic, and the sequence*

$$\cdots \to \pi_j(F) \xrightarrow{i_*} \pi_j(E) \xrightarrow{j_*} \pi_j(E, F) \xrightarrow{\partial} \pi_{j-1}(F) \to \cdots \tag{2}$$
$$\pi_j(E) \xrightarrow{p_*} \pi_j(M) \cong \pi_j(E, F).$$

is exact.

Remark. We indicate an alternative construction of the homomorphism $\partial\colon \pi_n(M) \to \pi_{n-1}(F)$ determined by this corollary, which does not involve the relative homotopy group $\pi_n(E, F)$. Let $f\colon D^n \to M$ be a map representing the element α of $\pi_n(M, x_0)$, and so sending the boundary S^{n-1} of the disc to the point x_0. Let a_0 be a fixed point of the boundary S^{n-1} of the disc, and denote by $[a_0, a]$ the chord in D^n joining a_0 to any other point a of the boundary S^{n-1}; then $f[a_0, a]$ will be a closed path in M beginning and ending at x_0. Lift this path to a covering path in the total space of the fibre bundle, beginning at the point y_0 (where $p(y_0) = x_0$); the terminal point b of this path will also be a point of the fibre $p^{-1}(x_0)$ above x_0. We now define a map $\hat{f}\colon S^{n-1} \to F$ by setting $\hat{f}(a) = b$, and leave it as an exercise to show that this map is homotopic to $\partial\alpha$ where ∂ is the "boundary homomorphism" determined by Theorem 22.2.1.

From this alternative definition of the boundary homomorphism $\partial\colon \pi_n(M) \to \pi_{n-1}(F)$, it easily follows that in particular for a trivial fibre bundle ∂ is the zero homomorphism. (Verify this!)

We now apply the "fibre bundle exact sequence" (2) to the computation of certain homotopy groups. (Recall that in §22.2 from such homotopy exact sequences we were able to obtain information about the homotopy groups of covering spaces and path spaces.)

Examples. (a) Among the simplest principal fibre bundles are the following two:

(i) $\mathbb{R}P^3 \cong SO(3) \xrightarrow{p} S^2$ (with fibre $SO(2) \cong S^1$);
(ii) $S^3 \cong SU(2) \xrightarrow{p} S^2$ (with fibre S^1).

(For the diffeomorphisms $\mathbb{R}P^3 \cong SO(3)$, $S^3 \cong SU(2)$, see Example (b) of §2.2.) The first of these is the principal fibre bundle $SO(3) \to SO(3)/SO(2) \cong S^2$, with homogeneous base space (see §5.2(a)); alternatively it can be interpreted as the unit tangent bundle over the 2-sphere S^2. The fibre bundle (ii) is the analogous principal fibre bundle $SU(2) \to SU(2)/U(1) \cong S^2$, where $U(1)$ is identified with the subgroup of $SU(2)$ consisting of the diagonal matrices; this is known under the name of the *Hopf fibering*, or *Hopf bundle*. Since by the results of §22.2 we have $\pi_j(\mathbb{R}P^3) \simeq \pi_j(S^3)$ for $j > 1$, it follows that as far as the higher homotopy groups are concerned the fibre bundles (i) and (ii) yield the same information; we shall therefore confine our attention to (ii), for which the homotopy exact sequence (2) takes the form

$$\cdots \pi_j(S^1) \xrightarrow{i_*} \pi_j(S^3) \xrightarrow{p_*} \pi_j(S^2) \xrightarrow{\partial} \pi_{j-1}(S^1) \xrightarrow{i_*} \cdots .$$

Since by Example (a) of §22.2 we have $\pi_j(S^1)=\pi_{j-1}(S^1)=0$ for $j>2$, this yields the exact sequence

$$0 \to \pi_j(S^3) \xrightarrow{p_*} \pi_j(S^2) \to 0 \qquad (j>2),$$

whence $\pi_j(S^3) \simeq \pi_j(S^2)$ for all $j>2$. Thus in particular $\pi_3(S^2) \simeq \pi_3(S^3) \simeq \mathbb{Z}$. (That $\pi_n(S^n) \simeq \mathbb{Z}$ for $n \geq 1$ was noted towards the conclusion of §21.1.)

(b) We now turn to the principal fibre bundle known as the *(general) Hopf bundle*, which has fibre $F=S^1$, and projection map

$$p: S^{2n+1} \to \mathbb{C}P^n$$

defined as follows. We take the unit sphere S^{2n+1} in $\mathbb{R}^{2n+2}$ $(=\mathbb{C}^{n+1})$ to be given by the complex equation $\sum_{j=0}^{n} |z_j|^2 = 1$, and then define the action of the circle group S^1 on this sphere by

$$z \mapsto e^{i\varphi} z \qquad (z=(z_0, \ldots, z_n) \in S^{2n+1}, \; e^{i\varphi} \in S^1).$$

Since the orbits of this action are just the points of $\mathbb{C}P^n$ (essentially by definition of the latter—see §2.2), this gives the desired projection map p.

As already noted (in part) in the preceding example, we know (see the end of §21.1) that

$$\pi_1(S^1) \simeq \pi_{2n+1}(S^{2n+1}) \simeq \mathbb{Z},$$
$$\pi_j(S^1)=0 \;\text{ for } j>1 \;\text{ and }\; \pi_q(S^{2n+1})=0 \;\text{ for } q<2n+1.$$

Hence from the exact sequence (2) as it applies to the (general) Hopf bundle we deduce that (verify it!)

$$\pi_2(\mathbb{C}P^n) \simeq \mathbb{Z}, \qquad \pi_j(\mathbb{C}P^n) \simeq \pi_j(S^{2n+1}) \;\text{ for } j>2$$

whence in particular $\pi_{2n+1}(\mathbb{C}P^n) \simeq \mathbb{Z}$.

(c) We next consider the tangent n-frame bundle on the sphere S^n, which may be identified (see §5.2, Example (a)) with the principal fibre bundle given by

(i) $SO(n+1) \to SO(n+1)/SO(n) \cong S^n$, with fibre $SO(n)$ and homogeneous base space.

We shall consider also the associated fibre bundle of k-frames given by

(ii) $V_{n+1,k+1} \to S^n$ (with fibre $V_{n,k}$).

(Recall from §5.2, Example (c) that $V_{n,k}$ is by definition the manifold whose points are the orthonormal k-frames in an n-dimensional Euclidean (i.e. inner product) space, and that $V_{n+1,k+1} \cong SO(n+1)/SO(n-k+1)$, whence the definition of the fibre bundle (ii) as associated with that given by (i).) In the case $k=1$, this becomes the fibre bundle with projection $V_{n+1,2}\to S^n$ and fibre $V_{n,1}\cong S^{n-1}$, i.e. the fibre bundle of unit tangent vectors on S^n.

To begin with we consider the exact sequence (2) as it applies to the fibre bundle given by (i). That exact sequence takes in this case the form

$$\cdots\pi_{j+1}(S^n)\xrightarrow{\partial}\pi_j(SO(n))\xrightarrow{i_*}\pi_j(SO(n+1))\xrightarrow{p_*}\pi_j(S^n)\to\cdots,$$

whence for $j<n-1$ (using $\pi_i(S^m)=0$ for $i<m$; see §21.1) we infer that $\pi_j(SO(n))\simeq\pi_j(SO(n+1))$. On the other hand for $j=n-1$ we obtain the exact sequence

$$\begin{array}{ccccccc}\pi_n(S^n) & \xrightarrow{\partial} & \pi_{n-1}(SO(n)) & \xrightarrow{i_*} & \pi_{n-1}(SO(n+1)) & \xrightarrow{p_*} & \pi_{n-1}(S^n),\\ \wr\wr & & & & & & \| \\ \mathbb{Z} & & & & & & 0\end{array}$$

from which we conclude that when $j=n-1$ the homomorphism i_* is an epimorphism (i.e. is "onto"), and has cyclic kernel $\partial(\pi_n(S^n))$. If the tangent bundle over the sphere S^n happens to be trivial, as for instance turns out to be the case when $n=3$ (since S^3 can be given a group structure making it a Lie group), then the boundary map ∂ in the above sequence is zero (see the observation following the remark above); hence in particular $SO(4)$ is topologically equivalent to $SO(3)\times S^3$, so that $\pi_j(SO(4))\simeq\pi_j(SO(3))\oplus\pi_j(S^3)$.

It is convenient at this point to deduce from Theorem 24.3.1 (applied to the fibre bundle (i)) the following proposition, of which we made substantial use in §23.

24.3.3. Proposition. *Let M be a manifold of dimension q. If $q<n$, then every map $M\to SO(n+1)$ is homotopic to a map $M\to SO(n)\subset SO(n+1)$; if $q<n-1$ then the inclusion map $SO(n)\to SO(n+1)$ determines a one-to-one correspondence*

$$[M,SO(n)]\leftrightarrow[M,SO(n+1)].$$

PROOF. Let $q<n$ and let $\tilde{f}\colon M\to SO(n+1)$ be any map. The projection of this map $f=p\tilde{f}\colon M\to S^n$ (with p as in (i)) is contractible to a constant map, i.e. is homotopic to a map to a single point of S^n (see the remark towards the end of §17.5); let $F=\{f_t\}\colon M\times I\to S^n$ be a homotopy effecting this, i.e. such that $f_0=f$ and $f_1(M)=s_0\in S^n$. Then the covering homotopy $\tilde{F}=\{\tilde{f}_t\}$ with $\tilde{f}_0=\tilde{f}$, guaranteed by Theorem 24.3.1, deforms $\tilde{f}$ to a map $\tilde{f}_1\colon M\to SO(n)=p^{-1}(s_0)$. This proves the first assertion of the proposition.

Now suppose $q<n-1$, and let $\tilde{f}_0,\tilde{f}_1\colon M\to SO(n)$ be two maps homotopic as maps to $SO(n+1)$, via a homotopy $\tilde{F}\colon M\times I\to SO(n+1)$. The projection $F=p\tilde{F}\colon M\times I\to S^n$ of this homotopy will map the boundary $(M\times 0)\cup(M\times 1)$ to the point $s_0\in S^n$. Since $M\times I$ has dimension $q+1<n$, there is a homotopy Φ_t deforming the map $p\tilde{F}=F=\Phi_0\colon M\times I\to S^n$, to the constant map $M\times I\to s_0$, throughout which the base $M\times 0$ and lid $M\times 1$ of the cylinder continue to be mapped to the single point s_0. The covering homotopy

$\Phi_t: M \times I \to SO(n+1)$ then yields at time $t=1$ a homotopy between $\tilde{f}_0$ and $\tilde{f}_1$ during which the image of M is contained in the single fibre $SO(n) = p^{-1}(s_0)$. The desired one-to-one correspondence now follows from this together with the first assertion of the proposition. □

We now turn to the fibre bundle given by (ii), with fibre $V_{n,k}$. Consider first the case $k=1$:

$$p: V_{n+1,2} \to S^n \quad \text{(with fibre } V_{n,1} \cong S^{n-1}).$$

For this fibre bundle the homotopy exact sequence (2) takes the form

$$\cdots \xrightarrow{p_*} \pi_j(S^n) \xrightarrow{\partial} \pi_{j-1}(S^{n-1}) \to \pi_{j-1}(V_{n+1,2}) \to \pi_{j-1}(S^n) \to \cdots .$$

If $j \leq n-1$, then $\pi_{j-1}(S^{n-1}) = 0 = \pi_{j-1}(S^n)$, whence $\pi_{j-1}(V_{n+1,2}) = 0$. If $j=n$, then $\pi_j(S^n) \simeq \mathbb{Z} \simeq \pi_{n-1}(S^{n-1})$, $\pi_{j-1}(S^n) = 0$, and we obtain the exact sequence

$$\begin{array}{ccccc} \pi_n(S^n) & \xrightarrow{\partial} & \pi_{n-1}(S^{n-1}) & \to \pi_{n-1}(V_{n+1,2}) \to 0. \\ \wr\wr & & \wr\wr & \\ \mathbb{Z} & & \mathbb{Z} & \end{array}$$

Thus in order to deduce from this the structure of $\pi_{n-1}(V_{n+1,2})$, we need to find ∂. To this end we consider a tangent vector field ξ on the sphere S^n with exactly one singular point, which we denote by s_0. (This singular point will of necessity be degenerate in view of the results of §15.2; its index, in the sense of Definition 14.4.2, can be seen to be zero if n is odd and 2 (up to sign) if n is even.) In terms of this vector field we define a map $\tilde{f}$ by

$$\tilde{f}: S^n \setminus \{s_0\} \cong D^n \to V_{n+1,2},$$

$$\tau \mapsto \left(\tau, \frac{\xi}{|\xi|}\right),$$

where τ is any unit vector in $\mathbb{R}^{n+1}$ save that one whose tip is at s_0, and ξ is evaluated at the tip of τ. We leave to the reader the verification that the map $p\tilde{f} = f: D^n \to S^n$ can be extended to the boundary ∂D^n by defining $f(\partial D^n) = s_0$, and has degree 1 (so that f represents a generator of $\pi_n(S^n)$), and also the verification of the fact that restriction of the closure of $\tilde{f}$ to the boundary ∂D^n yields a map from ∂D^n to the fibre $p^{-1}(s_0) = V_{n,1} \cong S^{n-1}$, which has degree equal to the index of the vector field at the singular point s_0. It then follows from the above direct construction of the image of f under the boundary homomorphism (see the remark following Corollary 24.3.2), that as a homomorphism $\mathbb{Z} \to \mathbb{Z}$, ∂ is just multiplication by the integer equal to the index of the vector field ξ at the singular point s_0. Hence

$$\pi_{n-1}(S^{n-1})/\partial\pi_n(S^n) \simeq \begin{cases} \mathbb{Z} & \text{if } n \text{ is odd,} \\ \mathbb{Z}_2 & \text{if } n \text{ is even,} \end{cases}$$

so that

$$\pi_{n-1}(V_{n+1,2}) \simeq \begin{cases} \mathbb{Z} & \text{if } n \text{ is odd,} \\ \mathbb{Z}_2 & \text{if } n \text{ is even,} \end{cases}$$

and, as we saw earlier,

$$\pi_j(V_{n+1,2})=0 \quad \text{for } j<n-1.$$

By considering in succession the fibre bundles given by

$$p\colon V_{n+1,k+1}\to S^n \quad (\text{with fibre } V_{n,k}),$$

one readily infers from the exact sequence (2) that, more generally,

$$\pi_j(V_{n+1,k+1})=0 \quad \text{for } j<n-k,$$

and that $\pi_{n-k}(V_{n+1,k+1})$ is cyclic. (Verify this!)

The fibre bundles over spheres for the unitary and symplectic groups:

$$U(n)\to S^{2n-1} \quad (\text{with fibre } U(n-1)),$$

$$Sp(n)\to S^{4n-1} \quad (\text{with fibre } Sp(n-1)),$$

can be exploited in a similar way. (That S^{2n-1} can be realized as the homogeneous space $U(n)/U(n-1)$ was noted in §5.2(e).) The exact sequence (2) for the first of these fibre spaces readily yields

$$\pi_j(U(n))\simeq\pi_j(U(n-1)) \quad \text{for } j<2n-2,$$

and for the second

$$\pi_j(Sp(n))\simeq\pi_j(Sp(n-1)) \quad \text{for } j<4n-2.$$

Thus the homotopy groups $\pi_j(SO(n))$ for $j<n-1$, the groups $\pi_j(U(n))$ for $j<2n-2$, and the groups $\pi_j(Sp(n))$ for $j<4n-2$ are independent of n; they are for this reason denoted simply by $\pi_j(SO)$, $\pi_j(U)$, $\pi_j(Sp)$, and (as for the homotopy groups of spheres) termed *stable*.

(d) We next consider the unit tangent bundle over a closed orientable surface (i.e. over a sphere-with-g-handles):

$$p\colon E\to M_g^2 \quad (\text{with fibre } S^1).$$

Thus the points of E are the pairs (x, τ) where $x\in M_g^2$ and for each x, τ runs over the unit tangent vectors to M_g^2 at x. For $g=0$ we have, as noted already in Example (a) above, $E\simeq SO(3)$, while for $g=1$ it turns out that $E\cong S^1\times M_g^2=S^1\times T^2=T^3$; hence we shall restrict attention to the cases $g\geq 2$. The homotopy exact sequence for our fibre bundle has the form

$$\cdots\to\pi_i(S^1)\xrightarrow{i_*}\pi_i(E)\xrightarrow{p_*}\pi_i(M_g^2)\xrightarrow{\partial}\pi_{i-1}(S^1)\to\cdots.$$

For $i>1$ we have $\pi_i(S^1)=\pi_i(M_g^2)=0$, since the universal covering spaces for S^1 and M_g^2 are contractible, i.e. homotopically equivalent to a point (see Corollary 22.2.2 and Corollary 20.11). Hence $\pi_i(E)=0$ for $i>1$. For $i=1$ we have the short exact sequence

$$0\to\underset{\underset{\mathbb{Z}}{\wr\wr}}{\pi_1(S^1)}\xrightarrow{i_*}\pi_1(E)\xrightarrow{p_*}\pi_1(M_g^2)\to 0. \tag{3}$$

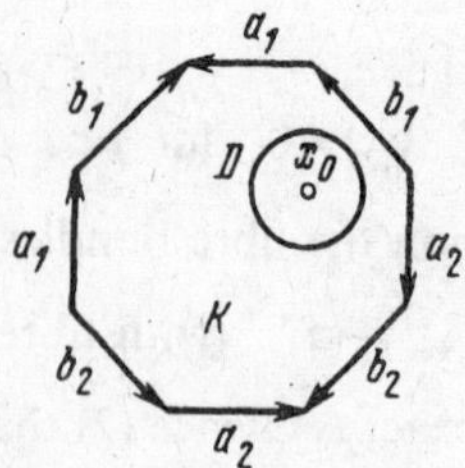

Figure 90

Denote the natural generator of the infinite cyclic group $\pi_1(S^1)$ by τ, and the canonical generators of the group $\pi_1(M_g^2)$ by $a_1, \ldots, a_g, b_1, \ldots, b_g$; the latter satisfy the relation

$$a_1 b_1 a_1^{-1} b_1^{-1} \cdots a_g b_g a_g^{-1} b_g^{-1} = 1, \tag{4}$$

and all relations on these generators are consequences of this one. The paths a_j, b_j determine "canonical cuts" of the surface M_g^2, as a result of which the surface is transformed into the $4g$-gon Q_{4g} depicted in Figure 90 (cf. Figures 61, 76). Write $i_*(\tau) = \bar{\tau} \in \pi_1(E)$, and choose elements $\bar{a}_1, \ldots, \bar{a}_g$, $\bar{b}_1, \ldots, \bar{b}_g$ in $\pi_1(E)$ such that $p_*(\bar{a}_j) = a_j$, $p_*(\bar{b}_j) = b_j$. By the exactness of the sequence (3) we have that $i_*\pi_1(S^1)$ is a normal subgroup of $\pi_1(E)$, and that $\pi_1(E)$ is generated by its elements $\bar{\tau}, \bar{a}_j, \bar{b}_j$. In view of the normality of the subgroup generated by $\bar{\tau}$, and the fact that $\pi_1(M_g^2)$ is defined by the single relation (4), it follows that $\pi_1(E)$ is presented by the defining relations

(1) $\bar{a}_j \bar{\tau} \bar{a}_j^{-1} = \bar{\tau}^{\alpha_j}$,
(2) $\bar{b}_j \bar{\tau} \bar{b}_j^{-1} = \bar{\tau}^{\beta_j}$,
(3) $\bar{a}_1 \bar{b}_1 \bar{a}_1^{-1} \bar{b}_1^{-1} \cdots \bar{a}_g \bar{b}_g \bar{a}_g^{-1} \bar{b}_g^{-1} = \bar{\tau}^{\gamma}$,

where the integers $\alpha_j, \beta_j (j = 1, \ldots, g)$ and γ are to be determined.

We shall now show that in fact $\alpha_j = \beta_j = 1$ and $\gamma = 2 - 2g$. To see that $\alpha_j = \beta_j = 1$, observe first that by §17.2 the conjugation $\bar{\tau} \mapsto \bar{a}_j \bar{\tau} \bar{a}_j^{-1}$ amounts to "parallel transporting" the fibre S^1 (representing the element $\bar{\tau}$ of $\pi_1(E)$) along the path $a_j = p(\bar{a}_j)$ in the base M_g^2. The orientability of M_g^2 ensures that the resulting self-diffeomorphism of S^1 (as the set of unit tangent vectors to the (single) end-point of the path a_j) is orientation-preserving, whence $\alpha_j = \beta_j = 1$ for all j.

We now indicate why it is that $\gamma = 2 - 2g$. Let ξ be any (tangent) vector field on the surface M_g^2 with exactly one zero (i.e. singular point) x_0, not lying on any of the paths a_j, b_j, and write $n(x) = \xi/|\xi|$ for $x \neq x_0$. Remove from the surface M_g^2 a small disc D with centre x_0 (as shown in Figure 90). The field $n(x)$ determines a map from $K = Q_{4g} - D \cong S^1 \times I$ to E (defined by $x \mapsto (x, n(x))$), and if we take the lifts $\bar{a}_j, \bar{b}_j$ to E of the paths a_j, b_j, to be those obtained by means of the vector field $n(x)$ (i.e. if we take $\bar{\gamma}(t) = (\gamma(t), n(\gamma(t)))$ for $\gamma(t)$ any of the paths a_j, b_j), then this map $K \to E$ serves as a homotopy

$$\bar{a}_1 \bar{b}_1 \bar{a}_1^{-1} \bar{b}_1^{-1} \cdots \bar{a}_g \bar{b}_g \bar{a}_g^{-1} \bar{b}_g^{-1} \sim n|_{\partial D} \sim \bar{\tau}^{\gamma}$$

between the product path (the boundary of Q_{4g}) and the image of the boundary ∂D of D. It follows that γ is the degree of the Gauss map of ∂D determined by the vector field $n|_{\partial D}$, which is, by Definition 14.4.2, just the index of the singular point s_0. Since by Hopf's theorem (not quite in the form we gave it in §15.2) that index is $2-2g$, we conclude that $\gamma=2-2g$, as claimed.

(e) We conclude with the particular example known as the "Hopf quaternion fibre bundle", which is defined as follows. We first endow $\mathbb{R}^{4n+4}$ with the structure of the $(n+1)$-dimensional quaternion space $\mathbb{H}^{n+1}$ with quaternion co-ordinates $q_0, \ldots, q_n$. In terms of these co-ordinates the unit sphere S^{4n+3} in $\mathbb{R}^{4n+4}$ is given by the equation $\sum_{\alpha=0}^{n}|q_\alpha|^2=1$. There is a natural (left) action of the group $SU(2)=Sp(1)\cong S^3$ of unit quaternions q ($|q|=1$) on this sphere, defined by

$$q(q_0, \ldots, q_n)=(qq_0, \ldots, qq_n).$$

The resulting orbit space $S^{4n+3}/SU(2)$ is clearly just the quaternion projective space $\mathbb{H}P^n$ (see Exercise 6 of §2.3); hence we obtain a principal fibre bundle with group (and fibre) $SU(2)$, and projection

$$S^{4n+3}\rightarrow S^{4n+3}/SU(2)\cong \mathbb{H}P^n.$$

Since $\mathbb{H}P^1\cong S^4$ (§2.3, Exercise 6), when $n=1$ this becomes a fibre bundle over S^4 with total space S^7:

$$S^7\rightarrow S^4 \quad \text{(with fibre } S^3\text{)};$$

this is what is called the *Hopf quaternion fibre bundle*. The exact sequence (2) for this fibre bundle, breaks up into segments of the form

$$0\rightarrow\pi_i(S^7)\rightarrow\pi_i(S^4)\rightarrow\pi_{i-1}(S^3)\rightarrow 0,$$

since, in view of the fact that the embedding of the fibre S^3 in S^7 is homotopic to a constant map, the inclusion homomorphisms $i_*: \pi_i(S^3)\rightarrow\pi_i(S^7)$ are all zero. It follows at once that $\pi_7(S^4)$ is an infinite group.

24.4. The Classification of Fibre Bundles

By way of leading up to the classification theorem we first consider the following principal fibre bundles over the Grassmannian manifolds $G_{n,k}$ and their analogues (see §5.2):

(a) projection $V_{n,k}\rightarrow G_{n,k}$, fibre $O(k)$, where $G_{n,k}$ is the Grassmannian manifold whose points (the reader will recall) are the k-dimensional planes in $\mathbb{R}^n$ passing through the origin, and the projection is defined via the realizations (given in §5.2) of $V_{n,k}$ and $G_{n,k}$ as the homogeneous spaces $O(n)/O(n-k)$ and $O(n)/(O(k)\times O(n-k))$ respectively;

(b) projection $V_{n,k}\rightarrow \hat{G}_{n,k}$, fibre $SO(k)$, where $\hat{G}_{n,k}$ is the manifold whose points are the oriented k-dimensional planes in $\mathbb{R}^n$ passing through the origin (so that the manifold $\hat{G}_{n,k}$ forms a 2-sheeted covering space for $G_{n,k}$), and the projection is defined analogously to that of the fibre bundle (a);

(c) projection $V^{\mathbb{C}}_{n,k} \to G^{\mathbb{C}}_{n,k}$, fibre $U(k)$, where $G^{\mathbb{C}}_{n,k}$ is the manifold of complex k-dimensional planes in $\mathbb{C}^n$ passing through the origin, $V^{\mathbb{C}}_{n,k}$ is the manifold of unitary k-frames in $\mathbb{C}^n$, and the projection is analogous to the preceding ones;

(d) projection $V^{\mathbb{H}}_{n,k} \to G^{\mathbb{H}}_{n,k}$, fibre $Sp(n)$, where $G^{\mathbb{H}}_{n,k}$ is the manifold of k-dimensional quaternion planes in $\mathbb{H}^n$ through the origin, $V^{\mathbb{H}}_{n,k}$ is the manifold of orthogonal quaternion k-frames, and the projection is analogous to the preceding ones.

In Example (c) of the preceding subsection we showed that $\pi_j(V_{n,k})=0$ for $j<n-k$; a similar argument shows that the analogous result holds over $\mathbb{C}$ and $\mathbb{H}$. Thus

$$\begin{aligned} \pi_j(V_{n,k})&=0 \quad \text{for } j<n-k, \\ \pi_j(V^{\mathbb{C}}_{n,k})&=0 \quad \text{for } j<2(n-k), \\ \pi_j(V^{\mathbb{H}}_{n,k})&=0 \quad \text{for } j<4(n-k). \end{aligned} \tag{5}$$

If we now fix k and let $n\to\infty$ (i.e. take the union of the ascending sequence $V_{n,k} \subset V_{n+1,k} \subset \cdots$, and of each of its analogues), then from (5) it follows that the resulting (infinite-dimensional) total spaces $V_{\infty,k}$, $V^{\mathbb{C}}_{\infty,k}$, $V^{\mathbb{H}}_{\infty,k}$ have all their homotopy groups zero (and consequently are contractible, i.e. homotopically equivalent to a point; verify this!). These constitute important examples of the following concept, crucial to the classification of fibre bundles.

24.4.1. Definition. A principal fibre bundle $E\to B_G$ (where we now admit "infinite-dimensional manifolds") with structure group G, is called *universal* for the group G if E is contractible (or, equivalently, if all $\pi_j(E)$ are zero). (It can be shown that universal bundles always exist, and that the base B_G of a universal G-bundle is unique up to a homotopy equivalence.) If $\pi_j(E)=0$ for all $j\le n+1$, the fibre bundle is said to be *n-universal* for G.

The classification theorem (whose proof we omit) then asserts the following: *There is a natural one-to-one correspondence between the set of (equivalence classes of) principal fibre bundles with given base M and structure group G, and the set $[M, B_G]$ of homotopy classes of maps from M to the base B_G of a universal fibre bundle for G.* (If $\dim M<n$, then in this statement B_G may be replaced by the base space of any n-universal fibre bundle.)

In fact each principal fibre bundle with structure group G and base M can be obtained from a map $f: M\to B_G$, as the "induced fibre bundle" corresponding to that map; we now define this important concept. Given a fibre bundle with projection $p: E\to M$, fibre F, and structure group G, together with a map $f: M'\to M$, the corresponding *induced fibre bundle* (induced via the map f) has the same fibre F and group G, and has projection $p': E'\to M'$ defined as follows: if the structure of the original fibre bundle is determined by a covering $M=\bigcup_\alpha U_\alpha$ together with transition functions $\lambda_{\alpha\beta}: F\times U_{\alpha\beta}\to F\times U_{\alpha\beta}$, $U_{\alpha\beta}=U_\alpha\cap U_\beta$, then the structure of the induced fibre bundle (with fibre F, group G and base M') is defined by taking the regions $U'_\alpha=f^{-1}(U_\alpha)$ as local co-ordinate neighbourhoods covering M', and as transition functions the maps $\lambda'_{\alpha\beta}: F\times U'_{\alpha\beta}\to F\times U'_{\alpha\beta}$, given by

$$\lambda'_{\alpha\beta}(y, x)=(\acute{T}^{\alpha\beta}(x)y, x), \qquad y\in F,$$
$$\acute{T}^{\alpha\beta}(x)=T^{\alpha\beta}(f(x)).$$

(Thus the co-ordinate neighbourhoods $U'_\alpha\subset M'$, and in a certain sense the maps $\lambda'_{\alpha\beta}$, are just the complete preimages (under the inducing map f) of the corresponding entities defining the structure of the original fibre space $p: E\to M$.) This completes the definition of the induced fibre bundle $p': E'\to M'$. (Note incidentally the obvious map $E'\to E$ covering the map $f': M'\to M$ of the bases.)

Returning to the classification theorem we see from it that the problem of classifying the fibre spaces (with prescribed structure group G) over the sphere S^q reduces to that of determining the set $[S^q, B_G]$ (where the "classifying space" B_G is as before the base space of a universal fibre bundle for G), or, equivalently, to the determination of the homotopy group $\pi_q(B_G)$. Now it follows from the fibre bundle exact sequence (2) as applied (infinite-dimensionality notwithstanding) to a universal fibre bundle for G, that $\pi_j(G)\simeq\pi_{j+1}(B_G)$ (verify this!). Hence considering in particular the above examples of universal fibre bundles we have:

$$\pi_j(O(k))\simeq\pi_{j+1}(G_{\infty,k}); \qquad \pi_j(SO(k))\simeq\pi_{j+1}(\hat{G}_{\infty,k});$$
$$\pi_j(U(k))\simeq\pi_{j+1}(G^{\mathbb{C}}_{\infty,k}); \qquad \pi_j(Sp(k))\simeq\pi_{j+1}(G^{\mathbb{H}}_{\infty,k}).$$

Similarly, from (5) and the exact sequences corresponding to the initial four examples, we obtain

$$\pi_j(SO(k)))\simeq\pi_{j+1}(\hat{G}_{n,k}) \quad \text{provided } j+1<n-k,$$

and analogous isomorphisms for $\pi_j(O(k))$, $\pi_j(U(k))$, $\pi_j(Sp(k))$. These isomorphisms and the preceding ones then yield

$$\pi_j(G_{\infty,k})\simeq\pi_j(G_{n,k}), \qquad \pi_j(\hat{G}_{\infty,k})\simeq\pi_j(\hat{G}_{n,k}) \quad \text{if } j<n-k,$$
$$\pi_j(G^{\mathbb{C}}_{\infty,k})\simeq\pi_j(G^{\mathbb{C}}_{n,k}) \quad \text{if } j<2(n-k),$$
$$\pi_j(G^{\mathbb{H}}_{\infty,k})\simeq\pi_j(G^{\mathbb{H}}_{n,k}) \quad \text{if } j<4(n-k).$$

We conclude this particular discussion by listing universal fibre bundles and noting the relevant homotopy groups $\pi_j(B_G)$ in a few simple cases.

(a) If $G=O(1)\simeq\mathbb{Z}_2$, then we may take

$$B_G=\lim_{n\to\infty}(S^n/\mathbb{Z}_2)=\lim_{n\to\infty}\mathbb{R}P^n=\mathbb{R}P^\infty,$$

(where, as already noted, the "limit" is the direct limit, i.e. union, of the sequence $\mathbb{R}P^1 \subset \mathbb{R}P^2 \subset \cdots$). It is easy to see that here we have $\pi_1(\mathbb{R}P^\infty) \simeq \mathbb{Z}_2$ and, for $j>1$, $\pi_j(\mathbb{R}P^\infty)=0$.

(b) Let $G=\mathbb{Z}_m$, the cyclic group of order m, acting on each odd-dimensional sphere S^{2n+1} (given in $\mathbb{C}^{n+1}$ by the equation $\sum_{\alpha=0}^{n}|z_\alpha|^2=1$) according to the formula

$$a(z_0, \ldots, z_n)=(e^{2\pi i/m}z_0, \ldots, e^{2\pi i/m}z_n) \qquad (a^m=1),$$

where a is any particular generator of $\mathbb{Z}_m$. The corresponding B_G is given by

$$B_G = \lim_{n\to\infty} S^{2n+1}/\mathbb{Z}_m = \lim_{n\to\infty} L_{(m)}^{2n+1} = L_{(m)}^{\infty}.$$

(The orbit spaces $L_{(m)}^{2n+1}=S^{2n+1}/\mathbb{Z}_m$ are called *lens spaces.*) Here (as indeed quite generally for any discrete group) covering space theory shows that for $j>1$, we have $\pi_j(B_G)=0$. (Use Corollary 22.2.2 or alternatively the above-mentioned general isomorphism $\pi_j(G) \simeq \pi_{j+1}(B_G)$.)

(c) Consider $G=U(1)\simeq SO(2)\cong S^1$ acting on the sphere S^{2n+1} (given in $\mathbb{C}^{n+1}$ as in the preceding example) according to the formula

$$(z_0, \ldots, z_n) \mapsto (e^{i\varphi}z_0, \ldots, e^{i\varphi}z_n).$$

The base of the corresponding universal fibre bundle for G is then

$$B_G = \lim_{n\to\infty} S^{2n+1}/S^1 = \lim_{n\to\infty} \mathbb{C}P^n = \mathbb{C}P^\infty.$$

Since $\pi_1(S^1)\simeq\mathbb{Z}$ and for $j>1$, $\pi_j(S^1)=0$, it follows (again from $\pi_j(G)=\pi_{j+1}(B_G)$) that

$$\pi_j(B_G)=\pi_j(\mathbb{C}P^\infty)=0 \quad \text{for } j>2,$$
$$\pi_2(\mathbb{C}P^\infty)\simeq\mathbb{Z}.$$

(d) For the group $G=SU(2)\simeq Sp(1)\cong S^3$ acting on each sphere S^{4n+3} (defined in $\mathbb{H}^{n+1}$ by the equation $\sum_{\alpha=0}^{n}|q_\alpha|^2=1$) according to the formula

$$(q_0, \ldots, q_n) \mapsto (qq_0, \ldots, qq_n),$$

where q denotes any unit quaternion (i.e. point of S^3), the base space of the corresponding universal fibre bundle is

$$B_G = \lim_{n\to\infty} S^{4n+3}/S^3 = \lim_{n\to\infty} \mathbb{H}P^n = \mathbb{H}P^\infty.$$

(The universality of the fibre bundle here follows from the fact that the homotopy groups $\pi_j(S^{4n+3})$ are trivial for $j<4n+3$.)

The classification of G-bundles over the sphere S^n can also be achieved directly, without invoking the concept of a universal fibre bundle. However while the above classification procedure applies to more general fibre bundles (with E, M, F more general topological spaces, and G a topological transformation group), we shall now make definite use in particular of our assumption that G is a Lie group of transformations of F. As noted earlier (see §24.1) it suffices to classify the principal fibre bundles. We begin by describing the principal fibre bundles over the disc D^n.

24.4.2. Lemma. *Every principal fibre bundle over the disc D^n with Lie structure group G is trivial.*

PROOF. Let $p: E \to D^n$ be the projection map of a principal G-bundle, and choose arbitrarily a G-connexion on this fibre bundle, i.e. for each pair of points $x_0, x_1 \in D^n$ and each path γ joining them, choose a transformation $\varphi_\gamma: F_{x_0} \to F_{x_1}$ of the fibre $F_{x_0} \cong F_{x_1} \cong G$, satisfying the requisite conditions (see Definition 24.2.4), in particular the condition that all φ_γ belong to the transformation group G. (As noted earlier, the existence of G-connexions (on fibre bundles with Lie structure group G) will be established in §25.)

Now take x_0 to be the centre of the disc, and for each point $x \in D^n$ denote by γ_x the obvious "line segment" $[x_0 x]$ joining x_0 to x. It is then straightforward to verify that the map $\Phi: D^n \times F_{x_0} \to E$, defined by

$$\Phi(x, y) = \varphi_{\gamma_x}(y), \tag{6}$$

determines an equivalence (see 24.1.3) between the trivial fibre bundle and the given one. This completes the proof of the lemma. □

Suppose now that we have a principal G-bundle over S^n with projection $p: E \to S^n$. Writing S^n as the union of its upper and lower hemispheres D^n_+ and D^n_-, intersecting in the equator S^{n-1}, we obtain from the given fibre bundle over S^n two fibre bundles over D^n, with respective total spaces $p^{-1}(D^n_+)$ and $p^{-1}(D^n_-)$. Then each of the regions of E above D^n_+ and D^n_- can be re-co-ordinatized (as in the above lemma) with the co-ordinates of the direct product $D^n \times F$ (via maps defined as in (6)). Hence the structure of our original fibre bundle over S^n may now be regarded as determined by the two co-ordinate neighbourhoods $U_1 = D^n_+$ and $U_2 = D^n_-$, together with the appropriate transition function λ_{12} defined on $F \times U_{12}$, where $U_{12} = U_1 \cap U_2 = S^{n-1}$, or, equivalently, a map $T^{12}: S^{n-1} \to G$, in terms of which λ_{12} is given by

$$\lambda_{12}(y, x) = (T^{12}(x)y, x), \qquad x \in U_{12}, \quad y \in F.$$

24.4.3. Lemma. *Replacement of the map $T^{12}: S^{n-1} \to G$ by a map homotopic to it, yields a G-bundle equivalent to the original one.*

PROOF. We may assume that the homotopy in question has the form

$$T: S^{n-1} \times [-\varepsilon, \varepsilon] \to G,$$

where the restriction of T to $S^{n-1} \times \{-\varepsilon\}$ is T^{12}. Using this homotopy we construct a third G-bundle as follows. Within S^n extend the hemispheres D^n_- and D^n_+ respectively above and below the equator a distance ε, obtaining discs

$D^n_{\varepsilon,-}$ and $D^n_{\varepsilon,+}$ with intersection the cylinder $S^{n-1}\times[-\varepsilon,\varepsilon]$; this allows us to define a new fibre bundle (with the same projection map $p: E\to S^n$ as the original two) by taking $D^n_{\varepsilon,+}$ and $D^n_{\varepsilon,-}$ as its co-ordinate neighbourhoods and as transition function above their intersection the map determined by T. It is then not difficult to verify that this fibre bundle is equivalent to each of the original ones (with transition functions determined respectively by T^{12} and the map homotopic to it via T). (These equivalences follow in essence by taking as co-ordinate neighbourhoods for the new fibre bundle on the one hand the two discs intersecting in the $(n-1)$-sphere a distance ε below the equator, with transition function determined by T^{12}, and on the other hand the two discs intersecting in the $(n-1)$-sphere a distance ε above the equator.) This completes the proof of the lemma. □

We conclude that: *All G-bundles over S^n are determined up to equivalence by the elements of $\pi_{n-1}(G)$.* In view of this the following list, wherein the groups $\pi_{n-1}(G)$ are identified for $G=SO(k)$, $U(k)$, and $n=2, 3, 4$, allows the classification of G-bundles (for $G=SO(k)$, $U(k)$) over spheres S^n of dimension $n\le 4$. (Some of the isomorphisms listed were established earlier, the remainder we give without proof.)

$$\pi_i(U(1))\simeq\pi_i(SO(2))\simeq\begin{cases}\mathbb{Z} & \text{for } i=1,\\ 0 & \text{for } i>1.\end{cases}$$

$$\pi_i(SO(3))\simeq\begin{cases}\mathbb{Z}_2 & \text{for } i=1,\\ 0 & \text{for } i=2,\\ \mathbb{Z} & \text{for } i=3.\end{cases}$$

$$\pi_i(SO(4))\simeq\pi_i(SO(3))\oplus\pi_i(S^3)\simeq\begin{cases}\mathbb{Z}_2 & \text{for } i=1,\\ 0 & \text{for } i=2,\\ \mathbb{Z}\oplus\mathbb{Z} & \text{for } i=3.\end{cases}$$

For $q\ge 5$, we have

$$\pi_i(SO(q))\simeq\begin{cases}\mathbb{Z}_2 & \text{for } i=1,\\ 0 & \text{for } i=2,\\ \mathbb{Z} & \text{for } i=3.\end{cases}$$

Since (topologically) $U(q)\cong S^1\times SU(q)$, we have

$$\pi_i(U(q))\simeq\pi_i(S^1)\oplus\pi_i(SU(q)).$$

For $q\ge 2$, we have

$$\pi_i(SU(q))\simeq\begin{cases}0 & \text{for } i=1,\\ 0 & \text{for } i=2,\\ \mathbb{Z} & \text{for } i=3.\end{cases}$$

24.5. Vector Bundles and Operations on Them

We shall now investigate vector bundles in greater detail. (Recall from §24.1 that a *vector bundle* is a fibre bundle having as fibre real or complex n-space and as structure group a subgroup of the appropriate general linear, orthogonal or unitary group.) Since the essential structure of a fibre bundle with given projection map $p: E \to M$, is determined by its transition functions $\lambda_{\alpha\beta}$ defined above the intersections $U_{\alpha\beta} = U_\alpha \cap U_\beta$, or, equivalently, by the corresponding maps $T^{\alpha\beta}: U_{\alpha\beta} \to G$ satisfying the conditions (1) (namely $T^{\alpha\beta}(x) = (T^{\beta\alpha}(x))^{-1}$, and $T^{\alpha\beta}(x)T^{\beta\gamma}(x)T^{\gamma\alpha}(x) = 1$ on the intersection $U_{\alpha\beta\gamma} = U_\alpha \cap U_\beta \cap U_\gamma$), the "operations on fibre bundles" appropriate for consideration will be those which preserve the latter conditions. One example of such an operation is that of taking a (smooth) real or complex representation $\rho: G \to GL(n, \mathbb{R})$ or $G \to GL(n, \mathbb{C})$ (more particularly, orthogonal or unitary) or indeed any (smooth) homomorphism $\rho: G \to G'$. Such a homomorphism yields a new fibre bundle with structure group G' and new transition functions determined by $\rho(T^{\alpha\beta}) = \rho \circ T^{\alpha\beta}$ in place of the former $T^{\alpha\beta}$; we call this new fibre bundle a *representation* (via ρ) of the original one, and if the original fibre bundle is denoted by some symbol η say, then we denote the new one by $\rho(\eta)$. Another example is provided by the *direct* (or *Whitney*) *sum* $\eta_1 \oplus \eta_2$ of two vector bundles with fibres $\mathbb{R}^m$ and $\mathbb{R}^n$, groups G_1 and G_2, and projections $p_1: E_1 \to M$ and $p_2: E_2 \to M$ onto the same base M covered by the same co-ordinate neighbourhoods $U_\alpha \subset M$: this is the vector bundle over M with group $G_1 \times G_2$, fibre $\mathbb{R}^m \oplus \mathbb{R}^n = \mathbb{R}^{m+n}$, bundle space E the union over all $x \in M$ of the sets $p_1^{-1}(x) \oplus p_2^{-1}(x)$ (so that E is a subspace of $E_1 \times E_2$), projection p defined by $p(p_1^{-1}(x) \oplus p_2^{-1}(x)) = x$, and, finally, co-ordinate maps $\varphi_\alpha: U_\alpha \times \mathbb{R}^{m+n} \to p^{-1}(U_\alpha)$ defined by

$$\varphi_\alpha(x, y) = (\varphi_\alpha^1(x, y_1), \varphi_\alpha^2(x, y_2))$$

where $y = (y_1, y_2)$, $y_1 \in \mathbb{R}^m$, $y_2 \in \mathbb{R}^n$, and $\varphi_\alpha^1: U_\alpha \times \mathbb{R}^m \to p_1^{-1}(U_\alpha)$, $\varphi_\alpha^2: U_\alpha \times \mathbb{R}^n \to p_2^{-1}(U_\alpha)$, are the co-ordinate maps of η_1, η_2 respectively. Yet another important operation is the *tensor product* $\eta_1 \otimes \eta_2$ of vector bundles η_1, η_2 (with the same base M, etc., as above), defined as the vector bundle over M with the same group $G_1 \times G_2$, with fibre $\mathbb{R}^m \otimes \mathbb{R}^n = \mathbb{R}^{mn}$, bundle space E the union over all $x \in M$ of the sets $p_1^{-1}(x) \otimes p_2^{-1}(x)$, and with appropriate projection p and co-ordinate maps φ_α (whose definition we leave to the reader). (Note that those definitions extend to complex vector bundles.)

In fact, as the latter examples suggest, a quite general argument shows that corresponding to each of the familiar operations on vector spaces (i.e. on the fibres of vector bundles) there is a naturally corresponding operation on vector bundles. We now list the most important of these operations:

(i) the *determinant*, $\det \eta$, of a real (or complex) vector bundle η; this is the line bundle (i.e. with one-dimensional fibre) whose transition functions are given by the maps $\det T^{\alpha\beta}: x \mapsto \det(T^{\alpha\beta}(x))$, on the regions of overlap $U_{\alpha\beta} \subset M$;

(ii) the *dual bundle* η^* with fibre the dual space of linear forms on the fibre of η, and transition functions determined by $(T^{\alpha\beta})^*: x \mapsto (T^{\alpha\beta}(x))^*$, where $(T^{\alpha\beta}(x))^*$ denotes the dual of the transformation $T^{\alpha\beta}(x)$.

(iii) the *complex conjugate* $\bar{\eta}$ of a complex vector bundle η.

(iv) the *complexification* $c(\eta)$ of a real vector bundle η (with fibre space the complexification $\mathbb{C} \otimes_{\mathbb{R}} F$ of the original fibre, and structure group $G \times G$ acting in the natural way), and the *realization* $r(\eta_1)$ of a complex vector bundle η_1; as for the corresponding operations on vector spaces we have $cr(\eta_1) = \eta_1 \oplus \bar{\eta}_1$ and $rc(\eta) = \eta \oplus \eta$ (verify this!);

(v) the (k-fold) *tensor power* $\eta \otimes \cdots \otimes \eta$ of a vector bundle η, and its (k-fold) *exterior power* $\Lambda^k \eta$ (the skew-symmetric part of the tensor power), and (k-fold) *symmetric power* $S^k \eta$ (the symmetric part of the tensor power) (cf. §18.1 of Part I).

In terms of vector bundles and the above operations on them, several earlier concepts now re-emerge as special cases of the following single concept.

24.5.1. Definition. A *cross-section* of a fibre bundle with projection $p: E \to M$, is a (smooth) map $\psi: M \to E$ satisfying $p\psi(x) = x$ for every $x \in M$. (Thus a cross-section is a map ψ defined on M and taking at each point $x \in M$ a value in the fibre F_x above x; in particular a cross-section of the trivial bundle is essentially just an ordinary "scalar" function or map from the base M to the fibre F.)

If τ is the tangent bundle of some manifold M, then a cross-section of τ is what we have hitherto called a vector field on M, and similarly a cross-section of the dual bundle τ^* is just a covector field on M. More generally a tensor (field) of type (k, l) on M can now be regarded as a cross-section of the vector bundle

$$\underbrace{\tau \otimes \cdots \otimes \tau}_{\text{upper indices}} \otimes \underbrace{\tau^* \otimes \cdots \otimes \tau^*}_{\text{lower indices}};$$

thus in particular a general tensor of type $(0, k)$ on M is from this point of view a cross-section of the tensor power $\otimes^k \tau^*$, and a differential k-form on M is a cross-section of the exterior power $\Lambda^k \tau^*$. A symmetric bilinear form on vectors (e.g. a metric (g_{ij})) becomes a cross-section of the symmetric square $S^2\tau^*$. (Note incidentally that, essentially by Theorem 18.2.2 of Part I, if the base M is n-dimensional then the vector bundle $\Lambda^n \tau^*$ coincides with the determinant of τ (see (i) above), and also that the triviality of this bundle is equivalent to the orientability of M (verify this!).)

Among fibre bundles (especially vector bundles) over complex analytic manifolds, the *complex analytic fibre bundles*, defined as those with complex analytic transition functions, are of particular importance. Such for instance are the tangent bundles of complex analytic manifolds, and the vector bundles obtained from these by means of the above operations. We remark that "algebraic fibre bundles" over complex algebraic varieties (especially compact subvarieties of $\mathbb{C}P^n$) are defined similarly. (Note that by "Chow's theorem" every complex analytic submanifold of $\mathbb{C}P^n$ is an algebraic variety, i.e. the set of common zeros of a finite set of homogeneous polynomials in $n+1$ variables.) For instance the general Hopf bundle, which we considered earlier (see Example (b) of §24.3) without bringing in its complex analytic structure (taking $U(1)\cong SO(2)\cong S^1$ as bundle group and S^1 as fibre), furnishes (once recast in suitable form) an important example of an algebraic, complex line bundle over $\mathbb{C}P^n$, with bundle group (as for all complex line bundles) the multiplicative group $\mathbb{C}^*$ of non-zero complex numbers; the bundle space E is the set of pairs (l, x) where l ranges over the points of $\mathbb{C}P^n$ in homogeneous co-ordinates, i.e. regarded as straight lines in $\mathbb{C}^{n+1}$, and x ranges over the points in l, and the projection is defined by $p: (l, x)\mapsto l$; the fibre is thus $\mathbb{C}$.

The (equivalence classes of) complex analytic line bundles over a complex manifold form an abelian group under the operation of tensor product, which, equipped with a suitable complex structure, is called the "Picard variety" of the manifold. The connected component of the identity of this group can be shown to be a complex torus.

In the first two of the following exercises $\tau(M)$ denotes the (complex) tangent bundle of a complex manifold M, and η denotes the Hopf bundle over $\mathbb{C}P^n$.

Exercises

1. Prove that $\tau(\mathbb{C}P^n)\oplus 1 = \underbrace{\eta\otimes\cdots\otimes\eta}_{n+1}$, where 1 denotes the trivial complex line bundle over $\mathbb{C}P^n$.

2. Prove that

$$\Lambda^n\tau(\mathbb{C}P^n)=\det \tau(\mathbb{C}P^n)=\eta^{n+1},$$

where the equals sign denotes equivalence.

(Note that these exercises have (simpler) analogues for the real line bundle $\eta_{\mathbb{R}}$ over $\mathbb{R}P^n$ with group $O(1)\simeq\mathbb{Z}_2$. (The total space of $\eta_{\mathbb{R}}$, defined analogously to the complex case, is called the "generalized (unbounded) Möbius band", diffeomorphic to $\mathbb{R}P^{n+1}$ with a point removed.))

3. Prove that $\xi^*=\xi^{-1}$ for every complex line bundle ξ.

24.6. Meromorphic Functions

An interesting class of "fibre bundles with singularities" is afforded by the families of level curves of meromorphic functions on compact complex manifolds M (in particular algebraic functions defined on projective algebraic varieties $M\subset\mathbb{C}P^q$). (Recall that by definition a *meromorphic function* is just a complex analytic map $f: M \to \mathbb{C}P^1 \cong \mathbb{C} \cup \{\infty\}$; this is equivalent to the usual function-theoretic definition of a meromorphic function as a complex-valued function analytic on M except for some points all of which are "poles" of the function (and which comprise $f^{-1}(\infty)$).)

On any local co-ordinate neighbourhood $U_\alpha \subset M$ with co-ordinates $z_\alpha^1, \ldots, z_\alpha^n$ (where n is the complex dimension of M) a meromorphic function $w = f(z)$ can, as usual, be written as $f(z_\alpha^1, \ldots, z_\alpha^n)$ and the induced map df of tangent spaces defined (see §1.2 where df is denoted by f_*); a *singular fibre* is then defined to be the complete preimage $f^{-1}(f(z_0))$, where $z_0 = (z_{\alpha 0}^1, \ldots, z_{\alpha 0}^n)$ is a point in (for instance) U_α such that $df|_{z=z_0} = 0$ (cf. §10.2). The non-singular fibres, denoted by $F_a = \{x \mid f(x) = a\}$, are then (non-singular) submanifolds of M.

The compactness of M and the analyticity of f together imply that if f is not constant, then it has only finitely many singular points, say $z_1, \ldots, z_m$. If we write $w_1 = f(z_1), \ldots, w_m = f(z_m)$ for the images (assumed distinct) of the singular points (i.e. its singular values), then over the planar region $U_f \subset \mathbb{R}^2$ defined by

$$U_f = [S^2 \cong \mathbb{C}P^1 \cong \mathbb{C} \cup \{\infty\}] \backslash \{w_1, \ldots, w_m\},$$

we have a smooth fibre bundle with fibre $F \cong F_w = f^{-1}(w)$, $w \in U_f$. As structure group of this bundle we may take the fundamental group $\pi_1(U_f)$, which is generated by (the homotopy classes of) the loops $a_1, \ldots, a_m$ indicated in Figure 91; each of these loops a_j determines in the normal way (see §19.1) a monodromy transformation (i.e. a self-map $\varphi_{a_j}: F \to F$, of the non-singular fibre). Our ultimate aim is to determine the effect of such transformations on the first homology group of F.

Consider the (in a sense typical) case where the singular points $z_1, \ldots, z_m$ are non-degenerate, i.e. the symmetric bilinear form $d^2f|_{z_j}$ is non-degenerate for each j (see §10.4); in this case the topological structure of the fibres in a sufficiently small neighbourhood U_j of each singular point z_j, is determined by the "quadratic part" of the function $f - f(z_j) = \Delta f$. This is a consequence of the existence in some neighbourhood of each z_j of a local co-ordinate system $z^1, \ldots, z^n$ (with z_j as origin, and the index j implicit) in which the function Δf (by virtue of Taylor's theorem and the fact that the bilinear form $d^2f|_{z_j}$ is

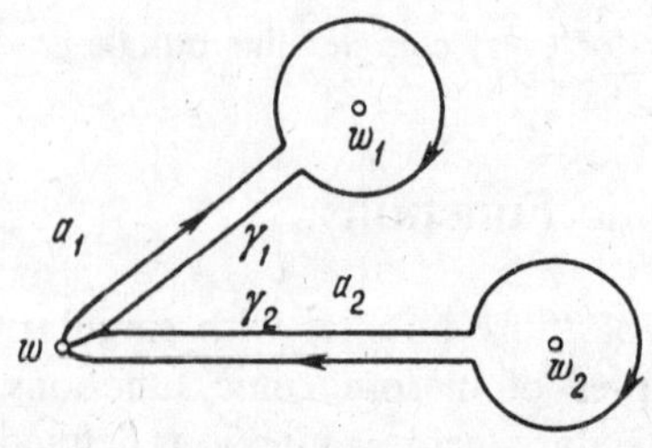

Figure 91

diagonal relative to a suitable system of co-ordinates) takes the form

$$\Delta f(z) = \sum_\alpha (z^\alpha)^2 + O(|z|^3);$$

hence in a sufficiently small neighbourhood $N_\varepsilon(z_j) = \{z \mid |z| < \varepsilon\}$ of z_j, as far as the topology of the fibres is concerned, the remainder $O(|z|^3)$ may be neglected:

$$\Delta f(z) \approx \sum_\alpha (z^\alpha)^2 = q(z).$$

The equation $q(z)=0$ (a complex cone) may thus be taken as defining topologically that part of the singular fibre through z_j within the ε-neighbourhood $N_\varepsilon(z_j)$, while for sufficiently small $\delta \neq 0$ (bounded in terms of ε) equations of the form $q(x) = \delta$ define those portions of nearby non-singular fibres in $N_\varepsilon(z_j)$. We denote these (portions of) so-called "quadrics" by K_δ; thus

$$K_\delta = \left\{ z \mid \sum_{\alpha=1}^{n} (z^\alpha)^2 = \delta \right\} \cap N_\varepsilon(z_j).$$

For each (sufficiently small) δ we now define a sphere $S_\delta^{n-1} \subset K_\delta$ as follows. If δ is real and positive set

$$S_\delta^{n-1} = \left\{ z \mid \sum_j (x^j)^2 = \delta,\ y^1 = \cdots = y^n = 0,\ z^\alpha = x^\alpha + iy^\alpha \right\};$$

if on the other hand $\delta = |\delta| e^{i\varphi}$ with $0 < \varphi < 2\pi$, we take S_δ^{n-1} to consist of those points of K_δ satisfying $\tilde{y}^\alpha = 0$ for $\alpha = 1, \ldots, n$, where the $\tilde{y}^\alpha$ are given by

$$z^\alpha e^{-i\varphi/2} = \tilde{z}^\alpha = \tilde{x}^\alpha + i\tilde{y}^\alpha,$$
$$\sum_\alpha (z^\alpha)^2 = \delta = |\delta| e^{i\varphi}.$$

(Note that in terms of the new co-ordinates $\tilde{z}^\alpha$, K_δ is defined by $\sum (\tilde{z}^\alpha)^2 = |\delta|$.) In view of the equations defining S_δ^{n-1}, it is clear that for δ sufficiently small S_δ^{n-1} will indeed be a whole sphere contained wholly within K_δ.

Exercise

Show that K_δ is diffeomorphic to the tangent bundle on the sphere S^{n-1} (consisting, as the reader will recall, of pairs (s, τ) where $s \in S^{n-1}$ and τ ranges over the tangent space to S^{n-1} at the point s).

If we now let $\delta \to 0$, then the non-singular fibre K_δ "collapses" onto the singular fibre K_0, in a manner expressed precisely by a map (or family of maps)

$$\varphi_\delta \colon K_\delta \to K_0.$$

Under this map the sphere S_δ^{n-1} is sent to a point; thus it "vanishes" so to speak, and for this reason it (or strictly speaking the family of spheres S_δ^{n-1}) is called a *vanishing cycle* of the singular point.

We wish to investigate the monodromy transformation $K_\delta \to K_\delta$ determined by the path $\alpha_j(t) = \delta e^{it}$, $0 \le t \le 2\pi$, around the singular point z_j; the fibre $K_{\delta e^{it}}$ deforms as this path is traced out, yielding finally, when $t = 2\pi$, a monodromy transformation $\sigma\colon K_\delta \to K_\delta$.

We shall from now on restrict ourselves to the case $n=2$. We begin by considering the purely quadratic function (on $\mathbb{C}P^1 \times \mathbb{C}P^1$) given by

$$w = f(z) = (z^1)^2 + (z^2)^2 = u^2 + v^2, \qquad u = z^1, \quad v = z^2,$$

since (as may be gathered from the above) this special case provides the key to the solution in the general case of our problem (which incidentally is that of determining the action of the monodromy group on the first homology group $H_1(F_w)$—see below). Here for each δ the quadric surface K_δ has equation $u^2 + v^2 = \delta = |\delta| e^{i\varphi}$, and the corresponding sphere S_δ^1 is the circle defined in K_δ by the equations $\operatorname{Im} \tilde{u} = \operatorname{Im} \tilde{v} = 0$, where $\tilde{u} = ue^{-i\varphi/2}$, $\tilde{v} = ve^{-i\varphi/2}$. Each non-singular (part-) fibre K_δ is easily seen to be (diffeomorphic to) the cylinder $S_\delta^1 \times \mathbb{R}^1$ (cf. the above exercise).

The map of fibres $K_{|\delta|} \to K_{|\delta| e^{it}}$, determined in the usual way via liftings of the path $\alpha(t) = |\delta| e^{it}$, $0 \le t \le 2\pi$, where we may suppose that initially δ is real and positive ($\delta = |\delta|$), can be written as

$$u \to ue^{it/2}, \qquad v \to ve^{it/2}. \tag{7}$$

Hence putting $t = 2\pi$ we obtain the desired monodromy transformation in the form

$$u \to ue^{i\pi} = -u, \qquad v \to ve^{i\pi} = -v.$$

Our immediate aim is now (for reasons given below) to apply a suitable deformation of the fibres, having the effect of changing the family of transformations $K_{|\delta|} \to K_{|\delta| e^{it}}$ (determined by traversal of the paths from $|\delta|$ to $|\delta| e^{it}$ for various δ) into a family of "traversal" transformations coinciding with (7) in some small neighbourhood of the vanishing cycle S_δ^1, but transforming "canonically" the region of K_δ outside some larger (though still small) neighbourhood of $S_\delta^1 \subset K_\delta$. (Here by "canonical" transformations we mean the diffeomorphisms between the various $K_\delta \backslash S_\delta^1$ defined by the "degenerate" maps $\varphi_\delta\colon K_\delta \to K_0$ (see above), each of which is one-to-one off the circle S_δ^1.) We need the family of traversal transformations to be of this form in order to be able to apply the local information in the neighbourhood of each singular point, framed in terms of quadrics, to the solution of the global problem of calculating the action of the monodromy group on the one-dimensional homology group $H_1(F_w) = \pi_1(F_w)/[\pi_1, \pi_1]$ of a non-singular fibre $F = F_w$, for an arbitrary map $f\colon M \to \mathbb{C}P^1$ of a compact, complex 2-manifold M (see the proof of Theorem 24.6.2 below for the details).

Thus, to repeat, we wish in effect to deform the family of traversal maps $K_{|\delta|} \to K_{|\delta| e^{it}}$ so that they coincide with the corresponding canonical maps between the manifolds $K_\delta \backslash S_\delta^1$ outside a small neighbourhood $U_\delta \supset S_\delta^1$, and with the maps defined by (7) inside a smaller neighbourhood V_δ (so that $K_\delta \supset U_\delta \supset V_\delta \supset S_\delta^1$).

With this aim in mind, we introduce convenient co-ordinates on each K_δ as follows. Observe first that since the singular fibre K_0 is defined by the equation $u^2+v^2=0$ or $u=\pm iv$, the manifold $K_0\backslash\{0\}$ falls into two disjoint identical pieces:

$$K_0\backslash\{0\}\cong(S^1\times\mathbb{R}^+)\cup(S^1\times\mathbb{R}^-).$$

On each of these pieces we introduce co-ordinates (ρ, θ) as follows:

$$\begin{aligned} &\rho>0, \qquad u=\rho e^{i\theta}, \qquad v=iu \qquad \text{(on the first piece);}\\ &\rho<0, \qquad u=-\rho e^{i\theta}, \qquad v=-iu \quad \text{(on the second piece).}\end{aligned} \tag{8}$$

For non-zero δ the manifold $K_\delta\backslash S_\delta^1$ has the same form:

$$K_\delta\backslash S_\delta^1\cong(S^1\times\mathbb{R}^+)\cup(S^1\times\mathbb{R}^-),$$

and the restriction of the map φ_δ: $K_\delta\to K_0$, is a diffeomorphism

$$K_\delta\backslash S_\delta^1\cong K_0\backslash\{0\}, \tag{9}$$

by means of which the co-ordinates ρ, θ can be introduced onto each $K_\delta\backslash S_\delta^1$. The co-ordinate θ is taken as starting from zero at the line of intersection of K_δ with the 3-dimensional hyperplane Im $u=0$. (The direction of increase of θ on the respective pieces is determined by (8) and the diffeomorphism (9).) We co-ordinatize the vanishing cycle of each non-singular fibre K_δ by extending the co-ordinates ρ, θ continuously from the first piece of K_δ to S_δ^1; we then have $\rho=0$ on each vanishing cycle. With this co-ordinatization of K_δ, the level curves of the form $\rho=$constant are the orbits under the action of the group of transformations of the form (cf. (7))

$$\begin{aligned} &u\to u\cos\theta+v\sin\theta \qquad (=ue^{i\theta} \text{ on the first piece}),\\ &v\to -u\sin\theta+v\cos\theta \quad (=ve^{i\theta} \text{ on the first piece}),\end{aligned}$$

(with similar formulae for the second piece). With respect to the obvious metric on K_δ, the vanishing cycle is the shortest of these orbits, and $|\rho|$ measures the distance of a point $(u, v)\in K_\delta$ from the vanishing cycle.

24.6.1. Lemma. *For every* $\varepsilon>0$ *the traversal maps* $K_{|\delta|}\to K_{|\delta|e^{it}}$ *can be arranged (via a suitable deformation of the fibres of the fibre bundle over* $\mathbb{R}^2\backslash\{0\}$ *defined by the function* $w=u^2+v^2$*) to be such that:*

(i) *for* $|\rho|>2\varepsilon$ *they are canonical (in the sense defined above);*
(ii) *for* $|\rho|<\varepsilon$ *they are as in (7);*
(iii) *the final* $(t=2\pi)$ *monodromy transformation*

$$\sigma\colon K_{|\delta|}\to K_{|\delta|}$$

has on both pieces of $K_{|\delta|}$ *the form (after replacing the co-ordinate* θ *by* $2\pi-\theta$ *on the second piece)*

$$\sigma\colon(\rho, \theta)\to(\rho, \theta+\theta(\rho)),$$

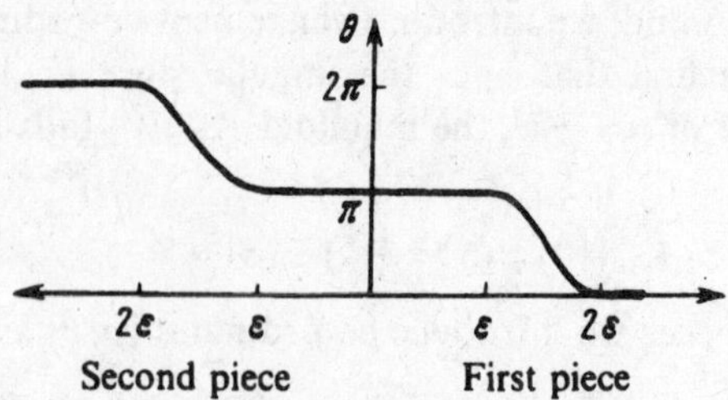

Figure 92

where the function $\theta(\rho)$ has graph as shown in Figure 92. (Thus $\theta(\rho)=\pi$ for $|\rho|\leq\varepsilon$; $\theta(\rho)=0$ for $\rho\geq 2\varepsilon$ (i.e. on the first piece) and $\theta(\rho)=2\pi$ for $\rho\leq -2\varepsilon$ (i.e. on the second piece), and as ρ varies from -2ε to 2ε the curve $(\rho, \theta(\rho))$ winds once round the cylinder $K_{|\delta|}$.)

PROOF. From the way in which the co-ordinate θ was introduced on $K_\delta\setminus S_\delta^1$ (via (8) and (9)), it follows that θ increases in opposite senses on the respective pieces, i.e. an increase in θ corresponds to rotations of the respective halves of the cylinder in opposite directions. Hence by (7) the transformation $K_{|\delta|}\to K_{|\delta|^{it}}$ due to the traversal of the path α from $|\delta|$ to $|\delta|e^{it}$ is given by

$$\theta\to\theta+\frac{\varphi}{2}\quad \text{(on the first piece)},$$

$$\theta\to\theta-\frac{\varphi}{2}\quad \text{(on the second piece)}.$$

If we now change the co-ordinate θ on the second piece, letting it vary from 2π to 0 instead of 0 to 2π (so that the new co-ordinate is 2π less the old), and if we then patch together (in the manner indicated in Figure 93) the resulting

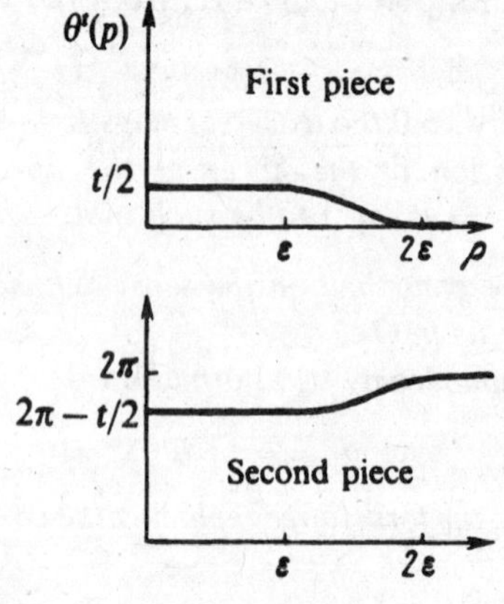

Figure 93

transformation restricted to the region $|\rho|<\varepsilon$, then we obtain a map $K_{|\delta|}\to K_{|\delta|e^{it}}$ given for $0\leq t\leq 2\pi$ by the formula

$$(\rho, \theta)\to(\rho, \theta+\theta^t(\rho))$$

where $\theta^t(\rho)$ has graph as shown in Figure 93. (The continuity of this map on the circle $S^1_\delta(\rho=0)$ follows from the fact that the (new) co-ordinate θ on the second piece now increases in the same sense around the cylinder as the co-ordinate θ on the first piece, but is out of phase with it by the amount 2π.) □

24.7. The Picard–Lefschetz Formula

We shall now apply the foregoing to the original (global) problem. Thus we have a complex analytic map $f: M\to\mathbb{C}P^1\cong S^2$, where M is a compact manifold of complex dimension 2, and all of the singular points $z_1,\ldots,z_m\in M$ of f (with corresponding singular values $w_j=f(z_j)$) are non-degenerate, and we wish to compute the action of the monodromy transformation determined by a closed path a_j, on the fibre

$$F=F_w=f^{-1}(w)\subset M$$

above a regular value $w\in S^2$ (where a_j and w are as in Figure 91).

For each $j=1,\ldots,m$, denote by

$$q_j\in H_1(F_w)=\pi_1/[\pi_1, \pi_1]$$

the cycle obtained by transporting the vanishing cycle S^1_δ corresponding to the singular point z_j (and some sufficiently small δ) from the fibre above $w_j-\delta$ containing it, to the fibre above the general regular value w, via the appropriate lifts of the path γ_j (indicated in Figure 91).

24.7.1. Theorem. *For each $j=1,\ldots,m$, the action on $H_1(F_w)$ of the monodromy transformation $\varphi_{a_j}: F_w\to F_w$, determined by the path $a_j=\gamma_j^{-1}\alpha_j\gamma_j$ (where $\alpha_j(t)=\delta e^{it}$ is the closed path around w_j) is given by the following "Picard–Lefschetz formula":*

$$\begin{gathered}(\varphi_{a_j})_*: H_1(F_w)\to H_1(F_w),\\ (\varphi_{a_j})_*(p)=p+(p\circ q_j)q_j,\end{gathered}\tag{10}$$

where p is any cycle (i.e. element of the group $H_1(F_w)$), q_j is the vanishing cycle (transported to the fibre F_w) of the singular point z_j, and $p\circ q_j$ is the intersection index of these cycles (or more precisely of any maps of circles in general position in F_w representing these cycles). It follows that the transformations $(\varphi_{a_j})_*$ preserve intersection indices; i.e. for any $p_1, p_2\in H_1(F_w)$,

$$p_1\circ p_2=[(\varphi_{a_j})_*p_1]\circ[(\varphi_{a_j})_*p_2].\tag{11}$$

Proof. Consider the cycle $\tilde{p}$ (the result of transporting p above γ_j^{-1}) in the fibre $F_{w_j-\delta}$, transversely intersecting the cycle $\tilde{q}_j = S^1_\delta$ (where $\delta < \varepsilon$ and ε is sufficiently small for the topological picture of the fibre bundle determined by f in the ε-neighbourhood of the singular point $z_j \in F_{w_j}$ to be determined by the quadratic form $(d^2f)_\varepsilon \approx \Delta f$ (see the beginning of the preceding subsection). In view of this and since it can be seen that the effect of traversal of the small circle α_j on the homology class of the cycle $\tilde{p}$ is determined by its effect in the vicinity of the singular point z_j (i.e. near the vanishing cycle contained in $F_{w_j-\delta}$), we may for our present purpose assume the monodromy transformation of $F_{w_j-\delta}$ to be as described in Lemma 24.6.1. It is then not difficult to see that in a neighbourhood of each point of (transverse) intersection of $\tilde{p}$ and S^1_δ, such a monodromy transformation changes $\tilde{p}$ by adding or subtracting a cycle homologous to S^1_δ according as the contribution of that point of intersection to the intersection index $\tilde{p} \circ S^1_\delta$ is $+1$ or -1. (The transformation in effect winds $\tilde{p}$ around the cylinder in one sense or the other exactly once for each point of intersection.) Hence the map of homology classes on $F_{w_j-\delta}$ induced by traversal of α_j is given by

$$[\tilde{p}] \mapsto [\tilde{p}] + (\tilde{p} \circ S^1_\delta)[S^1_\delta].$$

(In particular from $S^1_\delta \circ S^1_\delta = 0$ (or more directly) we have $[S^1_\delta] \mapsto [S^1_\delta]$.) Transporting all cycles from the fibre $F_{w_j-\delta}$ to the fibre F_w (above the path γ_j), we obtain the Picard–Lefschetz formula.

We now verify (11). Write briefly $\varphi = \varphi_{a_j}$; then we have

$$\varphi_* p_1 = p_1 + (p_1 \circ q_j) q_j,$$

$$\varphi_* p_2 = p_2 + (p_2 \circ q_j) q_j,$$

whence

$$\begin{aligned}(\varphi_* p_1) \circ (\varphi_* p_2) &= (p_1 + (p_1 \circ q_j) q_j) \circ (p_2 + (p_2 \circ q_j) q_j) \\ &= p_1 \circ p_2 + (p_1 \circ q_j)(q_j \circ p_2) + (p_1 \circ q_j)(p_2 \circ q_j) \\ &\quad + (p_1 \circ q_j)(p_2 \circ q_j)(q_j \circ q_j),\end{aligned}$$

and the latter expression is equal to $p_1 \circ p_2$ by virtue of the skew-symmetry of the intersection index: $p q = -q \circ p$ (see §15.1). This completes the proof of the theorem. □

EXERCISE

Find the analogue of the Picard–Lefschetz formula in dimensions $n > 2$. (The cases of even and odd n show an essential difference. What is it?)

We conclude by considering briefly the following special case. Let $M = \mathbb{C}P^2$ $(= \mathbb{C}^2 \cup \mathbb{C}P^1_\infty$, where $\mathbb{C}P^1_\infty$ is the one-dimensional complex projective line "at infinity" in $\mathbb{C}P^2$), and take our function to be that determined by a polynomial P_n in two variables:

$$P_n(z_1, z_2): \mathbb{C}^2 \to \mathbb{C}.$$

The fibre bundle with fibres the level surfaces $P_n(z_1, z_2) = \text{const.}$ can be extended in the usual way to a bundle with bundle space all of $\mathbb{C}P^2$ (and base $\mathbb{C}P^1$) using homogeneous co-ordinates u_0, u_1, u_2 where $z_1 = u_1/u_0$, $z_2 = u_2/u_0$.

EXERCISE

Find all singular fibres and compute the monodromy transformations in the hyperelliptic case $P_1(z_1, z_2) = z_1^2 - Q_n(z_2)$.

Here (i.e. in the hyperelliptic case) the non-singular fibres are just the orientable surfaces of genus g (this much follows from Theorem 2.1.4 and the classification theorem for surfaces), where $n = 2g+1$ or $2g+2$. Hence (see for instance §19.4) the fundamental group of a non-singular fibre is defined by the presentation on generators $a_1, b_1, \ldots, a_g, b_g$ with the single relation

$$a_1 b_1 a_1^{-1} b_1^{-1} \cdots a_g b_g a_g^{-1} b_g^{-1} = 1,$$

and the group $H_1 = \pi_1/[\pi_1, \pi_1]$ (isomorphic to $\mathbb{Z}^{2g}$) can be thought of as the $2g$-dimensional integral lattice (with generating "vectors" $[a_1], [b_1], \ldots, [a_g], [b_g]$) on which the monodromy transformations act by translation; this is a consequence of the Picard–Lefschetz formula and the following formulae for the intersection indices (valid for the natural choice of generating cycles a_i, b_i):

$$a_i \circ b_j = \delta_{ij}, \qquad a_i \circ a_j = b_i \circ b_j = 0.$$

§25. The Differential Geometry of Fibre Bundles

25.1. G-Connexions on Principal Fibre Bundles

Already in Chapter 6 of Part I we began what was in essence the local study of connexions and curvature on fibre bundles. In order to proceed to the study of connexions and curvature (and the associated topological invariants) in the more general context of a non-trivial fibre bundle over an arbitrary manifold rather than a region of Euclidean space, we now define these concepts anew in appropriate invariant (i.e. co-ordinate free) and more general form.

25.1.1. Definition. A *connexion* (or *G-connexion*) on a principal fibre bundle with total space E, base M, group G and projection $p\colon E \to M$, is a smoothly varying family of horizontal n-dimensional directions (where $n = \dim M$) invariant under the natural left action of the group G on E.

("Horizontal n-dimensional directions" on E were defined above (see Definition 24.2.1), as was the "natural left action" of G on E (see Theorem

24.1.1). We shall show below that the parallel transporting maps determined (as described in §24.2) by such an invariant family of horizontal directions necessarily belong to the group G, whence the equivalence of this definition of a G-connexion with the earlier one (Definition 24.2.4).)

The simplest differential-geometric method of obtaining a G-connexion on a fibre bundle is that which (analogously to the proof of Lemma 24.2.2) exploits the existence of a left-invariant metric (g_{ij}) on the manifold E (i.e. a metric with respect to which the transformations in G are all isometries). If such a metric exists (and we showed in §8.3 that it does exist at least in the case where the Lie group G is compact (and E also is compact)), then the G-connexion is defined by taking as horizontal direction at each point of E the n-dimensional subspace of the tangent space orthogonal to the fibre through that point.

An alternative method, convenient for the definition of curvature and for other applications of connexions on fibre bundles, consists in defining the field of horizontal directions on E by means of an equation of "Pfaffian" type, i.e. by means of a collection of differential 1-forms on E (or, equivalently, a single vector-valued 1-form). We shall now describe this method in detail.

Generalizing the definition given in Exercise 13, §24.7 of Part I, we define for each element τ in the Lie algebra $\mathfrak{g}$ of a Lie group G, a *right-invariant* tangent vector field ξ_τ on G, by

$$\xi_\tau(g) = -(R_g)_*\tau, \qquad g \in G,$$

where $R_g\colon y \mapsto yg$, $y \in G$, denotes right multiplication by g. Each such vector field is preserved by the maps of tangent spaces induced by the right multiplications in G (and conversely each tangent vector field which is right-invariant in this sense clearly has the above form ξ_τ for some $\tau \in \mathfrak{g}$. The map $\xi_\tau \mapsto \tau$ defines a linear bijection, which is in fact a Lie algebra isomorphism, between the Lie algebra of right-invariant tangent vector fields on G under the bracket operation (i.e. commutation—see §23.2 of Part I) on vector fields on the one hand, and the Lie algebra of G on the other. (This isomorphism was noted in the context of the classical matrix Lie groups in Part I, §24.3 and Exercise 13 of §24.7.) In view of this we may identify the Lie algebra of G with the Lie algebra of right-invariant vector fields on G.

In terms of this realization of the Lie algebra $\mathfrak{g}$ of G we now define a canonical 1-form ω_0 on G, with values in $\mathfrak{g}$ (so that it is strictly speaking a collection of 1-forms), as follows: if ξ is any right-invariant tangent vector field on G, then ω_0 evaluated at the tangent vector $\xi(y)$ to the point $y \in G$, is defined by

$$\omega_0(y, \xi(y)) = \xi$$

where, in the manner described above, ξ is regarded as an element of $\mathfrak{g}$. (We shall also use the suggestive notation $-(dg)g^{-1}$ for the form ω_0; here dg is intended to denote a typical tangent vector $\xi(g)$ (where ξ is any right-invariant vector field and g any element of G), and right multiplication by g^{-1}

the translation of $\xi(g)$ back to the identity, yielding $\xi(1)\in\mathfrak{g}$; see also below.) The form ω_0 has the following two properties:

(i) $\qquad \omega_0(y,\xi(y))\neq 0 \quad \text{if } \xi\neq 0;$

(ii) $d\omega_0(y,\xi_1(y),\xi_2(y))=\omega_0(y,[\xi_1,\xi_2](y))=[\omega_0(y,\xi_1(y)),\omega_0(y,\xi_2(y))]$
$$=[\xi_1,\xi_2].$$

The first of these is obvious, while the second follows directly by specializing Cartan's formula (25.2.3 of Part I).

We now examine the effect on the form ω_0 of left translations of the Lie group G. Via the induced map, a left translation $y\mapsto gy$ $(y,g\in G)$ sends each right-invariant vector field ξ on G to another such vector field $\eta=g_*\xi$, denoted also by $\mathrm{Ad}\,(g)\xi$. If G is a matrix Lie group then the induced map of tangent spaces sends the tangent vector $\xi(1)=y'(0)$ (where $y(t)$ is an integral curve of ξ) at the point $y(0)=1$, to the tangent vector $gy'(0)$ at the point g (where here the products are ordinary matrix products). On translating this tangent vector back to the identity, we obtain the vector $gy'(0)g^{-1}$; it follows that the right-invariant vector field $\mathrm{Ad}\,(g)\xi=g_*\xi$ is determined by $g\xi(1)g^{-1}$, just as ξ is determined by $\xi(1)=y'(0)$. Hence when G and $\mathfrak{g}$ are given in matrix form, then as the notation suggests (see §3.1), the transformation $\mathrm{Ad}\,(g)$ is determined by the corresponding inner automorphism of G, i.e. is given by $\xi\mapsto g\xi g^{-1}$, for all ξ (representing the elements of $\mathfrak{g}$), where the multiplication is ordinary matrix multiplication. (In this case the notation $\omega_0=-(dg)g^{-1}$ also takes on a more literal meaning.) The effect of a left translation $y\mapsto gy$ on the form ω_0 is given by

$$(g^*\omega_0)(y,\dot{\xi}(y))=\mathrm{Ad}\,(g)\omega_0(gy,(g_*\xi)(y)).$$

25.1.2. Definition. A *differential-geometric G-connexion* on a principal fibre bundle $p\colon E\to M$ with group G, is a 1-form ω on E taking its values in the Lie algebra $\mathfrak{g}$ of G, with the following two properties:

(i) the "normalization" property: restriction of the form ω to the fibre G yields the above-defined form $\omega_0=-(dg)g^{-1}$;

(ii) invariance property: under the natural left action of the group G on E (see Theorem 24.1.1) we have that

$$g^*\omega=\mathrm{Ad}\,(g)\omega=g\omega g^{-1}, \tag{1}$$

where we are now using the notation $g\omega g^{-1}$ (for $\mathrm{Ad}\,(g)\omega$), somewhat loosely for general (i.e. not necessarily matrix) Lie groups.

25.1.3. Lemma. (i) *Given any differential-geometric G-connexion ω on a principal fibre bundle $p\colon E\to M$ with group G, the equation $\omega=0$ defines a G-invariant family of horizontal directions on E (i.e. a G-connexion in the sense of Definition* 25.1.1).

(ii) *Conversely, any G-invariant family of horizontal directions on E is defined* *by an equation of the form $\omega=0$, for some differential-geometric G-connexion ω.*

PROOF. (i) Since the restriction of the given form ω to G is the form ω_0 which, as noted above, does not vanish on non-zero tangent vectors to G, it follows that at each point $y \in E$, the solutions of the equation $\omega(y, \tau)=0$ for the unknown tangent vector τ to E at y, form a subspace of dimension n (the dimension of the base M) of the tangent space T_yE to E at y, transversal to the tangent space to the fibre through y (i.e. modulo the latter, spanning the whole of T_yE). Hence the equation $\omega=0$ does indeed define a family of horizontal n-dimensional directions on E. The G-invariance of the family thus defined is immediate from the invariance property in Definition 25.1.2.

(ii) For the converse, given a G-invariant family of horizontal n-dimensional directions on E, we wish to construct from it a 1-form ω which at each point $y \in E$ is a linear map from T_yE to the Lie algebra $\mathfrak{g}$ of G, and has properties (i) and (ii) of Definition 25.1.2. Now since the structure group G acts on each fibre via the right multiplications R_g: $y \mapsto yg$, and since by its definition the form ω_0 on G is right-invariant:

$$R_g^*\omega_0=\omega_0,$$

it follows that the form ω_0 can be defined as a form on the fibre F_x above any point $x \in M (F_x=p^{-1}(x) \cong G)$ independently of the choice of local co-ordinate neighbourhoods of E with the appropriate direct-product structure (as in the definition of fibre bundle (in §24.1)). By its definition, the form ω_0, now regarded as a form on an arbitrary fibre F_x, determines at each point $y \in F_x$ an isomorphism between the tangent space T_yF to that fibre at y, and the Lie algebra $\mathfrak{g}$:

$$\omega_0: T_yF \to \mathfrak{g}.$$

The inherent transversality property of the given family $\{\mathbb{R}_y^n\}$ of horizontal n-dimensional directions is equivalent to the following direct decomposition of the tangent space T_yE at each $y \in E$:

$$T_yE=\mathbb{R}_y^n \oplus T_yF,$$

furnishing the projection

$$\pi: T_yE \to T_yF, \qquad \pi(\mathbb{R}_y^n)=0.$$

We may then take as the desired form ω the composite map $\omega_0\pi$:

$$T_yE \xrightarrow{\pi} T_yF \xrightarrow{\omega_0} \mathfrak{g}.$$

This ensures that the restriction of ω to each fibre F is ω_0. That the formula (1) holds for this ω is a consequence of the facts that under the natural left action g: $E \to E$, $g \in G$, we have firstly $g^*(\pi\omega_0)=\pi(g^*\omega_0)$ (in view of the G-invariance of the given family $\{\mathbb{R}_y^n\}$ of horizontal directions), and secondly that the set of values of the form ω_0 transforms according to the formula (1) (see the discussion immediately preceding Definition 25.1.2). □

25.1.4. Lemma. *Any principal fibre bundle can be endowed with a differential-geometric G-connexion (in the sense of Definition* 25.1.2).

PROOF. Suppose that the structure of the given principal fibre bundle $p: E \to M$, is given (in the usual way; see the definition in §24.1) by specifying local co-ordinate neighbourhoods U_α covering M, and diffeomorphisms $\varphi_\alpha: G \times U_\alpha \to p^{-1}(U_\alpha)$. We may assume (by choosing if necessary a finer covering of M by charts) that there exists a partition of unity $\{\psi_\alpha\}$ on M corresponding to this covering, i.e. smooth functions $\psi_\alpha: M \to \mathbb{R}$ such that $0 \le \psi_\alpha(x) \le 1$, $\sum_\alpha \psi_\alpha(x) = 1$, and $\psi_\alpha(x) \equiv 0$ for x outside U_α (see the remark following Theorem 8.1.3 for conditions on M ensuring the existence of a partition of unity). On each product $G \times U_\alpha \overset{\varphi_\alpha}{\cong} p^{-1}(U_\alpha)$, choose any differential-geometric G-connexion ω_α (for instance by taking the horizontal n-dimensional direction at each point $\varphi_\alpha(g, x)$ of $\varphi_\alpha(G \times U_\alpha)$ to be the tangent space to $\varphi_\alpha(g \times U_\alpha)$ at $\varphi_\alpha(g, x)$). The desired G-connexion ω on the whole of E may then be taken to be

$$\omega = \sum_\alpha (p^*\psi_\alpha)\omega_\alpha,$$

where the functions $p^*\psi_\alpha: E \to \mathbb{R}$ are defined for each $y \in E$ by $(p^*\psi_\alpha)(y) = \psi_\alpha(p(y))$, i.e. are obtained by "lifting" the ψ_α from the base. Clearly $p^*\psi_\alpha$ vanishes identically outside $p^{-1}(U_\alpha)$, so that the value of the form ω at a tangent vector τ to a point $y \in E$ is given by

$$\omega(y, \tau) = \sum_\alpha (p^*\psi_\alpha)(y)\omega_\alpha(y, \tau),$$

where it is understood that if $y \notin p^{-1}(U_\alpha)$, then since $(p^*\psi_\alpha)(y) = 0$, the contribution of the αth summand is zero. Since each ω_α yields ω_0 on restriction to the fibre G, the restriction of ω to any fibre F_x is

$$\sum_\alpha \psi_\alpha(x)\omega_0 = \omega_0.$$

That ω has property (ii) in Definition 25.1.2 is similarly immediate from the fact that each ω_α has that property (by definition). This completes the proof of the lemma. □

EXERCISE

Show that the set of connexions on a fibre bundle is path-wise connected.

Locally (i.e. above each local co-ordinate neighbourhood U_α of the base M, with co-ordinates $x_\alpha^1, \ldots, x_\alpha^n$ say) a G-connexion ω on E (or rather its restriction to the trivial principal bundle $E_\alpha = p^{-1}(U_\alpha) \overset{\varphi_\alpha}{\cong} F \times U_\alpha$ $(F = G)$) can be given in terms of a 1-form A defined on the base U_α (or any particular cross-section of E_α), taking its values in the Lie algebra $\mathfrak{g}$ of G, as follows: In terms of the co-ordinates (g, x) on E_α the relationship between the form $A = A_\mu\, dx_\alpha^\mu$ and the form $\omega|_{E_\alpha}$ is given by

$$\omega(g, x) = \omega_0(g) + gA_\mu(x)g^{-1}\, dx_\alpha^\mu, \tag{2}$$

where ω_0 is the above-defined canonical 1-form on G with values in $\mathfrak{g}$, and where the form of the term $gA_\mu(x)g^{-1}\,dx_\alpha^\mu=(\mathrm{Ad}\,g)A_\mu(x)\,dx_\alpha^\mu$ is determined by condition (1), which ω, as a G-connexion, must satisfy. In "invariant" (i.e. co-ordinate-free) notation (2) may be re-expressed as follows. The product structure of E_α is defined by a pair of "co-ordinate" maps

$$U_\alpha \xleftarrow{p} E_\alpha \xrightarrow{q} G=F$$

(yielding "co-ordinates" $(p(y), q(y))$ for each $y\in E_\alpha$); in terms of these maps the formula (2) may be re-written invariantly as

$$\omega|_{E_\alpha}=q^*(\omega_0)+qAq^{-1},$$

or

$$\omega(y)=\omega_0(q(y))+q(y)A(p(y))q(y)^{-1},\qquad y\in E_\alpha.$$

We wish to discover the transformation rule for the various 1-forms A, determined by a change of cross-section over the regions of overlap of the various U_α on which they are defined. Each cross-section (see Definition 24.5.1)

$$\psi\colon U_\alpha\to E_\alpha,\qquad p\psi=1,$$

of the fibre bundle E_α over U_α, determines a distinct co-ordinatization of the bundle E_α as a product, in the sense that for each $x\in U_\alpha$ we may take $\psi(x)(\in F_x)$, regarded as an element of G, as the origin of co-ordinates on F_x; a general point y of F_x will then be assigned, under the co-ordinatization determined by ψ, the co-ordinates (g, x) where $g\psi(x)=y$ (so that in particular $\psi(x)$ will be assigned co-ordinates $(1, x)$). Thus the corresponding projection map $q\colon E_\alpha\to G$, will for each $y\in E_\alpha$ be given by $q(y)=g\in G$, where g is defined by $g\psi(x)=y$, $y\in F_x$. It follows that two different cross-sections ψ_1 and ψ_2 of E_α will give rise to two different co-ordinatizations of E_α as a product, the distinction being determined for each point $x\in U_\alpha$ by the transformation $g(x)\in G$ defined by

$$\psi_1(x)=g(x)\psi_2(x). \tag{3}$$

Suppose now that the two cross-sections ψ_1, ψ_2 are cross-sections of the trivial bundle $E_{\alpha\beta}=p^{-1}(U_{\alpha\beta})$ where $U_{\alpha\beta}=U_\alpha\cap U_\beta$, and that as before the change of sections is achieved as in (3) by means of transformations $g(x)\in G$, $x\in U_{\alpha\beta}$, where now we are thinking of $g^{-1}(x)$ as given by the transition function $T^{\alpha\beta}\colon U_{\alpha\beta}\to G$. Then denoting by $q_i\colon E_{\alpha\beta}\to G$ the projection onto G corresponding to the co-ordinatization of $E_{\alpha\beta}$ as a product, afforded by ψ_i, $i=1, 2$, it follows that on the region $E_{\alpha\beta}$ the form ω (our original G-connexion on E) can be expressed in two ways:

$$\begin{aligned}\omega|_{E_{\alpha\beta}}&=q_1^*(\omega_0)+q_1A_\mu^{(1)}q_1^{-1}\,dx^\mu\\&=q_2^*(\omega_0)+q_2A_\mu^{(2)}q_2^{-1}\,dx^\mu.\end{aligned}\tag{4}$$

From the equality of these two expressions we deduce the following transformation rule (or "gauge transformation"; cf. §41 of Part I) for the components A_μ on (cross-sections above) each region of overlap $U_{\alpha\beta}$:

$$A_\mu^{(1)}(x)=g(x)A_\mu^{(2)}g^{-1}(x)-\frac{\partial g(x)}{\partial x^\mu}g^{-1}(x), \qquad g^{-1}(x)=T^{\alpha\beta}(x). \tag{5}$$

Thus this formula determines how the local "connexions" A are to be "glued" together to give the global G-connexion ω. (We invite the reader to carry out the detailed derivation of the formula (5) from (4).)

Note that if the cross-section ψ_2 can be chosen so that $A_\mu^{(2)}\equiv 0$, then the formula (5) simplifies to

$$A_\mu^{(1)}(x)=-\frac{\partial g(x)}{\partial x^\mu}g^{-1}(x).$$

Those G-connexions ω for which this can be achieved are called *trivial connexions*.

25.1.5. Theorem. *Let ω be any differential-geometric G-connexion on a principal fibre bundle p: $E\to M$ with group G, and let γ be a piecewise smooth path in the base M. Then each right translation of G determines a parallel transport of the fibre along the path γ.*

PROOF. Suppose the arc γ has the form $x=x(t)$, $a\le t\le b$. Corresponding to each point g_0 of the fibre $p^{-1}(x(a))\cong G$, we seek a horizontal path $\tilde{\gamma}$ in the total space, beginning at g_0 and covering γ. We may confine ourselves to the case where γ lies wholly in a single chart U_α of M (for which we have of course $p^{-1}(U_\alpha)\cong G\times U_\alpha$): in the general case the path $\tilde{\gamma}$ is obtained by piecing together the arcs above the U_α by means of the transition functions $\lambda_{\alpha\beta}$. In terms of the local co-ordinates (g, x), $g\in G$, $x\in U_\alpha$, for the space $E_\alpha=p^{-1}(U_\alpha)$, the desired curve $\tilde{\gamma}$ will be given by $(g(t), x(t))$ where $g(t)$ is yet to be determined. Since $\tilde{\gamma}$ is to be horizontal (i.e. to have horizontal tangent vector $(\dot{g}(t), \dot{x}(t))$), and since by Lemma 25.1.2 the horizontal directions are given by $\omega=0$, we must have, using (2), that

$$\omega(\dot{g}(t), \dot{x}(t))=-\dot{g}(t)g^{-1}(t)+\dot{x}^\mu(t)g(t)A_\mu(x(t))g^{-1}(t)=0.$$

Denoting by $B(t)$ the function with values in the Lie algebra $\mathfrak{g}$, defined by $B(t)=\dot{x}^\mu(t)A_\mu(x(t))$, we conclude that the desired function $g(t)$ is a solution of the ordinary linear differential equation

$$\dot{g}-gB=0.$$

Since the theory of ordinary differential equations guarantees (under conditions fulfilled by the present context) the existence and uniqueness of a solution $g(t)$ of this equation, defined for all $t\in[a, b]$ and satisfying the initial condition $g(a)=g_0\in G$, we at once infer that the desired parallel transport exists. It remains however to show that such parallel transports are determined by the right translations of the group G; this is immediate from the following lemma. □

25.1.6. Lemma. *Given a function $B: [a, b] \to \mathfrak{g}$, the (unique) solution $g(t)$, $a \le t \le b$, of the equation $\dot{g} - gB = 0$, satisfying the initial condition $g(a) = g_0 \in G$, has the form $g(t) = g_0 f(t)$, where $f: [a, b] \to G$ is independent of g_0.*

PROOF. If $B(t)$ does not depend on t, i.e. $B(t) = \text{const.} \in \mathfrak{g}$, then the equation $\dot{g} = gB$ has the explicit solution (with the prescribed initial condition) $g(t) = g_0 \exp[(t-a)B]$, where $\exp: \mathfrak{g} \to G$ is the exponential map (see §3.1), and this solution is of the required form.

If on the other hand $B(t)$ is not constant we can nonetheless exploit the above explicit solution in the following way. For arbitrary t we subdivide the interval $[a, t]$ into N small subintervals with end-points denoted by $a = t_0 < t_1 < \cdots < t_N = t$. We then have

$$\begin{aligned} g(t_1) &= g_0 + \dot{g}(t_0)(t_1 - t_0) + o(t_1 - t_0) \\ &= g(t_0) + g_0(t_1 - t_0)B(t_0) + o(t_1 - t_0) \\ &= g(t_0) \exp[(t_1 - t_0)B(t_0)] + o'(t_1 - t_0). \end{aligned}$$

Iterating this for the N intervals $[t_{i-1}, t_i]$, we obtain the following N approximations, valid to within $o(t_i - t_{i-1})$, $i = 1, \ldots, N$:

$$g(t_1) \approx g(t_0) \exp[(t_1 - t_0)B(t_0)];$$

$$\cdots\cdots\cdots\cdots\cdots$$

$$g(t_N) \approx g(t_{N-1}) \exp[(t_N - t_{N-1})B(t_{N-1})].$$

It follows that $g(t) = g_0 f(t)$ where

$$f(t) = \lim_{\substack{N \to \infty \\ |t_i - t_{i-1}| \to 0}} \exp[(t_N - t_{N-1})B(t_{N-1})] \cdots \exp[(t_1 - t_0)B(t_0)].$$

Since each factor of the expression under the limit sign belongs to G, so does the product, and therefore also the limit. This completes the proof of the lemma and with it the theorem. □

25.1.7. Corollary. *For differential-geometric G-connexions on principal fibre bundles, Lemma* 24.2.3 (*on the existence of* (*unique*) *horizontal covering paths*) *is valid without the assumption that the fibre be compact.*

We conclude by noting that the presence of a G-connexion allows the definition of the group of holonomies (cf. §19.1) of the bundle as the image under a homomorphism associating with the loops in $\Omega(x_0, M)$ certain (in general not all) right translations of G.

25.2. G-Connexions on Associated Fibre Bundles. Examples

We shall now discuss, in invariant terminology, connexions on arbitrary fibre bundles. Thus suppose we are given a fibre bundle $p_F: E_F \to M$ with structure group G acting as a group of transformations of the fibre F, and let $p: E \to M$ be the principal fibre bundle with which it is associated. Given a G-connexion ω on E we wish to define in terms of it an "associated" connexion ω_F on E_F. The form ω_F, which is to take its values in the tangent spaces at the points of F, is constructed as follows. We first define a form ω_F^0 on F by setting $\omega_F^0(y, \tau) = \tau$ for all $y \in F$ and all tangent vectors τ to F at y. We then require of the form ω_F that its restriction to each fibre F be just ω_F^0, whence it follows that in each region $p_F^{-1}(U_\alpha) \overset{\varphi_\alpha}{\cong} F \times U_\alpha$, in terms of the product co-ordinates (y, x), $y \in F$, $x \in U_\alpha$, $x = p_F(y)$, the form ω_F needs must be given (invariantly) by

$$\omega_F(y, x)(\eta) = \omega_F^0(y)(\eta_y) + A(y, x)(\eta_x) \tag{6}$$

where the form A is yet to be defined, and where η is an arbitrary tangent vector to $p_F^{-1}(U_\alpha) \subset E_F$ at (y, x), with components η_y and $\eta_x (\eta = \eta_y + \eta_x)$ in the tangent spaces to F and U_α respectively. In terms of co-ordinates x_α on the chart $U_\alpha \subset M$, we may write the form A as $A_\mu(y, x)\, dx_\alpha^\mu$.

It remains to define A, or, equivalently, the components $A_\mu(y, x)$. Observe first that each element of the Lie algebra $\mathfrak{g}$ of G naturally determines, via the given action of G on the fibre F, a tangent vector field on F (or in other words a (directional) differential operator on scalar functions defined on F); for if $g(t)$ is a one-parameter subgroup (see §3.1) of G with $g(0) = 1$, the identity transformation of F, then the vector field ξ on F determined by the one-parameter group $g(t)$ of diffeomorphisms (see §23.1 of Part I) is for each $y \in F$ given by

$$\xi(y) = \frac{d}{dt}[g(t)(y)]|_{t=0},$$

which is determined by the element $(d/dt)g(t)|_{t=0}$ of $\mathfrak{g}$. (To put it intuitively, each element of $\mathfrak{g}$ can be regarded as an infinitesimal transformation of F, naturally determining a tangent vector field on F.)

We need to define each of the components $A_\mu(y, x)$ as a tangent vector to F. In the notation of the preceding subsection, let $A_\mu(x)$ denote the "connexion" on U_α determined by the form ω on the principal bundle (the similarity of the notation $A_\mu(y, x)$ with this being intentional!); then for each $x \in U_\alpha$, $A_\mu(x)$ is an element of $\mathfrak{g}$, and so determines, in the manner just described, a tangent vector field ξ on F. We define $A_\mu(y, x)$, where $p_F(y) = x$, to be the value of ξ at y:

$$A_\mu(y, x) = \xi(y), \tag{7}$$

thereby completing the definition of ω_F in terms of ω.

Analogously to the case of the principal bundle (see Lemma 25.1.2), the equation $\omega_F=0$ (with ω_F defined invariantly by (6)) defines an n-dimensional horizontal direction $\mathbb{R}^n_x$ at each $x\in E_F$, the total space of the associated bundle.

We now examine in detail the case of vector bundles, i.e. fibre bundles whose fibre $F\cong\mathbb{R}^m$ has the additional structure of a vector space, on which the bundle group G acts linearly. In this case the elements of the Lie algebra $\mathfrak{g}$ become identified (in the manner described above) with linear vector fields ξ on F, and can therefore be regarded as linear transformations $A_\xi\colon\mathbb{R}^m\to\mathbb{R}^m$. Hence in terms of any basis for $\mathbb{R}^m$, yielding co-ordinates $\eta^1,\ldots,\eta^m$, A_ξ can be expressed as a matrix (a^i_j), and the field ξ is given at $\eta=(\eta^1,\ldots,\eta^m)$ by

$$\xi^i=a^i_j\eta^j.$$

It follows that by their definition (see (7)) each of the fields $A_\mu(\ ,x)$ (with value $A_\mu(\eta,x)$ at $\eta\in\mathbb{R}^m$) can in this context be put into matrix form:

$$A_\mu(\ ,x)=[A_\mu(\ ,x)^i_j]=(a^i_{j\mu}).$$

The upshot is that in a vector bundle with fibre $\mathbb{R}^m$ a G-connexion is defined (locally) by a matrix depending on x and μ:

$$[A_\mu(\ ,x)]^i_j=a^i_{j\mu},\qquad i,j=1,\ldots,m;\quad \mu=1,\ldots,n=\dim M,$$

or as a matrix-valued form $a^i_{j\mu}\,dx^\mu$.

The particular case where our vector bundle is the tangent bundle over M (which has fibre $\mathbb{R}^n$, i.e. $m=n$) was considered (though of course over less general base spaces) in Part I. In line with the terminology introduced there (see in particular §§28.2, 41.3 of Part I) we speak of the *torsion tensor* on M:

$$a^i_{j\mu}-a^i_{\mu j}=T^i_{\ \mu j}=-T^i_{j\mu},$$

and say that the connexion is *symmetric* if $T^i_{\ \mu j}\equiv 0$. Given a connexion on the tangent bundle over M, the corresponding parallel transport of the fibre along a path in M turns out to be a linear transformation (cf. Theorem 29.1.3 of Part I). In terms of local co-ordinates on each chart $U_\alpha\subset M$, a connexion on the tangent bundle determines operations of *covariant differentiation* on the sections ψ of the bundle:

$$\nabla_\mu\psi^i(x)=\frac{\partial\psi^i(x)}{\partial x^\mu}+a^i_{j\mu}(x)\psi^j(x),\tag{8}$$

and thence the operation of *directional covariant differentiation* $\nabla_\delta=\delta^\mu\nabla_\mu$ in the direction of any tangent vector $\delta=(\delta^1,\ldots,\delta^n)$ to the base M (cf. §29.1 of Part I). The fact that in general the operators ∇_μ do not commute amongst themselves leads to the *curvature form* $\Omega_{\nu\mu}=[\nabla_\nu,\nabla_\mu]$ (cf. §§30.1, 41.2 of Part I).

Given a smooth path $\gamma(t)$, $0\le t\le 1$, in M, and a connexion on the tangent bundle over M, the linear operator defined by the corresponding parallel transport of the fibre along the path from $\gamma(0)$ to $\gamma(1)$ is sometimes called (for reasons to be given below) the "chronological exponential operator" and denoted by

$$T\exp\left\{\int_0^1\left(-\nabla_{\dot\gamma(t)}+\frac{d}{dt}\right)dt.\right. \tag{9}$$

Here $T[A_1(t)A_2(t')\cdots]$ is the so-called "chronological product" of two or more non-commuting, time-dependent operators:

$$T[A_1(t)A_2(t')]=\begin{cases}A_1(t)A_2(t') & \text{for } t>t',\\ A_2(t')A_1(t) & \text{for } t<t'.\end{cases} \tag{10}$$

The expression (9), with an arbitrary time-dependent linear operator $A(t)$ in place of $-\nabla_{\dot\gamma(t)}+d/dt$, is defined as follows: subdividing the path $\gamma(t)$ into a variable number N of (small) subintervals with end-points $0=t_0<t_1<t_2<\cdots<t_N=1$, we set

$$\begin{aligned}T\exp\left\{\int_0^1 A(t)\,dt\right\}=\lim_{\substack{N\to\infty\\|t_i-t_{i-1}|\to 0}}&T\{\exp[(t_1-t_0)A(t_0)]\\ &\times\exp[(t_2-t_1)A(t_1)]\cdots\exp[(t_N-t_{N-1})A(t_{N-1})]\}.\end{aligned} \tag{11}$$

On putting $A(t)=-\nabla_{\dot\gamma(t)}+d/dt$, we obtain, as will be shown below, the linear operator defining parallel transport along $\gamma(t)$.

Exercises

1. Establish the following series representation for a continuously time-dependent linear operator $A(t)$:

$$\begin{aligned}T\exp\left\{\int_0^1 A(t)\,dt\right\}=1+\int_0^1 A(t)\,dt+\frac{1}{2}\int_0^1\int_0^1 T(A(t')A(t''))\,dt'\,dt''+\cdots\\ +\frac{1}{n!}\int_0^1\cdots\int_0^1 T(A(t_1)\cdots A(t_n))\,dt_1\cdots dt_n+\cdots.\end{aligned} \tag{12}$$

2. Show that the operator $T\exp\int_0^t A(\tau)\,d\tau=B(t)$ say, satisfies the equation

$$\frac{dB}{dt}=[A(t),B(t)], \tag{13}$$

and that the vector $\eta(t)=B(t)\eta_0$ satisfies the equation

$$\frac{d\eta(t)}{dt}=A(t)\eta(t). \tag{14}$$

Resuming the justification of our claim that the expression (9) is just the parallel transport operator, we recall first that as we formerly defined it (in §29.1 of Part I) the result of parallel transport is given by the vector (field) $\eta(t)$ determined by

$$\nabla_{\dot\gamma(t)}\eta(t)=0,\qquad \eta(0)=\eta_0,$$

or, equivalently, by

$$\frac{d\eta^i(t)}{dt}+a^i_{j\mu}(t)\dot{x}^\mu(t)\eta^j(t)=0, \qquad \eta(0)=\eta_0,$$

where $\gamma(t)=(x^1(t),\ldots,x^n(t))$. Thus the operator $\nabla_{\dot{\gamma}(t)}$ has the form $d/dt+a^i_{j\mu}(t)\dot{x}^\mu(t)$, so that $\eta(t)$ satisfies equation (14) with

$$A(t)=-\nabla_{\dot{\gamma}(t)}+\frac{d}{dt}=-a^i_{j\mu}(t)\dot{x}^\mu(t).$$

Hence by Exercise 2 above the operator (9) is the parallel transport operator, as claimed.

We now illustrate both this section and the preceding one by considering the simplest (but none the less important) sort of example, namely that of a fibre bundle with group $G=U(1)\cong S^1$ (or, equivalently, the isomorphic group $SO(2)$). The Lie algebra $\mathfrak{g}$ is in this case just $\mathbb{R}^1$ (with trivial commutation), and in terms of the angular co-ordinate φ, $0\leq\varphi<2\pi$, on $S^1=U(1)=\{e^{i\varphi}|0\leq\varphi<2\pi\}$, the 1-form ω_0 defined in the preceding subsection is given by $\omega_0=-(1/2\pi)\,d\varphi$ (where the $1/2\pi$ is merely a normalizing factor). The locally-defined forms $gA_\mu g^{-1}\,dx^\mu$ (i.e. defined on (the cross-sections above) each chart U_α of the base M) simplify to $A_\mu\,dx^\mu$, so that the corresponding G-connexion ω on our bundle is given by

$$\omega=\omega_0+gA_\mu g^{-1}\,dx^\mu=\frac{-1}{2\pi}\,d\varphi+A_\mu\,dx^\mu. \tag{15}$$

Of course as in the general case this G-connexion ω enjoys the invariance property (1), and the equation $\omega=0$ (which since ω is scalar-valued represents just a single equation of Pfaffian type) defines hyperplanes in E transverse to the fibres. By (5) above the gauge transformation is given by

$$A_\mu(x)\to A_\mu(x)-\frac{\partial g(x)}{\partial x^\mu}\,g(x)^{-1},$$

where here $(\partial g(x)/\partial x^\mu)g(x)^{-1}$, like $A_\mu(x)$, is for each x an element of $\mathfrak{g}=\mathbb{R}^1$. Putting $g(x)=e^{i\varphi(x)}$, we calculate that $(\partial g(x)/\partial x^\mu)g(x)^{-1}=i(\partial\varphi/\partial x^\mu)$, yielding the co-ordinate $\partial\varphi/\partial x^\mu\in\mathfrak{g}$ under the co-ordinatization of G by φ. Hence in the present context the gauge transformation reduces to the subtraction of a gradient term:

$$A_\mu\to A_\mu-\frac{\partial\varphi}{\partial x^\mu}.$$

Taking the group G in the form $U(1)=\{e^{i\varphi}|0\leq\varphi<2\pi\}$, we find that the covariant differential operator, operating on complex scalar fields (i.e. on cross-sections of fibre bundles with one-dimensional complex fibre $\mathbb{C}^1$ on which $U(1)$ acts by multiplication) has the form

$$\nabla_\mu=\frac{\partial}{\partial x^\mu}+iA_\mu(x), \tag{16}$$

where here $A_\mu(x)$ is real. (The factor i appears for the simple reason that the Lie algebra of $U(1)$, realized as above, consists of the purely imaginary complex numbers (which act on $\mathbb{C}^1$ via multiplication).) The "curvature form" of the connexion is obtained, as in the general case, by taking the commutators of the operators ∇_μ and ∇_ν for all μ and ν (cf. the conclusion of §25.3).

25.3. Curvature

We consider first the case of a complex analytic fibre bundle with one-dimensional fibre $\mathbb{C}^1$, structure group $G=\mathbb{C}^*$, and base an n-dimensional complex manifold M on each of whose charts U_α there are given local co-ordinates $z_\alpha^1, \ldots, z_\alpha^n$. The structure of the bundle is determined (as usual) by transition functions, in turn given by functions

$$T^{\alpha\beta}(z_\alpha^1, \ldots, z_\alpha^n) \neq 0, \tag{17}$$

defined on the regions of intersection $U_{\alpha\beta}=U_\alpha \cap U_\beta$, and taking their values in $G=\mathbb{C}^*$; these functions are, by assumption, complex analytic.

Digressing for a moment, we take the logarithms of the functions (17) (defined only to within an integer multiple of $2\pi i$):

$$a_{\alpha\beta}=\ln T^{\alpha\beta}(z^1, \ldots, z^n)(+2\pi im). \tag{18}$$

If we assume (as we shall both here and subsequently) that the regions of intersection $U_{\alpha\beta}$ are simply-connected (i.e. $\pi_1(U_{\alpha\beta})=1$), then we may take $a_{\alpha\beta}=\ln T^{\alpha\beta}(z)$ as defined unambiguously on each $U_{\alpha\beta}$, by fixing on a single branch of the logarithm. On the intersection $U_{\alpha\beta\gamma}=U_\alpha \cap U_\beta \cap U_\gamma$ of three charts we have (by §24.1(1)) $T^{\alpha\beta}T^{\beta\gamma}T^{\gamma\alpha}=1$, whence

$$a_{\alpha\beta}+a_{\beta\gamma}+a_{\gamma\alpha}=2\pi i n_{\alpha\beta\gamma},$$

for some integer $n_{\alpha\beta\gamma}$. (Note that in defining these integers we did not use the assumed analyticity of the function $T^{\alpha\beta}$.) We shall later show how to construct from the family $\{n_{\alpha\beta\gamma}\}$ (called a "cochain"), where $\alpha\beta\gamma$ runs over all triples for which $U_{\alpha\beta\gamma}$ is non-empty, a certain topological invariant of the fibre bundle. In the case that the base M is one-dimensional this topological invariant is defined simply as the residue modulo 2 of the integer

$$n=\sum_{\alpha<\beta<\gamma} n_{\alpha\beta\gamma}, \tag{19}$$

where the charts U_α have been arranged arbitrarily in a sequence $U_1, U_2, \ldots,$ and it is assumed that no four distinct U_α intersect (i.e. $U_{\alpha_1} \cap U_{\alpha_2} \cap U_{\alpha_3} \cap U_{\alpha_4}$ is empty for distinct $\alpha_1, \alpha_2, \alpha_3, \alpha_4$).

Exercise

Prove that the residue modulo 2 of the integer defined in (19) is indeed independent of the particular data (i.e. local co-ordinate neighbourhoods and transition functions) determining the structure of the fibre bundle (or, in other words, that it is preserved by fibre bundle equivalences). Calculate this invariant for the Hopf bundle η over $\mathbb{C}P^1$ and for its tensor powers η^k (see Example (a) of §24.3, and §24.5).

Resuming the main line of development of our topic, we first observe that the Lie algebra of the structure group $G=\mathbb{C}^+\cong S^1\times\mathbb{R}^+$ of our complex analytic line bundle is just the (commutative) Lie algebra $\mathbb{C}^1$; for if $w(t)=a(t)e^{i\varphi(t)}$, $a(t)>0$, is a curve in G such that $w(0)=1$, i.e. $a(0)=1$, $\varphi(0)=0$, then $\dot{w}(0)=\dot{a}(0)+i\dot{\varphi}(0)$. This also shows that the form ω_0 defined in §25.1 is given by $-d\ln w$, $w\in\mathbb{C}^*$. A G-connexion on our fibre bundle will by definition be given locally (i.e. in terms of the co-ordinates z_α, $\bar{z}_\alpha$ on the regions U_α of the base M) by

$$\omega=\omega_0+A^\alpha_\mu\,dz^\mu_\alpha+B^\alpha_\mu\,d\bar{z}^\mu_\alpha. \tag{20}$$

If in the present context we impose the additional requirement that the connexion be given locally in the form

$$\omega=\omega_0+d'f_\alpha=\omega_0+\frac{\partial f_\alpha}{\partial z^j}\,dz^j, \tag{21}$$

for some complex-valued function f_α (where d' (and the cognate operator d'' satisfying $d'+d''=d$) are as defined in Part I, §27.1), i.e. that

$$B_\mu(z)\equiv 0,\qquad A^\alpha_\mu(z)=\frac{\partial f_\alpha}{\partial z^\mu_\alpha}, \tag{22}$$

then by the formula (5) the difference $d'f_\alpha-d'f_\beta$ on the region of overlap $U_{\alpha\beta}=U_\alpha\cap U_\beta$ must be given by

$$d'f_\alpha-d'f_\beta=(dT^{\alpha\beta})(T^{\alpha\beta})^{-1}. \tag{23}$$

If the regions $U_{\alpha\beta}$ are simply-connected, then as above we can choose on each one a particular branch of the logarithm function to ensure that $\ln T^{\alpha\beta}$ is single-valued; then using $(dT^{\alpha\beta})(T^{\alpha\beta})^{-1}=d(\ln T^{\alpha\beta})$, equation (23) takes on the "gradient" form

$$d'f_\alpha-d'f_\beta=d\ln T^{\alpha\beta}=da_{\alpha\beta}, \tag{24}$$

where, as before, $a_{\alpha\beta}=\ln T^{\alpha\beta}$. Thus a G-connexion of the form (21) will be determined once we have chosen (not necessarily analytic) functions f_α (assuming they exist) satisfying (24).

Now since the functions $T^{\alpha\beta}$ are by assumption analytic, so also are the $a_{\alpha\beta}$, i.e. $d''a_{\alpha\beta}\equiv 0$ (by definition of analyticity; see Part I, §12.1). This and (24) together yield

$$d'd''f_\alpha-d'd''f_\beta\equiv-d''(d'+d'')a_{\alpha\beta}\equiv 0,$$

where we have used $(d')^2=(d'')^2=(d'+d'')^2=0$ and $d'd''=-d''d'$ (see Part I, Corollary 27.1.2). Hence the form $\Omega=d'd''f_\alpha$ is independent of the particular index α, i.e. this expression defines Ω as a form on the whole of M; we call Ω

the "curvature form" (or simply "curvature") determined by the connexion ω. Thus the curvature is given locally by

$$\Omega=\sum_{i,j}\frac{\partial^2 f_\alpha}{\partial z_\alpha^i\,\partial \bar{z}_\alpha^j}\,dz_\alpha^i\wedge d\bar{z}_\alpha^j.$$

In the 2-dimensional case (i.e. the case $n=1$ of one complex dimension) this becomes (on each chart U_α)

$$\Omega=\frac{\partial^2 f_\alpha}{\partial z_\alpha\,\partial \bar{z}_\alpha}\,dz_\alpha\wedge d\bar{z}_\alpha. \tag{25}$$

Since $\partial^2/\partial z\,\partial\bar{z}$ is just the real operator Δ (the Laplace operator), and $dz\wedge d\bar{z}=-2i\,dx\wedge dy$, it follows that for $n=1$ the real and imaginary parts of the form Ω are defined unambiguously, and moreover since $d\Omega\equiv 0$, as closed forms. It will subsequently be made clear that $\operatorname{Re}\Omega=d(\Omega_1)$ for some (real) form Ω_1, whence by the general Stokes formula (see §8.2)

$$\int_M \operatorname{Re}\Omega=0.$$

Finally we note the obvious fact, obtaining for every n, that if the functions f_α are analytic then $\Omega\equiv 0$.

We turn next to the case of a fibre bundle over a real manifold M, with structure group $G=U(1)\cong S^1$ and fibre $\mathbb{C}^1$. The transition functions on the regions of overlap here have the form $T^{\alpha\beta}(x)=e^{i\varphi_{\alpha\beta}(x)}$. It follows much as in our previous examination of fibre bundles with group $U(1)$ (at the end of the preceding subsection) that a G-connexion ω on such a bundle is determined by 1-forms ω_α on the charts U_α, whose differences on the regions of overlap $U_{\alpha\beta}$ are given by

$$\omega_\alpha-\omega_\beta=dq_{\alpha\beta}(x),$$

where $q_{\alpha\beta}(x)=i\varphi_{\alpha\beta}(x)=\ln T^{\alpha\beta}(x)$. The "curvature" Ω is then defined on each U_α by

$$2\pi i\Omega=d\omega_\alpha, \tag{26}$$

(where $2\pi i$ is simply a "normalizing" factor). Then (somewhat as in the previous case) since we have on each region of overlap $U_{\alpha\beta}$ that

$$d\omega_\alpha-d\omega_\beta=ddq_{\alpha\beta}\equiv 0,$$

we see that the curvature Ω is well defined by the formula (26).

The curvature form Ω may be defined alternatively, in invariant fashion, by the formula

$$p^*\Omega=\frac{1}{2\pi i}\,d\omega, \tag{27}$$

where $p\colon E\to M$ is the projection map of the fibre bundle. To verify that this definition of Ω is equivalent to that given by the formula (26), it suffices to recall (from the end of the preceding subsection) that locally, i.e. above each chart U_α, we have $\omega=\omega_0+p^*\omega_\alpha$, where $\omega_0=-(1/2\pi)\,d\varphi$, whence by (26) (and since $d^2=0$)

$$d\omega=dp^*\omega_0=p^*(2\pi i\Omega).$$

Having defined the curvature form for two rather special (though important) kinds of fibre bundles we now define curvature and investigate its properties in the general situation of an arbitrary (real or complex) fibre bundle $p: E\to M$ (with group G and fibre F), on which there is given a connexion ω. As has been noted before, via the n-dimensional horizontal direction $\mathbb{R}^n_y$ which it determines at each point $y\in E$, the connexion gives rise to the direct decomposition $T_yE=T_yF\oplus\mathbb{R}^n_y$ of the tangent space T_yE at y, and consequently to the projection map

$$H: T_yE\to\mathbb{R}^n_y, \qquad H(T_yF)=0. \tag{28}$$

25.3.1. Definition. Let ω_q be any q-form on the total space E of a fibre bundle endowed with a connexion ω. The *horizontal part* of the q-form ω_q is the q-form $H\omega_q$ whose value on a q-tuple $(\tau_1,\ldots,\tau_q)$ of tangent vectors to E at a point is given by

$$H\omega_q(\tau_1,\ldots,\tau_q)=\omega_q(H\tau_1,\ldots,H\tau_q). \tag{29}$$

It is clear from this definition that if one or more of the vectors τ_i is "vertical" (i.e. tangent to a fibre), then $H\omega_q(\tau_1,\ldots,\tau_q)=0$. Thus certainly the restriction of $H\omega_q$ to a fibre is always zero.

25.3.2. Definition. The *curvature form* Ω_E on the total space E of a (principal) fibre bundle endowed with a connexion ω, is defined by

$$\Omega_E=H\,d\omega. \tag{30}$$

25.3.3. Theorem. *If ω is a connexion on a principal fibre bundle and Ω_E the corresponding curvature form, then the following "structural equation" holds:*

$$d\omega+[\omega,\omega]=H\,d\omega=\Omega_E, \tag{31}$$

where the commutator $[\omega,\omega]$ is as defined below. The curvature form has the same invariance property as ω under the natural left action of the structure group G on the total space E:

$$g^*\Omega_E=\mathrm{Ad}\,(g)\Omega_E=g\Omega_E g^{-1}. \tag{32}$$

Before giving the proof we need to define the commutator of a pair of forms taking their values in a Lie algebra. (Recall that the forms ω and Ω_E figuring in the theorem take their values in the Lie algebra $\mathfrak{g}$ of G, or rather in the algebra of right-invariant vector fields on G, which we identified with $\mathfrak{g}$

(see §25.1).) Thus the *commutator* of a pair ω_p, ω_q of such forms, of ranks p and q respectively (or more generally the *product* of a pair of forms with values in any algebra equipped with a bilinear multiplication) is defined by

$$\binom{p+q}{p}[\omega_p, \omega_q](\tau_1, \ldots, \tau_{p+q}) = \sum_\sigma \operatorname{sgn} \sigma[\omega_p(\tau_{i_1}, \ldots, \tau_{i_p}), \omega_q(\tau_{j_1}, \ldots, \tau_{j_q})], \tag{33}$$

where $\tau_1, \ldots, \tau_{p+q}$ are tangent vectors at an arbitrary point of the underlying space (in our case E), $\binom{p+q}{p}$ is the binomial coefficient $(p+q)!/p!\,q!$, and σ is the permutation of indices given by

$$\sigma = \begin{pmatrix} 1 & 2\cdots p & p+1 & \cdots & p+q \\ i_1 & i_2\cdots i_p & j_1 & \cdots & j_q \end{pmatrix}. \tag{34}$$

In particular the commutator of a pair of 1-forms ω, $\hat{\omega}$ $(p=q=1)$ is defined by

$$[\omega, \hat{\omega}](\tau_1, \tau_2) = \tfrac{1}{2}([\omega(\tau_1), \hat{\omega}(\tau_2)] - [\omega(\tau_2), \hat{\omega}(\tau_1)]), \tag{35}$$

and if $\omega = \hat{\omega}$ this simplifies further to

$$[\omega, \omega](\tau_1, \tau_2) = \tfrac{1}{2}([\omega(\tau_1), \omega(\tau_2)] - [\omega(\tau_2), \omega(\tau_1)]) = [\omega(\tau_1), \omega(\tau_2)]. \tag{36}$$

Proof of the Theorem. To verify the structural equation (31) it suffices to show that it holds on each of the regions $E_\alpha = p^{-1}(U_\alpha)$ above the charts U_α. In terms of a local system of co-ordinates x_α^i (on U_α), the connexion is, in the notation of (2), given on the region $E_\alpha \subset E$ by the formula

$$\omega = \omega_0 + gA_\mu(x_\alpha)g^{-1}\, dx_\alpha^\mu = \omega_0 + gAg^{-1}, \tag{37}$$

where $A = A_\mu\, dx^\mu$ and $\omega_0 = -(dg)g^{-1}$. For convenience of calculation we assign to E_α, via the identification $\varphi_\alpha\colon G \times U_\alpha \to p^{-1}(U_\alpha) = E_\alpha$, the co-ordinates (g, x) of the product $G \times U_\alpha$. In terms of these co-ordinates a basis for the tangent space to E_α at the point $(1, x)$ is furnished by the basis vectors A_μ (interpreted as usual as elements of the Lie algebra of G) for the tangent space to G at 1, and $\partial_\mu = \partial/\partial x^\mu$ for the tangent space to U_α at x; a typical tangent vector to E_α at $(1, x)$ will then be represented in the form $(\xi^\mu A_\mu, \eta^\mu \partial_\mu)$.

Now by Lemma 25.1.3 the n-dimensional horizontal direction $\mathbb{R}^n_{(1,x)}$ at $(1, x)$ is the subspace of tangent vectors τ to E satisfying $\omega(\tau) = 0$. Since at the point $(1, x)$ we have by (37) that

$$\omega(A_\mu, \partial_\mu) = \omega_0(A_\mu) + A_\mu = -A_\mu + A_\mu = 0,$$

it follows that the vectors (A_μ, ∂_μ) form a basis for $\mathbb{R}^n_{(1,x)}$, and further that, by definition of the projection H,

$$H(0, \partial_\mu) = H(A_\mu, 0) = (A_\mu, \partial_\mu).$$

Hence

$$H\omega_0(A_\mu, 0) = \omega_0(H(A_\mu, 0)) = -A_\mu,$$
$$HA(0, \partial_\mu) = A(H(0, \partial_\mu)) = A_\mu,$$

so that we may write $H\omega_0 = -A$, $HA = A$. If we now calculate $d\omega$ from (37) (assuming if you like that G and $\mathfrak{g}$ are comprised of matrices) we obtain the following formula (appropriately interpreted):

$$d\omega = d\omega_0 + (dg)g^{-1}(gAg^{-1}) - (gAg^{-1})(dg)g^{-1} + g(dA)g^{-1}.$$

In view of the definition of the commutator of a pair of forms (see above), and the fact (established at the beginning of §25.1) that $d\omega_0 = [\omega_0, \omega_0]$, this becomes

$$d\omega = [\omega_0, \omega_0] - [\omega_0, gAg^{-1}] + [gAg^{-1}, \omega_0] + g\, dAg^{-1}, \tag{38}$$

whence, putting $g = 1$, applying H, and using $-H\omega_0 = A = HA$, we obtain

$$Hd\omega = [H\omega_0, H\omega_0] - [H\omega_0, HA] + [HA, H\omega_0] + H\, dA = [A, A] + dA. \tag{39}$$

Now since the operators H and d preserve the invariance property of ω (under the left action of G on E), so that $g^*(Hd\omega) = \mathrm{Ad}\,(g)H\, d\omega$, it follows from (39) that for all g (not just $g = 1$)

$$H\, d\omega = \Omega_E = g(dA + [A, A])g^{-1}. \tag{40}$$

From this and (37) we obtain finally

$$d\omega = -[\omega_0, \omega_0] - 2[\omega_0, gAg^{-1}] - g[A, A]g^{-1} + \Omega_E = \Omega_E - [\omega, \omega],$$

as required. The proof is completed by observing that the second statement of the theorem is a consequence of the aforementioned preservation of the invariance property of ω by the operators H and d. □

If we "lower" (i.e. project) the curvature form Ω_E onto the base M, we obtain as a result a form Ω, which in view of (39) is given locally (on each chart U_α of M) by

$$\Omega = DA = dA + [A, A] = \Omega_{\mu\nu}\, dx^\mu \wedge dx^\nu$$
$$= \left(\frac{\partial A_\nu}{\partial x^\mu} - \frac{\partial A_\mu}{\partial x^\nu} + [A_\mu, A_\nu]\right) dx^\mu \wedge dx^\nu, \tag{41}$$

whence it follows easily (as in §41.2 of Part I) that the coefficients $\Omega_{\mu\nu}$ are just the commutators $[\nabla_\mu, \nabla_\nu]$ of the appropriate covariant differential operators (cf. §25.2). Under a gauge transformation $g(x)$ we see from (40) that

$$\Omega \to g\Omega g^{-1}. \tag{42}$$

Applying the operators d and H to the form Ω_E on E we obtain (via the structural equation $\Omega_E = d\omega + [\omega, \omega]$) the "Bianchi identities" (cf. §30.5 of Part I)

$$\begin{aligned} d\Omega_E &= 2[\omega, \Omega_E], \\ Hd\Omega_E &= 0. \end{aligned} \tag{43}$$

For the projected form Ω it is also a straightforward calculation to show that on each chart U_α of the base M we have the further "Bianchi identity"

$$D\Omega = d\Omega + [A, \Omega] = 0. \tag{44}$$

25.4. Characteristic Classes: Constructions

For the one-dimensional complex fibre bundles $p: E \to M$ with (commutative) structure group $G = U(1) \cong S^1$ or $G = \mathbb{C}^*$, considered in the preceding two subsections, the structural equation (31) reduces to $\Omega_E = d\omega$ (and the Bianchi identities (43) reduce to the consequent closure of the curvature form, i.e. $d\Omega_E = 0$) since here the commutativity of the Lie algebra $\mathfrak{g}$ entails the vanishing of commutators of forms taking their values in $\mathfrak{g}$. For the same reason the equation (44) simplifies to $d\Omega = 0$; moreover since in this context $g\Omega g^{-1} \equiv \Omega$, the form Ω is unambiguously (i.e. gauge-invariantly) defined on M. (We shall refer to Ω also as the "curvature form", since Ω_E can be retrieved from it: $p^*\Omega = \Omega_E$.) Although as just noted Ω is here closed (and Ω_E exact) it can (and often does) happen that Ω is not exact (this will be discussed in the sequel). The difference between two such forms will always however be exact, as we shall now see.

25.4.1. Lemma. *If ω and $\bar{\omega}$ are two connexions on a fibre bundle with structure group $G = U(1) \cong S^1$ or $G = \mathbb{C}^*$, then the difference between the corresponding curvature forms Ω and $\bar{\Omega}$ on the base M, is an exact form; i.e. there is a form u such that*

$$\Omega - \bar{\Omega} = du.$$

Proof. We have $p^*\Omega = \Omega_E = d\omega$ and $p^*\bar{\Omega} = \bar{\Omega}_E = d\bar{\omega}$. From the formula (2) or (6) (with the commutativity of G taken into account) we obtain

$$\omega - \bar{\omega} = (\omega_0 + A_\mu\, dx^\mu) - (\omega_0 + \bar{A}_\mu\, dx^\mu) = (A_\mu - \bar{A}_\mu)\, dx^\mu,$$

which expresses $\omega - \bar{\omega}$ in the form p^*u with u a form on M. Hence $\Omega - \bar{\Omega} = du$, completing the proof. □

25.4.2. Corollary. *Let Ω and $\bar{\Omega}$ be as in the lemma and suppose that P is a two-dimensional, closed (i.e. compact and without boundary), oriented submanifold of the base M. Then the integrals $\int_P \Omega$ and $\int_P \bar{\Omega}$ are equal; consequently the quantity $\int_P \Omega$ is independent of the choice of connexion on the fibre bundle*

$p: E \to M$ (with group $U(1)$ or $\mathbb{C}^*$), and is thus a topological invariant of the bundle (and P).

PROOF. By the general Stokes formula (§8.2), since P has no boundary we have

$$\int_P (\Omega - \bar{\Omega}) = \int_P du = \int_{\partial P} u = 0. \qquad \square$$

Having prepared the ground somewhat with these particular groups, we now turn our attention to the general situation of a fibre bundle with structure group an arbitrary matrix Lie group. As before we suppose that a connexion ω is given on the fibre bundle, giving rise to a "curvature form" Ω on each chart U_α of M (with co-ordinates x^i), which transforms under gauge transformations as in (42):

$$\Omega \to g(x)\Omega g^{-1}(x).$$

It follows from this that since the trace of a matrix is unaffected by conjugation, so that in particular

$$\operatorname{tr} \Omega = \operatorname{tr}(g\Omega g^{-1}),$$

the scalar-valued 2-form $\operatorname{tr} \Omega$ is defined gauge-invariantly on the whole base M.

It is readily verified that $d(\operatorname{tr} \Omega) = \operatorname{tr}(d\Omega)$ (in fact that this holds for any matrix-valued form). From this and the Bianchi identity (44), namely $d\Omega = -[A, \Omega]$ (on each chart U_α of M), we deduce that the form $\operatorname{tr} \Omega$ is closed:

$$d(\operatorname{tr} \Omega) = \operatorname{tr}(d\Omega) = -\operatorname{tr}[A, \Omega] = 0,$$

where the final equality follows from the simple fact that the commutator of a pair of matrices has zero trace.

Given another connexion $\bar{\omega}$ on the fibre bundle, we obtain from the structural equation (31) that

$$d(\omega - \bar{\omega}) = d\omega - d\bar{\omega} = \Omega_E - \bar{\Omega}_E - [\omega, \omega] + [\bar{\omega}, \bar{\omega}],$$

whence, taking traces of the various matrix-valued forms appearing here, and using the fact that for such forms the taking of traces commutes with the operator d and with the pullback p^*, we deduce that

$$d(\operatorname{tr} \omega - \operatorname{tr} \bar{\omega}) = \operatorname{tr} \Omega_E - \operatorname{tr} \bar{\Omega}_E = p^*(\operatorname{tr} \Omega - \operatorname{tr} \bar{\Omega}).$$

From this and the local equations

$$\operatorname{tr} \omega = \operatorname{tr} \omega_0 + \operatorname{tr}(p^* A), \qquad \operatorname{tr} \bar{\omega} = \operatorname{tr} \omega_0 + \operatorname{tr}(p^* \bar{A}),$$

valid on each $U_\alpha \subset M$, we finally infer, as in the proof of Lemma 25.4.1, the exactness of the form $\operatorname{tr} \Omega - \operatorname{tr} \bar{\Omega}$:

$$\operatorname{tr} \Omega - \operatorname{tr} \bar{\Omega} = du$$

for some form u on M. We conclude that the integral of the 2-form $\operatorname{tr} \Omega$ over a given 2-dimensional, closed, oriented submanifold of M is a "topological invariant" (of the fibre bundle and the given submanifold of M) since it is independent of the connexion ω.

We next express our locally-defined, matrix-valued 2-form Ω in the usual differential notation, in terms of co-ordinates x^μ on each chart U_α of M:

$$\Omega=\Omega_{\mu\nu}\,dx^\mu \wedge dx^\nu=(q^i_j)_{\mu\nu}\,dx^\mu \wedge dx^\nu; \tag{45}$$

similarly the locally-defined, matrix-valued 1-form A (for which we may use the term "connexion", since this local form defines the connexion ω with which the fibre bundle is assumed to be endowed) is given in differential notation, on each U_α, by

$$A=A_\mu\,dx^\mu=(a^i_j)_\mu\,dx^\mu. \tag{46}$$

The differential forms (45) and (46) are linked by the following equations (see (41) or Part I, §41.2)

$$\Omega_{\mu\nu}=\frac{\partial A_\nu}{\partial x^\mu}-\frac{\partial A_\mu}{\partial x^\nu}+[A_\mu, A_\nu]. \tag{47}$$

In the preceding subsection (see (33)) we defined the "product" (actually commutator) of a pair of forms taking their values in a Lie algebra; we now define analogously the product (or "exterior product") $\omega_p \wedge \omega_q$ of a pair of matrix-valued forms with respect to the ordinary product of matrices (as opposed to the commutator), by setting

$$\begin{aligned}\binom{p+q}{p}&(\omega_p \wedge \omega_q)(\tau_1,\ldots,\tau_{p+q})\\ &=\sum_{\substack{i_1<\cdots<i_p\\ j_1<\cdots<j_q}}(\operatorname{sgn}\sigma)\omega_p(\tau_{i_1},\ldots,\tau_{i_p})\omega_q(\tau_{j_1},\ldots,\tau_{j_q}),\end{aligned} \tag{48}$$

where as before $\tau_1,\ldots,\tau_{p+q}$ are tangent vectors at an arbitrary point of the underlying space, σ is the permutation (34), and $\binom{p+q}{p}$ is the usual binomial coefficient. (This definition of the matrix-valued form $\omega_p \wedge \omega_q$ is the obvious generalization of the definition of exterior product for scalar-valued forms given in Part I, §18.3; however though associative (essentially because matrix multiplication is associative), the operation defined by (48) will not in general be skew-commutative (i.e. it will not in general be the case that $\omega_p \wedge \omega_q=(-1)^{pq}\omega_q \wedge \omega_p$).)

Armed with this "exterior product" of matrix-valued forms, we can now introduce further "characteristic classes" (the general definition of which we defer for the moment). The "characteristic classes" in question are defined by (cf. Part I, §42)

$$c_i=\operatorname{tr}(\Omega \wedge \cdots \wedge \Omega)=\operatorname{tr}\Omega^i, \qquad i\geq 1. \tag{49}$$

(Note that we have already considered the characteristic class $c_1=\operatorname{tr}\Omega$ (see above) without explicit mention of the name.) We now show that, as we have already established in the particular case $i=1$, the c_i are for all i gauge-invariant closed forms on the base M of our (principal) fibre bundle, yielding in a like manner certain topological invariants.

Under a gauge transformation determined by $g(x)$ on a region of intersection $U_{\alpha\beta}=U_\alpha \cap U_\beta$ of a pair of charts of M, we have

$$\Omega \to g\Omega g^{-1}, \qquad \Omega^i \to g\Omega^i g^{-1}, \qquad \operatorname{tr} \Omega^i = c_i \to c_i, \tag{50}$$

so that the form c_i is well defined on the whole base M.

To see that $c_i = \operatorname{tr} \Omega^i$ is closed, note first that by elementary properties of the operator d and the operation of taking traces of matrices, we have

$$d \operatorname{tr} \Omega^i = \operatorname{tr} d\Omega^i = \sum_{j=1}^{i} \operatorname{tr}(\Omega^{j-1} \wedge (d\Omega) \wedge \Omega^{i-j}).$$

That the last expression represents the zero $(2i+1)$-form is, as before, a consequence of Bianchi's identity $d\Omega = -[A, \Omega]$, and the vanishing of the trace of the commutator of a pair of matrices; we content ourselves with exemplifying the general argument by that for the case $i=2$, where the problem reduces to that of showing that $\operatorname{tr}(\Omega \wedge d\Omega)=0$. Here the component $(\Omega \wedge d\Omega)_{12345}$, for instance, of the 5-form $\Omega \wedge d\Omega$ will (after suitably grouping the terms in pairs and applying the above Bianchi identity) be a sum of expressions similar to

$$\text{const.} \times (\Omega_{12}[A_3, \Omega_{45}] + \Omega_{45}[A_3, \Omega_{12}]),$$

which can be re-expressed as

$$\text{const.} \times ([\Omega_{12} A_3, \Omega_{45}] + [\Omega_{45} A_3, \Omega_{12}]),$$

which clearly makes zero contribution to $\operatorname{tr}(\Omega \wedge d\Omega)$. (Here the $\Omega_{\mu\nu}$ and A_μ are as in (45) and (46).) In the general case a similar argument yields the desired conclusion: for all $i \geq 1$ the form c_i is closed.

Finally, if $\bar{\omega}$ is another connexion on the fibre bundle, then by an argument similar in essence to the earlier one (for the case $i=1$), it follows that

$$\operatorname{tr} \Omega^i - \operatorname{tr} \bar{\Omega}^i = du_i,$$

where

$$p^* u_i = \sum_{j=1}^{i} (-1)^j \operatorname{tr}(\Omega^{j-1} \wedge (A - \bar{A}) \wedge \Omega^{i-j}).$$

(We leave the detailed verification to the reader.) As earlier we conclude that, given a closed, oriented $2i$-dimensional submanifold P of M, the integral $\int_P c_i$ is for each i a topological invariant of the bundle (and P).

In the case $G = SO(2n)$ we can define further "Euler" characteristic classes χ_n (i.e. scalar-valued forms of rank $2n$ on the base M, enjoying certain desirable properties) by means of the following formula for the value of such a form at a $2n$-tuple $(\tau_1, \ldots, \tau_{2n})$ of tangent vectors at an arbitrary point of M:

$$\beta(n)\chi_n = \sum_{\substack{i_1<i_2\\ i_3<i_4\\ \cdots\cdots\\ i_{2n-1}<i_{2n}}} \chi[\Omega(\tau_{i_1}, \tau_{i_2}), \ldots, \Omega(\tau_{i_{2n-1}}, \tau_{i_{2n}})], \tag{51}$$

where σ is the permutation $\begin{pmatrix} 1 & 2 & \cdots & 2n \\ i_1 & i_2 & \cdots & i_{2n} \end{pmatrix}$, $\beta(n)$ is a numerical coefficient to be determined subsequently by certain normalizing requirements, and χ is a certain (scalar-valued) multi-linear form in n skew-symmetric, $2n \times 2n$ (variable) matrices $L^{(1)} = (l_{ij}^{(1)}), \ldots, L^{(n)} = (l_{ij}^{(n)})$ (i.e. elements of $so(2n)$), defined as follows: each skew-symmetric matrix $L^{(k)} = (l_{ij}^{(k)})$ determines an associated 2-form $l^{(k)} = l_{ij}^{(k)}\, du^i \wedge du^j$ on $\mathbb{R}^{2n}$ with co-ordinates $u_1, \ldots, u_{2n}$; χ is given in terms of these forms by the equation

$$l^{(1)} \wedge \cdots \wedge l^{(n)} = \chi[L^{(1)}, \ldots, L^{(n)}]\, du_1 \wedge \cdots \wedge du_{2n}. \tag{52}$$

(Thus χ is a general analogue of the so-called "Pfaffian".)

When $n = 1$, $\chi(L)$ is just the usual isomorphism between the Lie algebra of the group $SO(2) \cong S^1$ and the real line $\mathbb{R}^1$. (Note that for $G = U(1)$ $(\simeq SO(2))$, i.e. G in its complex guise, the form χ_1 was considered (implicitly) in §§25.2, 25.3, and at the beginning of the present subsection.)

When $n = 2$ the formula (51) becomes

$$\beta(2)\chi_2 = \frac{1}{4!}\, \varepsilon^{i_1 i_2 i_3 i_4}\Omega(\tau_{i_1}, \tau_{i_2}) \wedge \Omega(\tau_{i_3}, \tau_{i_4}). \tag{53}$$

Exercises

1. Prove that the form χ_n is as expected a well-defined, closed form on the base M of the fibre bundle with group $SO(2n)$. (The gauge-invariance of χ_n is a consequence of the invariance of χ under conjugation by elements of $SO(2n)$, which in turn follows from the transformation rule for forms of rank equal to the dimension of the ambient space (see Part I, Theorem 18.2.2).) Show further that if M is a Riemannian manifold of dimension $2n$ with metric g_{ij}, and the fibre bundle is the tangent bundle over M endowed with the symmetric connexion compatible with the metric, then (cf. Part I, §43):

$$\text{for } n = 1, \qquad \chi_1 = 2K\, d\sigma = 2K\sqrt{g}\, du \wedge dv,$$

where $g = \det(g_{ij})$ and K is the Gaussian curvature of M (and we have put $\beta(1) = \frac{1}{2}$); and

$$\text{for all } n \geq 1, \qquad \beta(n)\chi_n = \varepsilon^{i_1 \ldots i_n}\, \Omega_{i_1 i_2} \wedge \cdots \wedge \Omega_{i_{2n-1} i_{2n}}, \tag{54}$$

where $\Omega_{ij} = \sum_{k<l} R_{ijkl}\, dx^k \wedge dx^l$, R_{ijkl} denoting as usual the Riemannian curvature tensor on M (see Part I, §30).

2. Show that for fibre bundles with structure group $SO(n)$ the forms $c_{2i+1} = \operatorname{tr} \Omega^{2i+1}$ are globally exact, i.e. exact on the whole base (and therefore do not yield (when applicable) significant topological invariants). (Note also that in particular, since the Lie algebras of the groups $SO(n)$, $SU(n)$ consist of zero-trace matrices we have $c_1 = \operatorname{tr} \Omega \equiv 0$.)

For a fibre bundle with group $SO(4)$ and base M of dimension 4 the only (non-trivial) characteristic classes (arising as above) are χ_2 and $c_2=\mathrm{tr}(\Omega^2)$ (since the remaining c_i are trivial). If the bundle is as in the second part of Exercise 1 above, i.e. if it is the tangent bundle over a Riemannian 4-manifold M endowed with the connexion compatible with the metric, then it follows from (54) (and the formula for Ω_{ij} following it) that, with appropriate $\beta(2)$,

$$c_2 = -R^{ij}_{\lambda\varkappa}R_{ij\nu\mu}\, dx^\lambda \wedge dx^\varkappa \wedge dx^\nu \wedge dx^\mu, \tag{55}$$

$$\chi_2 = \frac{1}{4!}\,\varepsilon^{i_1i_2i_3i_4}R_{i_1i_2\nu\mu}R_{i_3i_4\lambda\varkappa}\, dx^\nu \wedge dx^\mu \wedge dx^\lambda \wedge dx^\varkappa.$$

The integrals $\int_M c_2$ and $\int_M \chi_2$ are clearly functionals of the metric g_{ij} given on M, with identically zero variational derivatives (i.e. stable under perturbations of the metric; see the exercise concluding §43 of Part I).

For the groups $SO(n)$ and $U(n)$ the above-defined characteristic classes are, for reasons which will appear in the next subsection, the most important. One can define characteristic classes b_j (analogous to the c_j) as forms of rank $2j$ on the bases of fibre bundles with structure group $Sp(n)$ (see Part I, §14.3), where $Sp(n)$ and its Lie algebra are realized as quaternion matrices (unitary and skew-Hermitian respectively). It turns out that only the b_{2i} are non-trivial. (We invite the reader to frame the definition of the b_j and prove the last statement.) These characteristic classes are however of lesser importance.

We shall next describe a general construction of characteristic classes. It may be deemed appropriate however to preface this description with the general definition: Given a Lie group G, a *differential-geometric characteristic class* for G is a class of closed, scalar-valued forms on the bases M of G-bundles, which are "uniformly defined" in terms of the connexions on such bundles, and have the further property that if one connexion (on any particular G-bundle) is replaced by another, then the corresponding form (representing the characteristic class) is altered only by the addition of an exact form. (It follows that the integral of (a representative form of) the characteristic class over a closed, oriented submanifold P of M of the appropriate dimension, will be a "topological invariant" of the bundle (and P).)

The promised general construction is as follows. As we well know, for each element g of an arbitrary Lie group G the inner automorphism $x \mapsto gxg^{-1}$, $x \in G$, determines an isomorphism $\mathrm{Ad}\, g$ of the Lie algebra $\mathfrak{g}$, which for a matrix group G (with corresponding matrix Lie algebra $\mathfrak{g}$) has the form

$$\mathrm{Ad}\, g\colon l \mapsto g\,lg^{-1}, \qquad l \in \mathfrak{g}.$$

25.4.3. Definition. A scalar-valued, symmetric, multilinear form $\psi[l_1, \ldots, l_m]$ in variables with provenance the Lie algebra $\mathfrak{g}$ of G, is said to be Ad-*invariant* if it is unchanged by the application of every map $\mathrm{Ad}\, g$, $g \in G$, i.e. if for all $g \in G$,

$$\psi[gl_1g^{-1}, \ldots, gl_mg^{-1}] = \psi[l_1, \ldots, l_m]. \tag{56}$$

Each such Ad-invariant form ψ gives rise to a characteristic class c_ψ defined as follows: if Ω denotes as usual the (local) curvature form on the base M (determined by a given connexion on the fibre bundle), then we set

$$c_\psi(\tau_1, \ldots, \tau_{2m}) = c_\psi(\Omega) = \sum_{\substack{i_1 < i_2 \\ \cdots\cdots \\ i_{2m-1} < i_{2m}}} \operatorname{sgn} \sigma \psi[\Omega(\tau_{i_1}, \tau_{i_2}), \ldots, \Omega(\tau_{i_{2m-1}}, \tau_{i_{2m}})], \tag{57}$$

where $(\tau_1, \ldots, \tau_{2m})$ is an arbitrary $2m$-tuple of tangent vectors at any point of M, and σ is the permutation $\begin{pmatrix} 1 \cdots 2m \\ i_1 \cdots i_{2m} \end{pmatrix}$. Clearly the Ad-invariance of ψ ensures that the scalar-valued form $c_\psi(\Omega)$ is well defined on the base M.

EXERCISE

Show that the form $c_\psi(\Omega)$ has the properties of a characteristic class, i.e. that it is closed, and up to the addition of an exact form is independent of the given connexion on the fibre bundle (so that integrals over closed, oriented submanifolds of M are topological invariants).

It is easy to relate this general construction to our previous special ones; for, in the case $G = SO(2n)$, if we take $\psi = \chi$ in the formula (57) (the Ad-invariance of χ forming part of Exercise 1 above), we obtain, to within a constant factor, the Euler characteristic class χ_n; and if for each $q \geq 1$ we take $\psi = \psi_q$ where

$$\psi_q(l_1, \ldots, l_q) = \sum_{i_1 < \ldots < i_q} \operatorname{tr}(l_{i_1} \cdots l_{i_q}), \tag{58}$$

then we obtain (again to within a constant factor) the characteristic class c_q.

25.4.4. Examples. (a) If we take G to be any of the (abelian) groups T^n or $\mathbb{R}^n$, then $\mathfrak{g} = \mathbb{R}^n$, and the operators $\operatorname{Ad} g$ are all trivial, so that every symmetric multi-linear form $\psi[l_1, \ldots, l_m]$ determines a characteristic class c_ψ. Hence in these cases for each m the totality of classes $c_\psi(\Omega)$ forms an algebra under the operations associating with each pair $c_\psi(\Omega)$, $c_{\hat\psi}(\Omega)$ the classes $c_{\psi+\hat\psi}(\Omega)$ and $c_{\psi\hat\psi}(\Omega)$ where $(\psi\hat\psi)[l_1, \ldots, l_m] = \psi[l_1, \ldots, l_m]\hat\psi[l_1, \ldots, l_m]$, and $\psi + \hat\psi$ is defined analogously. Note also that each ψ of rank m (whence also the corresponding $c_\psi(\Omega)$) can be identified in a natural way with a homogeneous polynomial of degree m in the n "elementary" forms $\psi_j(l) = \langle l, e_j \rangle$, where $\{e_1, \ldots, e_n\}$ is the standard basis for the Lie algebra $\mathfrak{g} = \mathbb{R}^n$, and $\langle\ ,\ \rangle$ denotes the Euclidean scalar product. If we denote the characteristic classes $c_{\psi_j}(\Omega)$ by $t_j(\Omega)$ (forms of rank 2 on the base of the underlying fibre bundle), then each class $c_\psi(\Omega)$, where $\psi = \psi[l_1, \ldots, l_m]$, can be expressed in the form

$$\sum \alpha_{i_1 \ldots i_q} t_{i_1}^{n_1} \cdots t_{i_q}^{n_q} = c_\psi(\Omega), \tag{59}$$

where $n_1 + \cdots + n_q = m$.

(b) $G=U(n)$. In this case (as we know from Part I, §24.2) the Lie algebra $\mathfrak{g}$ is comprised of all skew-Hermitian matrices. Here and in the succeeding examples we shall need the following general concepts and facts concerning them. We define a *Cartan subalgebra* of a Lie algebra to be a maximal commutative subalgebra. We shall make use of the result (whose proof we omit) that any Ad-invariant form on a Lie algebra is completely determined by its restriction to a Cartan subalgebra. (It can also be shown, at least when the ground field is $\mathbb{C}$, as it is here, that given any pair of Cartan subalgebras of a Lie algebra there is an automorphism of the Lie algebra sending one to the other.) For matrix Lie groups G, those automorphisms of the Lie algebra of the form $l \mapsto glg^{-1}, g \in G$, preserving a Cartan subalgebra, form a finite group called the *Weyl group* of G.

It is not too difficult to see that in the present case ($G=U(n)$) the subalgebra H of $\mathfrak{g}$ consisting of the diagonal skew-Hermitian matrices (i.e. diagonal matrices with purely imaginary diagonal entries) is a Cartan subalgebra. If we choose as a basis for this subalgebra the diagonal matrices l_j^0 with jth diagonal entry $(l_j^0)_{jj}=i$, and the rest zero, then the Weyl group turns out to be just the group S_n of all permutations of these basis elements (verify this!). If we denote by t_j the corresponding linear forms in the dual space $\mathfrak{g}^*$, i.e. defined by $t_j(l_k^0)=\delta_{jk}$, then in view of the fact that the Weyl group is S_n, the restriction of any symmetric Ad-invariant form ψ to the subalgebra $H \subset \mathfrak{g}$ will correspond in the natural way to a symmetric polynomial in $t_1, \ldots, t_n$. It is easy to see that for each k the symmetric Ad-invariant form on H defined by the (multiplicatively) basic symmetric polynomial

$$\psi_k^{(U)} = t_1^k + \cdots + t_n^k, \tag{60}$$

coincides with the restriction to H of the ψ_k defined in (58). (For example $\psi_1^{(U)} = t_1 + \cdots + t_n$, is clearly the trace operator on matrices in H, while for all $l, m \in H$, $\psi_2^{(U)}(l,m) = t_1(l)t_1(m) + \cdots + t_n(l)t_n(m) = \operatorname{tr}(lm)$.) Hence the polynomials $\psi_k^{(U)}$ can be extended from the Cartan subalgebra to the whole of $\mathfrak{g}$ via formula (58), and therefore correspond to the characteristic classes c_k.

(c) $G=SO(2n)$. The Lie algebra here consists of all $2n \times 2n$ skew-symmetric matrices. It can be shown that the "infinitesimal rotations" in the planes $\mathbb{R}_{12}, \mathbb{R}_{34}, \ldots, \mathbb{R}_{2n-1,2n}$ generate a Cartan subalgebra $H \subset \mathfrak{g}$, where the indices indicate the pair of standard basis vectors in $\mathbb{R}^{2n}$ determining the plane in question. Hence as a basis for H we may take the matrices $l_1^0, \ldots, l_n^0$ defined by

$$l_j^0 = \begin{pmatrix} \ddots & & & 0 \\ & 0 & 1 & \\ & -1 & 0 & \\ 0 & & & \ddots \end{pmatrix}, \tag{61}$$

where the only non-zero entries are

$$(l_j^0)_{2j-1,2j}=1, \qquad (l_j^0)_{2j,2j-1}=-1.$$

The reader might like to verify that the Weyl group of automorphisms of H of the form $l \mapsto glg^{-1}$, $g \in G$, is generated by the following transformations:

(i) all permutations of $l_1^0, \ldots, l_n^0$;
(ii) all simultaneous negations of pairs of the l_j^0; i.e. maps given by $l_i^0 \mapsto -l_i^0$, $l_k^0 \mapsto -l_k^0$, $l_j^0 \mapsto l_j^0$ for $j \neq i$, k (there being one such map for each pair i, k, $i \neq k$).

Hence the (symmetric) polynomials corresponding (as in the preceding example) to symmetric multilinear forms on H invariant under the Weyl group, are as follows:

$$\psi_q^{(SO)} = t_1^{2q} + \cdots + t_n^{2q}, \qquad q<n; \tag{62}$$
$$\tilde{\psi}_n^{(SO)} = t_1 \cdots t_n.$$

Much as in the preceding example, we see (via (58)) that the form $\psi_q^{(SO)}$ extends to the whole of $\mathfrak{g}$, yielding (to within a constant factor) the characteristic class c_{2q}. It is similarly easy to verify that the form $\tilde{\psi}_n^{(SO)}$ is essentially the restriction to H of χ (defined in (52)), and so corresponds to χ_n.

(d) $G=SO(2n+1)$. The Lie algebra $\mathfrak{g}$ again consists of all skew-symmetric matrices (here of degree $2n+1$). With $SO(2n)$ embedded in $SO(2n+1)$ in the most obvious way (so that $SO(2n)$ fixes the x^{2n+1}-axis), it turns out that the Cartan subalgebra H of the previous example (where G was $SO(2n)$) is also a Cartan subalgebra of our present Lie algebra $\mathfrak{g}=so(2n+1)$. (It is also, incidentally, a Cartan subalgebra for $U(n) \subset SO(2n)$, where the containment is defined via the realization map. In fact in all of our examples, including the commutative groups of Example (a) (where we had $H=\mathfrak{g}$ and trivial Weyl group), the Cartan subalgebras were of the same sort.) Thus we may take the same basis vectors $l_1^0, \ldots, l_n^0$ for H as before (or, more precisely, we obtain these basis vectors by augmenting those defined in (61) with a zero $(2n+1)$st row and column), and corresponding elementary forms t_j defined by $t_j(l_k^0)=\delta_{jk}$. However the Weyl group for $SO(2n+1)$ is larger than for $SO(2n)$; in addition to the generating transformations (i) and (ii) above, the Weyl group for $SO(2n+1)$ requires for its generation those transformations reversing the direction of a single l_i^0, each of which thereby induces a corresponding transformation of the space spanned by the t_n, given by $t_i \mapsto -t_i$, $t_j \mapsto t_j$, $j \neq i$. Hence among the symmetric forms on H invariant under the Weyl group, the multiplicatively basic ones (i.e. the algebra generators) will be just the following:

$$\psi_q^{(SO)} = t_1^{2q} + \cdots + t_n^{2q}.$$

As before these forms extend to the whole of $\mathfrak{g}$ to yield the characteristic classes c_{2q}. Thus in this as in the preceding two examples every symmetric form on the Cartan subalgebra, invariant under the Weyl group, extends to an Ad-invariant form on the whole Lie algebra.

25.5. Characteristic Classes: Enumeration

It turns out that for the groups $G=U(n)$, $SO(n)$ there are essentially no characteristic classes other than those constructed above; to put it more precisely, every "natural" or "covariant" construction (see below for the meaning of this term) which associates with each fibre bundle endowed with a connexion, a closed form on the base space M of the bundle with the property that integrals of the form over cycles (i.e. closed, oriented submanifolds of M) are topological invariants of the bundle (i.e. are not altered by changes of the connexion on the (same) fibre bundle), is "equivalent" to some polynomial in the characteristic classes c_i, χ_n (in the algebra of differential forms on M). (Closed forms a, b, of the same rank, on a manifold M are said to be *equivalent* if their difference is exact, i.e.

$$a=b+du$$

for some form u; if a and b are equivalent then their integrals over any cycle (see above) of the ambient manifold M coincide:

$$\int_P a=\int_P b,$$

for every cycle $P\subset M$.)

We shall now explain what is meant by a "natural" or covariant" construction. Earlier (see 21.4.2) we defined a "fibre bundle map" or "morphism" between two bundles with the same fibre and structure group, as a map $\bar{f}\colon E\to E'$ between the total spaces, which respects fibres (i.e. $fp=p'\bar{f}$, where $f\colon M\to M'$ is uniquely defined by this condition), and induces on each fibre F a diffeomorphism belonging to G. The reader will also recall (from §24.4) the construction of the "induced" fibre bundle $p\colon E\to M$, determined by a given fibre bundle $p'\colon E'\to M'$ and a map $f\colon M\to M'$, yielding a fibre bundle map $\bar{f}\colon E\to E'$. In connexion with this definition the "classification theorem" was stated (without proof): Every fibre bundle with base M and structure group G (e.g. a principal bundle) is induced from a universal (principal) fibre bundle $p_G\colon E_G\to B_G$ (i.e. one with contractible total space E_G; see Definition 24.4.1) via a map $M\to B_G$ unique up to a homotopy. (In §§24.3, 24.4 universal (and N-universal) fibre bundles were constructed for the groups $O(n)$, $SO(n)$ and $U(n)$, having smooth manifolds as bases for all N:

$$B_G=\hat{G}_{N,n} \quad \text{for} \quad SO(n), \quad N\to\infty;$$

$$B_G=G^{\mathbb{C}}_{N,n} \quad \text{for} \quad U(n), \quad N\to\infty;$$

$$B_G=\mathbb{C}P^N \quad \text{for} \quad U(1)=SO(2), \quad N\to\infty;$$

$$B_G=\mathbb{H}P^N \quad \text{for} \quad SU(2)=SP(1), \quad N\to\infty,$$

where the manifolds $\hat{G}_{N,n}$, $G^{\mathbb{C}}_{N,n}$ are as defined in §24.4.)

Suppose now that we have a fibre bundle map $\bar{f}\colon E\to E'$, giving rise (as above) to a map $f: M \to M'$ of bases, and that the bundle E' is endowed with a connexion ω'. Applying to the form ω' the restriction operation (i.e. pullback) $\bar{f}^*$ determined by the map $\bar{f}$, we obtain a form $\omega = \bar{f}^*\omega'$ on E which will in fact be a connexion on E. Since the operators d and H commute with the restriction operator $\bar{f}^*$, the corresponding curvature form Ω_E will be "covariant", i.e.

$$\bar{f}^*(\Omega'_E) = \Omega_E \qquad (\Omega'_E = H d\omega', \Omega_E = H d\omega).$$

From this and the details of their construction, it follows that all of the characteristic classes c_i, χ_n, and the general classes (see the preceding subsection) behave "naturally" or "covariantly" or, in the language of category theory, "functorially", with respect to bundle morphisms, by which we mean simply that for each Ad-invariant form ψ (see 25.4.3) we have

$$f^* c'_\psi = c_\psi, \tag{63}$$

where c'_ψ and c_ψ are respectively the characteristic classes of the bundles E' and E determined by the forms ω' and ω. (Of course as noted before (see the last exercise of the preceding subsection), the forms c'_ψ and c_ψ are closed, and, up to the addition of exact forms, independent of the choice of ω (and hence ω') on E (and E').)

25.5.1. Definition. Given a Lie group G, a closed form c (more precisely a class of forms) on the bases of fibre bundles is called a *topological characteristic class* for G if it has the following two properties:

(i) The form c is defined ("uniformly") for every principal fibre bundle with structure group G (and therefore in particular for every (base) manifold M);

(ii) Under bundle maps $\bar{f}: E \to E'$, the pullback operator f^* determined by the induced map $f: M \to M'$ of bases, acts on c according to the formula

$$f^*(c') = c + du,$$

for some exact form du. (Thus if we interpret c as the equivalence class of forms, defining two forms to be equivalent if they differ by an exact form, then the requirement is that c be covariant.)

It turns out that there are relatively few topological characteristic classes. In order to make this somewhat vague assertion more precise we require the notion of the *qth cohomology group* $H^q(M; \mathbb{R})$ of a manifold $M (q = 0, 1, 2, \ldots)$: the elements of $H^q(M; \mathbb{R})$ may be defined to be the equivalence classes a of real-valued closed forms $\tilde{a}$ $(d\tilde{a} = 0)$ on M, of rank q, (where two such forms are, as before, understood to be equivalent if they differ by an exact form, so that $\tilde{a}$ is equivalent to $\tilde{a} + du$); the additive operation on these classes is taken to be that induced by ordinary addition of forms. Thus the groups $H^q(M; \mathbb{R})$ are abelian (and $H^q(M; \mathbb{R}) = 0$ for $q > \dim M$); their direct sum

$H^*(M; \mathbb{R}) = \sum_q H^q(M; \mathbb{R})$ clearly forms an associative algebra if we take as the multiplicative operation (that induced by) the exterior product of (closed) forms. The following result provides a classification of topological characteristic classes.

25.5.2. Theorem. *Let B_G denote, as above, a classification space for G, i.e. the base of a universal principal G-bundle. Each element c of each cohomology group $H^q(B_G; \mathbb{R})$ determines a topological characteristic class (of rank q) defined (as it should be) on all principal G-bundles. Conversely, each topological characteristic class of rank q determines a unique element c of $H^q(B_G; \mathbb{R})$. (In other words there is a (natural) one-to-one correspondence between the elements of $H^q(B_G; \mathbb{R})$ and the topological characteristic classes of rank q.)*

PROOF. Given a topological characteristic class c as a closed form of rank q defined for all principal G-bundles (as in Definition 25.5.1) then in particular taking as bundle a universal one with base B_G, we shall have $c \in H^q(B_G; \mathbb{R})$.

For the converse, suppose that we are given an arbitrary element (i.e. cohomology class) $c \in H^q(B_G; \mathbb{R})$. By the classification theorem (see above or §24.4) each principal G-bundle η with base M is induced by a smooth map $f: M \to B_G$ unique up to a homotopy; if we define c on the (arbitrary) bundle η by setting

$$c\eta = f^*(c),$$

then the form $c(\eta)$ is closed (since $df^* = f^*d$) and so represents an element of $H^q(M; \mathbb{R})$. Since homotopic maps f_1, f_2 yield equivalent closed forms $f_1^*(c), f_2^*(c)$, i.e. $f_1^*(c) = f_2^*(c) + du$ (see Part III), this concludes the proof. □

25.5.3. Examples. In the following examples we identify (without proofs) the cohomology algebras of classification spaces B_G for various G.

(a) We begin by considering certain discrete groups G.

(i) $G = \mathbb{Z}$, $B_G = S^1$, $H^1(B_G; \mathbb{R}) = \mathbb{R}$, $H^q(B_G) = 0$ for $q > 1$.

(ii) $G = \mathbb{Z}_m$, m finite, with B_G the corresponding lens space (see §24.4, Example (b)); here $H^q(B_G; \mathbb{R}) = 0$ for all $q > 0$. (In fact for any finite group G, we have $H^q(B_G; \mathbb{R}) = 0$ for $q > 0$.)

(iii) $G = \mathbb{Z} \oplus \cdots \oplus \mathbb{Z}$ (n summands), $B_G = T^n$ (the n-dimensional torus); here $H^*(B_G; \mathbb{R})$ is the exterior algebra in n one-dimensional generators (i.e. rank-one forms).

(iv) $G = \pi_1(M_g^2)$ where M_g^2 is the closed, orientable surface of genus g; $B_G = M_g^2$; here $H^*(B_G; \mathbb{R})$ is the algebra on n one-dimensional (i.e. rank-one) generators $a_1, b_1, \ldots, a_g, b_g$ with relations $a_1 \wedge b_1 = \cdots = a_g \wedge b_g$, $a_i \wedge b_j = 0$ for $i \neq j$, $a_k \wedge a_l = b_k \wedge b_l = 0$.

(v) G the free group on p free generators; B_G the (punctured) plane $\mathbb{R}^2$ (with p points removed); here $H^1(B_G; \mathbb{R}) = \mathbb{R}^p$, $H^q(B_G; \mathbb{R}) = 0$ for $q > 1$.

(b) In this our second example we consider the various abelian groups

$G=\mathbb{R}^k\times T^m$. In the case $G=\mathbb{R}^k$ the space B_G is just a single point (or a space contractible to a point). For $G=T^1=S^1=U(1)\simeq SO(2)$, we have $B_G=\mathbb{C}P^\infty$ (see above). More generally for $G=\mathbb{R}^k\times T^m=\mathbb{R}^k\times S^1\times\cdots\times S^1$ (m factors S^1), we have

$$B_G\cong \underbrace{B_{S^1}\times\cdots\times B_{S^1}}_{m\ \text{factors}}=\underbrace{\mathbb{C}P^\infty\times\cdots\times\mathbb{C}P^\infty}_{m\ \text{factors}}$$

The cohomology algebra $H^*(B_G;\mathbb{R})$ then turns out to be the algebra of polynomials in m generators $t_1,\ldots,t_m$ from $H^2(B_G;\mathbb{R})$.

(c) For $G=U(n)$, we have $B_G=G^{\mathbb{C}}_{\infty,n}$ (see §24.4), and $H^*(B_G;\mathbb{R})$ is the algebra of polynomials in the n generators $c_1,\ldots,c_n$; $c_i\in H^{2i}(B_G;\mathbb{R})$. For $G=SU(n)$, $H^*(B_G,\mathbb{R})$ is the algebra of polynomials in the $(n-1)$ generators $c_2,\ldots,c_n$; $c_i\in H^{2i}(B_G;\mathbb{R})$.

(d) For $G=SO(2n)$, $H^*(B_G;\mathbb{R})$ is the algebra of polynomials in the generators $c_{2i}\in H^{4i}(B_G;\mathbb{R})$, $i=1,\ldots,n-1$, together with $\chi_n\in H^{2n}(B_G;\mathbb{R})$.

(e) For $G=SO(2n+1)$, $H^*(B_G;\mathbb{R})$ is the algebra of polynomials in the n generators $c_{2i}\in H^{4i}(B_G;\mathbb{R})$, $i=1,\ldots,n$.

From the last few of these examples (and Theorem 25.5.2 above) we deduce that for the important (compact) groups all the topological characteristic classes are (essentially) just the characteristic classes constructed earlier.

For a non-compact Lie group G the problem of classifying the topological characteristic classes (as cohomology classes of classifying spaces) reduces to the same problem for a maximal compact subgroup $K\subset G$; this is a consequence of the fact (whose proof we omit) that the base spaces B_K and B_G are homotopically equivalent (since this implies that $H^*(B_K;\mathbb{R})$ is (naturally) isomorphic to $H^*(B_G;\mathbb{R})$, whence by Theorem 25.5.2 the topological characteristic classes of the groups K and G coincide). If the Lie group G is semisimple (see §3.1) then we may also proceed as follows: Using the semisimplicity it can be shown that the complexification $\mathfrak{g}_{\mathbb{C}}=\mathbb{C}\otimes\mathfrak{g}$ of the Lie algebra $\mathfrak{g}$ is identifiable with the complexification $\mathfrak{g}'_{\mathbb{C}}$ of the Lie algebra $\mathfrak{g}'$ of some compact Lie group G' (called the *compact real form* of the complex Lie algebra $\mathfrak{g}'_{\mathbb{C}}=\mathfrak{g}_{\mathbb{C}}$). Given this, it is then clear that the constructions of differential-geometric characteristic classes from a connexion on a bundle (for instance as in the preceding subsection by means of "elementary" operations on the curvature form Ω) will lead to exactly the same results (locally on the base) for G' as for G (since it is essentially in terms of the Lie algebra $\mathfrak{g}$ that the form Ω on the base space was defined locally); thus for the groups G and G' we shall obtain locally the same closed forms, and consequent functionals of the connexion with identically zero variational derivative (i.e. integrals of the closed forms over submanifolds of the base). However these closed forms often turn out to be exact, in which case the corresponding functionals are "topologically trivial" (and the corresponding topological characteristic class is the zero class). For instance in the case $G=\mathbb{R}^k$ we have $G'=T^k$ for which, locally, the differential-geometric characteristic classes on any particular base space are the same; however for $G=\mathbb{R}^k$ (as might be expected from the opening sentence of this paragraph) all integrals of characteristic classes over cycles (i.e. closed, oriented submanifolds of the base) vanish. Similarly, although for $G=SO(p, q)$ the constructions of differential-geometric characteristic classes are (locally) exactly the same formally as for $G'=SO(p+q)$, yet non-trivial integrals of characteristic classes over cycles are again obtained only as for a maximal compact subgroup (here $SO(p)\times SO(q)\subset SO(p, q)$).

Thus, to repeat the above in the narrower (though important) context of the tangent bundle over a pseudo-Riemannian manifold of type (p, q) (where $p+q=\dim M$), although we can define locally on M the same differential-geometric characteristic classes as for Riemannian manifolds (where the metric is positive definite), with integrals over cycles of appropriate dimensions having, as in the Riemannian case, identically zero variational derivatives with respect to the metric, it will nonetheless often happen that many of these classes have no topological significance in the sense that their integrals over all cycles are zero. Consider for example the direct Lorentz group $SO(3, 1)=G$; its Lie algebra has the same complexification as the group $G'=SO(4)$ (and also the group $\hat{G}'=SU(2)\times SU(2)$). For G' (and also $\hat{G}'$) there are two generating differential-geometric characteristic classes of dimension (i.e. rank) 4, namely

$$c_2, \chi_2 \in H^4(B_{G'}; \mathbb{R}).$$

Thus by the above, for the group $G=SO(3, 1)$ there will be a corresponding pair of generating differential-geometric characteristic classes. However since on the other hand $SO(3)$, which is embeddable in $SO(3, 1)$ as a maximal compact subgroup, has only the one topological characteristic class of rank $\leqslant 4=\dim M$ (see Example (e) above)

$$c_2 \in H^4(B_G; \mathbb{R}),$$

it follows (from the opening sentence of the preceding paragraph) that $SO(3, 1)$ has also but one topological characteristic class.

Characteristic classes can be shown (using the method of universal bundles) to have the following important property: *The algebra of topological characteristic classes of a given group G has a multiplicative basis consisting of classes whose integrals over all cycles (i.e. closed, oriented submanifolds of the base manifolds of arbitrary G-bundles) are integer-valued.*

The validity of this general assertion follows, as we shall now show, from its truth for universal bundles: *The algebra $H^*(B_G; \mathbb{R})$ (where B_G is the base of a universal G-bundle) has an algebra basis of elements $d_1, \dots, d_k$ whose integrals are all cycles in B_G of appropriate dimensions, are integer-valued.* To see how the general statement follows from this, let M be the base of any G-bundle, let P be any closed, oriented q-dimensional manifold, and let $\varphi: P\to M$ be a smooth map. (The pair (P, φ) is often called a "singular cycle" of M.) By the classification theorem, each G-bundle over M is induced from the universal bundle over B_G by a map $f: M \to B_G$ (unique to within a homotopy). Composing the maps φ and f we obtain a singular cycle of dimension q in B_G:

$$(P, f\varphi); \qquad f\varphi: P \xrightarrow{\varphi} M \xrightarrow{f} B_G.$$

Corresponding to each basis element d_s of $H^*(B_G; \mathbb{R})$ we have a characteristic class $d_s' = f^*(d_s)$ on M. If d_s has rank $q = \dim P$, then integrating over the singular cycle (P, φ) of M, we obtain (using the analogue of the formula for change of variable of integration (see Part I, §22.1)):

$$\int_{(P,\varphi)} d_s' = \int_P \varphi^*(d_s') = \int_P \varphi^* f^*(d_s) = \int_{(P, f\varphi)} d_s,$$

which is an integer by assumption (being the integral over the cycle $(P, f\varphi)$ of the special basis element d_s).

25.5.4. Examples. (a) We first consider the commutative Lie groups $G = \mathbb{R}^1$ and $G' = SO(2) \simeq U(1)$. The curvature form on the base of any fibre bundle with G or G' as structure group is, as we know, a closed, scalar-valued form $\Omega = \Omega_{\mu\nu}\, dx^\mu \wedge dx^\nu$. The following question naturally arises: What conditions on a closed 2-form on the base of a G-bundle or G'-bundle will enable it to qualify as a curvature form? (In the language of physics this is equivalent to the question of the possibility of defining globally a "vector potential" A on the total space, essential for quantization.) It turns out that the precise conditions (different for the two groups G and G') are as follows (in both cases we omit the proof of sufficiency; the necessity follows from various earlier remarks):

(i) For the group $G = \mathbb{R}^1$, a closed 2-form Ω on the base M of a G-bundle is a curvature form if and only if we have $\int_P \Omega = 0$ for every 2-dimensional cycle $P \subset M$ (or, equivalently, if and only if $\Omega = dA$ for some 1-form on the base M; cf. the concluding sentence of the proof of Theorem 25.5.2).

(ii) For the group $G' = SO(2) \simeq U(1)$, a closed form Ω on M is a curvature form if and only if Ω can be normalized so that the integrals of Ω over all 2-cycles $P \subset M$ are integer-valued. (The vector-potential A (ω in earlier notation) will then be definable globally above M as a 1-form on the total space E, satisfying $dA = \Omega_E$; see the beginning of the preceding subsection.)

In the physical context of an electromagnetic field, the role of the curvature form Ω is played by the electromagnetic field tensor F:

$$F_{\mu\nu} = \Omega_{\mu\nu}, \qquad d(F_{\mu\nu}\, dx^\mu \wedge dx^\nu) = 0,$$

defined on a region of Minkowski space $\mathbb{R}^4_{3,1}$. (Here the closure of the curvature form is equivalent to Maxwell's equations (see Part I, §25.2).) If physical reality is such that the electrodynamics of a physical situation may be "compact" (in the sense that the group is $SO(2)$ rather than $\mathbb{R}^1$), then as was shown by Dirac, "magnetic monopoles" are possible. To see this consider the stationary (i.e. time-independent) situation of a magnetic field of strength $\Omega = F_{\mu\nu}\,dx^\mu \wedge dx^\nu$ (where now $\mu, \nu = 1, 2, 3$) in, for instance, a region of the form $U = \mathbb{R}^3 \setminus \{x_0\}$, i.e. $\mathbb{R}^3$ with the single point x_0 removed, where the field has a singularity at x_0. If $S^2 = S^2_\rho \subset U$ is the sphere in U given by the equation $\sum_{\mu=1}^3 (x^\mu - x_0^\mu)^2 = \rho^2 > 0$, then in view of (ii) above we may have

$$\int_{S^2} \Omega = n$$

for a non-zero integer n. Physically this means that the magnetic field flux through the surface S^2 can have non-zero (integer) values (i.e. there can exist a "magnetic monopole" at x_0), without there arising thereby any conflict with the possibility of defining a vector-potential (and quantizing the field in accordance with the general principles of the quantum theory of gauge fields). There may exist several monopoles, at points $x_0, \ldots, x_k \in \mathbb{R}^3$; there will then be a corresponding collection of "independent" cycles in the region $\mathbb{R}^3 \setminus \{x_0, \ldots, x_k\}$.

(b) As our second (and final) example we consider fibre bundles over the sphere S^k (for various k), with various structure groups G. By Lemma 24.4.3, for each k, G, the G-bundles over S^k are determined (up to equivalence) by the elements of $\pi_{k-1}(G)$.

(i) $k = 1$, $G = O(n)$, $SO(n)$, $U(n)$. Since the groups $SO(n)$ and $U(n)$ are connected, all fibre bundles over S^1 with these as structure groups will be trivial. However since $\pi_0(O(n)) \simeq \mathbb{Z}_2$, for $G = O(n)$ there does exist a non-trivial G-bundle over S^1. We shall later (in Chapter 7) encounter bundles with one-dimensional base (actually $\mathbb{R}^1$) as homogeneous models in the general theory of relativity. There is however no curvature theory for such bundles.

(ii) $k = 2$. In this case, for the group $G = SO(2)$ we have one G-bundle for each integer $m \in \pi_1(SO(2)) \simeq \mathbb{Z}$; for positive m these are just the Hopf bundle η (with fibre $\mathbb{C}^1$; see §25.3), and its tensor powers η^m (see §24.5). If Ω is the (suitably normalized) curvature form on S^2 determined by any connexion on η^m, then we shall have

$$m = \int_{S^2} \Omega,$$

i.e. the integer m is given by the integral of Ω over the base S^2. An alternative characterization of m may be obtained in the following way. Since $S^2 \setminus \{\infty\} \cong \mathbb{C}^1 \cong \mathbb{R}^2$, above the region $S^2 \setminus \{\infty\}$ any bundle over S^2 is trivial (cf. Lemma 24.4.2), whence it follows that the bundle is determined by specifying a connexion $A = A_\mu\,dx^\mu = A_z\,dz + A_{\bar z}\,d\bar z$ on $S^2 \setminus \{\infty\} \cong \mathbb{C}$, with the "boundary" condition that as $|x| \to \infty$ the connexion becomes (asymptotically) trivial, i.e. for some $g(x)$ (see below)

$$A_\mu \sim -\frac{\partial g(x)}{\partial x^\mu} g^{-1}(x) \quad \text{as } |x| \to \infty.$$

If the structure group G is $U(1)$ ($\simeq SO(2)$), then $g(x)$ has the form $e^{i\varphi(x)}$, whence $(\partial g/\partial x^\mu)g^{-1} = i(\partial\varphi/\partial x^\mu)$ (cf. end of §25.2); the map $g(x)$ (and hence also $\varphi(x)$) is defined only asymptotically as $|x| \to \infty$ on the rays with direction $x/|x|$, regarded as comprising S^1, and then $g(x)$ is given by

$$g\colon S^1 \to S^1, \qquad x \mapsto e^{i\varphi(x)} \quad (|x| \text{ large}).$$

The degree of this map can then be seen to coincide with m (if the bundle in question is that determined by the integer $m \in \mathbb{Z} \simeq \pi_1(SO(2))$).

(iii) $k=3$. Since $\pi_2(G)=0$ for all the Lie groups we have encountered (in fact for all Lie groups!), all fibre bundles over S^3 are topologically trivial. Any connexion yielding zero curvature form (i.e. any trivial connexion; cf. Part I, §41.2) will for some map $g\colon S^3 \to G$, be given by

$$A_\mu = -\frac{\partial g(x)}{\partial x^\mu} g^{-1}(x).$$

Thus here each homotopy class of maps $S^3 \to G$, i.e. each element of $\pi_3(G)$, determines a "homotopy class" of trivial connexions (cf. Lemma 24.4.3), though all fibre bundles are as noted, trivial. (Recall that: $\pi_3(SO(2))=0$; $\pi_3(SO(3)) \simeq \pi_3(SU(2)) \simeq \pi_3(SU(n)) \simeq \pi_3(SO(m)) \simeq \mathbb{Z}$ for $n \geq 3$, $m \geq 5$; $\pi_3(SO(4)) \simeq \mathbb{Z} \oplus \mathbb{Z}$.)

(iv) $k=4$. There are many inequivalent G-bundles over S^4, and a large collection of topological invariants. Since $S^4 \setminus \{\infty\} \cong \mathbb{R}^4$, each bundle over S^4 is determined (as in (ii) above) by specifying a connexion A_μ on the region $\mathbb{R}^4$ (above which the bundle is trivial), with the boundary condition

$$A_\mu(x) \sim -\frac{\partial g(x)}{\partial x^\mu} g^{-1}(x) \quad \text{as } |x| \to \infty,$$

for some (asymptotically defined) map $g(x)$ (which determines the bundle). The map $g(x)$ is as before defined asymptotically on the rays $x/|x|$ which may be regarded as points of S^3; thus each G-bundle is determined by a map $g\colon S^3 \to G$, so that each homotopy class of such maps (i.e. each element of $\pi_3(G)$) forms a topological invariant of the bundle.

Of particular interest are the groups $G=SU(2)$, $SO(4)$, $SO(3)$. For $G=SO(4)$ there are exactly two basic, integer-valued characteristic classes (while for each of $SU(2)$, $SO(3)$ there is only one), namely:

$$\begin{aligned} c_2 &= \int_{\mathbb{R}^4} \operatorname{tr}(F_{\mu\nu}F_{\lambda\varkappa})\, dx^\mu \wedge dx^\nu \wedge dx^\lambda \wedge dx^\varkappa; \\ \chi_2 &= \int_{\mathbb{R}^4} \operatorname{tr}(F_{\mu\nu}({}_*F_{\lambda\varkappa}))\, dx^\mu \wedge dx^\nu \wedge dx^\lambda \wedge dx^\varkappa, \end{aligned} \tag{64}$$

where $F_{\mu\nu}=\partial A_\mu/\partial x^\nu-\partial A_\nu/\partial x^\mu+[A_\mu, A_\nu]$. (For the definition of the operator $_*$ see Part I, §19.3. We leave to the reader the verification of the validity of the above expression for χ_2.)

EXERCISE

Show that for $G=SO(4)$ the total space E of a G-bundle over S^4 with fibre S^3, for which $\chi_2=1$ (but c_2 is arbitrary) is homotopically equivalent to the sphere S^7. (Such a bundle will be principal for $G=SU(2)$, but merely associated for $G=SO(4)$.)

Remark. It has been shown by Milnor that the total spaces E of certain of these bundles (namely those for which $c_2, \chi_2=1$) are homeomorphic, but not diffeomorphic, to the sphere S^7.

§26. Knots and Links. Braids

26.1. The Group of a Knot

The concept of the fundamental group has an important application in the theory of knots and links in 3-dimensional space. We define a *knot* to be a smooth, closed curve $\gamma=\gamma(t)$ $(0\le t\le 2\pi,\ \gamma(0)=\gamma(2\pi))$ in (Euclidean) $\mathbb{R}^3$, which does not intersect itself, and for all t has non-zero velocity vector; such a curve may indeed be "knotted" in $\mathbb{R}^3$ (see Figure 94).

By an *isotopy of a knot* γ we shall mean a "motion" of the knot through the ambient space $\mathbb{R}^3$ achieved by means of a deformation (i.e. homotopy) of the identity map of $\mathbb{R}^3$, in the class of self-diffeomorphisms of $\mathbb{R}^3$.† A knot is defined to be *trivial* if there is such an isotopy bringing it into the position of

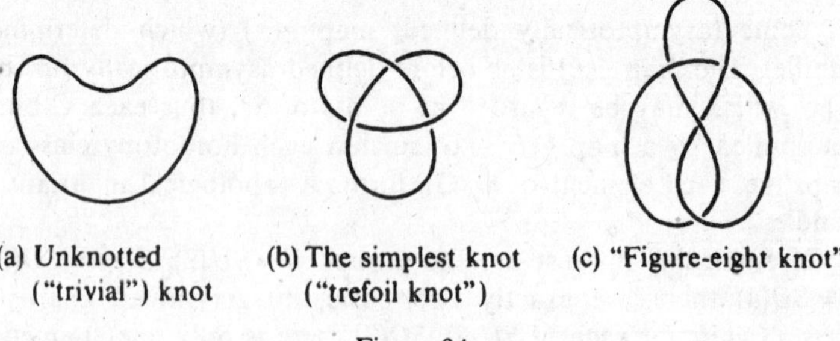

(a) Unknotted ("trivial") knot (b) The simplest knot ("trefoil knot") (c) "Figure-eight knot"

Figure 94

† Usually an isotopy of a knot is defined as a deformation (in the class of smooth embeddings) of the embedding of the circle (in space) which defines the knot. However it is then normally shown that such a deformation of the embedded circle extends to a deformation (of the requisite kind) of the whole space.

the particular (trivial) knot $\tilde{\gamma}=\{z=0, x^2+y^2=1\}$. It will be convenient to assume that our knot γ lies in $S^3 \supset \mathbb{R}^3$; this obviously does not affect the knot, nor does it require any essential modification of the definition of an isotopy of the knot (since a self-diffeomorphism of S^3 permutes the 2-dimensional simple closed surfaces in S^3, so that, by enclosing the knot in the interior of some region with such a surface as boundary, we can change any isotopy of S^3 so that it fixes some point of S^3 while determining essentially the same "motion" of the knot). For suitably small $\varepsilon>0$ consider an ε-neighbourhood (i.e. "tubular" neighbourhood) U_ε of the knot γ; the boundary ∂U_ε is a torus T^2, and $U_\varepsilon \cong D_\varepsilon^2 \times S^1$ where D_ε^2 represents a disc normal to γ of radius ε. If we remove from S^3 the interior of the tubular neighbourhood U_ε, we are left with manifold-with-boundary $V_\gamma \subset S^3$ whose boundary $\partial V_\gamma = \partial U_\varepsilon$ is diffeomorphic to the torus T^2. It is clear that V_γ is homotopically equivalent to the open region $S^3 \backslash \gamma = W_\gamma$, the *complement* of the knot.

26.1.1. Definition. The fundamental group $\pi_1(W_\gamma)=\pi_1(V_\gamma)$ is called the *knot group* of the knot γ.

We mention two simple properties of the knot group $\pi_1(W_\gamma)=\pi_1(V_\gamma)$:

(i) If the knot γ is trivial then $\pi_1(W_\gamma)\simeq \mathbb{Z}$. (This follows from the homotopy equivalence of $V_\gamma \subset S^3$ (or $W_\gamma \subset S^3$) where $\gamma=\{z=0, x^2+y^2=1\}$, with the circle S^1; see §17.5, Example (d).)

(ii) By definition of the regions V_γ and W_γ, the knot group $\pi_1(V_\gamma)=\pi_1(W_\gamma)$ (and in fact, to within a diffeomorphism, the topology of these regions) is unchanged by isotopies of the knot. Hence in particular the isomorphism $\pi_1(W_\gamma)\simeq \mathbb{Z}$ represents a necessary condition for the knot γ to be trivial. (Note that in fact this is also a sufficient condition; this however is a difficult result.)

We now describe the algorithm for calculating the knot group $\pi_1(W_\gamma)$. We first project the knot in a direction d onto a plane $\mathbb{R}^2$ (or "screen") perpendicular to d. We may suppose (by applying an isotopy of the knot if need be) that the knot is in general position with respect to the direction d; i.e. that all points of self-intersection of the image $\tilde{\gamma}$ on the screen $\mathbb{R}^2$ are double, and the angles of intersection are non-zero. Thus on the screen our image $\tilde{\gamma}$ is a planar, oriented graph (with orientation determined by that of $\gamma=\gamma(t)$, $0\le t\le 2\pi$) having exactly 4 edges incident with each vertex. Besides the orientation of $\tilde{\gamma}$ we need to indicate at each vertex (i.e. point of intersection) which of the intersecting arcs is "above" the other; we do this by labelling the "upper" (i.e. overcrossing) arc with a plus sign (+), and the "lower" with a minus sign (−). (In Figure 95 at the intersection B on the screen the preimage of the arc labelled + is indeed above the preimage of the one labelled −, if we regard the screen as horizontal with the knot suspended above it.) Taking the point ∞ as base point for the purpose of calculating $\pi_1(W_\gamma)$, the representative closed paths may be assumed to approach the knot from ∞ along the

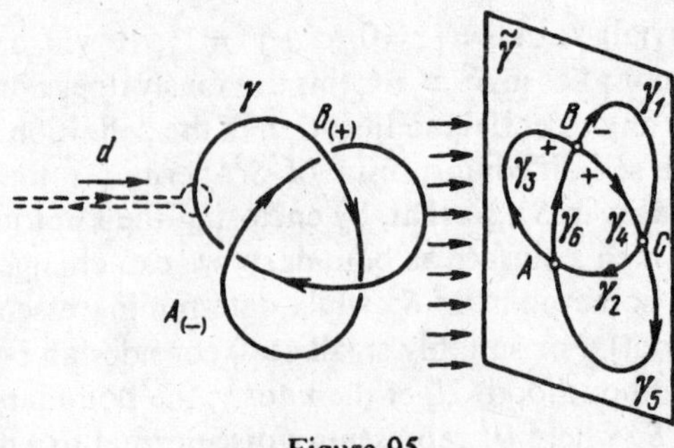

Figure 95

direction d perpendicular to the screen $\mathbb{R}^2$ (from the left in Figure 95). We now number the arcs on the screen in an order determined by tracing out the knot in the direction given by its orientation. (Thus in Figure 95 we have on the screen three vertices A, B, C and arcs $[B_{(-)}C_{+}]=\gamma_1$, $[C_{(+)}A_{(-)}]=\gamma_2$, $[A_{(-)}B_{(+)}]=\gamma_3$, $[B_{(+)}C_{(-)}]=\gamma_4$, $[C_{(-)}A_{(+)}]=\gamma_5$, $[A_{(+)}B_{(-)}]=\gamma_6$, in order as γ is traced out.) With each indexed arc γ_j we associate in the following way a loop a_j representing a generator of $\pi_1(W_\gamma)$: the path a_j comes from ∞ along the direction d almost to the midpoint of the arc γ_j, goes around that arc and returns to ∞; in Figure 95 the path a_3 is indicated, corresponding to the arc $\gamma_3=[A_{(-)}B_{(+)}]$.

It is easy to see (intuitively at least) that the paths a_j do indeed generate the group $\pi_1(W_\gamma)$. We now show how to obtain a full set of defining relations on these generators. With each vertex there are incident four arcs $\gamma_{j_1}, \gamma_{j_2}, \gamma_{j_3}, \gamma_{j_4}$, where we may suppose that the arc γ_{j_2} immediately follows γ_{j_1} (as the knot is traced out) and γ_{j_4} follows γ_{j_3}, so that $j_2=j_1+1$, $j_4=j_3+1$. We may also suppose that the pair of arcs with indices j_1, j_2 is "above" the pair with indices j_3, j_4 (or, in terms of Figure 95. that the segment comprised of the (preimages of) γ_{j_1} and γ_{j_2} lies to the left of that comprised of γ_{j_3} and γ_{j_4}). It is then easy to see that corresponding to the vertex we are considering we obtain the following relations: for the upper segment

$$a_{j_1}=a_{j_2}=a_{j_1+1}; \tag{1}$$

and for the lower segment

$$a_{j_4}=a_{j_3+1}=a_{j_1}^{-1}a_{j_3}a_{j_1}. \tag{2}$$

(Verify this!) (Note that there is some ambiguity here in that we have not specified the orientation of the a_j, so that the a_j are defined only to within replacements of the form $a_j \to a_j^{-1}$; however the relations (1) and (2) are not altered in any fundamental way by such replacements.)

EXERCISE

Show that the set of all relations of the form (1), (2) (obtained by considering in turn all vertices of the image of the knot on the screen) comprise a full set of defining relations for the knot group (i.e. that every relation is a consequence of these).

Note that it follows from the form of the relations (1), (2) that the commutator quotient group of the knot group (i.e. the knot group abelianized) is always isomorphic to $\mathbb{Z}$.

Example. For the trefoil knot (Figure 95) we obtain the following defining relations:

$$a=a_3=a_4, \quad a_1=a_3^{-1}a_6a_3 \quad \text{(vertex } B\text{)},$$
$$b=a_1=a_2, \quad a_5=a_1^{-1}a_4a_1 \quad \text{(vertex } C\text{)},$$
$$c=a_5=a_6, \quad a_3=a_5^{-1}a_2a_5 \quad \text{(vertex } A\text{)};$$

or,

$$b=a^{-1}ca, \qquad c=b^{-1}ab, \qquad a=c^{-1}bc.$$

26.2. The Alexander Polynomial of a Knot

The group of a knot often turns out to be rather complex. There is however a simpler invariant (actually defined in terms of the group of the knot) which though a considerably coarser invariant than the group, nonetheless provides a relatively straightforward test successfully distinguishing many knots. We shall now define this invariant. Let $a_1, \ldots, a_n$ be generators of our knot group (for instance as defined above), and let $r_i(a_1, \ldots, a_n)=1$ $(i=1, \ldots, m)$ be defining relations for the group in terms of these generators. Let G denote the free group with free generators $a_1, \ldots, a_n$ (so that now we regard the knot group as the quotient of G by the normal subgroup generated by the elements $r_i(a_1, \ldots, a_n)$, $i=1, \ldots, m$). For each free generator a_i we define the *free differential operator* $\partial/\partial a_i$ on the free group G by

$$\frac{\partial a_j}{\partial a_i}=\delta_{ij}, \qquad \frac{\partial a_i^{-1}}{\partial a_i}=-a_i^{-1}, \qquad \frac{\partial}{\partial a_i}(bc)=\frac{\partial b}{\partial a_i}+b\frac{\partial c}{\partial a_i}.$$

It is clear from this that the value of the derivative of any element of G is an element of the group ring $\mathbb{Z}[G]$, i.e. is a formal linear combination of elements of G with integer coefficients. We now form the $m \times n$ matrix $(\partial r_i/\partial a_j)$, with entries from the integral free group ring, and replace each generator occurring in these entries by a variable t, the same for all generators (thus a_j^k is replaced by t^k for all j, k). This yields an $m \times n$ matrix whose entries are polynomials in t and t^{-1} with integer coefficients; the *Alexander polynomial* of the knot is then the highest common factor $\Delta(t)$ of all $(n-1)\times(n-1)$ minors of this matrix, defined to within a factor of the form $\pm t^k$, k any integer.

EXERCISES

1. Prove that if the groups of two knots (or more generally any two finitely presented groups) are isomorphic, then (to within a factor of the form $\pm t^k$) their Alexander polynomials $\Delta(t)$, $\Delta'(t)$ either coincide, or satisfy $\Delta'(t)=\Delta(t^{-1})$.

2. Show that the Alexander polynomial of the trefoil knot is $\Delta(t)=1-t+t^2$.
3. Calculate the Alexander polynomial of the figure-eight knot, depicted in Figure 94(c), and deduce that that knot is not equivalent to the trefoil knot (i.e. isotopic neither to the trefoil knot nor to its reflection).

26.3. The Fibre Bundle Associated with a Knot

Since, as was noted above, $H_1(W_\gamma)\simeq\mathbb{Z}$, the embedding $\partial V_\gamma\cong T^2\to V_\gamma\sim W_\gamma$ (where $\sim$ denotes homotopy equivalence), determines a homomorphism (cf. §21.2)

$$H_1(\partial V_\gamma)\simeq H_1(T^2)\simeq\mathbb{Z}\oplus\mathbb{Z}\to\mathbb{Z}\simeq H_1(W_\gamma).$$

It follows that there is a generator $\bar{\gamma}$ on the torus $\partial V_\gamma\cong T^2$ (i.e. member of a generating pair for $\pi_1(\partial V_\gamma)$) which is homologous to zero in the complement V_γ of the knot γ. Clearly the generator $\bar{\gamma}$ is represented by the path traced out by the tip of a variable vector normal to γ and of constant length ε as its tail traces out γ (i.e. the path consisting of all end-points of vectors of a suitable normal vector field on γ of constant length ε). Consider any smooth map $\varphi\colon T^2\to S^1$, with the property that $\bar{\gamma}$ is the complete inverse image $\varphi^{-1}(s_0)$ of a regular value $s_0\in S^1$ of the map φ (see §10.2), and satisfying the further condition that its restriction to a representative of the other (meridional) generator of $\pi_1(\partial V_\gamma)$ has degree one as a map $S^1\to S^1$ (see §13.1). By the exercise below any such map φ can be extended from $T_2\cong\partial V_\gamma$ to the whole of the knot-complement V_γ, yielding a smooth map

$$\tilde{\varphi}\colon V_\gamma\to S^1,\qquad \tilde{\varphi}|_{\partial V_\gamma}=\varphi.$$

The complete preimage $\tilde{\varphi}^{-1}(s_0)$ of the regular point s_0 will then be a 2-dimensional bordered surface P with boundary $\partial P=\bar{\gamma}$ on the torus $\partial V_\gamma\cong T^2$. By letting the radius ε of the tubular neighbourhood U_ε of the knot (of which V_γ is actually the complement) approach zero, we obtain a surface P in $\mathbb{R}^3$ (or S^3) with boundary the knot γ itself.

EXERCISE

Prove that the above map $\varphi\colon T_2\to S^1$ can indeed be extended to all of V_γ. (*Hint*. Use the facts that $\bar{\gamma}$ is homologous to zero, that $H_1(V_\gamma)\simeq\mathbb{Z}$, and that $\pi_i(S^1)=0$ for $i>1$. Decompose V_γ as a "cell complex" (consisting of "0-cells" (points), "1-cells" (each homeomorphic to $[0,1]$, with boundary consisting of 0-cells), 2-cells (topological closed discs with boundaries consisting of 1-cells), and 3-cells (topological closed balls with boundaries consisting of 2-cells)), and extend the map φ first to the 1-skeleton (made up all 1-cells) (this is easy), then from the 1-skeleton to the 2-skeleton (this step requires some analysis), and thence to the 3-cells (invoking the fact that $\pi_2(S^1)=0$).)

26.3. Definition. The *genus of a knot* γ is the smallest possible genus that a non-self-intersecting surface P in $\mathbb{R}^3$ (or S^3) with boundary γ, can have.

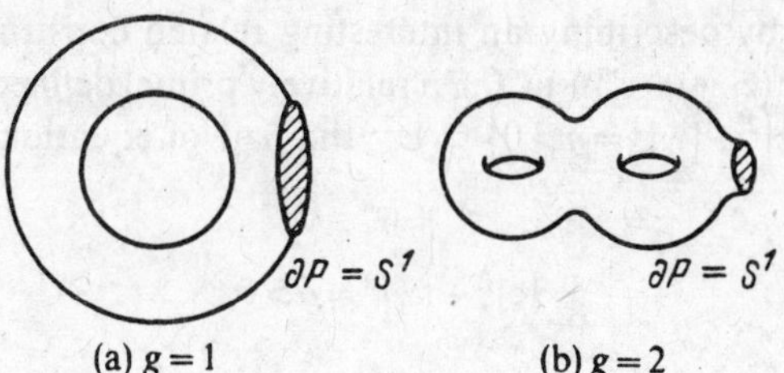

(a) $g=1$ (b) $g=2$

Figure 96

(Note that the *genus* of a closed surface, which by the classification theorem for surfaces is a "connected sum" of copies of T^2 or of $\mathbb{R}P^2$, is the number of such components; if the surface has boundary γ then its genus is defined to be that of the closed surface obtained by attaching a disc to γ.)

For many of the simplest knots γ the knot-complement V_γ, with boundary $\partial V_\gamma \cong T^2$, turns out to be a fibre bundle (with group taken to comprise all self-diffeomorphisms of the fibre) over a circle, with the property that above the toroidal boundary ∂V_γ the bundle φ: $T^2 \to S^1$ is the trivial one, having as fibres the small circles (of radius ε) encircling the knot once and lying in planes perpendicular to it.

Reversing the order in which the ingredients of this topological picture are given, we arrive at the following construction: We start with a smooth fibre bundle p: $V \to S^1$ over the circle having as fibre P a bordered surface of genus $g \geq 0$ with boundary S^1 (see Figure 96), such that the restriction of p to the boundary $\partial V \cong T^2$ defines the trivial bundle $\partial V \cong T^2 = S^1 \times S^1 \to S^1$. We then take the solid torus $D^2 \times S^1$ and identify its boundary (via the diffeomorphism $\partial V \cong T^2$) with that of V, thereby obtaining a closed 3-manifold M. If $M \cong S^3$ then V will indeed be a knot-complement, namely the complement of the knot $\gamma = 0 \times S^1$ running along the middle of the solid torus.

We consider the two simplest cases:

(i) If $g=0$, then $P=D^2$, $V=S^1 \times D^2$, and the resulting knot γ is trivial;

(ii) If $g=1$ (as in Figure 96(a)), then the manifold V is obtained as the product of the surface P by the interval $[0, 1]$, with identifications of the form $(x, 0) = (h(x, 1)(x \in P)$ for some self-homeomorphism h: $P \to P$, which is not orientation-reversing on the boundary S^1 (in order that ∂V be, as required, the trivial bundle over S^1: $\partial V \cong S^1 \times S^1$). The map h will induce on the group $H_1(P \cup D^2) \simeq H_1(T^2) \simeq \mathbb{Z} \oplus \mathbb{Z}$ an automorphism of the form $a \mapsto ma + nb$, $b \mapsto la + kb$, $mk - nl = 1$, where a, b are generators of $H_1(T^2)$.

Exercise

With reference to (ii), calculate the groups $\pi_1(V)$ and $\pi_1(V \cup (D^2 \times S^1))$. Show how to choose the identification map h so that $M = V \cup (D^2 \times S^1)$ is a sphere S^3 (in which case the construction yields a knot $\gamma \subset S^3$). Show in particular how to obtain the trefoil knot in this way (by taking $m=2$, $n=3$).

We conclude by describing an interesting related construction. Consider the polynomial $f(z, w)=z^m+w^n$ (m, n relatively prime) defined on $\mathbb{C}^2$, and the 3-sphere $S_\delta^3=\{|z|^2+|w|^2=\delta>0\}\subset\mathbb{C}^2$; the pair of equations

$$\begin{aligned} z^m+w^n&=0,\\ |z|^2+|w|^2&=\delta>0, \end{aligned} \tag{3}$$

define a curve $\gamma\subset S_\delta^3$.

EXERCISE

Show that the relative primality of m and n implies that the curve γ defined by (3) is connected (and non-self-intersecting), and is therefore a knot.

We now define a fibre bundle

$$p\colon S_\delta^3\backslash\gamma\to S^1,$$

with total space the knot-complement $W_\gamma=S_\delta^3\backslash\gamma$, and base S^1, by setting

$$p(z, w)=\frac{f(z, w)}{|f(z, w)|},\qquad f(z, w)\neq 0. \tag{4}$$

EXERCISES

1. Verify that the map (4) has rank one at every point, and, as claimed, defines a fibre bundle over S^1. Calculate the genus of the fibre.
2. Prove that the knot γ defined by the equations (3) is a "torus knot" $\gamma\subset T^2\subset S_\delta^3$ (i.e. that γ can be represented as a non-self-intersecting closed curve on the torus T^2, as for instance in Figure 97), which in the homology group $H_1(T^2)$ represents the element $ma+nb$, where a, b are the natural generators.
3. Show that the figure-eight knot (see Figure 94(c)) is not a torus knot.

26.4. Links

We now turn to the consideration of the more general concept of a "link" in $\mathbb{R}^3$ or S^3. A *link* is defined to be a collection of embedded (possibly knotted)

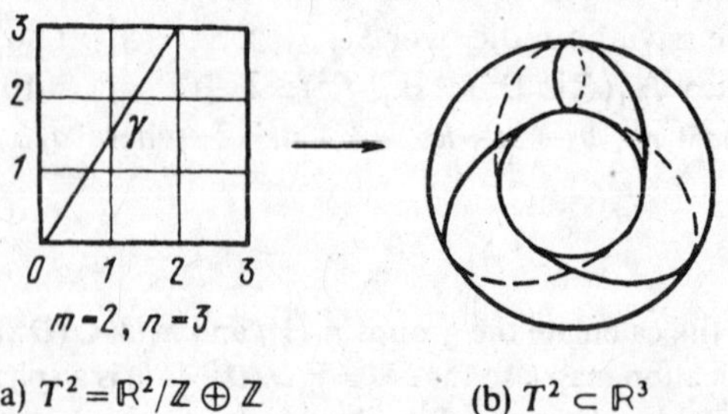

(a) $T^2=\mathbb{R}^2/\mathbb{Z}\oplus\mathbb{Z}$ (b) $T^2\subset\mathbb{R}^3$

Figure 97

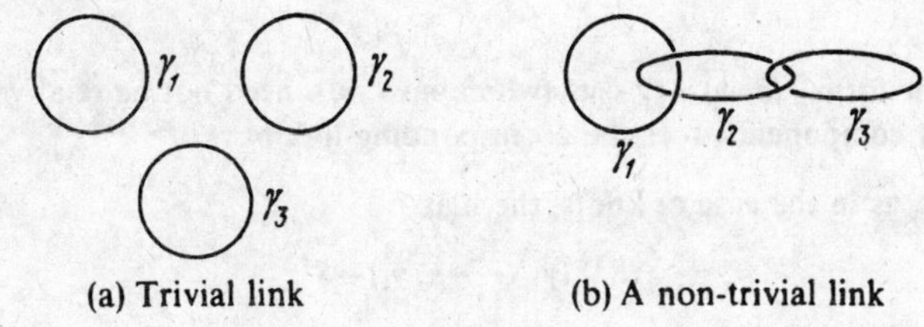

(a) Trivial link (b) A non-trivial link

Figure 98

circles $\gamma_1, \ldots, \gamma_k \subset S^3$, which are pairwise non-intersecting, non-self-intersecting, and have nowhere vanishing tangent vectors. (Examples of trivial and non-trivial links are depicted in Figure 98.)

As for knots, so also for links is the fundamental group $\pi_1(S^3 \backslash (\gamma_1 \cup \cdots \cup \gamma_k))$ of the complement of a link a natural and useful invariant; we call it the *group of the link*. One calculates the group of a link using the same algorithm as for knots, i.e. by projecting the link suitably onto a "screen" and choosing generators and computing the relations between them very much as described in §26.1.

EXERCISE

Calculate the group of each of the three links (for all of which $k=2$) shown in Figure 99.

We encountered earlier another invariant of a link, namely the collection of "linking coefficients" $\{\gamma_i, \gamma_j\}, i \neq j$ (see §15.4). However even for $k=2$ this invariant fails to distinguish inequivalent links; in Figure 99(c) (where each of the two component circles is individually unknotted) we have $k=2$, $\{\gamma_1, \gamma_2\}=0$ as for the trivial link (a), yet, as computation of the group of the link shows, its components cannot be "unlinked".

Interesting examples of links are afforded by pairs of equations of the form

$$f(z, w)=0, \qquad |z|^2+|w|^2=\delta>0, \tag{5}$$

where f is a polynomial.

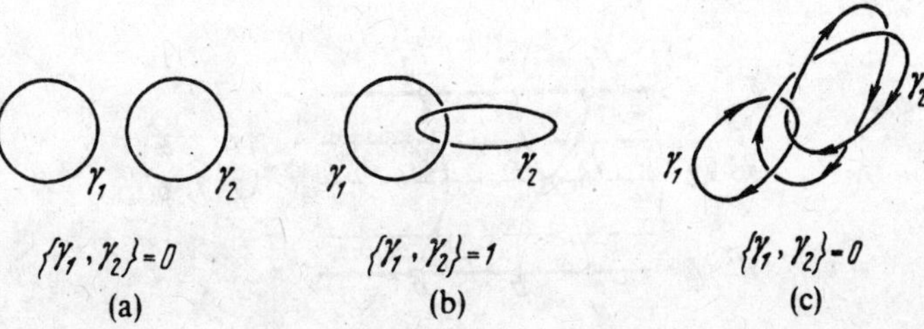

(a) (b) (c)

Figure 99

EXERCISES

1. If f has the form $f(z, w)=z^m+w^n$ (where now m, n need not be relatively prime), how many components does the corresponding link have?

2. Show that, as in the case of knots, the map

$$S^3\backslash(\gamma_1\cup\cdots\cup\gamma_k)\rightarrow S^1$$

given as in (4), defines a fibre bundle. Calculate the genus of the fibre. Find the group of such a link. Examine the particular examples (i) $f(z, w)=z^3+w^6$, (ii) $f(z, w)=z^2w+w^4$.

26.5. Braids

Finally in this section we consider "braids" and their associated groups. Let $P_1, \ldots, P_n$ be n (fixed) points in $\mathbb{R}^2$, and consider the product space $\mathbb{R}^2\times I$ where $I=[0, 1]$.

26.5.1. Definition. An *n-braid* b is a collection of n smooth arcs $\gamma_1, \ldots, \gamma_n$ in $\mathbb{R}^2\times I$ which do not intersect one another, are non-self-intersecting, have nowhere vanishing tangent vectors ($\dot{\gamma}_j\neq 0$), are transverse to the planes $\mathbb{R}^2\times t$, and begin and end at the points $P_1, \ldots, P_n$ on the respective planes $\mathbb{R}^2\times 0$ and $\mathbb{R}^2\times 1$, i.e.

$$\gamma_j(0)=(P_j, 0), \qquad j=1, \ldots, n,$$

$$\gamma_j(1)=(P_{\sigma(j)}, 1), \qquad j=1, \ldots, n,$$

where $\sigma=\sigma(b)$ is a permutation of $\{1, \ldots, n\}$. (See Figure 100.) A *pure* or *unpermuted n-braid* is one for which σ is the identity permutation $\sigma(j)=j$, $j=1, \ldots, n$.

For each n the isotopy classes of n-braids form a group B_n, called the *braid group on n threads*, under the operation defined as follows: the product K_1K_2 of two braids K_1 and K_2 is obtained by laying the lower plane $\mathbb{R}^2$ of K_1 on top of the upper plane of K_2 (so that the points corresponding to $P_1, \ldots, P_n$ in the two planes coincide, and the threads go from the lower plane of K_2 through to the upper plane of K_1; see Figure 101(b)); the *inverse* of a braid K is then essentially just the braid obtained from K by tracing its threads out in

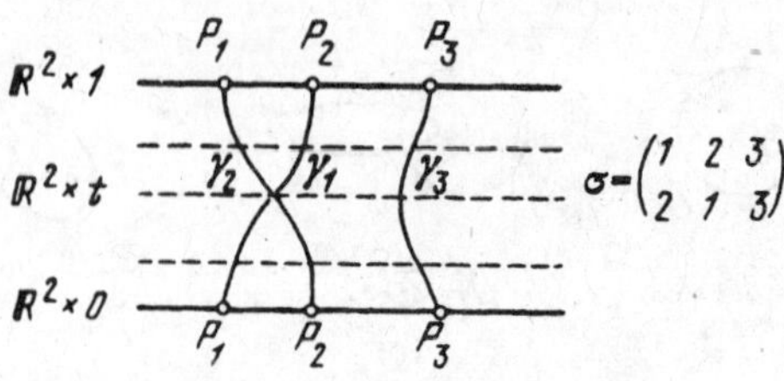

Figure 100

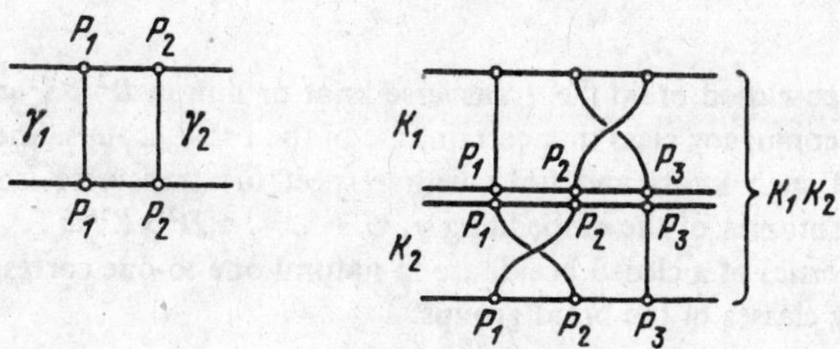

(a) Trivial braid (b) The product of braids K_1 and K_2

Figure 101

the direction opposite to that determined by t, and the *identity* or *trivial braid* is as shown (for $n=2$) in Figure 101(a).

Clearly the map defined by

$$b \mapsto \sigma(b),$$

is an epimorphism from the braid group B_n onto the symmetric group S_n of permutations of n symbols; its kernel obviously consists precisely of the pure braids. It is easy to see that B_n is generated by the $n-1$ braids $\beta_1, \ldots, \beta_{n-1}$ defined by Figure 102; these correspond in the obvious way to the $n-1$ transpositions

$$\sigma_i = \begin{pmatrix} \cdots & i & i+1 & \cdots \\ \cdots & i+1 & i & \cdots \end{pmatrix}, \qquad i=1, \ldots, n-1,$$

in S_n. One readily verifies the relations

$$\begin{aligned} \beta_i\beta_{i+1}\beta_i &= \beta_{i+1}\beta_i\beta_{i+1}, && i=1, \ldots, n-2, \\ \beta_i\beta_j &= \beta_j\beta_i, && |i-j|>1; \quad i,j=1, \ldots, n-1. \end{aligned} \tag{6}$$

EXERCISE

Prove that the relations (6) form a full set of defining relations for the braid group B_n.

Also of interest is the concept of a *closed braid*: This is defined as a knot or link "transversely" embedded in the solid torus $D^2 \times S^1 \subset \mathbb{R}^3$, or, to be more precise, as a collection of pairwise non-intersecting, individually non-self-intersecting, smooth closed curves $\gamma_1, \ldots, \gamma_k$ in the region $D^2 \times S^1$ of $\mathbb{R}^3$, with the further property that for no t, $0 \le t \le 2\pi$, do the tangent vectors $\dot{\gamma}_j(t)$ either vanish or lie in the plane of the disc $D^2 \times t$; i.e. the knot or link is transverse to the sections $D^2 \times t$.

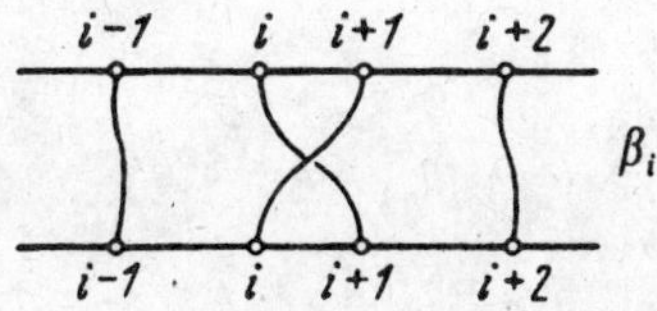

Figure 102

EXERCISE

Show that each closed braid (i.e. transverse knot or link in $D^2 \times S^1 \subset \mathbb{R}^3$) naturally determines a conjugacy class in a certain one of the braid groups. Show further that the classes of such knots and links with respect to "transverse isotopies" within $D^2 \times S^1$ (i.e. isotopies of the embedding $\gamma_1 \cup \cdots \cup \gamma_k \subset D^2 \times S^1 \subset \mathbb{R}^3$ preserving the defining properties of a closed braid) are in natural one-to-one correspondence with the conjugacy classes of the braid groups.

Remark. It can be shown that any knot or link in $\mathbb{R}^3$ can by means of an isotopy be brought into the form of an equivalent transverse knot or link in $D^2 \times S^1 \subset \mathbb{R}^3$ (i.e. closed braid). However this is not of great utility for the problem of classifying knots and links in $\mathbb{R}^3$, since in particular each knot is equivalent to many distinct closed braids.

The braid group arises in yet another interesting context. Consider the set U_n of all complex monic polynomials $f = z^n + a_1 z^{n-1} + \cdots + a_n$ of degree n, having n distinct roots. We indicate briefly how one proves that the group $\pi_1(U_n)$ (where the topology on U_n is defined by, for instance, the norm $|f| = \sum_{i=1}^n |a_i|$) is isomorphic to the braid group B_n.

Consider the space $V_n \subset \mathbb{C}^n = \mathbb{R}^{2n}$ defined by

$$V_n = \underbrace{\mathbb{R}^2 \times \cdots \times \mathbb{R}^2}_{n} \setminus \Delta,$$

where Δ consists of those complex n-tuples $(z_1, \ldots, z_n)$ with at least two of the z_j's equal.

EXERCISES

1. Prove that the group $\pi_1(V_n)$ is isomorphic to the subgroup of pure braids of some braid group.
2. Prove that the orbit space V_n/S_n ($\cong U_n$) has fundamental group isomorphic to the braid group B_n. (Here S_n denotes the symmetric group of degree n, acting on V_n according to the rule $(z_1, \ldots, z_n) \mapsto (z_{\sigma(1)}, \ldots, z_{\sigma(n)})$, $\sigma \in S_n$.)

CHAPTER 7

Some Examples of Dynamical Systems and Foliations on Manifolds

§27. The Simplest Concepts of the Qualitative Theory of Dynamical Systems. Two-dimensional Manifolds

27.1. Basic Definitions

What is a dynamical system?

27.1.1. Definition. A *dynamical system* (or as they say an *autonomous dynamical system*) on a manifold M, is a smooth vector field ξ on M.

In terms of local co-ordinates on M, a dynamical system ξ gives rise to the system of (autonomous) ordinary differential equations

$$\dot{x}^\alpha = \xi^\alpha(x^1, \ldots, x^n), \tag{1}$$

whose solutions are termed (as in Part I, §23.1) the *integral curves* or *integral trajectories* of the dynamical system; thus an integral trajectory is a curve $\gamma(t)$ on M whose velocity vector $\dot{\gamma}(t)$ coincides at each instant with $\xi(\gamma(t))$. In view of the appropriate existence and uniqueness theorems for systems of ordinary differential equations of the form (1), integral trajectories exist locally (if ξ is continuous), and are uniquely determined by initial conditions if ξ is smooth (as of course we normally assume it to be). On non-compact manifolds M it can happen that an integral trajectory "goes to infinity" in a finite amount of time, so that it is defined of course only for some finite interval of values of t. On the other hand on closed manifolds (i.e. compact manifolds (without boundary)), every trajectory can be extended without bound as far as the time is concerned, and is therefore defined for all values of t, $-\infty<t<\infty$.

Each vector field ξ on M determines a linear differential operator on real-valued functions f on M, namely the directional derivative of f in the direction of ξ at each point x of M (cf. the remark in §24.3 of Part I):

$$f\to\partial_\xi f=\xi^\alpha\frac{\partial f}{\partial x^\alpha}.$$

It is not difficult to see (cf. loc. cit.) that the exponential map applied to this operator, i.e. the operator defined by

$$\hat{S}_t=\exp(t\partial_\xi)=\sum_{k\geq 0}\frac{1}{k!}t^k(\partial_\xi)^k, \tag{2}$$

determines a shift of each function f along the integral trajectories of the given field ξ, i.e., for each $x\in M$

$$\hat{S}_t(f(x))=f(S_t(x)), \tag{3}$$

where $S_t(x)=\gamma(t)$, $\dot{\gamma}=\xi$, and $\gamma(0)=x$. Recall also (from Part I, §23.2) the definition of the commutator $[\xi,\eta]$ of two vector fields ξ,η on M, as the field given locally by the formula

$$[\xi,\eta]^\alpha=\xi^\gamma\frac{\partial\eta^\alpha}{\partial x^\gamma}-\eta^\gamma\frac{\partial\xi^\alpha}{\partial x^\gamma},$$

or

$$\partial_{[\xi,\eta]}=[\partial_\xi,\partial_\eta]. \tag{4}$$

We now introduce some of the general terminology associated with dynamical systems.

27.1.2. Definition. (i) The *limit sets* $\omega^+(\gamma)$ and $\omega^-(\gamma)$ of an integral trajectory $\gamma(t)$ of a dynamical system ξ, are the respective sets of all limit points of sequences of the form $\{\gamma(t_i)\}$, where $t_i\to\pm\infty$ (cf. §14.5). The union $\omega^+(\gamma)\cup\omega^-(\gamma)$, denoted by $\omega(\gamma)$, is called the *ω-limit set* of γ.

(ii) A *positive* (resp. *negative*) *invariant set* (or *invariant manifold*) of a dynamical system (1) is a subset (or manifold) $N\subset M$ with the property that for every point $x\in N$ the integral trajectory $\gamma(t)$ with $\gamma(0)=x$, lies in N for all $t\geq 0$ (resp. $t\leq 0$) (i.e. $S_t(x)\subset N$ for $t\geq 0$ (resp. $t\leq 0$)). Especially important are those subsets N which are both positively and negatively invariant; we call such sets (or submanifolds) simply *invariant*.

(iii) An invariant closed set $N\subset M$ is said to be *minimal* if N contains no proper (non-empty) invariant closed sets. (For example, a singular point x_0 (i.e. a point where the field vanishes: $\xi(x_0)=0$) will by itself form a minimal set, as will any periodic trajectory of the field ξ. We shall meet with more exotic examples of minimal sets in the sequel.)

(iv) We shall say that an integral trajectory is *captured* by a subset $N \subset M$ if for some t_0, $\gamma(t)$ lies in N for all $t \geq t_0$.
(v) A hypersurface $P \subset M$ is a *transversal* for a dynamical system ξ if the restriction of the field to P is nowhere tangent to P, and is a *closed transversal* if in addition P is a closed submanifold of the manifold M.

(In the latter case we have the following two possibilities: (1) the closed transversal P separates M into two parts W_1 and W_2 (i.e. $M = W_1 \cup W_2$, $W_1 \cap W_2 = P$), in which case one of the components W_1, W_2 will capture all integral trajectories of the field ξ, and the other component will capture all integral trajectories of the field $-\xi$ (verify this!); (2) the closed transversal P does not separate M into two components. Here of particular interest is the situation where for all points $x \in P$ the trajectory starting from x returns after a finite amount of time $t(x)$ (depending on the point) to the submanifold P (and intersects it); clearly the dependence on the point x of the function $t(x)$ giving the shortest time of return to P is smooth, so that the map $\psi: P \to P$ defined by $\psi(x) = \gamma(t(x))$ where $\gamma(0) = x$, will be a smooth map of manifolds.)

EXERCISE

With reference to the preceding remark, prove that the whole manifold M is diffeomorphic to the indicated quotient manifold:

$$M \cong P \times I(0, 1)/(x, 0) \sim (\psi(x), 1).$$

Deduce that M is diffeomorphic to a fibre bundle over S^1 with fibre P.

Besides dynamical systems the (cognate) qualitative theory of "one-dimensional foliations" on a manifold is also of great utility. A *one-dimensional foliation* on a manifold M is a field of directions (or "directors") $\xi(x)$ at each point, where the vector $\xi(x)$ is specified only up to a non-zero scalar multiple, so that in particular $\xi(x) \sim -\xi(x)$. (In other words the foliation is given by specifying at each point of M a one-dimensional (or possibly null) subspace of the tangent space to M at the point, varying smoothly with the point.) It is not difficult to see that locally, in some neighbourhood of a non-singular point x_0 of the foliation (i.e. a point x_0 such that $\xi(x_0) \neq 0$), it makes sense to express the foliation in the form (1), since for suitably well-behaved scalar-valued functions f on M the system $\dot{x} = \xi(x)$ has locally the same integral curves as the system $\dot{x} = f(x)\xi(x)$. (Multiplication by $f(x)$ corresponds to a change of time-scale $t = t(\tau)$ satisfying $dt(\tau)/d\tau = f(x(t(\tau)))$.) However globally the foliation may not be expressible in the form (1); in fact (as Figure 103 shows) it can even happen that there is no consistent time-direction definable. The reasons for considering one-dimensional foliations on a manifold, rather than vector fields, are various. For instance in the theory of "fluid crystals" the role of the "director" $\xi \sim -\xi$ at each point of the medium is filled appropriately by the axis of symmetry at the point of a certain axially symmetric rank-two tensor embodying the

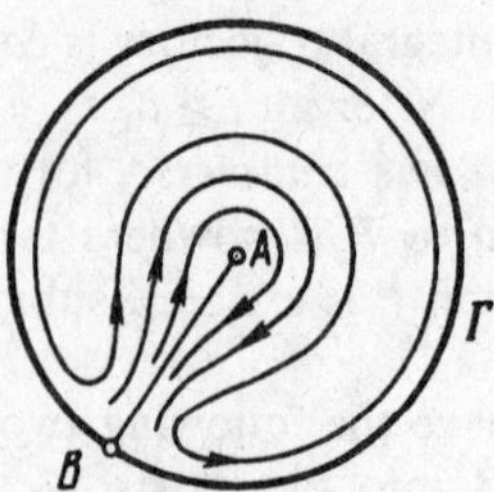

Figure 103. The points A and B are singular, with the point B on the boundary Γ a saddle; on the segment BA no time-direction is definable.

optical properties of the medium. One-dimensional foliations also arise naturally in studying the behaviour far from the origin of the integral trajectories of dynamical systems on $\mathbb{R}^n$, where the components of the right-hand side of (1) (i.e. of $\xi(x)$) are algebraic functions of the co-ordinates $x^1, \ldots, x^n$ of $\mathbb{R}^n$. For this purpose it becomes necessary to supplement $\mathbb{R}^n$ with points infinitely distant from the origin, i.e. to enlarge $\mathbb{R}^n$ to $\mathbb{R}P^n$; it turns out that such a dynamical system on $\mathbb{R}^n$ does extend to $\mathbb{R}P^n$, at the cost of foregoing the possibility of specifying a direction of flow of time. In the case $n=2$ for instance, a one-dimensional foliation on a 2-manifold M^2 will be given locally, on some region $U \subset M^2$ $(U \cong \mathbb{R}^2)$ with co-ordinates x, y, by a 1-form $\omega = P(x, y)\,dx + Q(x, y)\,dy$, since in this case the system (1) can be brought into the form

$$\omega = P(x, y)\,dx + Q(x, y)\,dy = 0. \tag{5}$$

A point (x, y) is clearly non-singular precisely if at least one of $P(x, y)$ and $Q(x, y)$ is non-zero. Suppose now that P and Q are polynomials of degree m. If we carry out the usual change to homogeneous co-ordinates u_0, u_1, u_2, given by

$$x = \frac{u_1}{u_0}, \qquad y = \frac{u_2}{u_0},$$

i.e. if we enlarge $U \cong \mathbb{R}^2$ in the usual way to $\mathbb{R}P^2$, then from the 1-form ω we obtain a 1-form Ω on $\mathbb{R}P^2$ given in terms of these co-ordinates by

$$\begin{aligned}\Omega = u_0^{m+2}\omega = u_0^{m+2} P\left(\frac{u_1}{u_0}, \frac{u_2}{u_0}\right)(u_0\,du_1 - u_1\,du_0)\\ + u_0^{m+2} Q\left(\frac{u_1}{u_0}, \frac{u_2}{u_0}\right)(u_0\,du_2 - u_2\,du_0).\end{aligned} \tag{6}$$

The equation $\Omega = 0$ then defines a one-dimensional foliation on all of $\mathbb{R}P^2$.

Exercise

Show that the line at infinity $\mathbb{R}P^1 \subset \mathbb{R}P^2$ (given by $u_0 = 0$) is an integral trajectory of the one-dimensional foliation $\Omega = 0$. Show also that this foliation is such that no

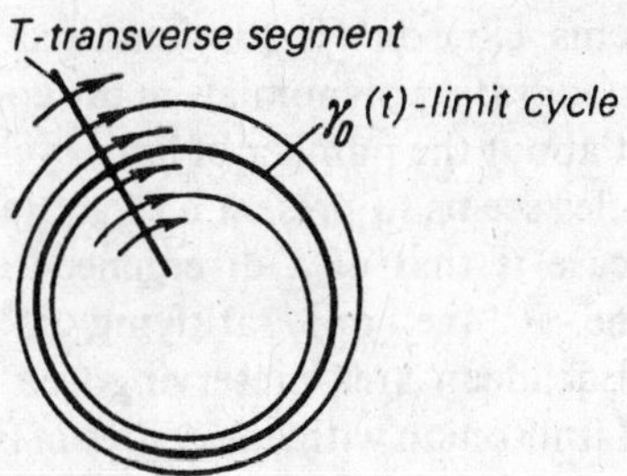

Figure 104. The *Poincaré function* $f: T \to T$ is defined for each $\tau \in T$ in terms of the trajectory γ starting at $\gamma(0)=\tau$: we set $f(\tau)=\gamma(t)$, where $\gamma(t) \in T$ is the first point where the curve γ again meets T. In particular if $\tau \in T$ is on the limit cycle γ_0, then $f(\tau)=\tau$.

consistent direction of flow of time can be chosen. Investigate the singular points at infinity of the foliation $\Omega=0$, in the case $m=2$.

27.1.3. Definition. A *limit cycle* of a dynamical system (or one-dimensional foliation) on a 2-dimensional manifold, is a periodic integral trajectory with the property that in some sufficiently small neighbourhood of it there are no other periodic trajectories. (It follows that the situation near a limit cycle will be essentially as shown in Figure 104.)

The situation where integral trajectories of a dynamical system on $\mathbb{R}^2$ enter a closed disc D^2 containing a "repellor" (i.e. a source), across the transversal $\Gamma=\partial D^2$, was considered in §14.5, where we proved the Poincaré–Bendixson theorem; that theorem in fact applies to one-dimensional foliations on the sphere S^2, where, as in Figure 105, the north pole for instance is a repelling singular point (i.e. a source). From the equality between the Euler characteristic of a closed orientable manifold and the sum of the indices of the singular points of a vector field in general position on the manifold (see Theorem 15.2.7), it follows that every such vector field on the sphere S^2 must have a source or sink, which fact, together with the Poincaré–Bendixson theorem, implies that limit cycles always exist for one-dimensional foliations on the 2-sphere. This represents the only significant result of any generality concerning the existence of limit cycles. Even in the

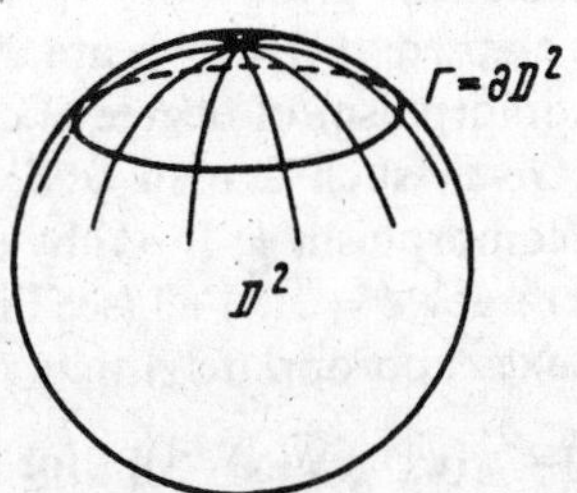

Figure 105

case of dynamical systems $\xi(x)$ on $\mathbb{R}^2$ (or foliations on $\mathbb{R}P^2$) where the components of $\xi(x)$ are quadratic polynomials in the co-ordinates x, y on $\mathbb{R}^2$, there is no general result about the number of limit cycles; in each individual case finding the limit cycles seems to present a separate non-trivial problem. A special (degenerate) case is that of a divergence-free dynamical system $\xi=(\xi^1, \xi^2)$ on the plane $\mathbb{R}^2$ (i.e. one satisfying $\partial\xi^\alpha/\partial x^\alpha=0$), where the transformations S_t are Euclidean area-preserving (see Part I, conclusion of §23.2); such a system is Hamiltonian with one degree of freedom and therefore an "energy" integral exists (see §28 below and Part I, §23.2, Example (a), for the meaning of this), and the system can be completely integrated.

27.2. Dynamical Systems on the Torus

From the classification theorem for surfaces together with the above-mentioned equality between the Euler characteristic of a closed oriented manifold and the sum of the indices of any vector field in general position on the manifold (Theorem 15.2.7), it follows that the torus T^2 is the only closed, orientable surface on which a nowhere vanishing vector field can be defined. One is led to consider such systems on the torus by for instance the question of the qualitative nature of the solutions of certain differential equations with periodic coefficients. Thus a differential equation of the form

$$\dot{x}=f(x, t), \tag{7}$$

where f is periodic in both arguments (i.e. $f(x+1, t)=f(x, t)=f(x, t+1)$) defines on the torus T^2, with co-ordinates x (modulo 1) and t (modulo 1), the dynamical system

$$\dot{x}=f(x, t), \qquad \dot{t}=1, \tag{8}$$

so that our original problem concerning the differential equation (7), is transformed into that of investigating the behaviour of the solutions of the system (8) on the torus. The system (8) has the obvious closed transversals $S^1 \subset T^2$ (see Definition 27.1.2(v)) each defined by an equation of the form $t=t_0$ (x arbitrary). Such a transversal does not separate the torus into two components, and clearly the trajectory $\gamma(t)=(x(t), t)$ starting at any point (x, t_0) of the closed transversal S^1 given by $t=t_0$, returns to that transversal after one unit of time has elapsed: $t(x)=1$; we are thus led naturally to a map $\psi: S^1\to S^1$ (in fact a diffeomorphism of degree 1) defined by $\psi(x)=\gamma(t_0+1)$, $\gamma(t_0)=(x, t_0)\in S^1 \subset T^2$. Given such a map $\psi: S^1\to S^1$, we define a corresponding real-valued diffeomorphism $\tilde{\psi}: \mathbb{R}\to\mathbb{R}$ by the conditions $\tilde{\psi}(x)\equiv\psi(x)$ mod 1 and $\tilde{\psi}(x+1)=\tilde{\psi}(x)+\deg\psi=\tilde{\psi}(x)+1$ (see Figure 106). We may clearly suppose (by shifting the axes appropriately) that $\tilde{\psi}(0)>0$. Write

$$\tilde{\psi}_n(x)=\underbrace{\tilde{\psi}(\tilde{\psi}(\cdots\tilde{\psi}(x)\cdots))}_{n\text{ times}} \quad \text{for } n>0,$$

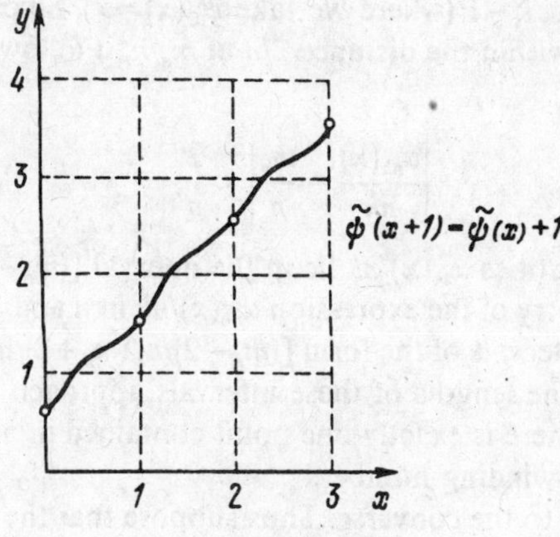

Figure 106

and $\tilde{\psi}_{-n}=(\tilde{\psi}_n)^{-1}$, the function inverse to $\tilde{\psi}_n$, and consider the expression

$$(\tilde{\psi}_n(x)-x)/n \tag{9}$$

(whose numerator in essence represents (geometrically speaking) the angle through which the point x is revolved by n applications of the map ψ).

27.2.1. Lemma.

(i) *The quantity* $(\tilde{\psi}_n(x)-x)/n$ *has a limit as* $n\to\infty$ *which is independent of the point* x.

(ii) *This limit, called the "winding number" of the map* ψ, *is rational if and only if* ψ *possesses a "periodic point", i.e. a point* x_0 *satisfying, for some integer* n, $\psi^n(x_0)=x_0$ *(or, equivalently,* $\tilde{\psi}_n(x_0)=x_0+m$ *for some integer* m*).*

PROOF. (i) Write briefly $\alpha_n(x)=\tilde{\psi}_n(x)-x$; thus, as noted above, $\alpha_n(x)$ is essentially the angle through which x is revolved by n successive applications of the diffeomorphism ψ. For any $x_1, x_2 \in \mathbb{R}$, we have

$$|\alpha_n(x_1)-\alpha_n(x_2)|<1; \tag{10}$$

this follows easily from the definition of $\tilde{\psi}(x)$ when $|x_1-x_2|<1$, and thence for any x_1, x_2 using the periodicity of $\alpha_n(x)$. If for each n we write m_n for the integer part of $\alpha_n(0)$, i.e. if we define m_n to be the integer satisfying the inequalities $m_n\le\alpha_n(0)<m_n+1$, then from (10) we infer that $|\alpha_n(x)-m_n|<2$, whence

$$\left|\frac{\alpha_n(x)}{n}-\frac{m_n}{n}\right|<\frac{2}{n} \quad \text{for all } x. \tag{11}$$

A direct verification from the definition of $\alpha_n(x)$ shows that

$$\alpha_{nk}(x)=\alpha_n(x)+\alpha_n(\tilde{\psi}_n(x))+\alpha_n(\tilde{\psi}_{2n}(x))+\cdots+\alpha_n(\tilde{\psi}_{n(k-1)}(x)),$$

whence it follows that $\alpha_{nk}(x)/nk$ is the arithmetic mean of the k quantities

$\alpha_n(\tilde{\psi}_{ni}(x))/n$, $i=0,\ldots,k-1$ (where we take $\tilde{\psi}_0(x)=x$). Since by (11) each of these k numbers is within the distance $2/n$ of m_n/n, it follows that for all k, n and x,

$$\left|\frac{\alpha_{nk}(x)}{nk}-\frac{m_n}{n}\right|<\frac{2}{n},$$

i.e. for all k the quantities $\alpha_{nk}(x)/nk$ lie in the interval $[(m_n-2)/n, (m_n+2)/n]$. This and the symmetry of the expression $\alpha_{nk}(x)/nk$ in n and k, together imply that every pair of intervals of the form $[(m_n-2)/n, (m_n+2)/n]$ has non-empty intersection. Since the lengths of these intervals approach zero as $n\to\infty$, it easily follows that there is exactly one point contained in all of them, which point is the desired winding number.

(ii) We now turn to the converse. Thus suppose that the winding number of ψ is $\alpha=m/n$ where m and n are integers. We define a function $\Delta(x)$ by

$$\tilde{\psi}_n(x)=x+m+\Delta(x).$$

In view of the periodicity of $\Delta(x)$ (i.e. in essence the compactness of S^1), if $\Delta(x)>0$ for all x, then $\Delta(x)$ must be bounded away from zero, say $\Delta(x)>\Delta_0>0$. However then

$$\tilde{\psi}_n(x)\geq x+m+\Delta_0,$$

whence, in view of the fact that $\tilde{\psi}(x)$ is increasing,

$$\tilde{\psi}_{nk}(x)\geq x+km+k\Delta_0 \quad \text{for } k>0,$$

yielding in turn

$$\frac{\alpha_{nk}(x)}{nk}=\frac{\tilde{\psi}_{nk}(x)-x}{nk}\geq\frac{m}{n}+\frac{\Delta_0}{n}.$$

Letting $k\to\infty$, we obtain $\alpha\geq m/n+\Delta_0/n>m/n$, contradicting the hypothesis that $\alpha=m/n$. A similar argument disposes of the possibility that $\Delta(x)\leq\Delta_0<0$. Hence $\Delta(x_0)=0$ for some point x_0 which will then be periodic for ψ. □

Clearly there will exist periodic solutions of the system (8) if and only if the map ψ possesses periodic points; hence in view of the above lemma the existence of periodic solutions of the system (8) on the torus is equivalent to the rationality of the winding number α of the map ψ.

We now consider the situation where the winding number α is irrational, i.e. where the system (8) has no periodic solutions. We begin with a corollary of (the proof of) Lemma 27.2.1.

27.2.2. Corollary. *If the winding number of the map ψ is irrational then for each x, and arbitrary N, the order of the points $x, \psi(x), \psi^2(x), \ldots, \psi^N(x)$ around the circle S^1 is the same as that obtained by replacing ψ by the rotation of S^1 through the angle α.*

PROOF. In the proof of part (ii) of Lemma 27.2.1, we showed essentially that $\tilde{\psi}_n(x) > x+m$ for all x if and only if $\alpha > m/n$, i.e. if and only if $x+n\alpha > x+m$ for all x, which shows that for each x the correspondence $x+n\alpha \pmod 1) \leftrightarrow \psi^n(x)$, $n=0, 1, \ldots, N$, does indeed preserve order around the circle. □

In the following lemma, the "limit sets" $\omega^{\pm}(x)$ are defined for each $x \in S^1$ (where S^1 is as before the circle defined by $t=t_0$) as the sets of limit points of the sequences of points $\psi^n(x)$ as $n \to \pm\infty$; obviously these sets are just the intersections $\omega^{\pm}(\gamma) \cap S^1$ where γ is the integral trajectory of the system (8) on the torus, passing through the point x at time $t=t_0$.

27.2.3. Lemma. *If the winding number of ψ is irrational then for each point x of S^1 $(t=t_0)$, the limit sets $\omega^+(x)$ and $\omega^-(x)$ just defined, coincide, are invariant under the diffeomorphism ψ, and are independent of the point x.*

PROOF. We begin by establishing the invariance of each of the sets $\omega^{\pm}(x)$ under the respective maps $\psi^{\pm 1}$. For each $y \in \omega^{\pm}(x)$ there is by definition of the sets $\omega^{\pm}(x)$ a sequence $n_1, n_2, \ldots$ of integers such that $n_k \to \pm\infty$ and $\psi^{n_k}(x) \to y$. It is then immediate that the sequence $\psi^{\pm 1}(\psi^{n_k}(x)) = \psi^{n_k \pm 1}(x)$ converges to the point $\psi^{\pm 1}(y)$, so that $\psi^{\pm 1}(y) \in \omega^{\pm}(x)$, as required.

For the justification of the remaining assertions of the lemma, we need to establish the following auxiliary fact: If we denote by a and $\bar{a}$ the arcs into which the two image points $\psi^n(x)$, $\psi^m(x)$ $(m \neq n)$ of our arbitrarily given point x, separate the circle (as shown in Figure 107), then for every $y \in S^1$, each of the "half-orbits" $\{\psi^q(y) \mid q \geq 0\}$ and $\{\psi^q(y) \mid q \leq 0\}$ contains points of both arcs. To see this consider to begin with the half-orbit $\{\psi^q(y) \mid q \geq 0\}$, and suppose for instance that $m > n$ and $y \in \bar{a}$. Form the sequence of arcs

$$a, \psi^{n-m}(a), \psi^{2(n-m)}(a), \ldots, \psi^{s(n-m)}(a), \ldots, \qquad s>0; \tag{12}$$

clearly these arcs are successively contiguous (i.e. a joins $\psi^m(x)$ and $\psi^n(x)$; $\psi^{n-m}(a)$ joins $\psi^n(x)$ and $\psi^{2n-m}(a)$, etc., as indicated in Figure 107), and their successive (common) end-points comprise a monotonic sequence of points of the circle (i.e. proceed around the circle in a single direction). It is not difficult to see that the arcs of the sequence (12) cover the circle; for if they did not then

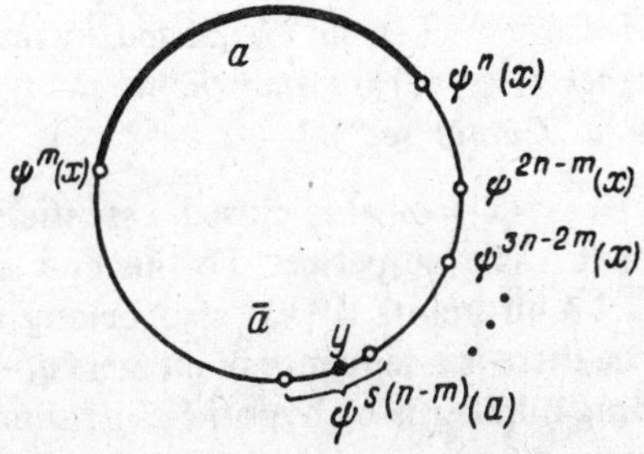

Figure 107

their end-points, i.e. the points of the form $\psi^{s(n-m)}\psi^n(x)$, $s=0, 1, \ldots$, would form a monotonic, "bounded" sequence of points on the circle (i.e. a monotonic sequence on the arc of the circle which they do cover), and would therefore converge. Their limit would then however represent a fixed point of the map ψ^{n-m}, which in view of the assumed irrationality of the winding number of ψ, is ruled out by Lemma 27.2.1. Thus the images of the arc a in (12) do cover the circle, so that $y \in \psi^{s(n-m)}(a)$ for some s, i.e. $\psi^{s(m-n)}(y) \in a$. The argument for the other half-orbit $\{\psi^q(y) | q \leq 0\}$ being essentially identical, this completes the proof of the fact that for every $y \in S^1$ each of the arcs a and $\bar{a}$ contains points (and hence infinitely many points) of both of the above half-orbits of y.

The remaining assertions of the lemma now follow easily. We content ourselves with showing that for any two points $x, y \in S^1$, we have $\omega^+(x) \subset \omega^-(y)$; the reverse inclusion, and the equality $\omega^+(x) = \omega^+(y)$ are established by very similar arguments. Thus suppose $x_0 \in \omega^+(x)$, and let q_k be a sequence of integers tending to ∞ such that $\psi^{q_k}(x) \to x_0$. In view of the above "auxiliary fact" each arc $a_k = [\psi^{q_k}(x), \psi^{q_k+1}(x)]$ contains infinitely many points of the half-orbit $\{\psi^q(y) | q \leq 0\}$; let s_k be a sequence of integers tending to $-\infty$ such that $\psi^{s_k}(y) \in a_k$. Since the lengths of the arcs a_k approach zero as $k \to \infty$, we must have $\psi^{s_k}(y) \to x_0$ as $k \to \infty$, whence $x_0 \in \omega^-(y)$. Hence $\omega^+(x) \subset \omega^-(y)$, as claimed. □

Before proceeding to the main consequences of the above two lemmas, we remind the reader of the definitions of a few relevant concepts from elementary point-set topology. A point y of a topological space T is a *limit point of a subset* $X \subset T$ (as opposed to a sequence) if every open set containing y contains a point of X other than y. A subset $X \subset T$ is then defined to be *closed* if it contains all of its limit points (which turns out to be equivalent to the requirement that its complement be open), and to be *perfect* if it coincides with the set of its limit points (or equivalently, if it has no isolated points). Finally a subset $X \subset T$ is *nowhere dense* in T if X has empty interior, i.e. if every non-empty open set of T contains a point outside X.

27.2.4. Theorem. *If the winding number of the map ψ is irrational then the limit set $\omega(x) = \omega(y)$ (defined prior to Lemma 27.2.3, and, by that lemma, independent of the point and whether $n \to \pm\infty$) is either the whole of S^1, or else a nowhere dense, perfect set (i.e. a "Cantor set").*

PROOF. The set $\omega(x) = \omega^{\pm}(x) = \omega(y)$ is closed, essentially by definition. It is not difficult to see that it is even perfect. To this end let x_0 be any point of $\omega(x)$. By Lemma 27.2.3 all points $\psi^q(x_0)$ also belong to $\omega(x)$, and $\omega^+(x_0) = \omega(x_0) = \omega(x)$. Hence there is a sequence of integers $q_k \to \infty$ such that $\psi^{q_k}(x_0) \to x_0$. Since the winding number is by hypothesis irrational, by Lemma 27.2.1 for no q_k can we have $\psi^{q_k}(x_0) = x_0$; hence the arbitrary point $x_0 \in \omega(x)$ is a limit point of $\omega(x)$, i.e. $\omega(x)$ is perfect.

What are the possibilities for perfect subsets $\omega(x) \subset S^1$? Either $\omega(x) = S^1$ (the case where the perfect subset is everywhere dense) or there are points of S^1 outside $\omega(x)$. If $\omega(x)$ contains some open arc b of S^1 then since for each point $x_0 \in b$ we have $\omega(x_0) = \omega(x)$, there will certainly exist an integer $m > 0$ such that also $\psi^m(x_0) \in b$; denote by a the closed sub-arc of b with end-points x_0 and $\psi^m(x_0)$. By arguing as in the preceding lemma it follows that the successive arcs of the sequence

$$a, \psi^m(a), \ldots, \psi^{sm}(a), \ldots$$

are contiguous and together cover the whole circle, whence $\omega(x) = S^1$. Hence either $\omega(x) = S^1$, or $\omega(x)$ contains no open arcs of S^1, i.e. is nowhere dense. □

Remark. There do in fact exist examples of C^1-maps $\psi: S^1 \to S^1$, arising from systems of the form (8) on the torus (with the right-hand side $f(x, t)$ of class C^1) for which the limit set $\omega(x)$ is a Cantor set. However if the map $\psi: S^1 \to S^1$ is of class C^2 (which it certainly will be for instance if $f(x, t)$ is a (real) analytic function of two variables, in particular a trigonometric polynomial), then there is a theorem of A. Denjoy to the effect that this cannot happen: *If the map ψ is of class C^2, and its winding number is irrational, then the limit set $\omega(x)$ coincides with the whole circle S^1 (i.e. the integral trajectories of the system* (8) *are everywhere dense on the torus T^2*). This result (in dimension 2) notwithstanding, the phenomenon of limit sets of trajectories of non-trivial dynamical systems which are Cantor sets, turns out to be inescapable in general, since in dimensions ≥ 3 such limit sets may occur even when the components of the right-hand side of the dynamical system are algebraic functions.

27.2.5. Theorem. *Suppose that the winding number α of the map ψ arising from a system of the form* (8) *on the torus, is irrational. If the limit set $\omega(x)$ of ψ is the whole circle S^1 (i.e. if the integral trajectories of the system are everywhere dense on the torus), then the map $\psi: S^1 \to S^1$ is "topologically equivalent" to a rotation; more precisely, there exists a homeomorphism $h: S^1 \to S^1$ (not necessarily smooth in general) such that*

$$h\psi h^{-1}(x) = x + \alpha, \qquad x \in S^1. \tag{13}$$

Proof. By Corollary 27.2.2, for each $x \in S^1$, the points $x_n = \psi^n(x)$ are placed around the circle in the same order as the points $n\alpha$ (mod 1) (the points of an orbit under the rotation of the circle through the angle α). In view of this, and since by hypothesis the points x_n form an everywhere dense subset of S^1 (i.e. a set whose limit points exhaust the circle), it follows that we can extend (by continuity) the map h defined by

$$h(x_n) = n\alpha \ (\text{mod } 1),$$

to a self-homeomorphism of S^1. It is easy to verify that this homeomorphism satisfies (13). □

Remarks. 1. We emphasize that the homeomorphism h serving to "linearize" the map $\psi: S^1 \to S^1$, may not in general be smooth; it can fail to be smooth for ψ of arbitrary smoothness class, or even (real) analytic. The question of the degree of smoothness of h is not, as it turns out, an easy one.

2. Throughout the present subsection we have been preoccupied with dynamical systems on the torus, in which case as we have seen the whole surface may represent a minimal set (see Definition 27.1.2(iii)). For orientable surfaces of genus > 1 on the other hand, the following result is known: *A minimal set of a vector field of class C^2 on an orientable surface of genus > 1 either consists of a single singular point, or is a periodic integral trajectory.* This extends the above-mentioned theorem of Denjoy in the sense that it excludes Cantor sets. (However for dynamical systems of class C^1 it is easy to construct examples having minimal sets which are Cantor.)

§28. Hamiltonian Systems on Manifolds. Liouville's Theorem. Examples

28.1. Hamiltonian Systems on Cotangent Bundles

Variational problems on arbitrary manifolds are posed exactly as in the (special) case of Euclidean space (see Part I, §31.1). Thus (in the case of one-dimensional variational problems) one starts with a *Lagrangian*, i.e. a scalar-valued function $L(x, v)$ of the points x of the manifold M and the tangent vectors v to M at the point x, and, exactly as in Part I, §31.1, one shows that the corresponding *extremals*, i.e. smooth arcs $\gamma(t)$ joining a given pair of points of M, which extremize the *action*

$$S = \int_{\gamma(t)} L(x, \dot{x})\, dt, \qquad x = \gamma(t), \qquad v = \dot{\gamma} = \dot{x},$$

satisfy a condition which locally takes the form of a second-order system of *Euler–Lagrange equations* on M:

$$\dot{p}_\alpha = \frac{d}{dt}\left(\frac{\partial L}{\partial v}\right) = \frac{\partial L}{\partial x^\alpha}. \tag{1}$$

Furthermore it can be shown just as in Part I, §33.1, that provided the "Legendre transformation" is "strongly non-singular", i.e. the local system of equations $p_\alpha = (\partial L/\partial v^\alpha)\,(x, v)$ has a unique solution of the form $v^\alpha = v^\alpha(x, p)$, then the Euler–Lagrange equations (1) can be transformed into the equivalent local *Hamiltonian* form

$$\dot{p}_\alpha = -\frac{\partial H}{\partial x^\alpha}, \qquad \dot{x}^\alpha = \frac{\partial H}{\partial p_\alpha}. \tag{2}$$

The Hamiltonian system (2) can clearly be regarded as a dynamical system on the cotangent bundle $T^*(M)$, whose points are just the pairs (x, p) where x varies over M, and for each x, p varies over the space of covectors at x; it is thus clear that $T^*(M)$ has dimension $2n$ where $n = \dim M$. (In Part I, §§33, 34, we called $T^*(M)$ "phase space".)

The differential form $\Omega = \sum_\alpha dx^\alpha \wedge dp_\alpha$ is closed (in fact even exact, since $\Omega = d\omega$ where $\omega = p_\alpha\, dx^\alpha$), is defined globally on the whole of $T^*(M)$, and is clearly non-degenerate. (As in Part I, §34.1, it follows from this non-degeneracy that $\Omega^n = \Omega \wedge \cdots \wedge \Omega$ is proportional to the volume element determined by the "skew-symmetric metric" on $T^*(M)$ afforded by the given form Ω.) Regarding Ω as defining a skew-symmetric metric on $T^*(M)$, we obtain from it a non-degenerate, *skew-symmetric scalar product* of vectors:

$$\langle \xi, \eta \rangle = -\langle \eta, \xi \rangle = J_{ab} \xi^a \eta^b, \tag{3}$$

where $a, b = 1, \ldots, 2n$, and $\Omega = J_{ab}\, dy^a \wedge dy^b$, the co-ordinates $y^1, \ldots, y^{2n}$ on $T^*(M)$ being given by $y^\alpha = x^\alpha$, $y^{n+\alpha} = p_\alpha$, $\alpha = 1, \ldots, n$. In terms of these co-ordinates the Hamiltonian system (2) takes the "skew-symmetric gradient" form

$$\dot{y}^a = J^{ab} \frac{\partial H}{\partial y^b}, \qquad J^{ab} J_{bc} = \delta^a_c. \tag{4}$$

As in Part I, §34.2 we define the *Poisson bracket* $\{f, g\}$ of two functions f, g defined on phase space $T^*(M)$, as the (skew-symmetric) scalar product of their "gradients" (see (4) or (6)) with respect to the co-ordinates $y^1, \ldots, y^{2n}$:

$$\{f, g\} = J^{ab} \frac{\partial f}{\partial y^a} \frac{\partial g}{\partial y^b} = \langle \nabla f, \nabla g \rangle. \tag{5}$$

By virtue of the fact that Ω is closed, it follows that the space of functions on $T^*(M)$ forms a Lie algebra with respect to the Poisson bracket (cf. Part I, Theorem 34.2.4). It can also be shown (essentially as in Part I, §34.2) that the "gradient" of the Poisson bracket $\{f, g\}$ of two functions f, g, is the negative of the commutator of their "gradients":

$$f \to \nabla f = \left(J^{ab} \frac{\partial f}{\partial y^b} \right), \qquad g \to \nabla g = \left(J^{ab} \frac{\partial g}{\partial y^b} \right), \tag{6}$$
$$\{f, g\} \to \nabla \{f, g\} = -[\nabla f, \nabla g].$$

The extension of these concepts and results (introduced in the context of Euclidean space in Part I, §§33, 34) to general manifolds, is automatic in view of their local character.

28.2. Hamiltonian Systems on Symplectic Manifolds. Examples

Before introducing the general concept of a symplectic manifold and its accompanying Hamiltonian system, we make the following two motivating remarks:

(i) Any closed 1-form $\omega = H_a\, dy^a$ on the manifold $T^*(M)$ determines via the system $\dot{y}^a = J^{ab} H_b$ a local group of canonical transformations S_t, since locally, i.e. on each chart (diffeomorphic to $\mathbb{R}^{2n}$), a closed 1-form is the differential of some function (although globally there may not exist a single-valued function of which ω is the differential).

(ii) There are interesting systems of Hamiltonian type which happen not to be defined on manifolds of the form $T^*(M)$, and where the analogue of the form Ω (see the preceding subsection), though closed, is not exact.

28.2.1. Definition. A *symplectic* manifold is an even-dimensional manifold M endowed with a non-degenerate, closed 2-form Ω given locally by

$$\Omega = J_{ab}\, dy^a \wedge dy^b, \qquad \det(J_{ab}) \neq 0.$$

Corresponding to each closed 1-form $\omega = H_a\, dy^a$ on M the form Ω determines a *gradient* or *Hamiltonian system* $\dot{y}^a = J^{ab} H_b$; each closed 1-form ω is locally the gradient of a real-valued function H on M, called a *Hamiltonian*: $\partial H/\partial y^a = H_a$. (In connexion with this definition it is appropriate to remind the reader of Darboux' Theorem (see Part I, §34.2) according to which the above form Ω can locally be brought into the canonical form $\sum_i dx^i \wedge dp_i$ by means of a suitable co-ordinate change.)

It is easy to see that the Poisson bracket

$$\{f, g\} = J^{ab} \frac{\partial f}{\partial y^a} \frac{\partial g}{\partial y^b}$$

of pairs of functions on a symplectic manifold continues to enjoy in this more general context the same properties as before (see above).

28.2.2. Examples. (a) A non-degenerate, closed form Ω can be defined on any 2-dimensional orientable manifold endowed with a Riemannian metric (see §8.2): any 2-form proportional to the element of area with respect to such a metric will serve; such a form is obviously non-degenerate, and is closed for the simple reason that it has maximal rank. A wider class of examples is afforded by Kählerian manifolds: Recall from Part I, §27.2 that a complex manifold is called *Kählerian* if it comes endowed with an Hermitian metric $g_{\alpha\beta}$ with the property that the form $\Omega = (i/2) g_{\alpha\beta}\, dz^\alpha \wedge d\bar{z}^\beta$, which is a real 2-form on the realized manifold, is closed. (See also Part I, §29.4.) As in Lemma 34.1.2 of Part I it follows that the form Ω^n, where n is the complex dimension of the given complex manifold, is proportional (with non-zero constant factor of proportionality) to the form representing the volume element. (Verify this!) (Cf. in particular the example in Part I, §27.2.) Hence if the manifold is closed (i.e. compact and without boundary), then the form Ω is not exact. Thus we see that Hamiltonian systems are defined in a natural way on Kählerian manifolds. Riemann surfaces (the simplest of which is $\mathbb{C}P^1 \cong S^2$), once equipped with the metric induced from their universal cover (L^2 in most cases; see §20), become Kählerian manifolds. We end this example by leaving it as an exercise for the reader to show that Hamiltonian systems on (2-dimensional) surfaces are precisely those systems for which the corresponding transformations S_t are area-preserving (cf. Part I, Corollary 34.3.2).

(b) Another interesting class of examples is afforded by semisimple Lie algebras (see §3.1), equipped with an Ad-invariant scalar product. Consider for example the Lie algebra $so(n)$ (of the Lie group $SO(n)$), consisting of all skew-symmetric $n \times n$ matrices. Since $so(n)$ is a linear space, as a manifold it is identifiable with $\mathbb{R}^{n(n-1)/2}$, so that its tangent vectors can be identified (in the usual way for the spaces $\mathbb{R}^m$) with its points. We define a 2-form Ω on pairs A, B of "tangent vectors" to $so(n)$ (identified with the appropriate elements of $so(n)$) at each point $C \in so(n)$, by (cf. §6.4(7))

$$\Omega_C(A, B) = \langle [A, B], C \rangle = \mathrm{tr}(C[A, B]).$$

Recall that, as has been noted before, for each $q \in SO(n)$, the operator Ad q on $so(n)$ is given in terms of matrix multiplication by the formula

$$C \mapsto qCq^{-1}, \qquad q \in SO(n), \qquad C \in so(n). \tag{7}$$

EXERCISES

1. Verify that the form Ω on $so(n)$ is Ad-invariant, i.e. is invariant under transformations of the form (7).
2. Prove that the restrictions of the form Ω to the orbits of $so(n)$ under the action of $SO(n)$ defined by (7), yield non-degenerate, closed forms (at least on those orbits of largest possible dimension).

In the particular case $n = 3$, we have $so(3) = \mathbb{R}^3$ and commutation is just the familiar cross product $[\xi, \eta]$, $\xi, \eta \in \mathbb{R}^3$ (see Part I, §24.2). At each $x \in \mathbb{R}^3$ the form Ω is in this case given by

$$\Omega_x(\xi, \eta) = \langle x, [\xi, \eta] \rangle,$$

or

$$\Omega = \text{const.} \times (x^1\, dx^2 \wedge dx^3 - x^2\, dx^1 \wedge dx^3 + x^3\, dx^1 \wedge dx^2).$$

This form is invariant under rotations of $\mathbb{R}^3$, and its restriction to each sphere $\sum_a (x^a)^2 = \text{const.}$ is non-degenerate.

The "Euler equations" of classical mechanics for the behaviour of a freely rotating rigid body in 3-space, have the form

$$\dot{M} = [M, \omega], \qquad \omega = \omega(M); \tag{8}$$

here ω is the angular velocity vector, M is the angular momentum, and Ω and M are linked by the equation $M = J\omega + \omega J$, where $J = (J_{ij}) = (\lambda_i \delta_{ij})$, $\lambda_i > 0$, is the "moment tensor of inertia" of the rigid body. In view of this we say of an equation of the form (8) on the Lie algebra $so(n)$, that it is of "Euler type".

EXERCISES

1. Show that an equation of Euler type on $so(n)$ defines a Hamiltonian system on each of the orbits of maximal dimension under the action (7) of $SO(n)$ on $so(n)$.

2. Prove that the orbits of maximal dimension are Kählerian manifolds.

28.3. Geodesic Flows

From a geometrical point of view the most important Hamiltonian systems are those known as "geodesic flows": A *geodesic flow* on a smooth manifold M endowed with a Riemannian metric $g_{\alpha\beta}$, is the Hamiltonian system $\dot{p}_\alpha = -\partial H/\partial x^\alpha$, $\dot{x}^\alpha = \partial H/\partial p_\alpha$, on the tangent bundle $T(M)$ of M, with Hamiltonian given locally by

$$H(x, p) = \tfrac{1}{2} g^{\alpha\beta} p_\alpha p_\beta, \qquad g^{\alpha\beta} g_{\beta\gamma} = \delta^\alpha_\gamma. \tag{9}$$

(This Hamiltonian arises from the Lagrangian $L = \frac{1}{2} g_{\alpha\beta} v^\alpha v^\beta$ whose extremals are the geodesics parametrized by the natural parameter (see Part I, §§31.2, 33.3). Note also that the above Hamiltonian system (with Hamiltonian (9)) can be regarded as being defined on $T(M)$ (as here) rather than $T^*(M)$ (as in §28.1), by simply raising the index: $p^\beta = g^{\beta\alpha} p_\alpha$.)

As in §33.3 of Part I it follows from Maupertuis' principle (extended to arbitrary smooth manifolds) that, in particular, a particle on a manifold M with metric $g_{\alpha\beta}(x)$, in a field of potential $U(x)$, $x \in M$, and confined to a fixed energy level $E = H(x, p) = \frac{1}{2}\langle p, p\rangle + U(x)$, moves along the geodesics of the new metric

$$\tilde{g}_{\alpha\beta}(x, E) = \text{const.} \times (E - U(x)) g_{\alpha\beta}(x), \tag{10}$$

so that the investigation of this physical situation is in this way subsumed under the general study of geodesic flows. (Note however that, as was observed in Part I, §33.3, the parameter arising in solving the extremal problem with respect to this new metric will not in general be natural (in the sense of being a constant multiple of distance travelled on M). In view of this, when Maupertuis' principle is exploited in this way, the resulting geodesic flow is of interest only as a one-dimensional foliation on M (rather than a vector field; see §27.1).)

We shall in the remainder of this subsection restrict our attention to geodesic flows on closed manifolds M (endowed with a Riemannian metric). Corresponding to each fixed energy level

$$E = H(x, p) = \tfrac{1}{2}\langle p, p\rangle = \tfrac{1}{2} g^{\alpha\beta} p_\alpha p_\beta = \tfrac{1}{2} g_{\alpha\beta} p^\alpha p^\beta$$

(where the final equality results from raising indices), we obtain a dynamical system (derived from (2) by raising indices in the first equation) on the (compact) bundle over M of tangent vectors of constant length $\sqrt{2E}$. (This bundle manifold is clearly a fibre bundle over M with fibre S^{n-1} where $n = \dim M$.) Two facts particularly relevant to the qualitative theory of such

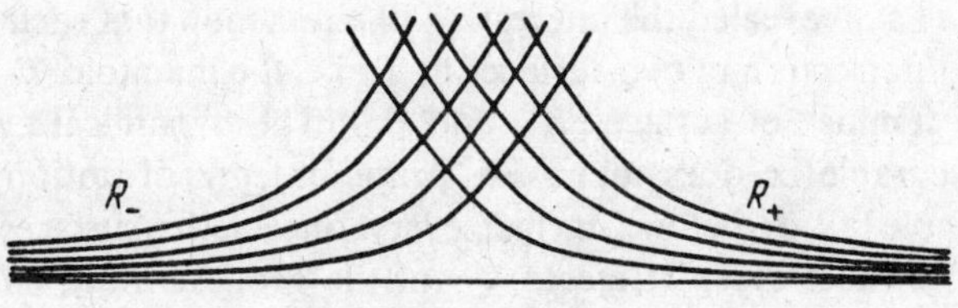

Figure 108

systems are the following: (i) they have no critical points; and (ii) their periodic trajectories may be investigated using the topological theory of critical points, applied to the length functional defined on the space of all closed curves in the bundle manifold. A case of particular interest is that of compact manifolds having negative curvature in every 2-dimensional direction (i.e. intuitively speaking, manifolds in which at every point the geodesics with coplanar tangent vectors diverge locally). For the sake of simplicity we here (briefly) describe the situation only for 2-dimensional surfaces endowed with a Riemannian metric of constant negative Gaussian curvature. At the fixed energy level $H(x, p)=1$, the geodesic flow on such a surface M can be represented by a dynamical system (i.e. a vector field ξ) on the unit tangent bundle T_1 of M. It turns out that corresponding to each conjugacy class of elements of the fundamental group $\pi_1(M)$, there is precisely one periodic trajectory. A characteristic feature of metrics of negative curvature is the "exponential" behaviour of their geodesics: if $\gamma(t)$ denotes any particular integral trajectory of the field ξ on T_1 (i.e. in essence a geodesic on M), then the collection of geodesics which approach $\gamma(t)$ "exponentially quickly" as $t \to +\infty$ forms a surface in T_1 (or in M) containing $\gamma(t)$, which surface we denote by $R_+(\gamma)$; the surface $R_-(\gamma)$ is defined analogously (for $t \to -\infty$) (see Figure 108). (The surfaces $R_\pm(\gamma)$ can be defined more precisely as the projections of those surfaces in the universal covering space (the Lobachevskian plane L^2) of M, consisting of the geodesics in L^2 converging on the respective "end-points at infinity" of γ (i.e. points on the boundary of the disc in the Poincaré model of L^2; see Figure 109). It is at least intuitively clear that the intersection $R_-(\gamma) \cap R_+(\gamma)$ of these surfaces is then precisely the geodesic

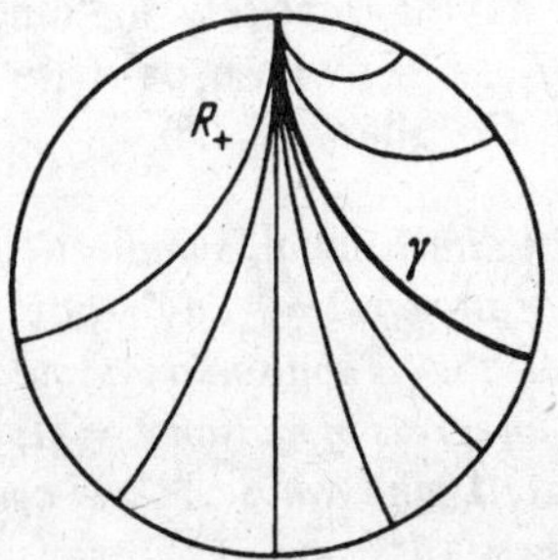

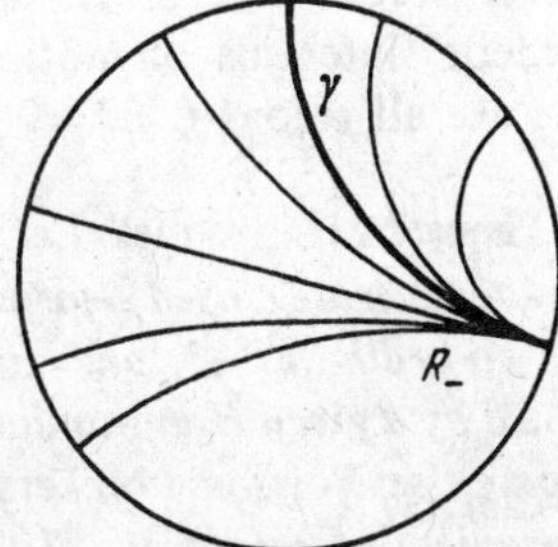

Figure 109

γ, so that we have revealed the interesting phenomenon that each trajectory $\gamma \subset T_1$ is the intersection of two surfaces in T_1, i.e. the manifold T_1 is "foliated" by the two families of surfaces R_+ and R_-. The dynamical system is not however integrable (i.e. does not have a "prime" integral of motion—see Part I, §23.2, Example (a), or below); in fact every one of the surfaces $R_+(\gamma)$ and $R_-(\gamma)$, and "almost every" trajectory γ of the geodesic flow, constitutes an everywhere dense subset of T_1. (We omit the proof of these facts.) The two families R_+, R_- of surfaces in T_1 furnish a rather curious example of a "2-dimensional foliation", which concept we shall consider in greater detail below (in §29).

Exercise

Prove that the "leaves" $R_+(\gamma)$ and $R_-(\gamma)$ of the foliations are topologically either planes $\mathbb{R}^2$ or cylinders $S^1 \times \mathbb{R}^1$. (The appropriate topology on each of the leaves $R_+(\gamma)$ and $R_-(\gamma)$ is that having as open base the collection of all finite intersections of the connected components of the subsets of $R_+(\gamma)$ (resp. $R_-(\gamma)$) open in the relative topology on $R_+(\gamma) \subset T_1$ (resp. $R_-(\gamma) \subset T_1$).)

28.4. Liouville's Theorem

Geodesic flows sometimes admit "spare" integrals of motion. (Recall from Part I, §§23.2, 34.2 that an "integral of the field" or "integral of motion" is a function on the manifold carrying the field, which is constant along each integral trajectory.) This will occur if the metric on the manifold M has a non-discrete group of isometries (as for example when M is a homogeneous space, or a surface of revolution in Euclidean space), and in certain other special cases. Of course if there are such integrals then no trajectory can be everywhere dense on the bundle manifold T of tangent vectors to M of constant length (given by $E=\frac{1}{2}\langle p, p\rangle=\text{const.}$). The analogous situation arises also in more general Hamiltonian systems having extra integrals of motion, i.e. integrals of motion other than the energy. The following theorem of Liouville is concerned with such systems, or, more precisely, with Hamiltonian systems with n degrees of freedom (i.e. on $\mathbb{R}^{2n}$, or more generally on any symplectic $2n$-dimensional manifold M endowed with a non-degenerate closed 2-form Ω; see §28.2), having precisely n "functionally independent" integrals of motion $H=f_1, f_2, \ldots, f_n$, whose pairwise Poisson brackets are all zero: $\{f_\alpha, f_\beta\}=0$ (cf. Part I, Corollary 34.2.6).

28.4.1. Theorem (Liouville). *Let M be a $2n$-dimensional, symplectic manifold with non-degenerate closed 2-form Ω given locally by $\Omega=J_{ab}\, dy^a \wedge dy^b$, in terms of local co-ordinates y^a, and suppose that the Hamiltonian system on M determined by a given Hamiltonian H has n integrals of motion $f_1=H, f_2, \ldots, f_n$ whose pairwise Poisson brackets are all zero, and whose "skew-symmetric" vector-gradients $\xi_i=(\xi_i^a)=(J^{ab}(\partial f_i/\partial y^b))$, where $(J^{ab})=(J_{ab})^{-1}$, are (pointwise) linearly independent. We then have that:*

(i) *Every (n-dimensional) level surface defined by n equations of the form* $f_1=a_1,\ldots,f_n=a_n$ *(where the* a_j *are constants), is non-singular, and each of its connected components can be obtained as a quotient group (and quotient space at the same time) of* $\mathbb{R}^n$ *by a (discrete) lattice of rank* $\leq n$*; it follows in particular that the non-singular, compact (connected) surfaces among these level surfaces are n-dimensional tori.*

(ii) *Given any compact, connected level surface (as defined in* (i)), *there is a neighbourhood (in M) of that surface in which there can be defined co-ordinates* $s_1,\ldots,s_n;\ \varphi_1,\ldots,\varphi_n$ $(0\leq\varphi_i<2\pi)$ *("action-angle" co-ordinates) with the following three properties:* (1) *in terms of these co-ordinates we have* $\Omega=\sum_\alpha ds_\alpha\wedge d\varphi_\alpha$, *i.e.* $\{s_\alpha,\ s_\beta\}=\{\varphi_\alpha,\ \varphi_\beta\}=0$, $\{s_\alpha,\ \varphi_\beta\}=\delta_{\alpha\beta}$; (2) $s_\alpha=s_\alpha(f_1,\ldots,f_n)$ *(i.e. the* s_α *are defined in terms of the functions* $f_1,\ldots,f_n$), *and the* φ_α *define co-ordinates on each surface (in the neighbourhood in question) of the form* $f_j=\text{const.}$ $(j=1,\ldots,n)$; (3) *in terms of these co-ordinates the original system takes on the equivalent form*

$$\dot{f}_\alpha=0\Leftrightarrow\dot{s}_\alpha=0,\qquad\dot{\varphi}_\alpha=\omega_\alpha(s_1,\ldots,s_n),\tag{11}$$

in the neighbourhood of the given surface.

PROOF. (i) In view of the linear independence of the gradients of the f_i, $i=1,\ldots,n$, each set of equations $f_1=a_1,\ldots,f_n=a_n$ defines a smooth (non-singular) n-dimensional submanifold $M^n(a_1,\ldots,a_n)$ say, of M. By Part I, §7.2, for each i the gradient $(\partial f_i/\partial x^a)$ is at each point perpendicular (in the Euclidean sense) to the level surface $f_i=\text{const.}$ through that point; it follows from this together with the condition

$$0=\{f_i,f_j\}=J^{ab}\frac{\partial f_i}{\partial y^a}\frac{\partial f_j}{\partial y^b}=\xi_i^b\frac{\partial f_j}{\partial y^b},$$

that the vector fields ξ_i are all tangential to the surface $M^n(a_1,\ldots,a_n)$. From the condition $\{f_i,f_j\}=0$ we also infer (via Theorem 34.2.4 of Part I) that the fields ξ_i are pairwise commuting: $[\xi_i,\xi_j]=0$. It follows that the n linearly independent fields ξ_i can be used to generate an action of the group $\mathbb{R}^n$ on M, which restricts to an action of $\mathbb{R}^n$ on the submanifold $M^n(a_1,\ldots,a_n)$ (and a neighbourhood of it). Choosing as "initial" point x_0 any point of $M^n(a_1,\ldots,a_n)$, we define a lattice $\{d\}$ to consist of all vectors $d\in\mathbb{R}^n$ which fix x_0 under this action of $\mathbb{R}^n$. It is then not difficult to see that $\{d\}$ is indeed a lattice in $\mathbb{R}^n$ (i.e. a discrete subgroup) and it is at least intuitive that (assuming M^n connected) we have $M^n(a_1,\ldots,a_n)\cong\mathbb{R}^n/\{d\}$ (cf. §5.1). Being a lattice, $\{d\}$ will be spanned by $k\leq n$ vectors; obviously the quotient space $\mathbb{R}^n/\{d\}$ will be compact if and only if $k=n$ (in which case $M^n(a_1,\ldots,a_n)\cong\mathbb{R}^n/\{d\}\cong T^n$).

(ii) Given a level surface $M^n(a_1,\ldots,a_n)\cong T^n$ (as in (i)), we first of all choose an origin of co-ordinates $x_0=x_0(\alpha_1,\ldots,\alpha_n)$ on each level surface $M^n(\alpha_1,\ldots,\alpha_n)\cong T^n$ (defined by $f_1=\alpha_1,\ldots,f_n=\alpha_n$) in a neighbourhood of $M^n(a_1,\ldots,a_n)$, such that $x_0(\alpha_1,\ldots,\alpha_n)$ depends smoothly on $\alpha_1,\ldots,\alpha_n$. It is clear that on any particular level surface $M^n(\alpha_1,\ldots,\alpha_n)$ in this neighbourhood, there exist linear combinations $\eta_j = \sum_{i=1}^n b_j^i \xi_i$ of the fields ξ_i, with the property that the co-ordinates defined on $\mathbb{R}^n$ (acting on the level surface $M^n(\alpha_1, \ldots, \alpha_n) \cong T^n$, as in Part (i) of the proof) by means of the fields η_j, are the usual angles $0 \leq \tilde{\varphi}_j < 2\pi$, $j = 1, \ldots, n$ (with $\tilde{\varphi}_j = 0$ at x_0). Here the coefficients b_j^i depend on the (variable) $\alpha_1, \ldots, \alpha_n$, i.e. on the particular values of $f_1, \ldots, f_n$ defining the surface; thus

$$\eta_j = b_j^i(f_1, \ldots, f_n)\xi_i,$$

so that our neighbourhood of the torus $M^n(a_1, \ldots, a_n)$ is co-ordinatized by $f_1, \ldots, f_n, \tilde{\varphi}_1, \ldots, \tilde{\varphi}_n$. Note that the matrix of Poisson brackets of these co-ordinates is

$$\begin{pmatrix} \{f_i, f_j\} = 0 & \{f_i, \tilde{\varphi}_j\} \\ \{\tilde{\varphi}_i, f_j\} & \{\tilde{\varphi}_i, \tilde{\varphi}_j\} \end{pmatrix} = \begin{pmatrix} 0 & A(f) \\ -A^{\mathrm{T}}(f) & B(f) \end{pmatrix}, \tag{12}$$

where $\det A \neq 0$ at each point of the neighbourhood (essentially since for each j the gradient $(\partial \tilde{\varphi}_j / \partial y^a)$ is parallel to the field η_j, which is a linear combination of the $\xi_i = (J^{ab}(\partial f_i / \partial y^b))$).

We now introduce the "action" co-ordinates $s_i = s_i(f_1, \ldots, f_n)$, $i = 1, \ldots, n$. In the special case of phase space $\mathbb{R}^{2n}$ with canonical co-ordinates $q_1, \ldots, q_n$, $p_1, \ldots, p_n$, the s_i are defined for each torus $M^n(\alpha_1, \ldots, \alpha_n)$ by (cf. Part I, §35.1)

$$s_i = \frac{1}{2\pi} \oint_{\gamma_i} \sum_{k=1}^n p_k\, dq_k, \qquad i = 1, \ldots, n, \tag{13}$$

where γ_i is the ith generating cycle of the torus:

$$\gamma_i = \gamma_i(\tilde{\varphi}_i), \qquad 0 \leq \tilde{\varphi}_i \leq 2\pi, \qquad \tilde{\varphi}_j = \text{const.} \quad \text{for } j \neq i.$$

We leave it to the reader as an exercise to verify (using $\{f_i, f_j\} = 0$) that these action co-ordinates satisfy $\{s_i, s_j\} = 0$ for all i, j, and also that they are canonically related to the angular co-ordinates $\tilde{\varphi}_1, \ldots, \tilde{\varphi}_n$:

$$\{s_i, \tilde{\varphi}_j\} = \delta_{ij}, \qquad i, j = 1, \ldots, n.$$

Returning now to the more general situation of our given $2n$-dimensional simplectic manifold M and the closed 2-form Ω defined on it, we observe first that the form Ω vanishes on the torus $M^n(a_1, \ldots, a_n)$ (since in view of the condition $\{f_i, f_j\} = 0$, the value of the form Ω on any pair ξ_i, ξ_j of the above-defined basic tangent vector fields to $M^n(a_1, \ldots, a_n)$, is zero (cf. also Part I, Corollary 35.1.6)), whence it follows that in some neighbourhood of this torus Ω is exact: $\Omega = d\omega$. The action co-ordinates s_i are then defined for each torus $M^n(\alpha_1, \ldots, \alpha_n)$ in that neighbourhood, analogously to (13):

$$s_i = \frac{1}{2\pi} \oint_{\gamma_i} \omega, \qquad i = 1, \ldots, n. \tag{14}$$

Finally we set

$$\varphi_i = \tilde{\varphi}_i + b_i(s_1, \ldots, s_n), \tag{15}$$

where the b_i (which vanish at x_0) are chosen so that the condition $\{\varphi_i, \varphi_j\}=0$ is fulfilled. (This is always possible in view of the fact that $\{s_i, \tilde{\varphi}_j\}=\delta_{ij}$; we again leave the details to the reader.) Thus on each level surface $f_1=\alpha_1, \ldots, f_n=\alpha_n$ in our neighbourhood, the co-ordinates $\varphi_1, \ldots, \varphi_n$ are obtained from the earlier $\tilde{\varphi}_i$ by means of a change of scale (ensuring $0 \le \varphi_i < 2\pi$).

The co-ordinates $s_1, \ldots, s_n; \varphi_1, \ldots, \varphi_n$ on our neighbourhood have thus been constructed to satisfy

$$\{s_i, s_j\}=\{\varphi_i, \varphi_j\}=0, \qquad \{s_i, \varphi_j\}=\delta_{ij}, \tag{16}$$

as required. In view of (13) we also have that the Hamiltonian $H=f_1$ is at each point determined solely by the "action" co-ordinates:

$$H=f_1(s_1, \ldots, s_n). \tag{17}$$

That the original Hamiltonian system takes the form (11) in terms of the co-ordinates $s_1, \ldots, s_n; \varphi_1, \ldots, \varphi_n$ is a consequence essentially of the fact that the f_j are integrals of motion (i.e. $f_j=$const. along each trajectory) and the definition of the s_j. □

28.5. Examples

The first three of the following examples represent instances of Liouville's theorem. (Two of these we met with earlier (in Part I, §§31, 32, 34)).

(a) Consider a Hamiltonian system of one degree of freedom (i.e. $n=1$), with Hamiltonian $H(x, p)$, and suppose that the surface (i.e. curve) $H(x, p)=E(=\text{const.})$ in 2-dimensional phase space, is compact (as in Figure 110). Then as canonical "action-angle" co-ordinates s, φ in a neighbourhood of that curve, we may take $s(E)=\oint_{H=E} p\, dx$ (the "truncated" action; see Part I, §33.3), and as φ the obvious angular co-ordinate scaled so that $\Omega = ds \wedge d\varphi$. Clearly then $H=H(s)$.

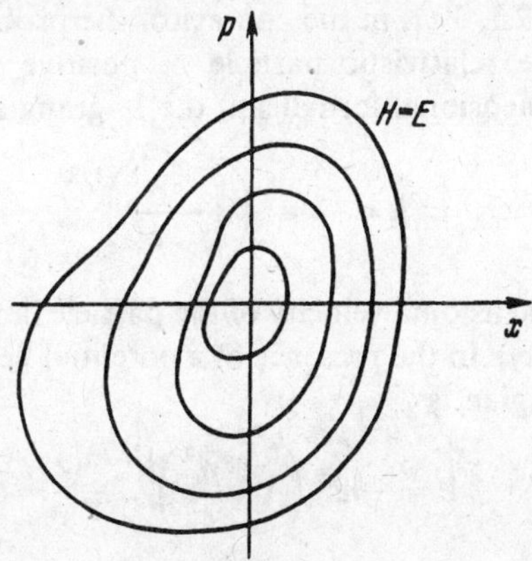

Figure 110

(b) The situation of a particle in a spherically symmetric potential field $U(r)$ in Euclidean 3-space with Euclidean co-ordinates x, y, z, where $r^2 = x^2 + y^2 + z^2$, was considered in Part I, §32.2, Example (d), and §34.2, Example (a), where it was observed that the angular momentum $M = [\bar{x}, p] = (M_x, M_y, M_z)$ is constant on each extremal, so that M_x, M_y, M_z represent integrals of motion of the appropriate Hamiltonian system. It is easy to check (using various of the results of Part I, §34.2) that the three integrals of motion

$$f_1 = H, \qquad f_2 = M_z, \qquad f_3 = |M|^2 = M_x^2 + M_y^2 + M_z^2, \tag{18}$$

satisfy the hypotheses of Liouville's theorem. (Verify that $\{f_i, f_j\} = 0.$)

(c) In Part I, §31.1, we investigated the geodesics on a surface of revolution (about the z-axis) in Euclidean $\mathbb{R}^3$, in terms of cylindrical co-ordinates r, φ, z. In this context we have the two integrals of motion

$$f_1 = H = \tfrac{1}{2} g_{ij} v^i v^j \qquad (x^1 = r, x^2 = \varphi), \text{ and } f_2 = p_\varphi. \tag{19}$$

Recall that as a consequence of the corresponding conservation laws we deduced (in Part I, §31.1) Clairaut's theorem, and the complete integrability of the geodesic flow on such a surface.

In our next (and last) two examples, we examine two relativistic problems with (as in (b) above) spherically symmetric Lagrangians, namely that of finding the trajectories of a charged particle moving through a central force field with potential α/r in STR (the Special Theory of Relativity), and the problem of the motion of a test particle of positive mass in a Schwarzschild gravitational field in GTR (the General Theory of Relativity) (see Part I, §39.1). (Other examples of systems with Lagrangians having a high degree of symmetry are afforded by the metrics on the 2-sphere and the Lobachevskian plane (where of course the trajectories are the geodesics on these surfaces; see Part I, §29.5, Exercises 6 and 7).)

(d) In Part I, §§32.1, 39.1, in the context of the problem of describing the behaviour of a free relativistic particle of positive mass m in STR, we considered (in 3-dimensional formalism) the Lagrangian

$$L^{(2)}_{\text{free}} = -mc^2 \left(1 - \frac{w^2}{c^2}\right)^{1/2},$$

where w is the 3-dimensional velocity of the particle (and we have written w^2 for $|w|^2$ for simplicity). In the presence of a potential field of the form α/r, the appropriate Lagrangian is

$$L = -mc^2 \left(1 - \frac{w^2}{c^2}\right)^{1/2} - \frac{\alpha}{r}. \tag{20}$$

It follows (much as in Part I, §32.1) that the energy, or in other words the

Hamiltonian, of the particle is given by

$$H(x, p)=c\sqrt{p^2+m^2c^2}+\frac{\alpha}{r}, \tag{21}$$

where $p=(p_1, p_2, p_3)$ is the 3-dimensional momentum (p^2 denoting $|p|^2$), and $x=(x^1, x^2, x^3)$ (x^1, x^2, x^3 being the spatial Minkowski co-ordinates on $\mathbb{R}^4_{1,3}$).

Since the angular momentum is conserved (cf. Part I, §32.2, Example (e)), the motion of the particle will be confined to a plane in space, which we may suppose co-ordinatized by x^1, x^2, or by the corresponding polar co-ordinates r, φ (where r has the same meaning as before). In terms of the latter co-ordinates we have $w^2=\dot{r}^2+r^2\dot{\varphi}^2$, whence $p_\varphi=\partial L/\partial\dot{\varphi}=r^2\dot{\varphi}$ (cf. Part I, §32.1, Example (b′)); writing $M=P_\varphi$, it follows that $p^2=p_r^2+M^2/r^2$ ($p=(p_r, p_\varphi)$), whence

$$H=c\sqrt{p_r^2+\frac{M^2}{r^2}+m^2c^2}+\frac{\alpha}{r}=\text{const.} \tag{22}$$

In the case of an attractive field (i.e. $\alpha<0$), if $Mc>|\alpha|$ then the expression in (22) approaches ∞ as $r\to 0$ whence it follows (using the constancy of H) that, as in the analogous classical situation, the particle cannot reach the centre of attraction ($r=0$). (If on the other hand $Mc<|\alpha|$ then under suitable conditions the particle may fall into the origin.)

For the exact solution of spherically symmetric, planar problems of this sort, it is convenient for technical reasons to use the Hamilton–Jacobi equation (introduced in Part I, §35.1)

$$-\frac{\partial S}{\partial t}=H\left(x, \frac{\partial S}{\partial x}\right), \tag{23}$$

where (in the general situation of n degrees of freedom) S is defined on the Lagrange surface Γ^{n+1} given by the equations $p_i=\partial S/\partial x^i$, $E(=p_{n+1})=-\partial S/\partial t$ in extended phase space with co-ordinates x, p, E, t ($E=p_{n+1}$, $t=x_{n+1}$), by means of integrals $\int \tilde{L}\,dt=\int p\,dx-E\,dt$ over paths from a fixed point Q to a variable point P of Γ^{n+1} (see Part I, loc. cit.).

In our present special context we have $n=2$ and co-ordinates r, φ, t; p_r, p_φ, E. We seek a solution of (23) in the form

$$S=-Et+M\varphi+f, \qquad f=f(r, M, E). \tag{24}$$

(Thus we are taking our Lagrange surfaces each to be defined by equations of the form $E=$const., $M=$const., p_r as in (26) below.) Given such a solution each integral trajectory $r(\varphi)$ will be defined by an equation of the form $\partial S/\partial M=$const., while the dependence of r on t will be given by an equation of the form $\partial S/\partial E=$const. To find f explicitly, observe first that since $p_r=\partial S/\partial r$, we have from (22) that

$$E=c\sqrt{\left(\frac{\partial S}{\partial r}\right)^2+\frac{M^2}{r^2}+m^2c^2}+\frac{\alpha}{r}, \tag{25}$$

whence

$$p_r = \frac{\partial S}{\partial r} = \sqrt{\frac{1}{c^2}\left(E - \frac{\alpha}{r}\right)^2 - \frac{M^2}{r^2} - m^2c^2}. \tag{26}$$

From this and (24) we obtain finally

$$S = -Et + M\varphi + \int \sqrt{\frac{1}{c^2}\left(E - \frac{\alpha}{r}\right)^2 - \frac{M^2}{r^2} - m^2c^2}\, dr, \tag{27}$$

whence we derive the equation of the trajectory in the form

$$\frac{\partial S}{\partial M} = \varphi + \frac{\partial f}{\partial M} = \text{const.} \tag{28}$$

(As noted above the dependence of r on t is for each trajectory given by an equation of the form $\partial S/\partial M = \text{const.}$)

Exercises

1. Prove that for $\alpha < 0$, $Mc < |\alpha|$, the above solutions are spirals which reach the origin in a finite amount of time.
2. Show that if $\alpha < 0$ and $E < mc^2$, then in general the trajectories are not closed. Find the correction to the elliptic orbits of classical mechanics.

(e) As our final example we consider the problem of describing the motion of a free particle of positive mass $m > 0$ in the Schwarzschild metric. In Part I, §39.2, we obtained that metric (as a stationary, spherically symmetric solution of Einstein's equation) in terms of the co-ordinates t, r, φ, θ, in the form

$$ds^2 = \left(1 - \frac{a}{r}\right)c^2\,dt^2 - \frac{1}{1 - \frac{a}{r}}\,dr^2 - r^2(d\theta^2 + \sin^2\theta\, d\varphi^2). \tag{29}$$

The Hamiltonian (corresponding to the usual Lagrangian $L = \frac{1}{2}g_{ab}\dot{x}^a\dot{x}^b$) is $H = \frac{1}{2}g^{ab}p_a p_b (= \dot{x}^a(\partial L/\partial \dot{x}^a) - L$: see Part I, §33.1); here $a, b = 0, 1, 2, 3$, and $x^0 = ct$, $x^1 = r$, $x^2 = \varphi$, $x^3 = \theta$. Since the motion is again planar we may, by choosing the spherical co-ordinates suitably, assume $\theta = \pi/2$, whereupon in (29) the term in $d\theta^2$ drops out, and the remaining (i.e. non-zero) g^{ab} are given by

$$g^{00} = (g_{00})^{-1} = \left(1 - \frac{a}{r}\right)^{-1}, \qquad g^{11} = g^{rr} = (g_{rr})^{-1} = 1 - \frac{a}{r},$$

$$g^{22} = g^{\varphi\varphi} = (g_{\varphi\varphi})^{-1} = -\frac{1}{r^2}.$$

In terms of such co-ordinates the Hamilton–Jacobi equations $\partial S/\partial x^a = p_a$, together with the constancy of the Hamiltonian along each trajectory, yield

$$\frac{\partial S}{\partial x^a}\frac{\partial S}{\partial x^b}g^{ab}=\text{const.}$$
$$=\frac{1}{1-\frac{a}{r}}\left(\frac{\partial S}{c\partial t}\right)^2-\left(1-\frac{a}{r}\right)\left(\frac{\partial S}{\partial r}\right)^2-\frac{1}{r^2}\left(\frac{\partial S}{\partial \varphi}\right)^2=g^{ab}p_ap_b. \tag{30}$$

(Note that we have used here the "truncated" Hamilton–Jacobi equations (see Part I, §35.1) in view of the fact that the time t functions essentially as one of our x-co-ordinates.) The constant in (30) may be taken to be m^2c^2 since we expect that, as in the case of flat (Minkowski) space (see Part I, §§32.2, 39.1), the 4-momentum of the free particle will lie on the mass surface $g^{ab}p_ap_b=m^2c^2$. As in the preceding example, we seek a solution of (30) in the form

$$S=-Et+M\varphi+f(r,M,E), \tag{31}$$

where $M=p_\varphi$, and $E=cp_0$ (the usual energy of the particle in 3-dimensional formalism ($E>mc^2$)), and then the functions $r(t)$, $r(\varphi)$ defining the trajectory will be given by equations of the form

$$\frac{\partial S}{\partial E}=\text{const.},\qquad \frac{\partial S}{\partial M}=\text{const.} \tag{32}$$

Substituting the expression (31) for S in (30) (where the constant is now taken to be m^2c^2), we obtain

$$\frac{\partial f}{\partial r}=\left[\left(\frac{E^2}{c^2\left(1-\frac{a}{r}\right)}-m^2c^2-\frac{M^2}{r^2}\right)\frac{1}{\left(1-\frac{a}{r}\right)}\right]^{1/2} \tag{33}$$

From this and the second of the equations (32) we obtain the equation of the trajectory in the form

$$\varphi=\int\frac{M}{r^2}\left[\frac{E^2}{c^2}-\left(m^2c^2+\frac{M^2}{r^2}\right)\left(1-\frac{a}{r}\right)\right]^{-1/2}dr. \tag{34}$$

Remark. On letting m go to zero in (34) we obtain the equation for the path of a particle of zero mass (the path of a light ray for instance) in the Schwarzschild metric.

The way in which r varies with t along a trajectory is given by the first of the equations (32); thus from that equation together with (33) we obtain

$$t=\int\left[\left(\frac{E^2}{c^2\left(1-\frac{a}{r}\right)}-m^2c^2-\frac{M^2}{r^2}\right)\frac{1}{\left(1-\frac{a}{r}\right)}\right]^{-1/2}\frac{E}{c^2\left(1-\frac{a}{r}\right)^2}dr,$$

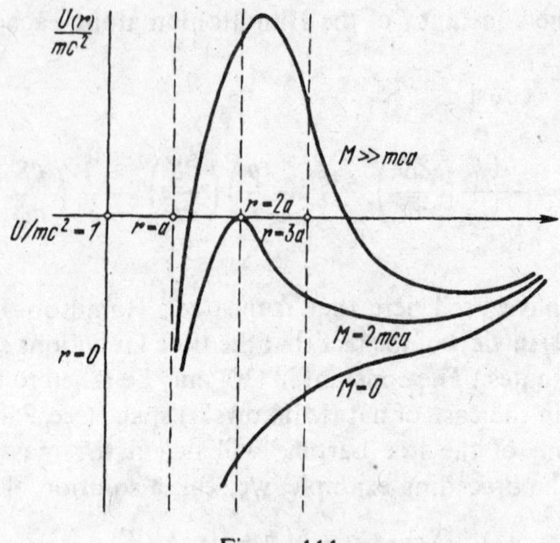

Figure 111

whence

$$\left(1-\frac{a}{r}\right)^{-1}\frac{dr}{cdt}=\frac{1}{E}[E^2-U^2(r)]^{1/2}, \tag{35}$$

where

$$U(r)=mc^2\left[\left(1-\frac{a}{r}\right)\left(1+\frac{M^2}{m^2c^2r^2}\right)\right]^{1/2} \qquad (m\neq 0). \tag{36}$$

The quantity $U(r)$ is called the "effective potential" at a given angular momentum M; the inequality $U(r)\leq E$ defines for each given M and E the possible regions in which the motion can take place (i.e. the possible values of r). From Figure 111, in which the graph of $U(r)/mc^2$ is indicated for various M, we see that, depending on M, the effective potential may have a maximum when r is of the order of $2a$. From the definition ((36)) of $U(r)$ it is immediate that $U(r)/mc^2\to 1$ as $r\to\infty$. From these facts (in conjunction with (35)) it follows in particular that it is possible for the particle to be "captured" by the gravitational field, i.e. for $r(t)$ to be such that $r(-\infty)=\infty$ while $r(+\infty)$ is finite.

§29. Foliations

29.1. Basic Definitions

We precede the definition of a "foliation" with that of an "integrable distribution" on a manifold.

29.1.1. Definition.

(i) A *k-dimensional distribution* on an n-dimensional manifold M is a smooth field of k-dimensional tangential directions on M, i.e. a map associating with each point $x \in M$ a k-dimensional subspace of the tangent space $T_x M$.

(ii) A distribution on a manifold M is said to be *integrable* if through each point of the manifold M there is a k-dimensional "integral" surface in M, i.e. a k-dimensional submanifold of M whose tangent space at each of its points is the k-dimensional subspace assigned to that point by the given distribution.

(iii) A manifold M is said to have a *k-dimensional foliation* defined on it if it is "foliated" into k-dimensional surfaces, i.e. if for each point of M there is specified precisely one smooth k-dimensional submanifold passing through that point, in a manner depending smoothly (or continuously) on the points of the manifold. (These specified surfaces are called the *leaves* of the foliation.) It is further required that in some neighbourhood of each point of M there can be introduced co-ordinates $x^1, \ldots, x^k$, $y^1, \ldots, y^{n-k}$ with the properties that the level surfaces $y^1 = a_1, \ldots, y^{n-k} = a_k$ are just the leaves of the foliation in that neighbourhood (one leaf for each $(n-k)$-tuple $a_1, \ldots, a_{n-k}$), and that $x^1, \ldots, x^k$ are local co-ordinates for each leaf. (We remark that foliations most often arise as integrable distributions.)

29.1.2. Examples. (a) We have already encountered (in §27.1) one-dimensional foliations, defined in terms of non-vanishing vector fields or fields of (one-dimensional) directions. A smooth one-dimensional distribution is always integrable, yielding a foliation, in view of the local theorem on the existence and uniqueness of solutions of systems of ordinary differential equations. Thus smooth one-dimensional distributions always give rise to (unique) foliations.

(b) Suppose there is given on a complex n-dimensional manifold M a (non-vanishing) complex vector field (or field of complex one-dimensional directions). If the vector field is holomorphic (i.e. if the map $M \to \mathbb{C}^n$ defining the field is complex analytic; see §4.1), then as in the preceding example the associated distribution is always integrable, yielding a one-dimensional complex foliation on the complex manifold M. Consider for example the case $n=2$, and the vector field on $\mathbb{C}^2$ defined by the complex differential equation

$$\frac{dz}{dw} = \frac{P(z, w)}{Q(z, w)}, \tag{1}$$

where P and Q are polynomials both of degree m; this equation, or its equivalent formulation $Q\,dz - P\,dw = 0$, defines (away from the common zeros of P and Q) a one-dimensional complex distribution, and thence, as just observed, a one-dimensional complex foliation on $\mathbb{C}^2$. By introducing in the usual way homogeneous co-ordinates u_0, u_1, u_2 by means of the substitutions

$$z=\frac{u_1}{u_0}, \qquad w=\frac{u_2}{u_0},$$

we can extend this to a foliation of the complex manifold $\mathbb{C}P^2 \supset \mathbb{C}^2$, namely to the foliation defined by the form $\omega = Q\,dz - P\,dw$ (cf. §27.1 where the analogous extension from $\mathbb{R}^2$ to $\mathbb{R}P^2$ was carried out). As already noted, the points where $P=Q=0$ will be "singular" points of the foliation, i.e. strictly speaking the foliation is defined only on the complement in $\mathbb{C}P^2$ of these points.

EXERCISE

Find the singular points on the complex line at infinity.

It is of interest to examine the arrangement of the leaves of such a complex foliation near a non-degenerate singular point (see §14.4 for the definition of non-degeneracy). The system defining the foliation may be written as

$$\begin{aligned} dz &= P(z, w)\,dt, & \dot{z} &= P, \\ dw &= Q(z, w)\,dt, & \dot{w} &= Q. \end{aligned} \tag{2}$$

We may clearly assume without loss of generality that the singular point in question is the origin $z=w=0$; thus $P(0,0)=Q(0,0)=0$. The non-degeneracy of the singular point $(0,0)$ means that the linear parts of P and Q, given by

$$\begin{aligned} \dot{z} &= az + bw + \cdots, \\ \dot{w} &= cz + dw + \cdots, \end{aligned} \tag{3}$$

satisfy $ad - bc \neq 0$. Assuming that the eigenvalues of the matrix $\begin{pmatrix} a & b \\ c & d \end{pmatrix}$ are distinct (which situation can clearly be achieved by means of an arbitrarily small perturbation of our system), we can, via a linear co-ordinate change, bring our system into the form

$$\begin{aligned} \dot{z} &= \lambda_1 z + \cdots, \\ \dot{w} &= \lambda_2 w + \cdots. \end{aligned} \tag{4}$$

Writing $\lambda = \lambda_1/\lambda_2$, we now turn our attention to the related purely linear system defined by $dz/dw = \lambda(z/w)$. (The question that arises here of the reducibility of the system (3) to purely linear form by means of a complex analytic co-ordinate change in some neighbourhood of the singular point $(0,0)$, is a difficult one, which we shall not enter into.) This linear system has the obvious general solution $z = aw^\lambda$, $a = \text{const.}$, and the two particular solutions $z \equiv 0$, w arbitrary; $w \equiv 0$, z arbitrary; hence the corresponding foliation has one leaf for each non-zero choice of the constant a, and the two additional leaves A, B defined respectively by $z \equiv 0$ and $w \equiv 0$. If we remove the singular point $(0, 0)$, we obtain a (non-singular) foliation of $\mathbb{C}^2 \setminus \{(0, 0)\}$, whose leaves A and B are not simply-connected (having each the topology of $\mathbb{C} \setminus \{(0, 0)\}$).

EXERCISE

Let $\gamma_1 \in \pi_1(A)$ and $\gamma_2 \in \pi_1(B)$ be generators of these fundamental groups. Show that both γ_1 and γ_2 are (represented by) limit cycles of the foliation (see Definition 27.1.3). Calculate the "holonomy representation" $\gamma_i \mapsto R_{\gamma_i}$ (see Example (c) below, and the next subsection).

(c) Consider a fibre bundle with base space M, structure group G, fibre F, total space E, and projection $p: E \to M$. In §24.2 we defined a connexion (or G-connexion) on a fibre bundle in terms of a family of horizontal n-dimensional "planes" $\mathbb{R}^n(y)$ (where $n = \dim M$) associated with the points $y \in E$; these "planes" were required to vary smoothly with the points of E, to be transverse to the fibre through y, and to satisfy a further condition involving the group G.

EXERCISE

Show that the distribution of horizontal n-dimensional directions afforded by a connexion on a fibre bundle, is integrable if and only if the curvature tensor of the connexion (see §25.3) is identically zero. (*Hint*. Use the integrability conditions below.)

It is immediate from the above-quoted definition of a connexion on a fibre bundle, that if the connexion determines a foliation of E, then the projection onto M of each leaf W of that foliation is locally a diffeomorphism, so that each leaf W constitutes a covering space for M. The monodromy group of each of these covering spaces is called in this context the "discrete group of holonomies" (cf. §19.1). If the base M is simply-connected, then it is its own universal cover, and each of the leaves W will be globally diffeomorphic to M.

Our examples concluded we now give two (equivalent) necessary and sufficient conditions for integrability of a k-dimensional distribution (to yield a foliation).

First formulation. In view of the final requirement of Definition 29.1.1(iii), the integrability problem is equivalent to that of solving (locally) a system of partial differential ("Pfaffian") equations of the form

$$\frac{\partial y^\alpha}{\partial x^\beta} = f^\alpha_\beta(x, y), \qquad \alpha = 1, \ldots, n-k, \quad \beta = 1, \ldots, k, \tag{5}$$

for the functions $y^\alpha(x^1, \ldots, x^k)$ defining (locally) the leaves of the foliation; here $x^1, \ldots, x^k, y^1, \ldots, y^{n-k}$ are local co-ordinates on the given manifold M, and the k vectors $(f^1_\beta, \ldots, f^{n-k}_\beta)$, $\beta = 1, \ldots, k$, form a basis for the k-dimensional subspace attached (via the given distribution) to the point $(x^1, \ldots, x^k, y^1, \ldots, y^{n-k})$ of M. The equality of mixed partial derivatives furnishes the desired integrability condition:

$$\frac{\partial^2 y^\alpha}{\partial x^\beta \partial x^\gamma} = \frac{\partial}{\partial x^\gamma} f^\alpha_\beta(x, y) = \frac{\partial}{\partial x^\beta} f^\alpha_\gamma(x, y). \tag{6}$$

Second formulation. Given a k-dimensional distribution on a manifold M, let ξ, η be two vector fields on M whose values $\xi(x)$, $\eta(x)$ at each point $x \in M$ lie in the specified k-dimensional subspace attached to the point x via the given distribution. If the distribution can be integrated to yield a foliation, then at each point x the fields ξ and η will be tangential to the leaf through x, whence so will their commutator $[\xi, \eta]$ (see Part I, §24.1). This condition (namely that the commutator $[\xi, \eta]$ of such vector fields should likewise have its values at each point in the k-dimensional "plane" attached to that point) turns out to be also sufficient for integrability.

Exercise

Prove that the above two formulations of the condition for integrability of a distribution are indeed equivalent.

We shall not give the proof of the sufficiency of each of these two conditions. (Their necessity is obvious.)

We conclude this subsection by considering the special case of $(n-1)$-dimensional foliations (i.e. of "codimension 1") of an n-manifold, i.e. foliations whose leaves each have dimension $k=n-1$. Such a foliation is given locally by a 1-form (or equation of "Pfaffian type"; cf. §25.1 or Part I, Lemma 34.1.2)

$$\omega = P_i(x)\, dx^i = 0. \tag{7}$$

(Note that of course at a non-singular point not all of the $P_i(x)$ will vanish.) If the form ω is closed then a distribution defined by an equation of the form (7) will be integrable, since then, as is well known, ω will be locally exact, i.e. $\omega = dH$ locally, for some scalar-valued function H, and the leaves of the foliation will be defined locally as the level surfaces $H=\text{const}$. If, more generally, there exists a nowhere-vanishing function $f(x)$ which is an "integrating factor" for ω, i.e. is such that the form $f(x)\omega$ is closed, then again the distribution defined by the Pfaffian equation $\omega=0$ is integrable since it is equivalent to $f(x)\omega=0$, which is integrable for the same reason as before.

Remark. It follows (to generalize further) that a foliation of codimension 1 can be defined on a manifold M with a given atlas of charts U_j, by means of equations $\omega_j=0$ (analogous to (7)) on the U_j (one equation for each U_j), where on the regions of overlap $U_i \cap U_j$ the forms ω_i, ω_j satisfy $f_{ij}(x)\omega_i = \omega_j$ for some function $f_{ij} \neq 0$ on $U_i \cap U_j$.

In connexion with the discussion preceding this remark, observe that if $d(f\omega)=0$, then $d\omega = -(df/f) \wedge \omega$, whence it follows that: (i) ω is a factor of $d\omega$ (with respect to the wedge product of forms); and (ii) $\omega \wedge d\omega = 0$.

Exercise

Show that the integrability condition (in either of the above formulations) for a (non-singular) distribution given by a Pfaffian equation $\omega=0$, is equivalent to each of the conditions (i) and (ii).

Note finally that in the case of 3-dimensional space $\mathbb{R}^3$ the form ω is in the customary manner written as a covector field $P=(P_\alpha)$, and $d\omega$ as curl P; in this notation the condition (ii) takes the form

$$\langle P, \operatorname{curl} P\rangle=0. \tag{8}$$

29.2. Examples of Foliations of Codimension 1

Let ω be a closed 1-form on a compact (connected) manifold M. As we have just seen (in the preceding subsection), the equation $\omega=0$ defines a foliation of codimension 1 on M. Given any basis of one-dimensional cycles $z_1,\ldots,z_q$ (i.e. (equivalence classes of) closed paths in M) for the group $\tilde{H}_1(M)=H_1/\text{Tors}$, where Tors denotes the subgroup of all torsion elements (see §19.3), the form ω determines a corresponding collection of "periods"

$$\oint_{z_j}\omega=a_j,\qquad j=1,\ldots,q.$$

If $\hat{M}\xrightarrow{p}M$ is the universal cover for M, then $\pi_1(\hat{M})=1$, whence it follows that the pullback $p^*(\omega)$ of the 1-form ω to $\hat{M}$ is (globally) exact, i.e. $p^*(\omega)=df$ for some function f. There is however a "lower" covering space $M_1\xrightarrow{p_1}M$, with non-trivial fundamental group, such that the form $p_1^*(\omega)$ is still exact; we now indicate the construction of an appropriate such covering space M_1. Thus let $A<\pi_1(M)$ be the subgroup of $\pi_1(M)$ consisting of all elements z of $\pi_1(M)$ such that

$$\int_z\omega=0;$$

it is easy to see that A contains the commutator subgroup of $\pi_1(M)$, so that A is certainly a normal subgroup. By §19.2, Exercise 1, there exists a covering space $M_1\xrightarrow{p_1}M$ such that

$$\pi_1(M_1)\simeq p_{1*}\pi_1(M_1)=A<\pi_1(M), \tag{9}$$

and, essentially by Theorem 19.2.3, the monodromy group B of this covering is isomorphic to the quotient group of $\pi_1(M)$ by A:

$$B=\sigma\pi_1(M)\simeq\pi_1(M)/A. \tag{10}$$

It follows readily from the definition of A that B is free abelian.

It is immediate from (9) (and the definition of A) that the form $p_1{}^*(\omega)$ on M_1 has all periods zero, so that it is (globally) exact: $p_1^*(\omega)=dg$ where g is a real-valued function on M_1. Clearly we may take g to be given by

$$g(x)=\int_{x_0}^{x} p_1^*(\omega), \tag{11}$$

where x_0 is any fixed point of M_1, and the integral, being by the above path-independent, may be taken along any path joining x_0 to x. The upshot is that by lifting the defining distribution to the covering space M_1, the original foliation has been transformed, as it were, into the family of level surfaces of the function $g(x)$ on M_1 (which furthermore as a covering space for M has free abelian monodromy group).

The first of the following examples represents a particular instance of this situation.

29.2.1. Examples. (a) Take M to be the n-dimensional torus T^n furnished with angular co-ordinates $\varphi^1, \ldots, \varphi^n$, $0 \le \varphi^j < 2\pi$, and consider any form of the type

$$\omega = b_i\, d\varphi^i, \qquad b_i = \text{const.} \tag{12}$$

Exercise

Show that the "minimal" covering $p_1\colon M_1 \to T^n$ for which $p_1^*(\omega)=dg$, has monodromy group isomorphic to $\mathbb{Z}\oplus\cdots\oplus\mathbb{Z}$ where the number of summands is equal to the rank of the set $\{b_1, \ldots, b_n\}$ over the field $\mathbb{Q}$ of rational numbers (i.e. the dimension of the vector space over $\mathbb{Q}$ spanned by the b_i).

(b) Let M be a compact Riemann surface (i.e. a (compact) one-dimensional complex manifold defined as in §4.2) and let ω be a holomorphic differential given locally by $\omega = f(z)\, dz$ where $f(z)$ is analytic (without poles). The differential $\operatorname{Re}\omega$ is then given locally by

$$\operatorname{Re}\omega = \operatorname{Re}(f(z)\, dz) = \operatorname{Re}(u+iv)(dx+i\, dy), \tag{13}$$

where $z=x+iy$, $f=u+iv$; the equation $\operatorname{Re}\omega=0$ clearly defines a one-dimensional foliation on the (realized) manifold M.

Exercise

Under the assumption that the singular points of this one-dimensional foliation are non-degenerate, show that they are all saddles, and deduce that they are equal in number to the Euler characteristic of the manifold M (see §§14, 15). Investigate the integral trajectories (i.e. the leaves) of this one-dimensional foliation when M is a hyperelliptic Riemann surface, i.e. is given by an equation of the form

$$w^2 = P_{2n+1}(z) = \prod_{\alpha=0}^{2n} (z-z_\alpha), \qquad z_\alpha \ne z_\beta. \tag{14}$$

On such a surface a holomorphic differential ω (without poles) can be shown to have the form

$$\omega = \frac{Q(z)\, dz}{\sqrt{P_{2n+1}(z)}}, \tag{15}$$

where $Q(z)$ is a polynomial of degree $\leq n-1$. Show that corresponding to almost every choice of the $2n+1$ (distinct) complex numbers z_α in (14), and almost every ω of the form (15), there exists an integral trajectory which is everywhere dense on M. Investigate the foliations defined by forms with poles, i.e. by forms ω as in (15) where now Q is a rational function. Given such a form ω, what singular points of the corresponding foliation arise from the poles of ω?

(c) We have already encountered (in §28.3) foliations with a more involved structure which lift (in the sense defined above) to families of level surfaces of functions defined on some (no longer "abelian") covering space. To recapitulate somewhat, let T_1 denote, as in §28.3, the unit tangent bundle over the closed surface M_g^2 (the sphere-with-g-handles) endowed with a metric of constant negative Gaussian curvature (inherited via its construction (for $g>1$) as the orbit space of its universal cover L^2 (the Lobachevskian plane) under the action of a discrete group of isometries of L^2; see Corollary 20.11). In §28.3, in the course of our discussion of geodesic flows, we defined two 2-dimensional foliations R_+ and R_- of T_1: each leaf of the foliation R_+ was defined to consist of all geodesics on M_g^2 which approach each other asymptotically as $t\to+\infty$ (and analogously for R_- with $t\to-\infty$). (Of course strictly speaking it is not the geodesics on M_g^2 which make up the leaves of R_+ and R_-, but rather the corresponding curves in T_1 obtained by pairing off each point of each geodesic with the unit tangent vector to the geodesic at that point.) Each pair of leaves, one from R_+ and one from R_-, then intersects in exactly one geodesic; this is made especially clear by lifting the geodesics on M_g^2 onto its universal cover, the Lobachevskian plane L^2, where each of the leaves of R_+ (for instance) becomes a collection of geodesics approaching a single point at infinity (i.e. a point on the boundary of the disc in the Poincaré model of L^2; see Figure 109). The corresponding covering space $\hat{T}_1 \xrightarrow{\hat{p}} T_1$, the unit tangent bundle over L^2, is not universal, since it is contractible to its fibre S^1, and so has fundamental group $\pi_1(\hat{T}_1)\simeq\mathbb{Z}$.

Exercise

Using the fact (mentioned in §28.3) that the conjugacy classes of $\pi_1(M_g^2)$ are in natural one-to-one correspondence with the closed geodesics, show that each of the foliations R_+ and R_- lifts to a family of level surfaces of a real-valued function on a covering space $T_2\to T_1$ only if T_2 covers $\hat{T}_1$. (Recall from §24.3, Example (d), that the group $\pi_1(T_1)$ has the presentation with generators $a_1,\dots,a_g,\ b_1,\dots,b_g,\tau$ and defining relations

$$\prod_{i=1}^{g} a_i b_i a_i^{-1} b_i^{-1} = \tau^{2-2g}, \qquad a_i\tau=\tau a_i, \qquad b_i\tau=\tau b_i. \tag{16}$$

From that example together with Theorem 19.2.3, it follows that the monodromy group of the covering $\hat{T}_1\to T_1$ is isomorphic to the group $\pi_1(T_1)/(\tau)\simeq\pi_1(M_g^2)$, which is non-abelian for $g>1$. (Here (τ) denotes the (normal) subgroup generated by τ.)

The foliations R_+ and R_- clearly have no singular points. Their leaves may have different topologies: it follows essentially from the contractibility of each leaf of R_+ to any geodesic comprising it (which can be seen by letting $t \to +\infty$), that a leaf will be topologically equivalent to $S^1 \times \mathbb{R}^1$ if among the geodesics defining it there is a periodic one, and to $\mathbb{R}^2$ otherwise. (Note that there are only countably many periodic geodesics in view of the above-mentioned one-to-one correspondence between such geodesics and the conjugacy classes of $\pi_1(M_g^2)$.)

Remark. As already noted in §28.3 the existence of foliations R_+ and R_- with the above properties turns out to be a characteristic feature of geodesic flows on compact manifolds of negative curvature (and for certain other spaces), and is of great importance for the general theory of dynamical systems, having as it does several remarkable consequences; we shall however not pursue this particular avenue further.

(d) We conclude our examples with a geometric construction of a 2-dimensional foliation (the "Reeb foliation") of the solid torus $D^2 \times S^1$, of which the boundary torus $T^2 = \partial(D^2 \times S^1)$ constitutes a single leaf. To this end consider first the solid cylindrical region $U \subset \mathbb{R}^3$ consisting of those points (x, y, z) for which $y^2 + z^2 \leq 1$, $-\infty < x < \infty$; thus $U \cong D^2 \times \mathbb{R}^1$. We foliate this solid cylinder as indicated in Figure 112: a single leaf is obtained by revolving about the x-axis an arc in the (x, y)-plane asymptotic to the lines $y = \pm 1$, and then all leaves (except for the boundary) are obtained by means of arbitrary translations of this one in the direction of the x-axis. (Clearly these leaves are invariant under the transformation $y \to -y$, $z \to -z$, $x \to x$.) On performing the identification $(x, y, z) \sim (x + 1, y, z)$ we obtain from this foliation of the solid cylinder the desired *Reeb foliation* of the solid torus $D^2 \times S^1$. Note that since the boundary $T^2 = \partial(D^2 \times S^1)$ is a leaf, we can obtain a foliation of the 3-sphere

$$S^3 = (D^2 \times S^1) \cup (S^1 \times D^2)$$

by identifying the boundaries of two such foliated solid tori.

We end this section by introducing two important topological invariants of foliations, namely "limit cycles" and "vanishing cycles", and examining these concepts as they pertain to some of the above examples of foliations. Let x_0 be a point of a leaf W of a k-dimensional smooth foliation of a compact n-dimensional manifold M, and let γ be a smooth closed path representing an

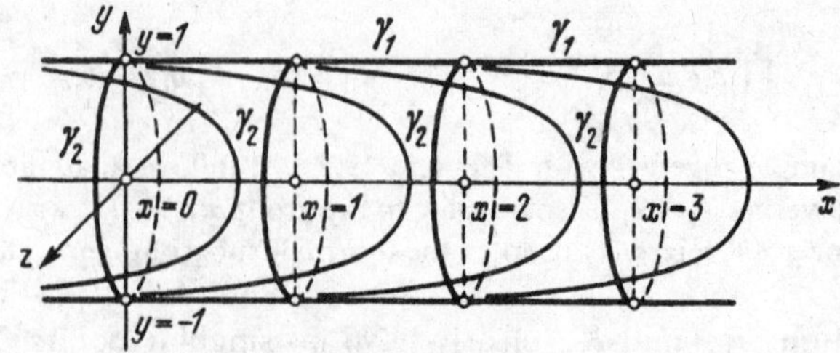

Figure 112

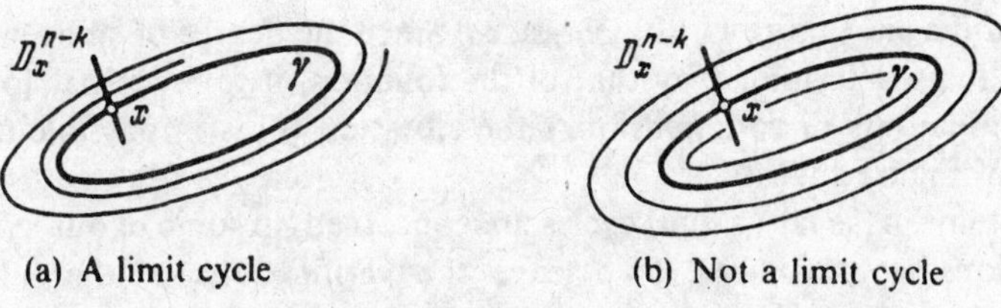

(a) A limit cycle (b) Not a limit cycle

Figure 113

element of $\pi_1(W, x_0)$. Corresponding to each point x on γ choose in M a (small) $(n-k)$-dimensional (closed) disc (or ball) D_x^{n-k} centred on x, which is transverse to γ and varies smoothly with x (as shown in Figure 113 for $n-k=1$). By intersecting the neighbouring leaves W_y with the family of discs D_x^{n-k} we obtain curves in terms of which there is defined a "transporting" map of the disc $D_{x_0}^{n-k}$ with centre x_0 (assumed sufficiently small) around γ:

$$R_\gamma: D_{x_0}^{n-k} \to D_{x_0}^{n-k}.$$

It is not difficult to see that, provided the discs are sufficiently small, the map R_γ is unchanged by deformations of the family of discs D_x^{n-k} or of γ, which leave x_0 and $D_{x_0}^{n-k}$ fixed. The correspondence $\gamma \mapsto R_\gamma$ therefore defines a map from the group $\pi_1(W, x_0)$ to the group of "germs" of self-maps of the disc $D_{x_0}^{n-k}$. (A *germ* of such maps is an equivalence class where two maps are equivalent if they are identical on some neighbourhood in $D_{x_0}^{n-k}$ of x_0; by resorting to germs of self-maps of $D_{x_0}^{n-k}$, the size of that disc is made irrelevant.) This representation of $\pi_1(W, x_0)$ is called the *holonomy group of the foliation on the leaf W*. If $\gamma \in \pi_1(W, x_0)$ is such that R_γ does not represent the identity map germ, we call γ a *limit cycle of the foliation* (see Figure 113). (Cf. Definition 27.1.3.)

In the case $k=n-1$, the disc $D_{x_0}^{n-k}=D_{x_0}^1=I$ is a path segment (i.e. a closed interval), separated into two pieces by the point x_0, as a consequence of which there arise in this case the two possibilities (a) and (b) illustrated in Figure 114. If the cycle γ is as in Figure 114(a), it is called a *two-sided limit*

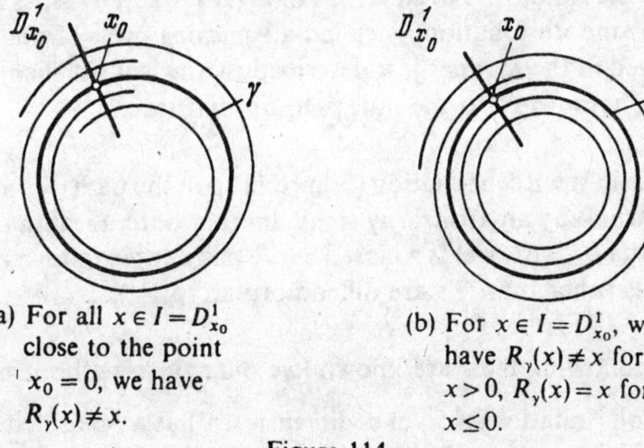

(a) For all $x \in I = D_{x_0}^1$ close to the point $x_0 = 0$, we have $R_\gamma(x) \neq x$.

(b) For $x \in I = D_{x_0}^1$, we have $R_\gamma(x) \neq x$ for $x > 0$, $R_\gamma(x) = x$ for $x \leq 0$.

Figure 114

cycle, and if as in Figure 114(b), *one-sided*. Since the degree of smoothness of the map R_γ is determined by that of the foliation, it follows that for (real) analytic foliations of codimension 1 the situation (b) is impossible (see the caption of Figure 114(b)).

The situation (as far as limit cycles are concerned) in some of our examples of foliations is as follows. (We leave the verifications of the following statements to the reader.)

(α) A non-singular foliation defined by a closed 1-form ω (see the conclusion of the preceding subsection) has no limit cycles.

(β) For the foliations R_+ and R_- of the unit tangent bundle over a compact surface (see Example (c) above), every leaf W is either simply-connected or else $\pi_1(W) \simeq \mathbb{Z}$, and in the latter circumstance the generator of the group $\pi_1(W)$ is a two-sided limit cycle.

(γ) Returning to the Reeb foliation of the solid torus $D^2 \times S^1$ (Example (d) above), let γ_1, γ_2 denote generators of the fundamental group $\pi_1(T^2) \simeq \mathbb{Z} \oplus \mathbb{Z}$ of the boundary torus T^2, as indicated in Figure 112. Here the element γ_1 is a one-sided limit cycle (from the inside), while γ_2 is not a (one-sided) limit cycle. On the other hand for the foliation of the 3-sphere $S^3 = (D^2 \times S^1) \cup (S^1 \times D^2)$ obtained by glueing together two Reeb-foliated solid tori, the cycles γ_1 and γ_2 are both one-sided limit cycles (since now γ_2 on one of the solid tori is identified with the γ_1 of the other). It follows that this foliation of the 3-sphere cannot be real analytic (although it may be infinitely differentiable).

Returning to the general situation of a k-dimensional smooth foliation of a manifold M, observe that if $\gamma \in \pi_1(W)$ (where W is any leaf of the foliation) is not a limit cycle (i.e. if R_γ represents the identity map), then it can be "nudged" onto leaves sufficiently close to it and still remain a closed curve (see Figure 113(b)). (If $k = n-1$ and γ is a limit cycle on one side then it can be moved intact onto neighbouring leaves on the other side.)

We single out for special attention from among such cycles the "vanishing" ones.

29.2.2. Definition. A non-trivial element γ of $\pi_1(W)$ (where W is, as above, any leaf of a given smooth foliation) is called a *vanishing cycle of the foliation* if when it is moved in the manner just described to any leaf sufficiently close to W, it becomes homotopic to the null path on that leaf.

For example in the Reeb foliation (Figure 112) on the path γ_2 on the leaf $W = T^2$, when shrunk by an arbitrarily small amount onto a leaf in the interior of the solid torus $D^2 \times S^1$, yields a closed curve homotopic to zero on that leaf (since all leaves other than T^2 are diffeomorphic to $\mathbb{R}^2$).

Remark. The following facts are known (we offer them without proof):

(i) Any smooth foliation of S^3 of codimension 1 has a one-sided limit cycle and is therefore not analytic.

(ii) Every smooth foliation of S^3 of codimension 1 possesses a closed leaf which is diffeomorphic to T^2 and bounds a region $D^2 \times S^1$ with the Reeb foliation induced on it.

(iii) If the universal covering space $\hat{M}$ of a 3-manifold M is neither contractible, nor diffeomorphic to $S^2 \times \mathbb{R}$, then every foliation of codimension 1 of M possesses (non-trivial) vanishing cycles.

(iv) If a foliation of codimension 1 of a 3-manifold M has no limit cycles, there is an abelian cover $P_1: M_1 \to M$ where $M_1 = W \times \mathbb{R}^1$, with the property that the foliation of M_1 obtained as above by "lifting" the given foliation of M, has as its leaves the surfaces $W \times \{\text{const.}\}$, whence it follows that $M \cong W \times (\mathbb{R}/\mathbb{Z} \oplus \cdots \oplus \mathbb{Z})$. Such foliations share the topological structure of foliations defined by closed non-degenerate 1-forms (cf. (α) above).

§30. Variational Problems Involving Higher Derivatives

30.1. Hamiltonian Formalism

On any manifold, in principle, one can formulate the variational problem of finding the extremal arcs of the functional $S = \int L\, dt$ arising from a Lagrangian L which is a (scalar-valued) function not only of the points x and the velocity $v = \dot{x}$, but also of the higher derivatives of x with respect to the time, i.e. where locally, on each chart of M with co-ordinates $x^1, \ldots, x^n$ ($n = \dim M$), the Lagrangian has the form $L = L(x, \dot{x}, \ddot{x}, \ldots, x^{(m)})$ for some $m > 1$. (Note that this generalizes the "one-dimensional variational problem" introduced in Part I, §31.1, in the (usual) sense that the extremals sought are one-dimensional (in contrast with §37.1 of Part 1).) The following lemma gives the generalized "Euler–Lagrange equations", which the extremals must satisfy in this more general situation.

30.1.1. Lemma (cf. Part I, Theorem 31.1.2). *The equation $\delta S = 0$ is equivalent to the following Euler–Lagrange equations*:

$$\sum_{q=0}^{m} (-1)^q \frac{d^q}{dt^q}\left(\frac{\partial L}{\partial v_q^i}\right) = 0, \tag{1}$$

where $v_q^i = d^q x^i/dt^q$, $i = 1, \ldots, n$, $q = 0, \ldots, m$.

PROOF. In the conventional notation of the calculus of variations we have

$$\delta S = \int \sum_{q=0}^{m} \left(\frac{\partial L}{\partial v_q^i}\, \delta v_q^i\right) dt, \tag{2}$$

where $\delta v_q^i = (d^q/dt^q)\ (\delta x^i(t))$, and we are assuming that the variation (or "perturbation") $\delta x^i(t)$ is differentiable infinitely often, and vanishes outside the relevant interval of values of t. Integrating by parts q times (for each q) the summand $(\partial L/\partial v_q^i)\delta v_q^i$ (analogously to the proof of Theorem 31.1.2 of Part I), we obtain from (2)

$$\delta S = \int \left[\sum_{q,i} (-1)^q \frac{d^q}{dt^q} \left(\frac{\partial L}{\partial v_q^i} \right) \right] \delta x^i(t)\, dt, \tag{3}$$

whence in turn, by exploiting the arbitrariness of the variation δx^i (again essentially as in the proof of Theorem 31.1.2 of Part I), the desired Euler–Lagrange equations (1) follow. □

Analogously to the earlier situation we call the entity $(\delta S/\delta x^i(t))$ defined by

$$\frac{\partial S}{\delta x^i(t)} = \sum_q (-1)^q \left(\frac{\partial L}{\partial v_q^i} \right)^{(q)}, \tag{4}$$

the *variational derivative.*

Remark. In Part I, §37.4 we showed, using Hilbert's variational principle, how Einstein's equations for the metric defining a gravitational field could be obtained as the Euler–Lagrange equations corresponding to the action

$$S = \int R\sqrt{-g}\, d^4x.$$

Here the scalar curvature R, expressed in terms of the gravitational metric $g_{ab}(x)$, involves the second derivatives of the g_{ab}. (We showed however (in the first part of the proof of Theorem 37.4.1 of Part I) that, essentially by adding a term to the Lagrangian $R\sqrt{-g}$ making zero contribution to the total variation δS, we could remove those second derivatives). We also considered earlier other Lagrangians involving the curvature, namely those arising from characteristic classes (defined in Part I, §42 as certain closed forms whose integrals have identically zero variational derivatives; see also §25.4 et seqq.); second derivatives will therefore also figure in these. (Of course in most of these cases the variational problems are higher dimensional in the sense that the manifolds comprising the appropriate variational classes have dimension > 1; see Part I, §37.1.)

Returning to the Euler–Lagrange equations (1), we observe that in general they will have order $2m$ where m is the number of derivatives figuring in L. It turns out that there is in the present more general context an analogue of the "Legendre transformation" (see Part I, §33.1) by means of which in the "non-singular" case the Euler–Lagrange equations (1) can be transformed into equivalent Hamiltonian form on a space of dimension $2mn$ (where n is as before the dimension of the original manifold M). (This is "Ostrogradskii's theorem".) Having in view certain interesting applications of the variational method to particular problems involving derivatives higher than the first, we shall prove Ostrogradskiĭ's theorem only in the case $n=1$. Thus we have but a single (local) co-ordinate, which we shall denote by u. We shall use x (rather than t) to denote the independent time variable, and shall accordingly denote the derivatives of u by $u', u'', \dots, u^{(m)}$; our Lagrangian (on the real line) will then be of the form

$$L=L(u, u', \dots, u^{(m)}). \tag{5}$$

We introduce canonical variables q_j and p_j ($j=1, \dots, m$) defined by

$$\begin{aligned} &q_1=u, q_2=u', \dots, q_m=u^{(m-1)};\\ &p_1=\frac{\partial L}{\partial u'}-\left(\frac{\partial L}{\partial u''}\right)'+\cdots+(-1)^{m-1}\left(\frac{\partial L}{\partial u^{(m)}}\right)^{(m-1)},\\ &p_2=\frac{\partial L}{\partial u''}-\left(\frac{\partial L}{\partial u'''}\right)'+\cdots+(-1)^{m-2}\left(\frac{\partial L}{\partial u^{(m)}}\right)^{(m-2)},\\ &\cdots\cdots\cdots\cdots\cdots\cdots\cdots\cdots\cdots\cdots\\ &p_m=\frac{\partial L}{\partial u^{(m)}}, \end{aligned} \tag{6}$$

and then set

$$H(p, q)=-L+u'p_1+u''p_2+\cdots+u^{(m)}p_m. \tag{7}$$

30.1.2. Definition (cf. Part I, Definition 33.1.1). A Lagrangian $L(u, u', \dots, u^{(m)})$ is said to be (*strongly*) *non-singular* if the equations (6) can be uniquely solved in the form

$$u=u(p, q), u'=u'(p, q), \dots, u^{(2m-1)}=u^{(2m-1)}(p, q).$$

30.1.3. Lemma. *If the Lagrangian L has the form*

$$L=a(u^{(m)})^2+\tilde{L}(u, u', \dots, u^{(m-1)}), \tag{8}$$

then it is non-singular.

PROOF. Observe first that the $u^{(\alpha)}$ with $\alpha=0, \dots, m-1$, are trivially expressible in terms of p and q since for these α by definition $q_{\alpha+1}=u^{\alpha}$. With L as in (8) the last of the equations (6) becomes

$$p_m=2au^{(m)},$$

from which $u^{(m)}$ can be obtained in terms of p_m. We next show how to express $u^{(m+1)}$ in terms of p and q. To this end consider

$$p_{m-1}=\frac{\partial L}{\partial u^{(m-1)}}-\left(\frac{\partial L}{\partial u^{(m)}}\right)'=\frac{\partial \tilde{L}}{\partial u^{(m-1)}}-p_m'=\frac{\partial \tilde{L}}{\partial u^{(m-1)}}-2au^{(m+1)},$$

whence we have

$$u^{(m+1)} = \frac{1}{2a}\left(\frac{\partial \tilde{L}}{\partial u^{(m-1)}} - p_{m-1}\right).$$

Since $\tilde{L}$ involves only those $u^{(\alpha)}$ with $\alpha < m$, this yields (on replacing $u, u', \ldots, u^{(m-1)}$ by $q_1, \ldots, q_m$ in $\partial\tilde{L}/\partial u^{(m-1)}$) an expression for $u^{(m+1)}$ in terms of p and q. An easy iteration of this procedure completes the proof. □

30.1.4. Theorem (Ostrogradskiĭ). *For non-singular Lagrangians $L(u, u', \ldots, u^{(m)})$ the Euler–Lagrange equation is equivalent to the Hamiltonian system*

$$q'_\alpha = \frac{\partial H}{\partial p_\alpha}, \qquad p'_\alpha = -\frac{\partial H}{\partial q_\alpha}, \qquad \alpha = 1, \ldots, m, \tag{9}$$

where $H(p, q) = -L + u'p_1 + u''p_2 + \cdots + u^{(m)}p_m$.

PROOF. For ease of calculation we restrict ourselves to the case $m=2$; thus $L = L(u, u', u'')$. The Euler–Lagrange equation is then

$$\frac{\partial L}{\partial u} - \frac{d}{dx}\frac{\partial L}{\partial u'} + \frac{d^2}{dx^2}\frac{\partial L}{\partial u''} = 0, \tag{10}$$

and the canonical variables are given by

$$\begin{aligned} q_1 &= u, & q_2 &= u', \\ p_1 &= \frac{\partial L}{\partial u'} - \frac{d}{dx}\frac{\partial L}{\partial u''}, & p_2 &= \frac{\partial L}{\partial u''}. \end{aligned} \tag{11}$$

Since the last of these equations (namely $p_2 = \partial L/\partial u''$) expresses p_2 as a function of q_1, q_2 and u'', it follows from the non-singularity assumption that u'' can be expressed in the form

$$u'' = f(q_1, q_2, p_2).$$

Hence the Hamiltonian has the form

$$\begin{aligned} H = p_1u' + p_2u'' - L(u, u', u'') &= p_1q_2 + p_2u'' - L(q_1, q_2, u'') \\ &= p_1q_2 + \Phi(q_1, q_2, p_2). \end{aligned} \tag{12}$$

From this and (11) it follows that

$$q'_1 = u' = q_2 = \frac{\partial H}{\partial p_1}, \qquad q'_2 = u'' = \frac{\partial H}{\partial p_2},$$

$$p'_2 = \frac{d}{dx}\left(\frac{\partial L}{\partial u''}\right) = -p_1 + \frac{\partial L}{\partial u'} = -\frac{\partial H}{\partial q_2}.$$

Finally from (10) and (11) we have $dp_1/dx = \partial L/\partial u$, whence $p'_1 = -\partial H/\partial q_1$. With this we have completed the deduction from (11) (and (12)) of the

Hamiltonian system (9) on the phase space with co-ordinates q_1, q_2, p_1, p_2. We leave the inference in the reverse direction to the reader. □

30.2. Examples

Consider the "Sturm–Liouville operator" $\mathscr{L} = -d^2/dx^2 + u(x)$, where the function $u(x)$ (the "potential") is smooth. The following two situations are of particular interest (see below):

(i) $u(x) \to 0$ as $|x| \to \infty$. (If the stronger condition $\int_{-\infty}^{\infty} |u(x)|(1+|x|)\,dx < \infty$ obtains, the potential is said to approach zero *rapidly*.)
(ii) $u(x+T) = u(x)$, the case of a *periodic potential*.

Consider, for the moment purely formally, the ordinary differential equation

$$\mathscr{L}\psi = \lambda\psi, \qquad \psi = \psi(x, \lambda), \tag{13}$$

where λ is a parameter (the "spectral" parameter). On making the substitution $\chi(x, \lambda) = i(d\ln\psi/dx)$ in (13), that equation is transformed into the "Riccati equation"

$$i\chi' + \chi^2 = \lambda - u. \tag{14}$$

Assuming λ large positive and writing $\lambda = k^2$, we seek a solution of the Riccati equation in the form of a (formal) power series in $\sqrt{\lambda} = k$:

$$\chi(x, k) = k + \sum_{n \geq 1} \frac{\chi_n(x)}{(2k)^n}. \tag{15}$$

To determine the functions $\chi_n(x)$ we substitute (15) into the Riccati equation, obtaining

$$\sum_{n \geq 1} \frac{i\chi_n'(x)}{(2k)^n} + k^2 + 2k \sum_{n \geq 1} \frac{\chi_n(x)}{(2k)^n} + \left(\sum_{n \geq 1} \frac{\chi_n}{(2k)^n} \right)^2 = k^2 - u,$$

whence, equating like powers of $2k$ on both sides, we are led to

$$\chi_1 = -u; \qquad i\chi_n' + \chi_{n+1} + \sum_{i=1}^{n-1} \chi_i \chi_{n-i} = 0, \qquad n \geq 1. \tag{16}$$

From this the following facts emerge concerning the χ_n:

(i) Each χ_n is a polynomial (with constant coefficients) in $u, u', \ldots, u^{(n-1)}$.
(ii) Each of the χ_{2q+1} is real, while each of the χ_{2q} is purely imaginary and an (exact) derivative (of a polynomial in $u, u', \ldots$). (This reflects the fact, following directly from Riccati's equation, that

$$\chi_{\text{Im}} = -\tfrac{1}{2}(\ln \chi_{\text{Re}})', \tag{17}$$

where $\chi = \chi_{\text{Re}} + i\chi_{\text{Im}}$.)

(iii) The first few odd-indexed χ_n calculate out (using (16)) as follows:

$$\chi_1 = -u, \qquad \chi_3 = u'' - u^2,$$
$$\chi_5 = -u^{(iv)} + 5(u')^2 + 6uu'' - 2u^3. \tag{18}$$

For each $q \geq -1$ we now cast χ_{2q+3} in the role of a Lagrangian L_q:

$$L_q(u, u', u'', \ldots) = \chi_{2q+3}. \tag{19}$$

It can be shown that all of these Lagrangians are non-singular (in the sense of Definition 30.1.2). Moreover for $q \geq 0$ they have the following remarkable property: for each q (≥ 0) there exists a differential operator A_q of order $2q+1$ having polynomials in $u, u', u'', \ldots$ as coefficients, with the property that the commutator $[\mathscr{L}, A_q] = \mathscr{L}A_q - A_q\mathscr{L}$ is the operator which simply multiplies by the scalar-valued function $f_q(u, u', u'', \ldots)$ given by

$$f_q(u, u', u'', \ldots) = \frac{d}{dx}\frac{\delta S_q}{\delta u(x)}, \qquad S_q = \int L_q\, dx; \tag{20}$$

thus we may write $[\mathscr{L}, A_q] = f_q$ with f_q given by (20). We shall not establish this in general here; in the cases $q = 0, 1$ it can be verified by direct calculation: from

$$[\mathscr{L}, A_0] = \mathscr{L}A_0 - A_0\mathscr{L} = f_0 = \frac{d}{dx}\frac{\delta S_0}{\delta u(x)}, \qquad S_0 = \int (u'' - u^2)\, dx,$$

we obtain

$$f_0 = -2u', \quad \text{whence} \quad A_0 = -2\frac{d}{dx};$$

similarly from

$$[\mathscr{L}, A_1] = \mathscr{L}A_1 - A_1\mathscr{L} = f_1 = \frac{d}{dx}\frac{\delta S_1}{\delta u(x)},$$
$$S_1 = \int (-u^{(iv)} + 5(u')^2 + 6uu'' - 2u^3)\, dx,$$

we obtain

$$f_1 = 2u''' - 12u'u, \quad \text{and thence} \quad A_1 = -8\frac{d^3}{dx^3} + 6\left(u\frac{d}{dx} + \frac{d}{dx}u\right).$$

(Verify that indeed $[\mathscr{L}, A_0] = f_0$ and $[\mathscr{L}, A_1] = f_1$.)

Exercise

Calculate χ_7 and thence verify that to within a constant factor

$$A_2 = 16\frac{d^5}{dx^5} - 20\left(u\frac{d^3}{dx^3} + \frac{d^3}{dx^3}u\right) + 30u\frac{d}{dx}u + 5\left(u''\frac{d}{dx} + \frac{d}{dx}u''\right).$$

In view of the general result (that for each $q \geq 0$ there is an operator A_q such that $[\mathscr{L}, A_q] = f_q$), the following "commutativity equation"

$$[\mathscr{L}, A_q + c_1 A_{q-1} + \cdots + c_q A_0] = 0 \tag{21}$$

reduces to an ordinary differential equation in the potential u. (The reason for considering such commutativity equations will emerge in the following subsections. We remark that already in the 1920's there had been brought to light a curious property of pairs of commuting operators, namely that they are related by an algebraic equation $R(\mathscr{L}, A) = 0$, determining in turn a Riemann surface.) From the form of the functions f_q given by (20) it is immediate that the commutativity equation (21) has the equivalent "Lagrangian" form

$$\frac{d}{dx}\frac{\delta S}{\delta u(x)} = 0, \quad \text{where} \quad S = S_q + c_1 S_{q-1} + \cdots + c_q S_0, \qquad S_j = \int \chi_{2j+3}\, dx. \tag{22}$$

It follows that to within an additive constant the commutativity equation is equivalent to the Euler–Lagrange equation

$$\frac{\delta \hat{S}}{\delta u(x)} = 0, \quad \text{where} \quad \hat{S} = S_q + c_1 S_{q-1} + \cdots + c_q S_0 + c_{q+1} S_{-1}, \tag{23}$$

(and where $S_{-1} = -\int u\, dx$). We investigate this equation in the cases $q = 0, 1$. It is immediate from (18) that the corresponding Lagrangians are given respectively by

$$(q=0) \quad L = L_0 + c_1 L_{-1} = u'' - u^2 - c_1 u,$$

$$(q=1) \quad L = L_1 + c_1 L_0 + c_2 L_{-1} = -u^{(\mathrm{iv})} + 5(u')^2 + 6uu'' - 2u^3 + c_1(u'' - u^2) - c_2 u,$$

so that the respective commutativity equations (in the form (23)) become

$$(q=0) \quad -c_1 - 2u = 0, \quad \text{or} \quad u = -c_1/2,$$

$$(q=1) \quad u'' = 3u^2 + c_1 u + c_2/2.$$

In the latter case, writing $v = u'$ we obtain

$$v\frac{dv}{dx} = \frac{d}{dx}\left(u^3 + \frac{c_1}{2}u^2 + \frac{c^2}{2}u\right),$$

whence $v^2/2 = u^3 + (c_1/2)u^2 + (c_2/2)u + d/2$, and then

$$x - x_0 = \int \frac{du}{\sqrt{2u^3 + c_1 u^2 + c_2 u + d}}, \tag{24}$$

which (to within a constant factor and an additive constant) defines u as a Weierstrass elliptic function $\wp(x)$. (Note that here by means of a change of variable of the form $u \to u + \text{const.}$, we may without loss of generality suppose $c_1 = 0$.)

In the next case, $q = 2$, it turns out to be convenient to recast the equation (23) in Hamiltonian form (in accordance with Theorem 30.1.3). To integrate the resulting system fully an additional integral (i.e. integral of motion) apart from the Hamiltonian H, is needed; it happens that the equation (22) possesses a "latent symmetry" which does permit the system to be fully integrated in this case also. We pursue this in more detail in the following subsection.

30.3. Integration of the Commutativity Equations. The Connexion with the Kovalevskaja Problem. Finite-zoned Periodic Potentials

We continue our investigation of the commutativity equation (22). Thus suppose $u = u(x)$ to be such that the differential operator

$$A = A_q + \sum_{i=1}^{q} c_i A_{q-i} = \text{const.} \times \frac{d^{2q+1}}{dx^{2q+1}} + \sum_{i=1}^{q} Q_i(u, u', \ldots) \frac{d^{2q+1-i}}{dx^{2q+1-i}}, \tag{25}$$

of order $2q+1$, commutes with the Sturm–Liouville operator $\mathscr{L} = -d^2/dx^2 + u(x)$. Let $c = c(x, x_0, \lambda)$, $s = s(x, x_0, \lambda)$ form a basis for the space of solutions of the Sturm–Liouville equation (13), satisfying the initial conditions

$$c = 1, \quad c' = 0, \quad s = 0, \quad s' = 1, \quad \text{when} \quad x = x_0. \tag{26}$$

Since A commutes with $\mathscr{L}$, it acts as a linear transformation on this solution whose matrix $\Lambda = (\alpha_{ij})$ relative to the chosen basis is given by

$$\begin{aligned} Ac &= \alpha_{11} c + \alpha_{12} s, \\ As &= \alpha_{21} c + \alpha_{22} s, \end{aligned} \tag{27}$$

where the entries α_{ij} depend only on x_0 and λ. By (26) $c|_{x_0} = 1$, $c'|_{x_0} = 0$, and since $c'' = (u(x) - \lambda)c$, we have further $c''|_{x_0} = -\lambda + u(x_0)$, $c'''|_{x_0} = (u'c + (u - \lambda)c')|_{x_0} = u'(x_0)$, and so on; thus for all n the dependence of $c^{(n)}|_{x=x_0}$ (and similarly $s^{(n)}|_{x=x_0}$) on λ is polynomial. It follows from this (and the definition of A in (25)) that Ac, As and their derivatives, evaluated at $x = x_0$, depend polynomially on λ, and therefore, in view of

$$\begin{aligned} (Ac)_{x=x_0} &= \alpha_{11}, & (As)_{x=x_0} &= \alpha_{21}, \\ \left(\frac{d}{dx} Ac\right)_{x=x_0} &= \alpha_{12}, & \left(\frac{d}{dx} As\right)_{x=x_0} &= \alpha_{22}, \end{aligned} \tag{28}$$

that the dependence of the α_{ij} in (27) on λ is likewise polynomial. This and the fact that the coefficients Q_i appearing in the operator A (see (25)) are polynomials in $u, u', u'', \ldots,$ together imply that the entries α_{ij} in the matrix Λ are polynomials in $\lambda, u(x_0), u'(x_0), u''(x_0), \ldots$. Explicit calculation of the matrix Λ_q of the operator A_q in the cases $q=0, 1$, yields respectively (using (28) together with the results of our calculations of A_0 and A_1 in the preceding subsection)

$$\Lambda_0=\begin{pmatrix} 0 & -2 \\ 2(\lambda-u(x_0)) & 0 \end{pmatrix},$$
$$\Lambda_1=\begin{pmatrix} -2u'(x_0) & -2u''(x_0)+4u^2(x_0)+4\lambda u(x_0)-8\lambda^2 \\ 4u(x_0)+8\lambda & 2u'(x_0) \end{pmatrix}. \tag{29}$$

(The calculation of the matrix Λ_2 of the operator A_2 $(q=2)$ is similarly straightforward, though more cumbersome.)

Returning to the general situation (with A as in (25)), we note first that the trace $\operatorname{tr}\Lambda$ is always zero, so that the characteristic polynomial of Λ has the form

$$\det(w\cdot 1-\Lambda(\lambda))=w^2+\det\Lambda=w^2+P_{2q+1}(\lambda),$$

where, as it turns out, P_{2q+1} is a polynomial in λ of degree $2q+1$ of the form

$$P_{2q+1}(\lambda)=\text{const.}\times\lambda^{2q+1}+\sum_{i=1}^{2q+1}\varphi_i(u(x_0), u'(x_0), u''(x_0), \ldots)\lambda^{2q+1-i}. \tag{30}$$

Since the characteristic polynomial is of course independent of the choice of basis for the solution space of the Sturm–Liouville equation $\mathscr{L}\psi=\lambda\psi$, it follows that the coefficients $\varphi_i(u(x_0), u'(x_0), \ldots)$ in (30) of $P_{2q+1}(\lambda)=\det\Lambda$, are independent of the point x_0, i.e. are unaffected by replacement in them of $u(x_0)$ by $u(x)$ (where $u(x)$ is some solution of the commutativity equation (22)); thus these coefficients furnish integrals of the commutativity equation.

In the case $q=1$, we have $\Lambda=\Lambda_1+c_1\Lambda_0$. As noted in the preceding subsection we may without loss of generality suppose $c_1=0$. This assumed, it follows from the expression for Λ_1 in (29) that

$$\begin{aligned} P_3(\lambda)=\det\Lambda=\det\Lambda_1&=64\lambda^3-(48u^3-16u'')\lambda \\ &\quad-(4(u')^2+16u^3-8uu''), \end{aligned}$$

yielding the two integrals

$$3u^2-u''=\text{const.}, \qquad (u')^2+4u^3-2uu''=\text{const.}$$

The first of these (in the form $u''=3u^2+c_1u+c_2/2$) was exploited in the preceding section to obtain an explicit solution of the commutativity equation. Putting $u''=3u^2+c_2/2$ in the second integral, we obtain

$(u')^2-2u^3-c_2u$, which is just the negative of the Hamiltonian

$$H=-(u')^2+2u^3+c_1u^2+c_2u=-L+u'p_1+u''p_2+u'''p_3+u^{(\mathrm{iv})}p_4 \quad (31)$$

(with $c_1=0$).

In the case $q=2$, we have $\Lambda=\Lambda_2+c_1\Lambda_1+c_2\Lambda_0$. Again we may without loss of generality arrange that $c_1=0$ by means of a transformation of the form $u \to u+\text{const.}$ (verify this!) It can be shown that in this case the coefficients of

$$P_5(\lambda)=\det\Lambda=\det(\Lambda_2+c_2\Lambda_0)$$

yield, in addition to the Hamiltonian $H=J_1$, a further non-trivial integral J_2. If we introduce variables γ_1, γ_2 defined by

$$\begin{aligned} u&=-2(\gamma_1+\gamma_2),\\ \tfrac{1}{8}(3u^2-u'')&=\gamma_1\gamma_2-\tfrac{1}{2}\sigma, \end{aligned} \quad (32)$$

where σ is the coefficient of λ^3 in the polynomial $P_5(\lambda)$ made monic, then the following assertion can be verified by means of elementary calculation.

30.3.1. Proposition. *On the level surface defined by the equations* $H(p_j, q_j)=\text{const.}$, $J_2(p_j, q_j)=\text{const.}$, *the commutativity equation* (22) *with* $q=2$ *takes the form*

$$\gamma_1'=\frac{2i\sqrt{P_5(\gamma_1)}}{\gamma_1-\gamma_2}, \qquad \gamma_2'=\frac{2i\sqrt{P_5(\gamma_2)}}{\gamma_2-\gamma_1}, \quad (33)$$

where γ_1 *and* γ_2 *are as in* (32).

The equations (33) have the same form as the Kovalevskaja equations for a spinning top in a gravitational field (see [31]); the potential $u(x)$ can be found in terms of theta-functions, and then angular variables introduced via "Abel's transformation" (see [26, 31]). In the terminology of Riemann surfaces we can say the following concerning the above proposition: It follows from the equations (33) that each level surface defined by the two integrals $H(=J_1)$ and J_2 (i.e. by equations of the form $H=\text{const.}$, $J_2=\text{const.}$) is a "Jacobian variety" $J(\Gamma)$ of a Riemann surface Γ of genus 2 (see [26]), i.e. a 2-dimensional complex abelian toroidal variety T^4 defined by a 4-dimensional (real) lattice in $\mathbb{C}^2=\mathbb{R}^4$ (and so determined by four independent (real) vectors). (The analogous statement in the case $q=1$ would be that a level surface given by $H=\text{const.}$, corresponding to the single integral H, is a Riemann surface of genus 1, i.e. a torus T^2; this is in fact obviously true since in view of (32), the equation $H=\text{const.}$ has the form $q_2^2=R_3(q_1)$ where R_3 is a polynomial of degree 3 (see §4.2).)

Thus in the case $q=2$ also, the Hamiltonian system defined by (22) (or (23)) is fully integrable. It follows from Liouville's theorem (28.4.1) that the open region of the space of solutions $u(x)$ of the commutativity equation (22) (with

$q=2$), corresponding to an open neighbourhood in phase space (with co-ordinates q_1, q_2, p_1, p_2) of a closed, connected component of a level surface $H=\text{const.}$, $J_2=\text{const.}$ (which by that theorem will be a torus T^2), will consist of periodic and almost periodic functions obtained essentially by restricting certain functions of two variables (defined on the neighbouring toroidal level surfaces) to trajectories of the Hamiltonian system defined by (22).

We now turn our attention to the Sturm–Liouville equation itself (i.e. to the ordinary differential equation $\mathscr{L}\psi=\lambda\psi$ where as usual $\mathscr{L}=-d^2/dx^2+u(x)$) in the case of periodic potential $u(x)$ (with period T). (Note that in this case it is often called instead the "Hill equation".) It is convenient for this purpose to introduce the operator $\hat{T}$ which "translates by the period" the solutions of $\mathscr{L}\psi=\lambda\psi$, i.e. is defined by $\hat{T}\psi(x)=\psi(x+T)$. It turns out to be appropriate to look (for each λ) for a pair of linearly independent solutions $\psi_\pm$ satisfying

$$\hat{T}\psi_\pm(x)=\psi_\pm(x+T)=\exp\{\pm ipT\}\psi_\pm(x), \tag{34}$$

and the further norm requirement that $\psi_\pm=1$ for $x=x_0$; for each λ such a pair of functions will then clearly constitute a pair of eigenfunctions (called *Floquet* (or *Bloch*) *functions*) for the translation operator $\hat{T}$.

EXERCISES

1. Deduce from the existence of such a pair of eigenfunctions that the operator $\hat{T}$ has determinant 1.
2. Show that for real λ the "quasi-momentum" $p=p(\lambda)$ (see (34)) is always either real or purely imaginary.

30.3.2. Definition. The set of (real) λ for which $p(\lambda)$ is real is called the *zone of stability* (or *permissible zone*) of the pair of eigenfunctions $\psi_\pm$. (Its complement, i.e. the set of λ for which $p(\lambda)$ is purely imaginary, is called the *zone of instability* (or *impermissible zone* or *zone of parametric resonance*).)

EXERCISE

Show that those λ for which $\psi_+(x,\lambda)=\psi_-(x,\lambda)$, are boundary points of the zone of stability.

For each pair of values of x, x_0 (considered as parameters) we may alternatively regard the pair $\psi_\pm(x, x_0, \lambda)$ as defining a single two-valued function of λ, with the boundary points of the zone of stability as branch points; in this way we are led to a Riemann surface Γ on which the pair $\psi_\pm(x, x_0, \lambda)$ is defined as a single-valued Floquet function of λ.

30.3.3. Definition. The potential $u(x)$ is said to be *finite-zoned* if the Riemann surface Γ determined by a Floquet function $\psi_\pm(x, x_0, \lambda)$ has finite genus. (We remark that the genus of Γ coincides with the number of finite "lacunae", i.e. intervals of finite length contained in the zone of instability with end-points which are boundary points of the zone of stability.)

We state without proof the following fact: The *finite-zoned potentials* $u(x)$ *are precisely the periodic solutions of the commutativity equations* (22). (The Riemann surface Γ corresponding in the above-defined manner to such a potential, is then given by $w^2 = -\det \Lambda = -P_{2q+1}(\lambda)$, in the notation of the preceding subsection (see in particular (30)), and for each x, x_0 the boundary points of the zone of stability are given by $\det \Lambda = 0$.)

This is a consequence of the fact (whose proof we also omit) that a pair of Floquet eigenfunctions $\psi_{\pm}(x, x_0, \lambda)$, regarded, as above, as a single-valued function defined on the Riemann surface Γ, is an eigenvector also of the operator A (or corresponding matrix Λ) defined in (25). (Note that for each x, x_0, the eigenvalues and eigenvectors of the matrix $\Lambda(\lambda)$ can clearly be regarded as having as domain of definition the Riemann surface given by

$$0 = P(w, \lambda) = \det(w \cdot 1 - \Lambda(\lambda)) = w^2 + \det \Lambda(\lambda).)$$

30.4. The Korteweg–deVries Equation. Its Interpretation as an Infinite-dimensional Hamiltonian System

The investigation of finite-zoned potentials (in connexion with the corresponding Sturm–Liouville equation—see above) turns out to be useful as an aid to finding solutions periodic in x, of the *Korteweg–deVries* (KdV) *equation* for dispersive waves in a non-linear medium, well known in mathematical physics:

$$\frac{\partial u}{\partial t} = 6u\frac{\partial u}{\partial x} - \frac{\partial^3 u}{\partial x^3} = \frac{\partial}{\partial x}\frac{\delta S_1}{\delta u(x)}, \qquad u = u(x, t). \tag{35}$$

(Here S_1 is, to within a constant factor, as in (20), and $\delta S_1/\delta u(x)$ denotes the "partial" variational derivative with t held fixed.) In view of (20) et seqq., we can rewrite the KdV equation in operator form as

$$\dot{u} = \frac{\partial \mathscr{L}}{\partial t} = [A_1, \mathscr{L}] = 6u\frac{\partial u}{\partial x} - \frac{\partial^3 u}{\partial x^3} = \frac{\partial}{\partial x}\frac{\delta S_1}{\delta u(x)}, \tag{36}$$

where

$$A_1 = -4\frac{\partial^3}{\partial x^3} + 3\left(u\frac{\partial}{\partial x} + \frac{\partial}{\partial x}u\right).$$

Exercise

Prove that the quantities S_{-1}, S_0, S_1, S_2 (defined essentially as in §30.2) are integrals of the equation (36), i.e. that the KdV equation implies that

$$\frac{\partial}{\partial t}S_q = 0 \quad \text{for} \quad q = -1, 0, 1, 2. \tag{37}$$

It is not difficult to see that the KdV equation has solutions of the form

$u(x-ct)$ (representing "cnoidal waves"), which necessarily satisfy $[A_1+c A_0, \mathscr{L}]=0$, or, equivalently (cf. (24))

$$y-y_0=\int \frac{du}{\sqrt{2u^3+cu^2+c_2u+d}}, \qquad y=x-ct. \tag{38}$$

(To see this, note first that since

$$\frac{\partial u}{\partial t}=-c\frac{du}{dy}, \qquad \frac{\partial u}{\partial x}=\frac{du}{dy}, \qquad \frac{\partial^3 u}{\partial x^3}=\frac{d^3u}{dy^3},$$

the KdV equation becomes

$$-c\frac{du}{dy}-6u\frac{du}{dy}+\frac{d^3u}{dy^3}=0,$$

or

$$\frac{d^2u}{dy^2}=3u^2+cu+\frac{c_2}{2},$$

whence follows (38) exactly as in the derivation of (24).) Thus such a solution $u(x-ct)\,(=u(y))$ will, to within a constant factor, be of the form $\wp(y)+\text{const.}$, where $\wp(y)$ is a Weierstrass elliptic function.

Exercise

Show that the KdV equation has rapidly decreasing (see §30.2) solutions of the form $u(x-ct)$ (i.e. rapidly decreasing for each t as $|x|\to\infty$). (The corresponding waves are called "solitons".) Find the analytic form of these solutions, and in particular the relationship between the speed c and the amplitude $\max_x|u|$. Find also solutions of the form $u(x-ct)$ having poles at real values of x. Show that in particular there is a rational stationary solution of the form

$$u(x)=\frac{2}{(x-a)^2}, \qquad c=0, \quad a=\text{const.}$$

(All of these solutions represent degenerate cases of the above-noted solution in terms of a Weierstrass function $\wp(y)$.)

Solutions of the KdV equation of greater complexity can be investigated as follows: One considers the commutativity equation $[\mathscr{L}, A_2+c_1A_1+c_2A_0]=0$, in the form (cf. (23))

$$0=\frac{\delta S}{\delta u(x)}=\frac{\delta(S_2+c_1S_1+c_2S_0+c_3S_{-1})}{\delta u(x)}, \tag{39}$$

and takes the solutions $u(x)$ of this equation as initial data for the following Cauchy initial-value problem:

$$u(x,0)=u(x),$$

$$\frac{\partial u}{\partial t}=6u\frac{\partial u}{\partial x}-\frac{\partial^3 u}{\partial x^3};$$

for each $u(x)$ satisfying (39), the solution $u(x, t)$ of the latter problem can then be investigated using the fact that $u(x, t)$ will satisfy (39) not just for $t=0$ but for all t.

EXERCISES

1. Deduce this fact from the preservation of S_q for $q=-1, 0, 1, 2$ (see (37)).
2. Show that if $u(x)$ is a solution of the equation (22) with $q=1$, then $3u(x)$ is a solution for $q=2$. (Recall from §30.2 that (to within an additive constant) $u(x)$ will be a constant multiple of the appropriate Weierstrass function $\wp(x)$ or some degenerate case of it, as for example $u(x)=2/x^2$.) Show that more generally equation (22) with $q=2$ has a solution of the form $u(x-a)+u(x-b)+u(x-c)$ for appropriate a, b, c; find the conditions needed on a, b, c.

The KdV equation (35) constitutes (in a certain sense to be made precise) an "infinite-dimensional Hamiltonian system" on the space of functions $u(x)$. We define the *Poisson bracket of two functionals* J, I on that space by means of the formula (cf. §28.1 (5), and see §30.5 below)

$$\{J, I\}=\int \frac{\delta J}{\delta u(x)} \frac{d}{dx} \frac{\delta I}{\delta u(x)}\, dx. \tag{40}$$

EXERCISE

Prove that this Poisson bracket satisfies the Jacobi identity, and that the corresponding Hamiltonian system has the form (cf. §28.1 (4))

$$\dot{u}=\frac{d}{dx} \frac{\delta J}{\delta u(x)}, \tag{41}$$

where J is some functional of u.

It is now immediate that the KdV equation can be interpreted as a Hamiltonian system in this sense (see §30.5 below). It turns out that the properties of this Hamiltonian system very much depend on the space of functions u on which the system is defined; the classes of functions of most interest are (as before—see §30.2) those of rapidly decreasing functions, and of periodic functions of fixed period T.

EXERCISES

1. Find the simplest "canonical" variables p_α, q_α, $\alpha=1, 2, \ldots$, among the elementary trigonometric functions, in terms of which the Hamiltonian system (41), defined on the space of periodic functions with period T, takes the form

$$\dot{p}_\alpha=-\frac{\partial H}{\partial q_\alpha}, \qquad \dot{q}_\alpha=\frac{\partial H}{\partial p_\alpha}, \qquad \alpha=0, 1, 2, \ldots; \quad H=J. \tag{42}$$

2. Show that the KdV equation is equivalent to the matrix equation

$$\frac{\partial \Lambda_1}{\partial x}-\frac{\partial \Lambda_0}{\partial t}=[\Lambda_0, \Lambda_1], \tag{43}$$

where the matrices Λ_0, Λ_1 are as in (29) (with x_0 replaced by x).

3. Show that the commutativity equation (22) with $q=2$, is equivalent to the matrix equation

$$\frac{d\Lambda}{dx}=[\Lambda_0, \Lambda], \tag{44}$$

where $\Lambda=\Lambda_2+c_1\Lambda_1+c_2\Lambda_0$, the matrices $\Lambda_0, \Lambda_1, \Lambda_2$ being as in (29) et seqq., except that in them x_0 is to be replaced by x. (It follows in particular that, as noted in §30.3, the coefficients of the polynomial $\det \Lambda=P_5(\lambda)$, are integrals of the commutativity equation (22) with $q=2$.)

4. Show that for the solutions $u(x)$ of the commutativity equation (22) with $q=2$ (or equivalently (44)), the dependence on the time t, as given by the KdV equation, is determined equivalently by the equation

$$\frac{d\Lambda}{dt}=[\Lambda_1, \Lambda]. \tag{45}$$

30.5. Hamiltonian Formalism of Field Systems

In the absence of thermodynamical non-reversibility, a physical system from which there is no energy-loss is said to be "conservative". The modern view is that among the systems of physical origin, the conservative ones should be Hamiltonian. (However as examples show the resulting Hamiltonian formalism will not always be "trivial", in the sense of reducing to Lagrangian formalism.) In this subsection we present a few results concerning the Hamiltonian formalism of "field systems". The basic concept here, as in the finite-dimensional case, is the important "Poisson bracket" operation. The underlying space will here be a function space consisting of C^∞-functions $u^j(x^1, \ldots, x^n)$ of n variables (cf. end of the preceding subsection). (We shall not specify precisely the domain of these functions, nor the boundary conditions on them; since all variations will be finite and we do not wish to become involved in a discussion of the appropriate boundary conditions, we shall instead make the conventional assumption that the integral of the total derivative of our functions over the whole of the space co-ordinatized by $x^1, \ldots, x^n$, is always zero.)

The Poisson bracket operation is defined between functionals of m "function-fields" $u^1, \ldots, u^m$ (i.e. C^∞-functions). We shall adopt the convenient formalism of the theoretical physicists wherein the result of the Poisson bracket operation is expressed in terms of "point-functionals", i.e. via the individual fields considered at each point x. (Thus for each point x and each u^i we have the functional whose value on (u^j) is the number $u^i(x)$.) Using this notation we now formally define a *Poisson bracket* $\{\ ,\ \}$ by means of a formula of the form

$$\{u^i(x), u^j(y)\}=F^{ij}(x, y) \tag{46}$$

(where the n-tuples x, y are regarded as independent "continuous variables") with the following properties: the Poisson bracket should be skew-symmetric, linear in each argument, and should satisfy Leibniz' condition ($\{fg, h\} = f\{g, h\} + g\{f, h\}$) and the Jacobi identity (cf. Part I, Theorem 34.2.2, and §28.1 above).

Consider an arbitrary functional of the form

$$J = \int P(u, \nabla u, \ldots)\, d^n y;$$

we shall now derive a formula for the Poisson bracket $\{u^i(x), J\}$. Note first that in view of the linearity condition we have

$$\left\{u^i(x), \int P\, d^n y\right\} = \int \{u^i(x), P(y)\}\, d^n y$$

(where we have written $P(y)$ for $P(u(y), \nabla u(y), \ldots)$, and

$$\left\{u^i(x), \frac{\partial}{\partial y^k} v(y)\right\} = \frac{\partial}{\partial y^k}\{u^i(x), v(y)\}.$$

In the following calculations we shall assume, purely for the sake of notational simplicity, that u is a single field (rather than an m-tuple of fields) and that $n=1$, i.e. that the space co-ordinatized by $x^1, \ldots, x^n$ is one-dimensional, co-ordinatized by $x^1 = x$. One can show (or at least infer from known facts) that the linearity condition and the Leibniz condition together imply that for some r,

$$\{u(x), P(y)\} = \{u(x), u(y)\} \frac{\partial P}{\partial u}(y) + \{u(x), u'(y)\} \frac{\partial P}{\partial u'}(y) + \cdots$$
$$+ \{u(x), u^{(r)}(y)\} \frac{\partial P}{\partial u^{(r)}}(y).$$

Thus

$$\{u(x), P(y)\} = \sum_{s \geq 0} \frac{\partial P}{\partial u^{(s)}}(y) \frac{\partial^s}{\partial y^s}\{u(x), u(y)\},$$

whence, using $(fg)' = f'g + fg'$, and our underlying assumption about integrals of total derivatives (see above), we obtain

$$\left\{u(x), \int P\, dy\right\} = \int \left\{\sum_{s \geq 0} \{u(x), u(y)\}(-1)^s \frac{\partial^s}{\partial y^s}\left(\frac{\partial P}{\partial u^{(s)}}\right)\right\} dy$$
$$= \int \{u(x), u(y)\} \frac{\delta J}{\delta u}\, dy,$$

where $\delta J/\delta u$ is (the appropriate generalization of) the variational derivative (cf., e.g. §29.2(4)). In the general case of m function-fields u^i and n variables x^j, a completely analogous argument yields

$$\{u^i(x), J\} = \int \{u^i(x), u^s(y)\} \frac{\delta J}{\delta u^s(y)} d^n y. \tag{47}$$

A further argument along similar lines yields the following formula for the Poisson bracket of a pair of functionals J_1, J_2:

$$\{J_1, J_2\} = \int \frac{\delta J_1}{\delta u^r(x)} \frac{\delta J_2}{\partial u^s(y)} \{u^r(x), u^s(y)\} d^n x\, d^n y. \tag{48}$$

(Note that, in the formalism of the calculus of variations, we have as usual

$$\delta J = \int \frac{\delta J}{\delta u^j(x)} \delta u^j(x)\, d^n x,$$

with the variations δu^j all restricted to being finite.)

30.5.1. Definition. A Poisson bracket as defined by (46) et seqq. is said to be *local* if each $\{u^i(x), u^j(y)\}$ can be expressed as a finite sum of the form

$$\{u^i(x), u^j(y)\} = F^{ij}(x, y) = \sum_k B_k^{ij}\, \partial_x^k\, \delta(x-y), \tag{49}$$

where the "local principle" prevails, i.e. the families B_k of coefficients depend on the values of the fields u and at most finitely many of their derivatives only at the point x; here $k=(k_1, \ldots, k_n)$, $k_j \geq 0$;

$$\partial_x^k = \partial_{1_x}^{k_1} \circ \cdots \circ \partial_{n_x}^{k_n}, \qquad \partial_{j_x} = \frac{\partial}{\partial x^j}; \quad \text{and} \quad \delta(x-y) = \prod_{i=1}^n \delta(x^i - y^i)$$

denotes as usual the Dirac δ-function, i.e. the kernel of the identity operator:

$$\int f(x)\delta(x-y)\, d^n x = f(y).$$

(In the present context we shall operate with this symbolism purely formally, i.e. algebraically, avoiding any discussion of function spaces.)

Assuming we have a local Poisson bracket given by (49), we shall now derive an expression for $\{J_1, J_2\}$ in terms of the differential operator

$$A = (A_x^{ij}) = \sum_k B_k^{ij}(u(x))\partial_x^k.$$

Beginning with (47), we have

$$\begin{aligned}
\{u^i(x), J\} &= \int \{u^i(x), u^j(y)\} \frac{\delta J}{\delta u^j(y)} d^n y \\
&= \int A_x^{ij} \delta(x-y) \frac{\delta J}{\delta u^j(y)} d^n y \\
&= A_x^{ij} \int \delta(x-y) \frac{\delta J}{\delta u^j(y)} d^n y \\
&= A_x^{ij} \frac{\delta J}{\delta u^j(x)},
\end{aligned}$$

where we have used the fact that the differential operator A_x^{ij} (regarded as operating on $\delta(x-y)\,(\delta J/\delta u^j(y))$) depends only on x and so can be taken outside the integral sign. Thus

$$\{u^i(x), J\} = A_x^{ij} \frac{\delta J}{\delta u^j(x)}, \tag{50}$$

whence we have immediately from (48) that

$$\{J_1, J_2\} = \int \frac{\delta J_1}{\delta u^i(x)} A_x^{ij} \frac{\delta J_2}{\delta u^j(x)}\, d^n x. \tag{51}$$

The verification of the Jacobi identity for an arbitrary Poisson bracket of the form (51) (determined by a general matrix differential operator $A = (A^{ij})$) is difficult. However the verification for the particular case where the coefficients of A are "constant" (in the sense of being independent of the fields u and their derivatives, though possibly depending (explicitly) on x) is easy; this case represents the direct analogue of the finite-dimensional case of a Poisson bracket with constant coefficients (see §28.1 above, or Part I, §34.2).

We now consider two of the simplest examples (with constant coefficients).

(a) *Lagrange brackets.* Suppose we are given fields $p_1, \ldots, p_n, q^1, \ldots, q^n$, with Poisson bracket defined by (cf. §28.4)

$$\begin{aligned} \{p_i(x), p_j(y)\} &= \{q^i(x), q^j(y)\} = 0, \\ \{q^j(x), p_i(y)\} &= \delta_i^j \delta(x-y). \end{aligned} \tag{52}$$

Such a bracket arises from non-singular functionals (cf. Definition 30.1.2) of the form

$$S(q) = \int \Lambda\left(q, \frac{dq}{dt}, \left(\frac{\partial q}{\partial x^\alpha}\right)\right) d^n x\, dt,$$

with

$$p_j(x) = \frac{\partial \Lambda}{\partial \dot{q}^j}, \qquad \left(\dot{q}^j = \frac{dq^j}{dt}\right).$$

(Analogously to §30.1 (see in particular (6), (7) et seqq.) this and what follows may readily be extended to Lagrangians involving higher derivatives than the first.) If we take as Hamiltonian the usual one

$$H = P_0 = \int T_0^0\, d^n x = \int (p_j \dot{q}^j - \Lambda)\, d^n x, \tag{53}$$

then (cf., e.g. Part I, §34.2) the Euler–Lagrange equations take on the explicit Hamiltonian form (verify!)

$$\begin{aligned} \dot{p}_j(x) &= \{p_j(x), H\} = -\frac{\delta H}{\delta q^j(x)}, \\ \dot{q}^j(x) &= \{q^j(x), H\} = \frac{\delta H}{\delta p_j(x)}. \end{aligned} \tag{54}$$

(Note that if instead we take as Hamiltonian any other component of the total momentum, namely (cf. Part I, Definition 37.2.2)

$$P_\alpha = \frac{1}{c}\int T^0_\alpha \, d^n x = \frac{1}{c}\int p_j \frac{\partial q^j}{\partial x^\alpha} \, d^n x, \tag{55}$$

then we obtain (in the analogous way) merely a field generating (via the Poisson bracket) the action of the group of translations of the fields in the direction of the x^α-axis. Verify this!)

(b) *The Korteweg–deVries* (KdV) *equation.* (Cf. §30.4.) Here we have but a single field $u(x)$ defined on one-dimensional space (and also dependent on the time t), with Poisson bracket given by

$$\{u(x), u(y)\} = \delta'(x-y) \tag{56}$$

(where δ' is the "derivative" of the δ-function, and so has the property

$$\int \delta'(x) f(x)\, dx = -\int \delta(x) f'(x)\, dx = -f'(0),$$

for every function f.) This bracket has non-trivial annihilator, i.e. there exist non-zero functionals I such that

$$\{u(x), I\} = 0;$$

in fact $I_0 = \int u(y)\, dy$ has this property, as is easily verified.

The momentum P is defined in this context by

$$P = \int \frac{u^2}{2}\, dx, \qquad \left(\text{whence } \{u(x), P\} = \frac{\partial u}{\partial x}\right); \tag{57}$$

thus, analogously to the preceding example, the momentum functional P is a generator (via the Poisson bracket) of the action of the group of translations of the x-axis.

A Hamiltonian of the form

$$H = -\int \left(\frac{u_x^2}{2} + u^3 + \frac{cu^2}{2}\right) dx, \qquad c = \text{const.}, \tag{58}$$

gives rise to the KdV equation (as the corresponding Hamiltonian system; cf. Part I, §34.2(22) et seqq.):

$$\dot{u}(x) = \{u(x), H\} = -\frac{\partial}{\partial x}\frac{\delta H}{\delta u(x)} = u_{xxx} + 6uu_x + cu_x. \tag{59}$$

Exercise

Show that each of the entities L_q defined in §30.2(19) yields a conservation law for the KdV equation; i.e. (to be explicit) that the KdV equation implies that the functionals $I_q = \int L_q\, dx$, satisfy

$$\dot{I}_q = 0.$$

Show also that $\{I_q, I_p\} = 0$, and $H = I_1 + cI_0$.

Returning to the general discussion, we note the following fundamental property of the Poisson bracket, non-trivial in the sense that (in the finite case as in the present more general context) it depends crucially on the Jacobi identity (as well as the other, simpler, defining conditions of a Poisson bracket). Let J_1, J_2 be any two functionals, and form from them the third $J_3=\{J_1, J_2\}$. These three functionals determine three "Hamiltonian" vector fields (or "currents") on the function space, via the formula

$$\frac{\partial u^j(x)}{\partial t_\alpha}=\{u_j(x), J_\alpha\}, \qquad \alpha=1, 2, 3;$$

it follows readily from the Jacobi identity for the Poisson bracket that the vector field commutator of the first two of these fields is equal to the third. (To see this write ∂_α for the operator $[\ , J_\alpha]$, and calculate $[\partial_1, \partial_2](u^j(x)) =\partial_1(\partial_2 u^j(x))-\partial_2(\partial_1 u^j(x))$.) Hence in particular if $\{J_1, J_2\}=0$, then the corresponding currents commute.

Note also that the *annihilator* of a Poisson bracket is defined to be the totality of functionals whose Poisson bracket with every functional, is zero. A Poisson bracket is *non-degenerate* if its annihilator is trivial (i.e. zero).

Interesting examples of Poisson brackets occur in hydrodynamics (fluid mechanics); we shall now consider some of these. To begin with we consider a Lie algebra L_n of vector fields on $\mathbb{R}^n$ and its subalgebra $L_n^0 \subset L_n$ consisting of those fields $w=w^i e_i$ in L_n with identically vanishing divergence: $\partial_i w^i \equiv 0$. By definition of the commutator of a pair of vector fields (see Part I, Definition 23.2.3), we have

$$[v, w]^\alpha=v^i\partial_i w^\alpha - w^i \partial_i v^\alpha,$$

where

$$v=v^i e_i, \qquad w=w^i e_i, \qquad [e_i, e_j]=0.$$

If in terms of the field components $v^i(x)$, $w^j(y)$ of an arbitrary pair of vector fields v, w, we define "structural constants" $c_{ij}^k(x, y, z)$ by

$$[v, w](z)=\iint c_{ij}^k(x, y, z)v^i(x)w^j(y)e_k\, d^n x\, d^n y, \tag{60}$$

then it is readily verified that

$$c_{ij}^k(x, y, z)=\delta_j^k\delta(x-z)\delta_i(x-y)-\delta_i^k\delta(y-z)\delta_j(y-x), \tag{61}$$

where $\delta_i(x)=(\partial/\partial x^i)\delta(x)$. Denoting by $p_j(x)$ the components of an element $p(x)$ of the dual space L_n^*, we have that the scalar $p_j(x)v^j(x)$ (i.e. real-valued function of the points of $\mathbb{R}^n$) is invariant under smooth changes of the coordinates x (see Part I, Example 17.2.2(a)). It follows that the "scalar product"

$$\int p_i(x)v^i(x)\, d^n x \tag{62}$$

will, from the tensorial point of view, be invariant under such co-ordinate changes, provided we regard each $p \in L_n^*$ as the "density" of a covector (rather than just a covector) which under a co-ordinate change $x = x(x')$ is multiplied by the Jacobian of the change. We shall in fact use the term *momentum density* for such p. Note that for each p, the scalar product (62) defines a functional on L_n (the totality of which functionals we also denote by L_n^*); the corresponding dual L_n^{0*} will then be a quotient of

$$L_n^*/((p_j = \partial_j \varphi)), \tag{63}$$

since for densities of the form $(p_j = \partial_j \varphi)$ we have

$$\int (\partial_j \varphi) v^j \, d^n x = -\int \varphi(\partial_j v^j) \, d^n(x) = 0$$

if $\partial_j v^j \equiv 0$, i.e. if $v \in L_n^0$.

The Poisson bracket of momentum densities is taken to be

$$\begin{aligned} \{p_i(x), p_j(y)\} &= \int c_{ij}^k (x, y, z) p_k(z) \, d^n z \\ &= p_j(x)\delta_i(x-y) - p_i(y)\delta_j(y-x). \end{aligned} \tag{64}$$

The hydrodynamical equations for an incompressible fluid flow, in the case of constant density ρ, arise from the Hamiltonian

$$H = \int \frac{|p|^2}{2\rho} \, d^n x, \tag{65}$$

where $p_i = \rho v^i$ ($v = v(x)$ being the velocity of the fluid at each point, assumed to satisfy $\partial_i v^i \equiv 0$), and where the metric on the phase space L_n^{0*} is taken to be Euclidean (so that $v^i = v_i$). These equations are generally formulated on the whole of L_n^* in the form

$$\dot{p}_i = \rho \dot{v}^i = \{p_i, H\} + \partial_i P = \rho v^k \partial_k v^i + \partial_i \left(\frac{\rho |v|^2}{2} + P \right), \tag{66}$$

where $P = P(x)$ is the pressure (which may from one point of view be taken to be defined by (66)).

For a compressible fluid on the other hand it is necessary to enlarge the algebra L_n by means of "internal variables". In the language of fields this means adjoining two new field variables, the "mass density" ρ and the "entropy density" s, with Poisson brackets given by

$$\begin{aligned} \{p_i(x), \rho(y)\} = \rho(x)\delta_i(x-y), \qquad \{p_i(x), s(y)\} = s(x)\delta_i(x-y), \\ \{s(x), s(y)\} = \{\rho(x), \rho(y)\} = \{s(x), \rho(y)\} = 0. \end{aligned} \tag{67}$$

Here the Hamiltonian has the form (assuming as before the Euclidean metric)

$$H = \int \left[\frac{|p|^2}{2\rho} + \varepsilon_0(\rho, s) \right] d^n x \qquad (p_i = \rho v^i),$$

where ε_0 is the energy density.

EXERCISES

1. Show that

$$\{v_i(x), v_j(y)\} = \frac{1}{\rho}(\partial_i v_j - \partial_j v_i)\delta(x-y).$$

2. Assuming $n=2$, consider "Clebsch variables" ψ, α satisfying

$$p_i = \rho\partial_i\psi + s\partial_i\alpha.$$

Show that

$$\{\rho(x), \psi(y)\} = \{s(x), \alpha(y)\} = \delta(x-y),$$

while all the other possible brackets vanish. Examine the question of the global introduction of Clebsch variables.

3. In the case $n=3$ consider field variables $\psi, \alpha, \beta, \gamma$ satisfying

$$p_i = \rho\partial_i\psi + s\partial_i\alpha + \beta\partial_i\gamma,$$

with (as the only non-zero brackets)

$$\{\rho(x), \psi(y)\} = \{s(x), \alpha(y)\} = \{\beta(x), \gamma(y)\} = \delta(x-y).$$

Show that this bracket accords with (67). Investigate the "gauge group" arising from the non-uniqueness of the Clebsch variables $\rho, \psi, s, \alpha, \beta, \gamma$.

4. Suppose $n=3$ and that the fluid is "barotropic", i.e. $\varepsilon_0 = \varepsilon_0(\rho)$ and the entropy plays no role as a field variable in the process under investigation. Examine the question of the global introduction of Clebsch variables $\rho, \psi, \alpha, \beta$ satisfying

$$p_i = \rho\partial_i\psi + \alpha\partial_i\beta.$$

Given any tensor field $T(x)$ on $\mathbb{R}^n$, one may extend the Poisson bracket to the field $T(x)$, i.e. define the brackets $\{p_i(x), T(y)\}$ (setting $\{T(x), T(y)\}=0$), in a manner suggested by the condition for a tensor field T to be "frozen in", which for each vector field w takes the form

$$\dot{T}(x) = L_w T(x),$$

where $L_w T$ is the Lie derivative of the tensor field in the direction of w. Thus for each Hamiltonian

$$H_w = \int w^i(x) p_i(x)\, d^n(x),$$

we set

$$\{T(x), H_w\} = L_w T(x). \tag{68}$$

This, together with the condition $\{T(x), T(y)\}=0$, then uniquely determines the extension of the Poisson bracket to include $T(x)$.

Of particular interest here is the case where T is a magnetic field (or, abstractly, a closed 2-form in $\mathbb{R}^3$):

$$T = (H_{ij}).$$

The mass density ρ and the entropy density s will here be 3-forms in $\mathbb{R}^3$, since under a transformation of co-ordinates they are multiplied by the Jacobian (see Part I, Theorem 18.2.3), and consequently the quotient $s\rho^{-1}$ will be a scalar in $\mathbb{R}^3$. It turns out that the Hamiltonian

$$H=\int\left(\frac{|p|^2}{2\rho}+\varepsilon_0(\rho, s)+\frac{|(H_{ij})|^2}{8\pi}\right)d^3x, \tag{69}$$

together with the above-defined bracket (see (68)), yield the equations of magnetic hydrodynamics in the situation where the field is "frozen in" the fluid.

The Poisson bracket defined in (64) is a special case of the more general "differential-geometric" bracket.

30.5.2. Definition. Given a family of fields $u^1(x), \ldots, u^m(x)$, we call a bracket operation of the form

$$\begin{aligned}\{u^i(x), u^j(y)\} &= g^{ij,\alpha}(u(x))\partial_\alpha\delta(x-y)\\ &\quad+b_k^{ij,\alpha}(u(x))u_\alpha^k\delta(x-y)\end{aligned} \tag{70}$$

(where $u_\alpha^k=\partial_\alpha u^k(x)$; $x=(x^1, \ldots, x^n)$; $\alpha=1, \ldots, n$; $i, j=1, \ldots, m$) a *homogeneous differential-geometric bracket*. An *inhomogeneous differential-geometric bracket* has the form

$$\begin{aligned}\{u^i(x), u^j(y)\} &= g^{ij,\alpha}(u(x))\partial_\alpha\delta(x-y)\\ &\quad+b_k^{ij,\alpha}(u(x))u_\alpha^k\delta(x-y)\\ &\quad+c^{ij}(u(x))\delta(x-y).\end{aligned} \tag{71}$$

Functionals of the form

$$H=\int h(u)\,d^nx,$$

where the "density" $h(u)$ (cf. (65), (69)) is independent of the fields u^i_α, are naturally said to be of "hydrodynamical type". In conjunction with a Poisson bracket of the form (70) such a Hamiltonian gives rise to a "system (of equations) of hydrodynamical type":

$$\dot{u}^i=\{u^i(x), H\}=a_j^{i,\alpha}(u)u_\alpha^j. \tag{72}$$

It is not difficult to see that the classes of Poisson brackets of the form (70), of functionals of hydrodynamical type, and of concomitant equations of the form (72), are each invariant under local changes $u=u(v)$ of the variables u, not involving any derivatives. The precise way in which the bracket (70) transforms is given by the following result, a straightforward consequence of the Leibniz condition.

30.5.3. Lemma. *Under local changes of variables of the form $u=u(v)$, the components $g^{ij,\alpha}$ (in the formula (70) for a Poisson bracket) transform (for each fixed α) like those of a tensor of type (2, 0) on the u-space, while for each fixed α for which $(g^{ij,\alpha})$ is non-degenerate, the components $b_k^{ij,\alpha}$ transform (after appropriate lowering of an index using the "metric" $g^{ij,\alpha}$) like Christoffel symbols; thus we may write*

$$b_k^{ij,\alpha}=g^{is,\alpha}\Gamma_{sk}^{j,\alpha}.$$

In the case where the x-space is one-dimensional (i.e. $n=1$, so that $g^{ij,\alpha}=g^{ij,1}=g^{ij}$ say) more detailed facts are known. (Their proof, which we omit, involves a certain amount of calculation.)

30.5.4. Theorem. *In the case $n=1$ and under the assumption that* $\det(g^{ij})$ *does not vanish, the expression* (70) *satisfies all the conditions of a Poisson bracket (including the Jacobi identity) if and only if the "metric" (g^{ij}) is symmetric, and (Γ_{sk}^j) is the symmetric connexion compatible with this metric, yielding zero curvature (see Part* I, §§29.3, 30).

30.5.5. Corollary. *In the case $n=1$, a Poisson bracket of the form* (70) *is determined (locally) by a single invariant, namely the signature of the "metric" $g^{ij}=g^{ji}$ $(\det(g^{ij})\neq 0)$. There exist local co-ordinates for the u-space in terms of which $g^{ij}\equiv$ const., $b_k^{ij}\equiv 0$.*

In the case $n=1$, it turns out that if a system of the form (72) is Hamiltonian and diagonalizable (i.e. the matrix $(a_j^i(u))$ can be diagonalized on the whole region of interest) then the system is integrable in a certain precise "Liouvillian" sense. About the subcase where in addition $m=2$ (the case of a 2-component system) several facts had already been discovered in the nineteenth century, beginning perhaps with Riemann; these are summarized in the following

EXERCISES

1. Prove that for functions $x=x(u^1, u^2)$, $t=t(u^1, u^2)$, the system (72) (defining a so-called "hodograph" transformation) is linear. (*Note.* The "hodograph" of a particle's motion is the curve described by the tip of the velocity vector of the particle, laid off from the origin.)

2. Show that for $m=2$ there is a local co-ordinate change $u=u(v)$ which diagonalizes the system (72).

For $m>2$ (i.e. for more than two fields) there has been discovered quite recently a generalization of the "hodograph" method, which, however, in contrast with the classical case $m=2$, relies heavily on the assumption that the system is Hamiltonian. Assuming as before that $n=1$, one considers a pair of commuting Hamiltonian systems of the form (72):

$$u^i_t = v^i_j(u)u^j_x,$$
$$u^i_t = w^i_j(u)u^j_x, \tag{73}$$

where (v^i_j) is diagonal, i.e. $v^i_j = \delta^i_j v^i(u)$, and where furthermore the diagonal entries v^i are all distinct. Then (w^i_j) is also diagonal, say $w^i_j = \delta^i_j w^i$ (verify!) The system of equations

$$w^i(u) = v^i(u)x + t \tag{74}$$

determines a family of functions (as solutions)

$$u^1(x, t), \ldots, u^m(x, t),$$

which, as it turns out, satisfy the original system $u^i_t = v^i_j(u)u^j_x$.

We conclude by stating a result for the inhomogeneous bracket (71). (Its proof under the stated assumptions is not complicated.)

30.5.6. Theorem. *Suppose* $n = 1$ *and* $\det(g^{ij,\alpha}) = \det(g^{ij}) \neq 0$, *and that we are given co-ordinates* u^k *in terms of which* $g^{ij} \equiv \text{const.}$, $b^{ij}_k \equiv 0$. *If the formula* (71) *defines a Poisson bracket, then in terms of such co-ordinates* u^k *the quantities* $c^{ij}(u)$ *must have the form*

$$c^{ij}(u) = c^{ij}_k u^k + c^{ij}_0, \tag{75}$$

where $c^{ij}_k \equiv \text{const.}$, $c^{ij}_0 \equiv \text{const.}$, *and where moreover the* c^{ij}_k *are the structural constants* (*see Part* I, §24.5) *of a semi-simple Lie algebra with Killing metric* g^{ij}, *and* (c^{ij}_0) *represents a cocycle on that Lie algebra, i.e. satisfies*

$$c^{is}_0 c^{jk}_s + c^{ks}_0 c^{ij}_s + c^{js}_0 c^{ki}_s = 0. \tag{76}$$

We shall not broach here the higher-dimensional case or other generalizations of this theory.

CHAPTER 8

The Global Structure of Solutions of Higher-Dimensional Variational Problems

§31. Some Manifolds Arising in the General Theory of Relativity (GTR)

31.1. Statement of the Problem

Geometrically speaking the fundamental problem of GTR is that of finding 4-dimensional manifolds M^4 endowed with a metric g_{ab} of signature $(+ - - -)$, satisfying Einstein's equation (see Part I, §§37.4, 39)

$$R_{ab} - \tfrac{1}{2} R g_{ab} = \frac{8\pi G}{c^4} T_{ab}, \tag{1}$$

where T_{ab} is the energy-momentum tensor of the medium. From the most general point of view there is only one restriction on the tensor T_{ab}, namely that the "energy density" be non-negative, which comes down to the condition that for every time-like vector $\xi = (\xi^a)$ (i.e. such that $g_{ab}\xi^a\xi^b > 0$) at each point of M^4, we have $T_{ab}\xi^a\xi^b \geq 0$. However in the present chapter we shall for the most part concern ourselves with the so-called "hydrodynamical energy-momentum tensor" of an isotropic dense medium (forming a "perfect fluid"), already introduced in Part I, §39.4, important in the investigation of the gravitational fields of macroscopic bodies:

$$T_{ab} = (p + \varepsilon) u_a u_b - p g_{ab}, \tag{2}$$

where $u = (u_a)$ is the 4-velocity ($\langle u, u \rangle = 1$), p is the (isotropic) pressure, and ε is the energy density (see loc. cit.).

A general analysis of the Einstein equation (1) presents problems of a complexity which might be termed transcendental. However in several special situations, where a metric is sought which possesses a large group G of isometries, equation (1) reduces, in suitable systems of local co-ordinates, to a comparatively simple form, allowing precise solutions to be found, or at least making qualitative investigation possible. In such cases there always arises the question of the extent to which any solution we have found is a full solution, i.e. whether it represents the whole manifold M^4 or just some region of it. The simplest, as well as most fundamental, of these special cases, where the purely co-ordinate approach yields only a region of M^4, is the relativistic analogue of the field of a point-mass, i.e. the "Schwarzschild solution", which we derived in Part I, §39.3.

We conclude this introduction with the following definition (in essence encountered several times in Part I).

31.1.1. Definition. A function $f(x)$ on M^4 is said to be respectively: (i) *time-like*; (ii) *space-like*; (iii) *light-like* (or *isotropic*) according as:

(i) $g^{ab}\dfrac{\partial f}{\partial x^a}\dfrac{\partial f}{\partial x^b} > 0$, i.e. $\langle \nabla f, \nabla f \rangle > 0$;

(ii) $g^{ab}\dfrac{\partial f}{\partial x^a}\dfrac{\partial f}{\partial x^b} < 0$, i.e. $\langle \nabla f, \nabla f \rangle < 0$;

(iii) $g^{ab}\dfrac{\partial f}{\partial x^a}\dfrac{\partial f}{\partial x^b} = 0$, i.e. $\langle \nabla f, \nabla f \rangle = 0$.

(Here $\nabla f = \operatorname{grad} f$; thus if for example $f(x) = x^a$, then $\langle \nabla f, \nabla f \rangle = g^{aa}$.)

We shall often denote the scalar square $\langle \nabla f, \nabla f \rangle$ by g^{ff}.

31.2. Spherically Symmetric Solutions

We begin with the definition

31.2.1. Definition. A manifold M^4 endowed with an Einstein metric g_{ab} is said to be *spherically symmetric* if it admits among its isometries those of the group $SO(3)$, having 2-dimensional orbits S^2. (Think of the simple case where the transformations in $SO(3)$ fix the "time"-co-ordinate, and rotate the "spatial" co-ordinates.)

The orbits S^2 of M^4 under the action of $SO(3)$ will of necessity be space-like (i.e. their non-zero tangent vectors $\xi = (\xi^a)$ at each point will all satisfy $g_{ab}\xi^a\xi^b < 0$) in view of the fact that each stationary subgroup of $SO(3)$ (i.e. each stabilizer in $SO(3)$ of a point) acts isotropically on the tangent space to the orbit S^2 through that point, while on the other hand the set of non-space-like tangent vectors (i.e. of time-like and isotropic vectors) to M^4 at each point, forming the 3-dimensional light cone together with its interior (i.e. given by $g_{ab}\xi^a\xi^b > 0$), does not contain a 2-dimensional subspace (essentially since g_{ab} has signature $(+ - - -)$).

We denote the orbit space $M^4/SO(3)$ by M^2; this will be a 2-dimensional "space of parameters" where the parameters in question index the orbits. We thus obtain a fibre bundle with the natural projection

$$p: M^4 \to M^2, \tag{3}$$

and with fibres $p^{-1}(x)$, $x \in M^2$, the orbits S^2. There is a naturally defined connexion (in fact a G-connexion; see §24.2), namely that afforded by the 2-dimensional planes at the points of M^4, orthogonal (with respect to the metric g_{ab}) to the fibres, i.e. to the orbits S^2 of the action of $SO(3)$. Let H_y denote the stationary group (i.e. the stabilizer in $SO(3)$) of an arbitrary point $y \in M^4$. From the fact that on its orbit the point y is isolated among the points fixed by H_y, together with the above-noted isotropic action of H_y on the tangent plane to the orbit through y, it follows that the set of fixed points of M^4 under the group H_y forms a 2-dimensional submanifold of M^4 orthogonal to the fibres; we have thus established the following:

31.2.2. Lemma. *The connexion on the fibre bundle* (3) *whose horizontal planes at each point are orthogonal* (*with respect to the metric* g_{ab}) *to the fibres* S^2, *is "trivial", i.e. as a 2-dimensional distribution on* M^4, *is integrable* (*in the sense of Definition* 29.1.1).

Given any metric of signature $(+ -)$ on the base M^2 of the fibre bundle (3), there always exist on some neighbourhood U of each point $x \in M^2$, "orthogonal" co-ordinates τ, R, i.e. co-ordinates in terms of which the metric takes the form

$$g_{00}\,d\tau^2 + g_{11}\,dR^2, \qquad g_{00} > 0, \qquad g_{11} < 0. \tag{4}$$

(There would seem to be little risk of confusing this R with the scalar curvature R in (1)!) Let θ, φ be the standard spherical co-ordinates on the unit sphere S^2; the usual spherical metric $d\Omega^2 = d\theta^2 + \sin^2\theta\,d\varphi^2$ is then invariant under $SO(3)$, and is essentially the only such invariant metric on S^2. It follows from this together with the above lemma that on the region $p^{-1}(U) \subset M^4$ there are co-ordinates τ, R, θ, φ in terms of which the Einstein metric has the form

$$g_{00}\,d\tau^2 + g_{11}\,dR^2 - r^2\,d\Omega^2, \qquad r = r(R, \tau), \tag{5}$$

where g_{00}, g_{11} depend only on τ and R, and the "radius" r is some measure, invariant under $SO(3)$, of the "size" of the orbit S^2 through each point. Thus the expression (5) gives the local form of an arbitrary $SO(3)$-invariant metric on M^4.

In the particular case of the Schwarzschild solution of the Einstein equation (1) with $T_{ab} \equiv 0$ (derived in Part I, §39.3), we took (putting the velocity of light $c = 1$ for simplicity)

$$\tau = t, \qquad R = r, \tag{6}$$

(where t is the time and r the radial co-ordinate, i.e. spatial distance from the "centre"), and obtained the solution in the form (see loc. cit.)

$$g_{00} = 1 - \frac{a}{r}, \qquad g_{11} = -\frac{1}{1 - a/r}, \qquad a = \text{const.} > 0. \tag{7}$$

This solution is physically significant only for $r > a$, i.e. in the "region of an external observer". Considering it formally for $r < a$ we see that, in the terminology of Definition 31.1.1:

(i) the function r (the radial co-ordinate) is time-like for $r < a$, i.e. $g^{rr} = g^{11} > 0$;
(ii) the function t (the time co-ordinate) is space-like for $r < a$, i.e. $g^{tt} = g^{00} < 0$.

As $r \to a$ from the outside (i.e. from above), we have $g^{rr} \to 0$, $g^{tt} \to \infty$, so that by (ii) the function t has no reasonable meaning for $r = a$, while by (i) if the function r on M^4 is meaningful, then it is isotropic (in the sense of 31.1.1).

It is in any case clear from (7) that the above co-ordinates t, r, θ, φ are not appropriate when $r = a$. However as we shall now see there is a choice of the co-ordinates τ, R, different from that yielding the Schwarzschild solution (7), whose region of application is larger than that of an "external observer". In solving the Einstein equation in empty space, i.e. the equation

$$R_{ab} - \tfrac{1}{2} R g_{ab} = 0, \tag{8}$$

or, equivalently, $R_{ab} = 0$ (see Part I, §37.4), we may assume without loss of generality that locally the co-ordinate τ can be chosen so that $g_{00} \equiv 1$; this can be shown to follow from the above-mentioned fact that any metric of signature $(+\,-)$ on M^2 takes the form (4) in terms of suitable local co-ordinates, and thence the form $d\tau^2 + g_{11}\, dR^2$, $g_{11} < 1$, after a further change of co-ordinates. Thus in terms of local co-ordinates τ, R, θ, φ we may suppose the metric given by

$$d\tau^2 + g_{11}\, dR^2 - r^2(d\theta^2 + \sin^2\theta\, d\varphi^2), \tag{9}$$

where r and g_{11} each depend only on R and τ. The solution of the system (8) (or equivalently the system $R_{ab} = 0$) for g_{11} and r (in (9)) in terms of τ and R, can be obtained without great difficulty. One first needs to calculate the Christoffel symbols Γ^a_{bc} and thence the Ricci tensor R_{ab} using the appropriate formulae (see Part I, Theorem 29.3.2 and §39.3(30)), in order to obtain the specific form taken by the equations (8) as they apply to a metric of the form (9). These calculations are facilitated by the introduction of the new parameters ν, λ, μ defined by

$$g_{00} = e^{\nu}, \qquad -g_{11} = e^{\lambda}, \qquad r^2 = e^{\mu}; \tag{10}$$

of course in the present context we have $g_{00} \equiv 1$ so that $v \equiv 0$. We shall give the results of these calculations below (in the more general situation where $v \not\equiv 0$ and $T_{ab} \not\equiv 0$; see the right-hand sides of equations (16), where differentiation with respect to τ is indicated by a dot, and with respect to R by a prime). It turns out that the solution for r in terms of τ and R is as follows (here η denotes a further convenient parameter defined by the two equations):

$$\frac{r}{a} = \frac{1}{2}\left(\frac{R^2}{a^2} + 1\right)(1 - \cos \eta),$$

$$\frac{\tau}{a} = \frac{1}{2}\left(\frac{R^2}{a^2} + 1\right)^{3/2} (\pi - \eta + \sin \eta), \tag{11}$$

$$0 \le \eta \le 2\pi.$$

The (local) structure of the resulting "Kruskal manifold" M^4 (as reflected in the function $r(R, \tau)$) is indicated in Figure 115. Note that the point $r = 0$, where the orbits under the group $SO(3)$ reduce to a point, is a singularity of the metric (i.e. of M^4). By drawing the light cones at various points of the (τ, R)-plane (we recommend that the reader attempt this exercise), the

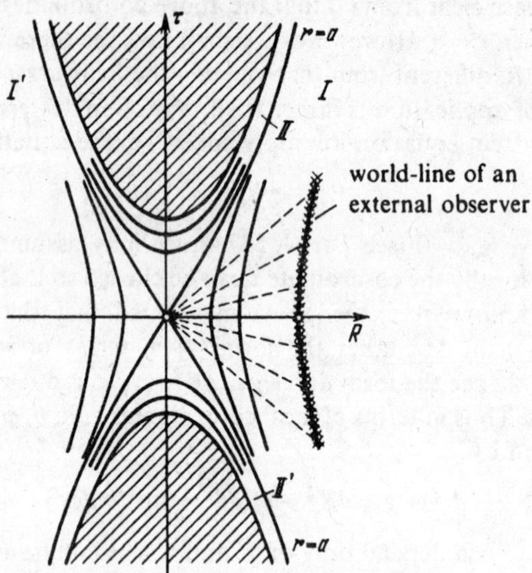

Figure 115. The (R, τ)-plane. The "meaningless" region $r \le 0$ is shaded. The solid curves represent the level curves $r(R, \tau) = \text{const.}$, and the dotted lines the level curves $t(R, \tau) = \text{const.}$ (Here t is the time co-ordinate in the Schwarzschild region $r > a$; see (6).) At $r = a$ we have $|t| = \infty$ except for the origin $R = \tau = 0$, where the time is undefinable since all of the level curves $t = \text{const.}$ converge there. At that point we have $\partial r/\partial \tau = \partial r/\partial R = 0$; it is in fact a saddle point. On the curves $r = a$ we have $\langle \nabla r, \nabla r \rangle = 0$. The regions I and I′ are isometric (under the map $R \leftrightarrow -R$), as are the regions II and II′ $(\tau \leftrightarrow -\tau)$.

following assertions can be verified. Note first that the world-line indicated in Figure 115, of an external observer in the region I (the "Schwarzschild" region) is given by

$$r = r_0 > a, \qquad -\infty < t < \infty, \qquad \theta = \theta_0, \qquad \varphi = \varphi_0. \tag{12}$$

Signals can reach the external observer from the region II′, but signals from the observer cannot enter that region (a "white hole"). On the other hand for the region II the reverse situation obtains: the observer can send signals into that region, but no signal can reach the observer from it. (Thus that the region II represents a so-called "black hole".) Communication between the regions I and I′ in either direction is impossible.

The solutions of Schwarzschild and Kruskal have an important application to the situation of a collapsing spherically symmetric mass (a collapsing star or "collapsar"). Consider the time-like world-line γ of each point of the star's boundary (i.e. for each fixed pair of values of θ and φ), of the form shown in Figure 116. Together all such world-lines (for all values of θ and φ) form a 3-dimensional surface $S_\gamma \subset M^4$, separating M^4 into two regions, an "external" region A and an "internal" region B (as in Figure 116). We expect the process of collapse of the star to be described by a spherically symmetric solution of Einstein's equation (1) with the following properties:

(i) in the external region A the metric g_{ab} should be that given by the Kruskal solution of $R_{ab} = 0$;
(ii) in the internal region B the metric g_{ab} should satisfy the Einstein equation (1) with T_{ab} the hydrodynamical energy-momentum tensor given by (2) where the pressure p and energy density ε are linked by an appropriate "state equation" $\varepsilon = \varepsilon(p)$;
(iii) the boundary of the star (represented by the curve γ in the (τ, R)-plane) should intersect (as $t \to +\infty$) the "horizon line" $r = a$ where $g^{rr} = 0$, without having done so earlier (as $t \to -\infty$).

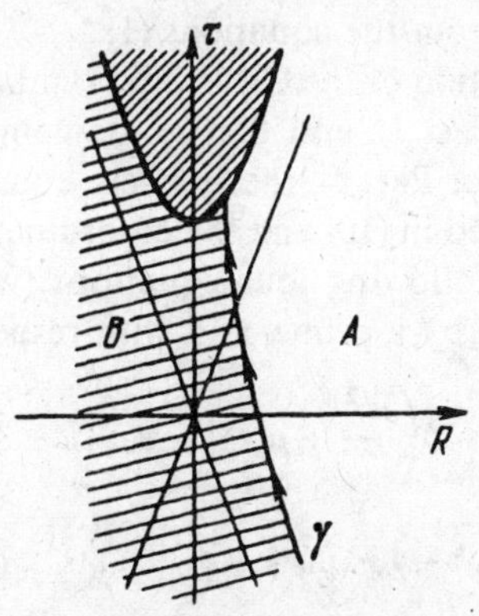

Figure 116

In view of (ii) we need to solve the Einstein equation in the region B, where $T_{ab} \neq 0$. As noted in Part I, §39.4, in terms of an attached co-ordinate system (i.e. a "co-moving" frame accompanying the star), i.e. one for which the 4-velocity satisfies $\langle u, u\rangle = 1$, $u_\alpha = 0$, $\alpha = 1, 2, 3$, we shall have (see (2))

$$T_0^0 = \varepsilon, \qquad T_\alpha^\alpha = -p\,(\alpha = 1, 2, 3), \qquad T_b^a = 0 \quad \text{for } a \neq b, \tag{13}$$

and in view of the assumed spherical symmetry of the star the metric will in terms of suitable such co-ordinates still have the form (5).

Remark. Generally speaking an attached co-ordinate system with the property that the "time"-co-ordinate τ is orthogonal to the co-ordinates R, θ, φ, will not be "synchronous", i.e. in general it cannot be arranged (as it was in Kruskal's solution above) that $g_{00} \equiv 1$.

Continuing with our solution in the region B, note first that the conservation law $\nabla_a T_b^a = 0$ (see Part I, §§37.2, 39.4) yields (in terms of the parameters ν, λ, μ defined as in (10))

$$\nu' = -\frac{2p'}{p+\varepsilon}, \qquad (\dot{\lambda} + 2\dot{\mu}) = -\frac{2\dot{p}}{p+\varepsilon}, \tag{14}$$

whence we obtain immediately

$$\begin{aligned} -g_{11}r^4 &= \exp\left\{-2\int \frac{d\varepsilon}{p+\varepsilon}\right\} \exp \varkappa(R), \\ g_{00} &= \exp\left\{-2\int \frac{dp}{p+\varepsilon}\right\} \exp \psi(\tau), \end{aligned} \tag{15}$$

where the right-hand side integrals are to be interpreted suitably, and where $\varkappa(R)$ and $\psi(\tau)$ are arbitrary functions. The latitude in choosing these functions reflects that remaining in the choice of the co-ordinates τ, R; by choosing $\varkappa, \psi$ appropriately, thereby finally fixing on the co-ordinates τ, R, the quantities p, ε can be eliminated from the equations (1).

After explicit calculation of the Christoffel symbols Γ_{bc}^a (using the formula in Theorem 29.3.2 of Part I), and thence the components R_{ab} of the Ricci tensor (using §39.3(30) of Part I), the Einstein equations (1), in terms of the parameters ν, λ, μ defined in (10) and the co-ordinates $x^0 = \tau$, $x^1 = R$, $x^2 = \theta$, $x^3 = \varphi$, reduce to the following four equations (where differentiation with respect to R is denoted by a prime, and with respect to τ by a dot):

$$\begin{aligned} -\frac{8\pi G}{c^4} T_1^1 &= \tfrac{1}{2}e^{-\lambda}\left(\frac{\mu'^2}{2} + \mu'\nu'\right) - e^{-\nu}(\ddot{\mu} - \tfrac{1}{2}\dot{\mu}\dot{\nu} + \tfrac{3}{4}\dot{\mu}^2) - e^{-\mu}; \\ -\frac{8\pi G}{c^4} T_2^2 &= \tfrac{1}{4}e^{-\lambda}(2\nu'' + \nu'^2 + 2\mu'' + \mu'^2 - \mu'\lambda' - \nu'\lambda' + \mu'\nu) \\ &\quad + \tfrac{1}{4}e^{-\nu}(\dot{\lambda}\dot{\nu} + \dot{\mu}\dot{\nu} - \dot{\lambda}\dot{\mu} - 2\ddot{\lambda} - \dot{\lambda}^2 - 2\ddot{\mu} - \dot{\mu}^2); \end{aligned} \tag{16}$$

$$\frac{8\pi G}{c^4} T_0^0 = -e^{-\lambda}\left(\mu'' + \tfrac{3}{4}\mu'^2 - \frac{\mu'\lambda'}{2}\right) + \tfrac{1}{2}e^{-\nu}\left(\dot{\lambda}\dot{\nu} + \frac{\dot{\mu}^2}{2}\right) + e^{-\mu};$$

$$\frac{8\pi G}{c^4} T_0^1 = 0 = \tfrac{1}{2}e^{-\lambda}(2\dot{\mu}' + \dot{\mu}\mu' - \dot{\lambda}\mu' - \nu'\dot{\mu}). \tag{16}$$

We define $a = a(\tau, R)$ by $\langle \nabla r, \nabla r \rangle = g^{rr} = 1 - a/r$. In the Schwarzschild solution (Part I, §39.3) we had $a = \text{const.} = 2MG/c^2$ where M is the mass of the body, and G Newton's universal gravitational constant. In the present more general context (of a hydrodynamical energy-momentum tensor which is not necessarily zero) the equation

$$a(\tau, R) = r(\tau, R) \tag{17}$$

defines the "horizon", where $\langle \nabla r, \nabla r \rangle = g^{rr} = 0$, i.e. where the function $r(\tau, R)$ becomes light-like.

Remark. The Einstein equations (16) (after elimination of $T_1^1 = T_2^2 = -p$ and $T_0^0 = \varepsilon$ by means of equations (15) with $\varkappa \equiv \psi \equiv 0$, and a further elimination of ν, so that they then involve only λ, μ (and their derivatives with respect to τ and R)) can be obtained alternatively via the following action of "2-dimensional field theory":

$$\tilde{S} = \int \tilde{\Lambda}\, dR\, d\tau,$$

where the Lagrangian $\tilde{\Lambda}$ is defined by

$$\tilde{\Lambda} = \tilde{\Lambda}(\lambda, \mu, \dot{\lambda}, \dot{\mu}, \lambda', \mu') = T_1 + T_2 + U,$$

$$T_1 = -\exp\left\{\frac{k-1}{2}\lambda + (k+1)\mu\right\}\left(\frac{\mu'^2}{2} + \mu'(k\lambda' + 2k\mu')\right),$$

$$T_2 = \exp\left\{-\frac{k-1}{2}\lambda + (1-k)\mu\right\}\left(\frac{\dot{\mu}^2}{2} + \dot{\mu}\dot{\lambda}\right), \qquad U = 2\exp\left\{\frac{k+1}{2}\lambda + k\mu\right\},$$

where k is constant. On calculating the formal energy-momentum tensor (as defined in Part I, §37.2(8)) on the 2-dimensional space co-ordinatized by τ and R, we obtain the following expressions for its components:

$$\tilde{T}_0^0 = -\tilde{\Lambda} + \dot{\lambda}\frac{\partial\tilde{\Lambda}}{\partial\dot{\lambda}} + \dot{\mu}\frac{\partial\tilde{\Lambda}}{\partial\dot{\mu}} = T_2 - T_1 - U,$$

$$\tilde{T}_1^1 = -\tilde{\Lambda} + \lambda'\frac{\partial\tilde{\Lambda}}{\partial\lambda'} + \mu'\frac{\partial\tilde{\Lambda}}{\partial\mu'} = T_1 - T_2 - U,$$

$$\tilde{T}_0^1 = \dot{\lambda}\frac{\partial\tilde{\Lambda}}{\partial\lambda'} + \dot{\mu}\frac{\partial\tilde{\Lambda}}{\partial\mu'} = \tfrac{1}{2}e^{-\lambda}(2\dot{\mu}' + \dot{\mu}\mu' - \dot{\lambda}\mu' - (k\lambda' + 2k\mu')\dot{\mu}),$$

$$\tilde{T}_1^0 = \lambda'\frac{\partial\tilde{\Lambda}}{\partial\dot{\lambda}} + \mu'\frac{\partial\tilde{\Lambda}}{\partial\dot{\mu}} = -T_0^1 e^{-(k-1)\lambda - 2k\mu}.$$

It can be shown that the Einstein equations (16) for a spherically symmetric body with hydrodynamical energy-momentum tensor, are equivalent to the Euler–Lagrange equations $\delta\tilde{S}=0$ arising from the Lagrangian $\tilde{\Lambda}$, subject to the constraints $\tilde{T}_0^1=\tilde{T}_1^0=0$. This avenue of approach (i.e. via the functional $\tilde{S}$) turns out to be convenient for the investigation of solutions independent of one or the other of τ and R. We note also that if $\dot{\lambda}\neq 0$, $\lambda'\neq 0$, $\dot{\mu}\neq 0$, $\mu'\neq 0$, then the equations $\tilde{T}_0^1=\tilde{T}_1^0=0$ can be solved, and thence the Einstein equations (16) reduced to the first-order system (23); we shall now indicate how to carry out this reduction, but via a more direct route.

From the definition of $a(\tau, R)$ above, we have

$$a=r(1-\langle\nabla r,\nabla r\rangle)=r\left(1-g^{00}\left(\frac{\partial r}{\partial\tau}\right)^2-g^{11}\left(\frac{\partial r}{\partial R}\right)^2\right),$$

whence, setting as usual $g_{00}=e^{\nu}$, $g_{11}=-e^{\lambda}$, $r^2=e^{\mu}$, we obtain

$$a=e^{3\mu/2}(e^{-\mu}-\tfrac{1}{4}(\dot{\mu})^2e^{-\nu}-\tfrac{1}{4}(\mu')^2e^{-\lambda}).$$

Using this together with the Einstein equations (16) (with $T_1^1=T_2^2=-p$, $T_0^0=\varepsilon$) one verifies by direct calculation the following two equations:

$$\frac{\partial a(\tau,R)}{\partial\tau}=-pr^2\frac{\partial r}{\partial\tau}\frac{8\pi G}{c^4},$$
$$\frac{\partial a}{\partial R}=\varepsilon r^2\frac{\partial r}{\partial R}\frac{8\pi G}{c^4}. \tag{18}$$

(Note that since ε is the energy density, the energy of the body when $\tau=\tau_0$ $=$ const. is given by

$$\Sigma=\int_{\tau=\tau_0}\varepsilon\sqrt{|g_{11}g_{22}g_{33}|}\,dR\,d\theta\,d\varphi=\int_{\tau=\tau_0}e^{\lambda/2}4\pi r^2\varepsilon\,dR;$$

on the other hand in view of the second of the equations (18) the quantity defined by

$$E_{\text{total}}=\int_{\tau=\tau_0}4\pi\varepsilon r^2\,dr=\int_{\tau=\tau_0}4\pi\varepsilon r^2\frac{\partial r}{\partial R}\,dR$$

is an integral of the system (18) and turns out to be in fact the Hamiltonian of the system determined by the Lagrangian $\tilde{\Lambda}$ under the supplementary conditions $\tilde{T}_0^1=\tilde{T}_1^0=0$ (see the above remark). Thus arises the so-called "gravitational energy defect": $E_{\text{total}}\neq\Sigma$.)

Consider first the case $p\equiv 0$ (obtaining for instance in a dust-cloud). It follows immediately from the first equation in (18), and the second equation of the pair (15) (linking ε and p to the metric), that in this case

$$a=a(R),\qquad g_{00}=\text{const.}, \tag{19}$$

whence in particular we may assume without loss of generality that $g_{00} \equiv 1$. It follows easily from the first equation in (15) that $g_{11} = -1/r^4\varepsilon^2$, whence, substituting for ε from the second equation in (18), we obtain

$$g_{11} = -\frac{(r')^2}{(a')^2}\left(\frac{8\pi G}{c^4}\right)^2.$$

Hence by definition of a (and invoking $g^{00} \equiv 1$),

$$1 - \frac{a}{r} = g^{00}(\dot{r})^2 + g^{11}(r')^2 = (\dot{r})^2 - (a')^2\left(\frac{8\pi G}{c^4}\right)^{-2}, \tag{20}$$

so that finally

$$\dot{r} = \pm\sqrt{1 - \frac{a(R)}{r} + (a'(R))^2\left(\frac{8\pi G}{c^4}\right)^{-2}}. \tag{21}$$

This equation can by means of a single "quadrature" be integrated to yield the well-known "Tolman solution", from which we conclude that the dust-like matter will either collapse or expand indefinitely. This conclusion is in fact clear from (20) since in the region outside the horizon (i.e. where $g^{rr} = 1 - a/r > 0$) we have from (20) that

$$\dot{r} = \pm\sqrt{g^{rr} + |g^{11}r'^2|} \neq 0,$$

so that, in the absence of singularities of the metric and the energy density $(0 < \varepsilon < \infty)$ outside the horizon, the sign of $\dot{r}$ cannot change.

We now turn to the more general case where the state equation has the form $p = k\varepsilon$, $k = \text{const.}$, $0 \leq k \leq 1$. From (15) with $\varkappa \equiv \psi \equiv 0$, we obtain

$$g_{00} = \varepsilon^{-2k/(k+1)}, \qquad g_{11} = -\frac{\varepsilon^{-2/(k+1)}}{r^4}. \tag{22}$$

The equations (18), together with (20) (modified using (22)), form the system

$$\begin{aligned} \dot{a} &= -pr^2\dot{r}\frac{8\pi G}{c^4}, \\ a' &= \varepsilon r^2 r'\frac{8\pi G}{c^4}, \\ 1 - \frac{a}{r} &= \dot{r}^2\varepsilon^{2k/(k+1)} - r'^2 r^4\varepsilon^{2/(k+1)}. \end{aligned} \tag{23}$$

From the last of these equations and the result of eliminating ε between the first two, we obtain

$$\begin{aligned} \dot{r} &= \pm\sqrt{\Phi(r, r', a, a')} = \pm\sqrt{(g^{rr} + |r'^2 r^4\varepsilon^{2/(k+1)}|)\varepsilon^{-2k/(k+1)}}, \\ \dot{a} &= -k\frac{a'}{r'}\sqrt{\Phi}, \qquad 0 \leq k \leq 1. \end{aligned} \tag{24}$$

It is clear that, apart from the sign of $\dot{r}$, any solution of this system will be completely determined by specifying $r(0, R)$ and $a(0, R)$. We see also that the same conclusion as in the earlier case $p = 0$ can be drawn here (and for essentially the same reason); namely, that since $\Phi \geq g^{rr} > 0$ in the region outside the horizon, the sign of $\dot{r}$ cannot change there; thus if neither the metric nor the energy density $\varepsilon\,(>0)$ possesses singularities outside the horizon, there will occur either monotonic collapse or expansion of the matter.

Remarks. 1. The last of the equations (23) takes on an especially simple form in the case of a state equation of "limiting rigidity" (i.e. when $k = 1$), namely

$$1 - \frac{a}{r} = \varepsilon(\dot{r}^2 - (r')^2 r^4). \tag{25}$$

2. It is easy to verify that the system of partial differential equations (23) is invariant under the group of "scaling transformations"

$$\begin{aligned} r &\to \xi r, & a &\to \xi a, & \tau &\to \alpha\tau, \\ R &\to \beta R, & \varepsilon &\to \gamma\varepsilon, & p &\to \gamma p, \end{aligned} \tag{26}$$

where $\alpha^2/\beta^2 = \xi^{-8k/(k+1)}$, $\gamma = \xi^{-2}$. If one looks for so-called "self-similar" solutions, i.e. solutions of (23) invariant under the one-parameter subgroup of this group defined by the further relations

$$\alpha = \beta^s, \qquad \xi = \beta^\gamma, \qquad \gamma = \frac{(1-s)(1+k)}{4k}$$

(where s is the parameter, determining $\xi, \alpha, \beta, \gamma$ via these relations and the earlier ones), which can be shown to have the form

$$r = R^\gamma r_1(\lambda), \qquad a = R^\gamma a_1(\lambda), \qquad \varepsilon = R^{-2\gamma}\varepsilon_1(\lambda), \qquad \text{where } \lambda = \frac{\tau}{R^s}, \tag{27}$$

then one finds that for the particular value $s = (1-k)/(1+3k)$ of the parameter, the system (23) reduces to the following (dynamical) system of ordinary differential equations (verify!):

$$\begin{aligned} &\frac{da_1}{d\lambda} = -k\varepsilon_1 r_1^2 \frac{dr_1}{d\lambda}\frac{8\pi G}{c^4}, \\ &\gamma a_1 - s\lambda\frac{da_1}{d\lambda} = \varepsilon_1 r_1^2\left(\gamma r_1 - s\lambda\frac{dr_1}{d\lambda}\right)\frac{8\pi G}{c^4}, \\ &1 - \frac{a_1}{r_1} = \left(\frac{dr_1}{d\lambda}\right)^2 \varepsilon_1^{2k/(k+1)} - \left(\gamma r_1 - s\lambda\frac{dr_1}{d\lambda}\right)^2 r_1^4 \varepsilon_1^{2/(k+1)}. \end{aligned} \tag{28}$$

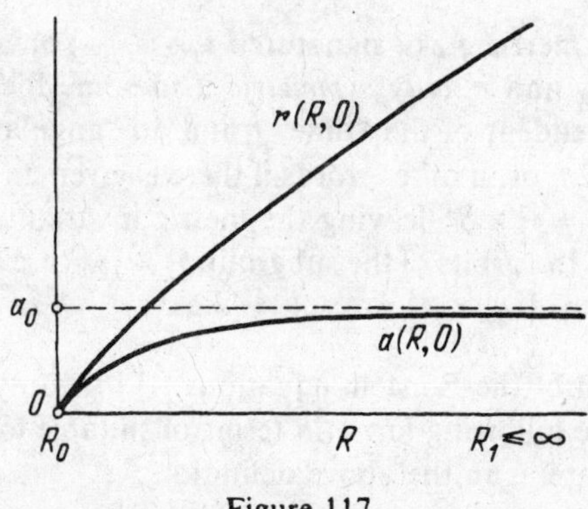

Figure 117

We remark also that the investigation of stationary solutions (i.e. solutions independent of τ), and solutions independent of R, also leads to dynamical systems.

We conclude this subsection by listing the conditions generally imposed in connexion with the boundary problem (at $\tau = 0$) for the system (23).

(i) It is required that the solutions satisfy the condition $r' > 0$ (corresponding to the requirement that the spherical fibres be ordered appropriately by their "radii" r), and $a' > 0$ (which follows from the first condition $r' > 0$ and the second of the equations (24) together with the general condition $\varepsilon > 0$).

(ii) As the (co-moving) co-ordinate R varies over its domain of variation $R_0 \leq R < R_1$ (where R_1 may be infinite), r is required to vary from 0 to ∞ $(0 \leq r < \infty)$ and a from 0 to a_0; thus $r(R_0, 0) = r(R_0) = a(R_0) = a(R_0, 0) = 0$, and $r \to \infty$, $a \to a_0$, as $R \to R_1$. Noting that the system (23) is equivalent to the Einstein equations (16) only under the conditions $\varepsilon \neq 0$, $r' \neq 0$, $\dot{r} \neq 0$ (and of course $p = k\varepsilon$), it then follows from the second equation in (23) that

$$a_0 = \frac{8\pi G}{c^4} \int_{R_0}^{R_1} \varepsilon r^2 \frac{\partial r}{\partial R}\, dR = E_{\text{total}} \times \text{const.} < \infty,$$

so that as $r \to \infty$ we must have $\varepsilon \to 0$ in such a way that this integral converges.

(iii) Finally, the requirement that when $\tau = 0$ the whole of the material body be observable, translates into the condition $r(R) = r(0, R) > a(R)$ for $R > R_0$ (see Figure 117).

31.3. Axially Symmetric Solutions

In this subsection we shall examine the stationary, axially symmetric solution of the Einstein equations (1), the "Kerr solution", which describes the gravitational field of a "rotating black hole".

31.3.1. Definition. A metric g_{ab} of signature $(+ - - -)$ on a manifold M^4 is said to be *stationary* and *axially symmetric* if in some local system of co-ordinates it is independent of the "time" t and an "angular" space-like co-ordinate φ, $0 \leq \varphi < 2\pi$; or in other words if there is given an action on M^4 of the abelian group $G = \mathbb{R} \times S^1$ leaving the metric invariant, and having the further property that the orbits of the subgroup $\mathbb{R} \times \{s_0\}$ are time-like and the orbits of the subgroup $\{t_0\} \times S^1$ are space-like.

The Kerr solution of the Einstein equations (1) in empty space ($T_{ab} \equiv 0$), turns out to have the following form (in terms of suitable local co-ordinates t, r, θ, φ, where φ, t are as in the above definition):

$$\begin{aligned} ds^2 = dt^2 - \Bigg[& dr^2 + 2a \sin^2\theta \, dr \, d\varphi + (r^2 + a^2) \sin^2\theta \, d\varphi^2 \\ & + \rho^2 \, d\theta^2 + \frac{2mr}{\rho^2} (dr + a \sin^2\theta \, d\varphi + dt)^2 \Bigg], \end{aligned} \tag{29}$$

where $m = \text{const.}$ (the mass of the rotating body), $\rho^2 = r^2 + a^2 \cos^2\theta$, $a = \text{const.}$, and we have put $c = 1$ and $G = 1$ for simplicity. Note also that, as we shall see below, the co-ordinate r may be negative. The local co-ordinates r, θ, φ (at $t = t_0$) are related to Cartesian (i.e. standard) co-ordinates x, y, z on the local co-ordinate neighbourhood $\mathbb{R}^3$ ($t = t_0$) by the equations

$$\begin{aligned} x &= \sqrt{r^2 + a^2} \sin\theta \cos\left(\varphi - \arctan \frac{a}{r}\right), \\ y &= \sqrt{r^2 + a^2} \sin\theta \sin\left(\varphi - \arctan \frac{a}{r}\right), \\ z &= r \cos\theta, \qquad 0 \leq \theta \leq \pi, \quad 0 \leq \varphi < 2\pi. \end{aligned} \tag{30}$$

It follows easily that in terms of the co-ordinates x, y, z, the level surfaces $r = \text{const.}$, $t = t_0$, are oblate spheroids, i.e. ellipsoids which depart from being spherical in being flattened in the z-direction. On the other hand each of the level surfaces defined by $\theta = \theta_0$, $t = t_0$, is a one-sheeted hyperboloid defined by

$$\frac{x^2 + y^2}{a^2 \sin^2\theta_0} - \frac{z^2}{a^2 \cos^2\theta_0} = 1. \tag{31}$$

If we introduce new co-ordinates t^*, φ^* by defining

$$\begin{aligned} dt^* &= dt - 2mr\frac{dr}{\Delta}, \\ d\varphi^* &= d\varphi + \frac{adr}{\Delta}, \\ & (\Delta = r^2 - 2mr + a^2), \end{aligned} \tag{32}$$

then in terms of the co-ordinates $t^*, r, \varphi^*, \theta$, the metric (29) takes the form (verify this!)

$$ds^2 = -\frac{\rho 2}{\Delta} dr^2 - \rho^2 \, d\theta^2 - (r^2 + a^2) \sin^2 \theta (d\varphi^*)^2$$
$$- \frac{2mr}{\rho^2} (a \sin^2 \theta \, d\varphi^* + dt^*)^2 + (dt^*)^2. \tag{33}$$

(Note that the group G still acts via transformations of the form $t^* \to t^* + \text{const.}, \varphi^* \to \varphi^* + \text{const.}$) It is easily verified that when $a = 0$ the metric (33) reduces to the Schwarzschild metric (see (7)), with $2m$ in the role of the constant (which is appropriate since we have taken $c = 1$, $G = 1$ for simplicity).

Exercise

Show that in the case $m = 0$ the metric (33) is equivalent to the Minkowski metric.

Inverting the matrix (g_{ab}) of the metric (33), we find that the non-zero components g^{ab} are as follows (with $c = 1$, $G = 1$):

$$\begin{gathered} g^{00} = \frac{1}{\Delta}\left(r^2 + a^2 + \frac{2mra^2}{\rho^2} \sin^2 \theta\right), \\ -g^{rr} = \frac{\Delta}{\rho^2}, \qquad -g^{\theta\theta} = \frac{1}{\rho^2}, \\ -g^{\varphi^*\varphi^*} = \frac{1}{\Delta \sin^2 \theta}\left(1 - \frac{2mr}{\rho^2}\right), \\ g^{\varphi^* t^*} = \frac{4mra}{\rho^2 \Delta}. \end{gathered} \tag{34}$$

As for the Schwarzschild solution, we define the horizon by means of the equation $g^{rr} = 0$, which in view of (34) is equivalent to the condition $\Delta = r^2 - 2mr + a^2 = 0$, i.e. to

$$r_\pm = m \pm \sqrt{m^2 - a^2}. \tag{35}$$

We consider now the implications of the two possibilities: (i) $a > m$ (in which case the roots $r_\pm$ are complex); and (ii) $a < m$ (when the roots $r_\pm$ will both be real positive).

(i) In the case $a > m$ (which may be described as that of "rapid rotation" since in fact ma is a measure, in some sense, of the angular momentum), Δ is positive for all real r, so that there is no horizon ($g^{rr} < 0$). From the first equation in (34) we see that $g^{00} > 0$. The metric (33) is defined (and has non-vanishing determinant) everywhere except where $\rho = 0$, i.e. both $r = 0$ and $\cos \theta = 0$; in view of the fact that these singularities are observable from the outside (i.e. from the region $0 < r < \infty$), they are called "naked singularities". In terms of the co-ordinates x, y, z defined by (30), they are given by

$$x^2 + y^2 = a^2, \qquad z = 0, \qquad t^* \text{ arbitrary.} \tag{36}$$

The equation $g^{\varphi^*\varphi^*} = 0$ is, by (34), equivalent to

$$r^2 - 2mr + a^2 \cos^2 \theta = 0 \quad \text{or} \quad r = m \pm \sqrt{m^2 - a^2 \cos^2 \theta}; \tag{37}$$

this equation defines the "ergosphere", inside which (i.e. at those points satisfying $r^2 - 2mr + a^2 \cos^2 \theta < 0$) we have $g^{\varphi^*\varphi^*} > 0$, so that the co-ordinate φ^* is time-like there (see Definition 31.1.1), while outside the ergosphere it is of course space-like. It is an easy calculation to verify that the region defined by

$$g_{\varphi^*\varphi^*} = -\sin^2 \theta \left(r^2 + a^2 + \frac{2mra^2 \sin^2 \theta}{\rho^2} \right) > 0 \tag{38}$$

is contained within the ergosphere so that those of the curves $t^* = \text{const.}$, $r = \text{const.}$, $\theta = \text{const.}$, φ^* variable (which from (30) we see to be closed) contained in the region (38), will be time-like. If on such a curve the constant value of θ satisfies $\cos \theta = 0$ (i.e. if $\theta = \pi/2$), then the (constant) value r_0 (< 0) of r on the curve must satisfy $r_0^2 < ma$; this follows easily from (38).

(ii) In the case $a < m$, the region (which we label I) of the external observer is given by

$$\text{(I)} \qquad r > m + \sqrt{m^2 - a^2} = r_+, \tag{39}$$

and we also have regions II and III given respectively by

$$\text{(II)} \qquad r_- < r < r_+, \tag{40}$$

where r is time-like, and

$$\text{(III)} \qquad -\infty < r < r_-, \tag{41}$$

where r is again space-like. By the first equation in (34), as $r \to r_+$ from within the region I, we have $g^{00} \to \infty$, so that an external observer's time t^* pertains only to that region. Of course by definition of $r_\pm$ (see (35)), the co-ordinate r is light-like on the horizon $r = r_\pm$. Since $r_\pm > 0$ the singularities of the metric (which as noted above occur at the points $r = 0$, $\cos \theta = 0$, t^*, φ^* arbitrary) all lie in the region III. In that region the curves defined by $r = r_0 < 0$, $\theta = \text{const.}$, $t^* = \text{const.}$, φ^* variable, are closed and time-like (verify this!).

The construction in this case of the full manifold M^4, by pasting together regions of types I, II and III, can be carried out in a manner modelled on the Kruskal manifold (see (11) and Figure 115). Thus to a region of type I "at" $r \to r_+$, can be joined two regions of type II (a "white hole" at $t \to -\infty$, and a "black hole" at $t \to +\infty$; cf. the Schwarzschild–Kruskal manifold as depicted in Figure 115). Similarly, to a region of type II we can join up (i) two regions of type III at $r \to r_-$, and (ii) two regions of type I at $r \to r_+$. Finally, to a region

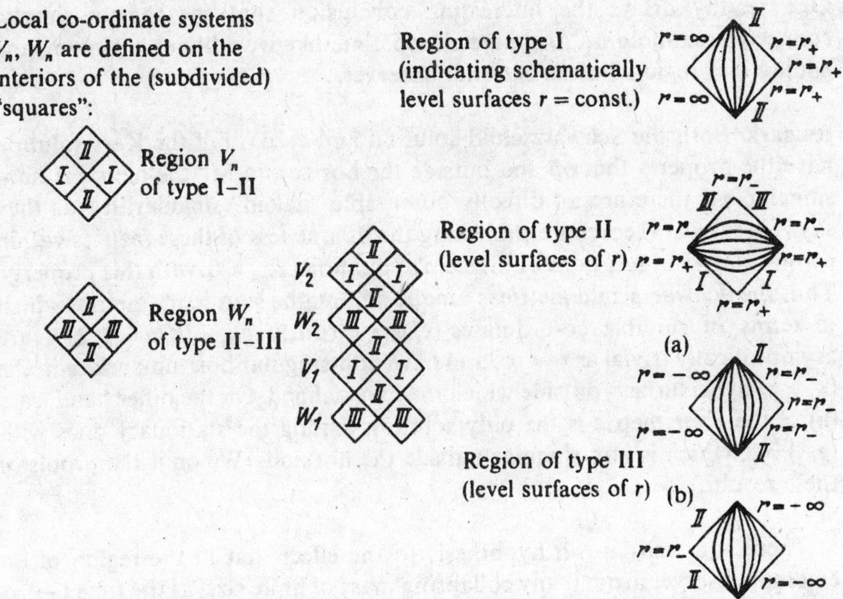

Figure 118

of type III we can join on two regions of type II at $r \to r_-$. (The permitted conjunctions are indicated schematically on the right in Figure 118.)

On each of the resulting composite regions of types I–II and II–III (as on the left in Figure 118), there will then be defined co-ordinates analogous to the Kruskal co-ordinates (see Figure 115). Although the requisite analysis is more difficult, it can be shown that these composite regions can also be pasted together, as shown schematically in the centre of Figure 118. Suitable iteration of such pastings yields a manifold M^4 with an obvious isometry B sending V_n to V_{n+1} and W_n to W_{n+1}, $n = 0, \pm 1, \ldots$, and we can obtain thence a manifold $\bar{M}^4$ by means of the usual identification of the points of M^4 under the action of the isometry B:

$$\bar{M}^4 = M^4/B.$$

The following is a list of the possible transitions or passages from one type of region to another in $\bar{M}^4$ along future-directed, time-like world-lines. (Here we use for instance "region (V_n, I)" to mean "some region of type I in the 'square' co-ordinatized by the system V_n", with reference to the diagram in the centre of Figure 118; and we use $\cong$ to denote identification under B.

$$\text{region } (V_n, \text{I}) \to \text{region } (W_{n+1}, \text{II}) \cong \text{region } (V_n, \text{II});$$
$$\text{region } (W_n, \text{II}) \to \text{region } (V_n, \text{I});$$
$$\text{region } (W_n, \text{III}) \to \text{region } (W_n, \text{II}) = \text{region } (V_n, \text{II}).$$

One readily draws the interesting conclusion that in the non-simply-connected manifold $\bar{M}^4$ there are closed, time-like world-lines beginning and ending in a region I of an external observer.

Remark. Both the Schwarzschild solution and case (ii) of the Kerr solution have the property that on and outside the horizon in M^4 the metric is non-singular, i.e. there are no directly observable "naked" singularities, as they say. There exist theorems establishing the uniqueness of these metrics within certain classes of solutions of Einstein's equations $R_{ab} = 0$, with this property. Thus the Schwarzschild metric is unique among the stationary metrics which in terms of suitable co-ordinates x, y, z, t satisfy $g_{0\alpha} = 0$, $\alpha = 1, 2, 3$, are asymptotically trivial as $r \to \infty$, and have non-singular horizon-surface in the (x, y, z)-hypersurface, outside which they are defined. On the other hand, case (ii) of the Kerr metric is the only solution among the stationary ones with $(g_{0\alpha}) \neq 0$, which is non-singular outside the horizon. (We omit the proofs of these results.)

There is a well-known hypothesis to the effect that in the region of an external observer around any collapsing mass of finite size, as the time $t \to \infty$ the metric will assume asymptotically the form of either the Schwarzschild metric or the Kerr metric (assuming that initially (at $t = 0$) the metric has no singularity).

31.4. Cosmological Models

Another type of problem of GTR is concerned with the construction and investigation of evolutionary models which might afford definite ideas of the manner of evolution of the global metric of the universe as a whole. Such 4-dimensional manifolds M^4 (with metric g_{ab} of signature $(+ - - -)$) are called *cosmological models*. In view of the difficulties arising from the impossibility of formulating universal boundary conditions for the Einstein equations for such a manifold, we shall restrict ourselves to the so-called homogeneous cosmological models, where it is assumed that at each moment of time the metric is in a certain sense the same at each point of space. We make this more precise in the following.

31.4.1. Definition. A 4-dimensional manifold M^4 endowed with a metric g_{ab} satisfying Einstein's equation (1) is called a *homogeneous cosmological model* if there is also given a group G of isometries whose orbits are all 3-dimensional and space-like.

Throughout the remainder of this section we shall choose our local co-ordinates x^0, x^1, x^2, x^3 so that the x^1-, x^2-, and x^3-axes (or, more precisely, the curves $x^0 = \text{const.}$, $x^2 = \text{const.}$, $x^3 = \text{const.}$, x^1 variable, etc.) are at each point tangential to the orbit of G through that point (so that the co-ordinates x^1, x^2, x^3 might then be appropriately termed "spatial"), and the x^0-axis, or "time" axis (more precisely each curve of the form $x^1 = \text{const.}$, $x^2 = \text{const.}$, $x^3 = \text{const.}$, x^0 variable) is transverse to the orbits of G. If in addition the conditions that $g_{00} \equiv 1$ and that the time co-ordinate $x^0 = ct$ be actually orthogonal to the orbits, are satisfied, we call the co-ordinate system *synchronous*. Our investigation of homogeneous models will largely be carried out in terms of synchronous co-ordinate frames. Note that since the orbits of the (Lie) group G are required to be 3-dimensional, it is clear that in the most general homogeneous models that group will have dimension 3 (i.e. G cannot have dimension < 3).

The totality of G-orbits (as points), i.e. the orbit-space, becomes, via the natural identification map p, a one-dimensional manifold N^1 (which might be called the "global time-axis") forming the base of the fibre bundle with projection map

$$p: M^4 \to N^1$$

sending each point of M^4 to its orbit; thus the fibres $p^{-1}(t_0)$ are the orbits. The local choice of a time axis, for instance one yielding a synchronous local frame, determines a connexion on the fibre bundle; the corresponding curvature will however be zero in view of the one-dimensionality of the base N^1 (see §§24.2, 25.3). In the case that G is 3-dimensional it is convenient to represent the metric on M^4 by means of a "tetrad", i.e. a quadruple of (pointwise linearly independent) left-invariant vector fields X_0, X_1, X_2, X_3 on M^4 (i.e. invariant under the map of tangent spaces induced by the (left) action of G on M^4, and not to be confused with left-invariant fields on G as defined in Part I, §24.3), with X_0 transverse to, and X_1, X_2, X_3 tangential to, the orbits of G, and with pairwise commutators satisfying

$$[X_0, X_\alpha] = 0, \qquad [X_\alpha, X_\beta] = c^\gamma_{\alpha\beta} X_\gamma, \tag{42}$$

where the coefficients $c^\gamma_{\alpha\beta}$ are the structural constants of the Lie algebra of G (independent of the time). (The classification of the 3-dimensional Lie algebras was given in Part I, §24.5.) It is intuitive that such a tetrad exists. We shall usually impose the further conditions (analogous to those defining a synchronous co-ordinate frame) that the field X_0 be orthogonal to the orbits, and that $\langle X_0, X_0 \rangle \equiv 1$; in this case the local synchronous co-ordinate system may clearly be chosen with the respective axes directed along the vector fields, and suitably scaled so that the components of the metric are given by:

$$g_{ab} = \langle X_a, X_b \rangle, \qquad g_{00} = 1, \qquad g_{0\alpha} = 0 \quad \text{for } \alpha = 1, 2, 3. \tag{43}$$

Note that in view of the homogeneity of the model the components $g_{\alpha\beta}(t)$ depend only on the time t (whose axis is by the above directed along the vector field X_0).

In terms of a synchronous local co-ordinate frame the Einstein equations (1), with, as is appropriate in the present context, the hydrodynamical energy-momentum tensor $T_{ab} = (p + \varepsilon)u_a u_b - p g_{ab}$, take the form of a system of second-order ordinary differential equations in the components $g_{\alpha\beta}(t)$, of which there are at most 6 distinct in view of the symmetry ($g_{\alpha\beta} = g_{\beta\alpha}$). We can eliminate u, ε and p from these equations (much as earlier) by means of the conservation law and the other natural conditions:

$$\nabla_a T^a_b = 0, \qquad \langle u, u\rangle = 1, \qquad p = p(\varepsilon). \tag{44}$$

(In fact, as in the latter part of §31.2 we shall take $p = k\varepsilon$, $0 \le k \le 1$, the cases $p = 0$ (see §31.2) and $p = \varepsilon/3$, in both of which we have $T^a_a = 0$, being of particular interest.) This done we end up with a dynamical system on the phase space of dimension 12 co-ordinatized by $g_{\alpha\beta}(t)$, $\dot g_{\alpha\beta}(t)$, or rather on that region of it determined by the following two conditions:

(i) $g_{\alpha\beta}\xi^\alpha\xi^\beta < 0$ for every non-zero 3-vector ξ (i.e. tangent vector to an orbit). (This is the condition that the orbits under G be space-like.)
(ii) $\varepsilon(g_{\alpha\beta}, \dot g_{\alpha\beta}) \ge 0$ (the condition that the energy density be non-negative).

We call this region the "physical region" of the phase space $\mathbb{R}^{12}$ co-ordinatized by the $g_{\alpha\beta}$, $\dot g_{\alpha\beta}$, and denote it by S. Thus in the context of a homogeneous cosmological model, Einstein's equations reduce to a dynamical system on the physical region S of the phase space $\mathbb{R}^{12}$. Each integral trajectory of the system represents a possible metric $g_{\alpha\beta}(t)$, or in other words a possible homogeneous cosmological model of the Einsteinian manifold M^4 with the prescribed 3-dimensional group G of isometries.

Of particular interest among these solutions $g_{\alpha\beta}(t)$, i.e. among the collection of (general) homogeneous models, are those for which there is a properly larger group of isometries of M^4 having the same orbits as G. For each solution $g_{\alpha\beta}(t)$ we denote by $\hat G$ ($\ge G$) the largest group of isometries of M^4 having the same orbits as G. Since the (Lie) group $\hat G$ acts transitively as a group of isometries of each orbit $M^3(t)$, the latter can be represented as the homogeneous space (see §5.1)

$$M^3(t) \cong \hat G/H, \tag{45}$$

where H is the stabilizer of a point of the orbit. The assumed 3-dimensionality of the (Lie) group G (and of the orbits) clearly implies that the intersection $G \cap H$ is discrete (i.e. zero-dimensional); in fact we have (essentially from (45)) that

$$\dim \hat G = 3 + \dim H.$$

Remark. It is theoretically possible for a homogeneous model to possess a group $\hat G$ of isometries of dimension > 3 with 3-dimensional space-like orbits, yet having no 3-dimensional subgroup G with the same orbits; we shall not pursue this, however. (In the general case, involving the largest number of parameters, and embracing the most interesting examples, such groups $\hat G$ do in fact turn out to contain such 3-dimensional subgroups G.)

EXERCISE

Investigate the subgroups of the isometry group of a 3-dimensional (Riemannian) manifold which do not contain 3-dimensional transitive subgroups.

If $\dim \hat{G} > 3$ then for each fixed t the corresponding metrics $g_{\alpha\beta}(t)$ on the homogeneous space (or orbit) $M^3(t)$ will form a proper subset of the set of all Einsteinian metrics $g_{\alpha\beta}(t)$ corresponding to the given group G, since the action of the (non-discrete) group $H < \hat{G}$, fixing a point of $M^3(t)$, entails additional restrictions on the metric $g_{\alpha\beta}(t)$, and also on $\dot{g}_{\alpha\beta}(t)$. If we regard H as acting on the tangent space $\mathbb{R}^3$ to $M^3(t)$ at the fixed point, then it follows from the condition $g_{\alpha\beta}\xi^\alpha\xi^\beta < 0$ (see (i) above) that $H \leq SO(3)$. Since (as is not difficult to see) the only (non-trivial) proper Lie subgroups of $SO(3)$ are all isomorphic to $SO(2)$, there can arise only the two situations

$$H \simeq SO(3) \quad \text{(complete isotropy)},$$

$$H \simeq SO(2) \quad \text{(axial isotropy in planes)},$$

in the first of which the group $\hat{G}$ will have dimension 6, and in the second dimension 4. We shall now examine in more detail the first of these situations.

31.5. Friedman's Models

The investigation of homogeneous and isotropic cosmological models ("Friedman's models"), where the group $\hat{G}$ defined above has dimension 6 (the case of full isotropy), presents no special difficulty, and yet is of fundamental significance for relativistic cosmology. Astronomical observations show that at the present stage of evolution of the universe the distribution of matter throughout the universe is such as to make it "on the average" homogeneous and isotropic, if the averaging is carried out over sufficiently large distances which are at the same time negligible in comparison with the "metagalactic", i.e. with that part of the universe observable at present. (The present size of the metagalactic is of the order of 10^{28} cm; as far as the isotropy of the distribution of matter is concerned it is appropriate to average over distances of the order of the size of galactic clusters, i.e. from 10^{25} to 10^{26} cm. We remark that investigation of the residual background radiation, whose energy density in the early stages of the evolution of the universe must have exceeded the density of matter, has so far revealed no anisotropy.)

It can be shown that there are only three homogeneous, isotropic, simply-connected, 3-dimensional Riemannian manifolds M^3 with isometry group $\hat{G}$ of dimension 6, namely the 3-sphere S^3, Euclidean space $\mathbb{R}^3$, and 3-dimensional Lobachevskian space L^3. (The 2-dimensional analogues S^2 and L^2 were considered in Part I, §§9, 10.) The metrics of these three spaces may be written respectively as follows:

$$dl^2(t) = a^2(t)\, dl_0^2 = a^2 \begin{cases} d\chi^2 + \sin^2\chi\, d\Omega^2\, (S^3), \\ d\chi^2 + \chi^2\, d\Omega^2\, (\mathbb{R}^3), \\ d\chi^2 + \sinh^2\chi\, d\Omega^2\, (L^3), \end{cases} \tag{46}$$

$$(d\Omega^2 = d\theta^2 + \sin^2\theta\, d\varphi^2),$$

where a^2 is the scaling or averaging factor. (Note that $d\Omega^2$ is the usual metric on the unit sphere S^2.) Hence in terms of a synchronous co-ordinate system the metric on the corresponding manifold M^4 has the form

$$ds^2 = c^2\, dt^2 - dl^2(t) = (dx^0)^2 - a^2(t)\, dl_0^2,$$

where dl_0^2 is given by one or another of the formulae in (46). If we change to a new time co-ordinate η defined by

$$c\, dt = a\, d\eta, \tag{47}$$

then the metric on M^4 becomes

$$ds^2 = a^2(\eta)(d\eta^2 - dl_0^2). \tag{48}$$

(Note incidentally that the metric of our homogeneous, isotropic model is thus "conformally flat" (i.e. conformally equivalent to the Minkowski metric; see Part I, §15). It is also clear from (48) that the metric has a singularity at $a = 0$.)

From the general formula for the components R_{ab} given in Part I, §39.3(30), applied to the metric (48), one finds (after some calculation and invoking the homogeneity assumption) that the Einstein equations reduce to the single equation

$$R_0^0 - \tfrac{1}{2}R = \frac{8\pi G}{c^4} T_0^0. \tag{49}$$

Since a non-zero spatial velocity at any point would contravene the prevailing assumption of complete isotropy at each time t, it follows that the 4-velocity must be trivial, i.e. $u \equiv (1, 0, 0, 0)$ (in terms of the old time-co-ordinate x^0), whence $T_0^0 = \varepsilon$. The conservation law $\nabla_a T_b^a = 0$ yields in the present context

$$3 \ln a = -\int \frac{d\varepsilon}{p + \varepsilon} + \text{const.} \tag{50}$$

Knowledge of the state equation $p = p(\varepsilon)$ together with (50) then allows $T_0^0 = \varepsilon$ to be eliminated from equation (49), which can then be completely solved for the function $a(t)$ (or rather for the functions $a(\eta)$ and $t(\eta)$) determining (via (48)) the metric ds^2. We now carry out this solution explicitly in the case $p = 0$ for each of the three possibilities S^3, L^3, $\mathbb{R}^3$ in turn. (Note that if $p = 0$ then $\varepsilon = \mu c^2$ where μ is the mass density, and it follows from (50) that $\varepsilon a^3 = \text{const.}$, so that $M = \mu a^3$ is an integral of the equation (49).)

(a) For the sphere S^3, calculation (using the formula for R_{ab} given in Part I, §39.3(30)) yields

$$R = -\frac{6}{a^3}(a + a''), \qquad R_0^0 = \frac{3}{a^4}(a'^2 - aa''),$$

where the prime denotes differentiation with respect to η. Hence the Einstein equation (49) (with $T_0^0 = \varepsilon$) becomes in this case

$$\frac{8\pi G}{c^4}\varepsilon = \frac{3}{a^4}(a'^2 + a^2). \tag{51}$$

Assuming $p = 0$, it follows, as noted above, that $\varepsilon a^3 = \text{const.}$, whence (as is easily verified) we obtain the solution

$$a = a_0(1 - \cos\eta), \qquad a_0 = \text{const.} \tag{52}$$

From this and (47) we then obtain

$$t = \frac{a_0}{c}(\eta - \sin\eta). \tag{53}$$

We shall now examine the behaviour of the mass density μ for small t (or η). From (51) and (52) we have

$$\mu = \frac{\varepsilon}{c^2} = \frac{3c^2}{4\pi G a_0^2}(1 - \cos\eta)^{-3},$$

while from (53) (and the Taylor series for the sine function) we obtain $t \sim a_0\eta^3/6c$. Since, similarly, $(1 - \cos\eta)^3 \sim \eta^6/8$, we deduce that

$$\mu \sim \frac{t^{-2}}{6\pi G} \quad \text{as } t \to 0. \tag{54}$$

(b) For Lobachevskian 3-space L^3, the Einstein equation (49) can be shown, by means of the analogous calculations of R_0^0 and R, to take the form

$$\frac{8\pi G}{c^4}\varepsilon = \frac{3}{a^4}(a'^2 - a^2),$$

whence we obtain (in the case $p = 0$)

$$a = a_0(\cosh\eta - 1), \qquad t = \frac{a_0}{c}(\sinh\eta - \eta), \qquad a_0 = \text{const.},$$

and we find that the mass density μ has the same asymptotic behaviour as before: $\mu \sim t^{-2}/6\pi G$ as $t \to 0$.

(c) If the orbits are Euclidean (i.e. $\mathbb{R}^3$) then the Einstein equation (49) (together with the usual condition (50)) gives

$$\frac{8\pi G}{c^4}\varepsilon = \frac{3}{a^2}\left(\frac{da}{dt}\right)^2, \qquad \mu a^3 = \text{const.},$$

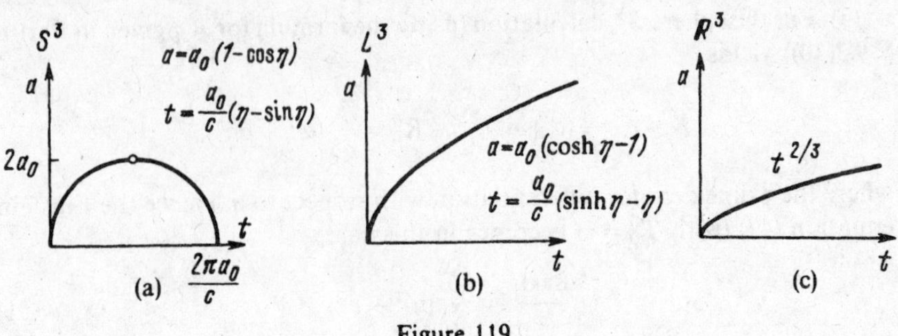

Figure 119

whence

$$a = \text{const.} \times t^{2/3}.$$

The asymptotic behaviour of μ as $t \to 0$ is then easily verified to be essentially as in the previous two cases; in fact

$$\mu = \frac{c^2}{6\pi G} t^{-2}.$$

The graph of a as a function of t is given in this and the other two cases in Figure 119. (Note that in all three cases the transformation $\eta \to -\eta$, $t \to -t$ preserves the solution.)

Remark. It is assumed that as $\varepsilon \to \infty$ (i.e. for small t) the state equation $p = \varepsilon/3$ is the appropriate one (rather than $p = 0$); the analogous formulae for this case can be obtained in a manner similar to the above, one difference being that the condition (50) yields $\varepsilon a^4 = \text{const.}$ (rather than $\varepsilon a^3 = \text{const.}$).

It is immediate from (48) that in our homogeneous isotropic model the paths of light rays satisfying $\theta = \text{const.}$, $\varphi = \text{const.}$, will be given by

$$\chi = \pm\eta + \text{const.} \qquad (\varphi = \text{const.}, \theta = \text{const.}). \tag{55}$$

Since the metric is not stationary, the frequency ω of a (monochromatic) light ray will not be an integral of the Einstein equation, i.e. will not be constant; it turns out that along a light ray we shall have rather $\omega a = \text{const.}$, as the following exercise shows.

EXERCISE

Consider Maxwell's equations $\partial S/\partial A_a = 0$ with respect to a metric of the form (48), where $a(\eta)$ is arbitrary. (Here S is as in the last exercise of §39.4 of Part I, and $A = (A^a)$ is the 4-potential (cf. Part I, §37.3).) Find (complex) solutions of the form

$$A(x, t) = e^{i\omega t} A(x), \qquad (c\,dt = a\,d\eta),$$

and show that $\omega a = \text{const.}$

Thus if ω_0 is the frequency of the light ray at the instant $\eta_0 = \eta - \chi$ of its emission, then at the (present) time η we shall have

$$\omega = \omega_0 \frac{a(\eta - \chi)}{a(\eta)}. \tag{56}$$

Astronomical observations show that the universe is expanding. The quantity

$$H = c\frac{a'(\eta)}{a^2} = \frac{d \ln a}{dt}, \tag{57}$$

known as the "Hubble constant" is directly observable; its reciprocal H^{-1}, which has the dimensions of time, has the present-day estimate

$$H \approx 13 \times 10^9 \text{ years } \pm 25\%.$$

From (57) and the solutions obtained above under the assumption $p = 0$ in each of the cases S^3, $\mathbb{R}^3$, L^3, it follows easily that

$$\begin{aligned} S^3&: t = \frac{a_0}{c}(\eta - \sin \eta) = \frac{1}{H} g(\eta) = \frac{1}{H} \frac{\sin \eta(\eta - \sin \eta)}{(1 - \cos \eta)^2}, \\ \mathbb{R}^3&: t = \tfrac{2}{3}H^{-1}, \\ L^3&: t = \frac{a_0}{c}(\sinh \eta - \eta) = \frac{1}{H} f(\eta) = \frac{1}{H} \frac{\sinh \eta(\sinh \eta - \eta)}{(\cosh \eta - 1)^2}. \end{aligned} \tag{58}$$

Since it can be shown that $0 < g(\eta) < \frac{2}{3}$ for $0 < \eta < 2\pi$ (at the endpoints of which interval $a = 0$), and $\frac{2}{3} < f(\eta) < 1$ for all η, we conclude that:

In the isotropic Friedman model of the universe (with $p = 0$) the time interval from the instant t_0 when $a = 0$, till the present, does not exceed the present value of H^{-1}.

Exercises

1. Show that for the state equation $p = \varepsilon/3$, this estimate of the age of the universe is still essentially valid (within the given framework of assumptions).
2. Verify (directly from the various forms taken by the Einstein equation in the above three cases, together with the definition (57) of H) that, assuming $p = 0$, the possibilities

$$\mu > \frac{3H^2}{8\pi G}, \qquad \mu = \frac{3H^2}{8\pi G}, \qquad \mu < \frac{3H^2}{8\pi G},$$

correspond respectively to orbits S^3, $\mathbb{R}^3$, L^3.

31.6. Anisotropic Vacuum Models

There naturally arises the question of the extent to which the most important of the conclusions obtained in the preceding subsection in the context of homogeneous and isotropic models, remain valid for more general models.

Various classes of "perturbations" of the isotropic models have been studied with just this question in mind. The only class of "large" perturbations of the isotropic models which has been thoroughly investigated up till now is that consisting of the general (non-isotropic) models; the two most important questions posed in connexion with this class are the following:

(i) Does there always exist a singularity, or is this specific to the isotropic models?
(ii) Is it possible to find sufficiently general initial conditions guaranteeing that the metric, as it evolves, will in some sense "isotropify" into its currently observed isotropic state?

Concerning the first of these questions, it is not difficult to establish the following fact: Write $g(t)=\det(g_{\alpha\beta}(t))$ $(\alpha, \beta=1, 2, 3)$, where g_{ab} is a metric of signature $(+---)$, in terms of synchronous local co-ordinates $x^0=ct$, x^1, x^2, x^3 on any homogeneous model M^4, so that in particular since $g_{00}=1$, $g_{0\alpha}=0$, the volume element on M^4 is given by $d\sigma=\sqrt{-g}\,d^4x$; it then follows from the Einstein equations (1) together with the condition that the energy density be non-negative (i.e. $T_0^0\geq 0$) that there is a value t_0 of the time co-ordinate such that $g(t_0)=0$. However, this does not necessarily imply that the metric g_{ab} on M^4 has a corresponding singularity, as we shall now see.

The above synchronous frame $x^0=ct, x^1, x^2, x^3$ has the property that the time axis is orthogonal to the "spatial hypersurfaces", each of which is an orbit under the given group G. We may instead choose a new "time" co-ordinate $\tilde{x}^0$ transverse, but not orthogonal, to the orbits, with respect to which the metric takes the form

$$\tilde{g}_{ab}=\begin{pmatrix} \tilde{g}_{00} & \tilde{g}_{01} & 0 & 0 \\ \tilde{g}_{10} & \tilde{g}_{11} & \tilde{g}_{12} & \tilde{g}_{13} \\ 0 & \tilde{g}_{21} & \tilde{g}_{22} & \tilde{g}_{23} \\ 0 & \tilde{g}_{31} & \tilde{g}_{32} & \tilde{g}_{33} \end{pmatrix}, \qquad \tilde{g}_{ab}=\tilde{g}_{ba}, \tag{59}$$

where $\tilde{g}_{00}\equiv 0$, $\tilde{g}_{01}\neq 0$. We can describe in greater detail how to make this choice appropriately, in terms of the corresponding new left-invariant vector fields $\tilde{X}_a$ satisfying the relations (42); if the integral curves of the field $\tilde{X}_0$ are for instance null geodesics (i.e. light geodesics), then we shall have $\tilde{g}_{00}\equiv 0$, and by choosing $\tilde{X}_2$ and $\tilde{X}_3$ orthogonal to $\tilde{X}_0$ and scaling $\tilde{X}_1$ appropriately, we can arrange that $\langle\tilde{X}_0, \tilde{X}_1\rangle=\tilde{g}_{01}=\text{const.}\neq 0$, $\tilde{g}_{02}\equiv\tilde{g}_{03}\equiv 0$. If at $t=t_0$ we then have $\tilde{g}_{11}=\tilde{g}_{12}=\tilde{g}_{13}=0$, the restriction of the metric to the orbit $M^3(t_0)$ will be singular, and on reverting to a synchronous frame the metric $(\tilde{g}_{ab})$ will change to one with zero determinant at t_0, even though $\det(\tilde{g}_{ab})\neq 0$. Points where this phenomenon occurs are called *fictional* (or *removable* or *co-ordinate*) *singularities* since they are accidents of a particular co-ordinate system. (Cf. the Schwarzschild as against the Kruskal co-ordinates in §31.2.) Fictional singularities may arise even in axially symmetric perturbations of the isotropic models.

The anisotropic homogeneous case differs from that of full isotropy in having several (as opposed to essentially just one) non-trivial "vacuum" solutions, i.e. under the assumption $\varepsilon = 0$. We shall now consider the simplest of these.

(a) Kasner's solution. Here $G = \mathbb{R}^3$, which has commutative Lie algebra (of type I in the classification given in Part I, §24.5, of the 3-dimensional Lie algebras). Thus the fields X_a (forming a "tetrad"—see §31.4) directed "parallel" to the respective co-ordinate x^a-axes, commute. The metric has the following form (putting $c = 1$ for simplicity):

$$ds^2 = dt^2 - \sum_{\alpha=1}^{3} t^{2p_\alpha}(dx^\alpha)^2. \tag{60}$$

For a metric of this form the following two conditions turn out to be equivalent to Einstein's equation $R_{ab} = 0$ in empty space:

$$R_{ab} = 0 \Leftrightarrow \sum_{\alpha=1}^{3} p_\alpha = 1, \qquad \sum_{\alpha=1}^{3} p_\alpha^2 = 1.$$

EXERCISE

Show that in the case $p_1 \equiv p_2 \equiv 0$, $p_3 \equiv 1$, the Kasner metric (60) transforms under a suitable co-ordinate change into the Minkowski metric (whence it follows that the singularity $t = 0$ is fictitious). How does the group $G = \mathbb{R}^3$ act in terms of the Minkowski coordinates?

(b) The Taub–Misner solution. Here $G = SU(2)$ (which has Lie algebra of type IX in the classification given in Part I, §24.5). It is assumed that with respect to some synchronous local co-ordinate system the matrix $g_{\alpha\beta}(t)$ is diagonal: $g_{\alpha\beta} = q_\alpha^2 \delta_{\alpha\beta}$. The Taub solution is then determined by the following two conditions:

(i) $q_1^2 = q_2^2$ (this is the condition of axial isotropy; thus the full group $\hat{G}$ of isometries is isomorphic to $SU(2) \times SO(2)$, and so is 4-dimensional).

(ii) $\varepsilon = 0$.

Under these conditions the Einstein equations $R_{ab} = 0$ are exactly integrable; in terms of our synchronous co-ordinate system the solution turns out to have the form (writing $a = q_1^2$, $b = q_2^2$)

$$\begin{aligned} a^2 = b^2 &= \frac{q}{2} \frac{\cosh(2q\tau + \delta_1)}{\cosh^2(q\tau + \delta_2)}, \\ c^2 &= \frac{2q}{\cosh(2q\tau + \delta_1)}, \qquad \frac{d\tau}{dt} = \frac{1}{abc}, \end{aligned} \tag{61}$$

where q, δ_1, δ_2 are constants.

For the space-like orbits $M^3(t)$ we have $M^3(t) \cong S^3 \cong SU(2)$. From (61) it is not difficult to see that as $t \to 0$, we have $\tau \to -\infty$, whence $c \to 0$; thus in one particular direction distances contract to zero as $t \to 0$. Hence in terms of our synchronous frame we obtain a corresponding projection map

$$(t \to 0), \qquad p\colon S^3 \to S^2 \cong M^3(0) \cong S^3/S^1 \cong SU(2)/SO(2),$$

with fibre S^1; this is just the Hopf fibre bundle over S^2 (see §24.3). It would appear, therefore, that the sphere $M^3(0)$ is singular in our Einsteinian manifold; however it is only fictitiously so since there is a pair of co-ordinate frames, with corresponding tetrads of fields $\tilde{X}_a^{\pm}$ satisfying (42), in terms of which the metric is non-singular at $t = 0$, having the form

$$(\tilde{g}_{ab}^{(\pm)}) = \begin{pmatrix} 0 & \pm 1 & 0 & 0 \\ \pm 1 & \tilde{g}_{11}^{(\pm)} & 0 & 0 \\ 0 & 0 & \tilde{g}_{22}^{(\pm)} & 0 \\ 0 & 0 & 0 & \tilde{g}_{33}^{(\pm)} \end{pmatrix} = (\langle \tilde{X}_a^{\pm}, \tilde{X}_b^{\pm} \rangle), \tag{62}$$

$$\tilde{g}_{11}^{(\pm)} = \tilde{q}_1^2 = \frac{At + 1 - 4B^2t^2}{B(4B^2t^2 + 1)}, \qquad \tilde{g}_{22}^{(\pm)} = \tilde{g}_{33}^{(\pm)} = Bt^2 + \frac{1}{4B}.$$

This is what is known as the Taub–Misner metric (or alternatively the Taub–NUT metric, for Newman, Unti and Tamburino).

EXERCISES

1. Show that the regions $t > 0$ of the Einsteinian manifolds endowed respectively with the metrics (61) and (62), are isometric.
2. Find the matrix $A(t)$ transforming the tetrad $(\tilde{X}_a^+)$ into $(\tilde{X}_a^-)$.
3. Show that on the orbit $M^3(0) \cong SU(2)$ (i.e. that corresponding to $t = 0$) the integral curves of each of the fields $\tilde{X}_1^{\pm}$ are closed, light-like curves. Show that the curves "parallel" to the former synchronous time co-ordinate $x^0 = ct$ (i.e. those defined by $x^1, x^2, x^3 = \text{const.}$) wind themselves as $t \downarrow 0$ (i.e. from within the region $t > 0$) onto these closed light-like curves, as onto limit cycles (thus constituting topologically infinite curves of finite length, revealing a kind of "incompleteness" of the indefinite metric).

Remark. The Taub–Misner metrics $\tilde{g}_{ab}^{\pm}$, which are both analytic continuations of the Taub metric (expressed in (61) in terms of a synchronous time co-ordinate t) beyond the region $t > 0$, have the interesting property that the change of co-ordinates in the region $t > 0$ transforming $\tilde{g}_{ab}^{+}$ into $\tilde{g}_{ab}^{-}$, can *not* be continued analytically throughout the whole manifold. In fact, the metrics $\tilde{g}_{ab}^{+}$ and $\tilde{g}_{ab}^{-}$ are not (analytically) equivalent on the whole of M^4 in view of the fact that there is only one isometry between them on the region $t > 0$, while the matrix $A(t)$ of the change from the tetrad $(\tilde{X}_a^+)$ to $(\tilde{X}_a^-)$ is not analytic at $t = 0$. We conclude from this that the Taub metric has essentially distinct analytic continuations beyond the fictionally singular surface $M^3(0)$.

The property of the Taub–Misner metric described in Exercise 3 above should not really cause any great surprise. To see why, consider the simple

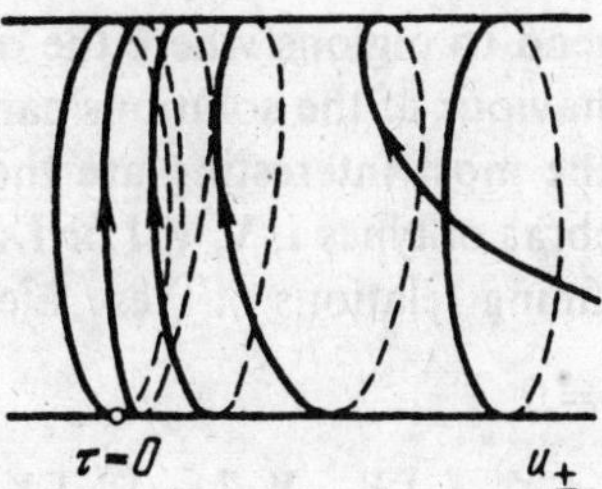

Figure 120. The (u, v)-cylinder, showing a curve $x = \ln u_- + v_- = \text{const.}$

example of a metric on a 2-dimensional manifold with co-ordinates x, τ, given in the region $\tau > 0$ by

$$ds^2 = d\tau^2 - \frac{\tau^2}{4} dx^2. \tag{63}$$

This metric admits the group G of transformations of the form $x \to x + x_0$; if we assume the co-ordinate x to be "cyclic", i.e. if we identify x with $x + 2\pi$, then $G = SO(2)$. In these, obviously synchronous, co-ordinates the metric is apparently singular when $\tau = 0$. However the following two co-ordinate changes enable us to extend the metric beyond the region $\tau > 0$ in two ways:

$$\tau = 2\sqrt{u_+}, \qquad x = \ln u_+ - v_+;$$

$$\tau = 2\sqrt{u_-}, \qquad x = \ln u_- + v_-.$$

In terms of these new co-ordinate systems the metric takes the form

$$ds_\pm^2 = \pm 2du_\pm \, dv_\pm - u_\pm (dv_\pm)^2,$$

so that the corresponding matrices $\tilde{g}_{ab}^{(\pm)}$ are given by

$$\tilde{g}_{ab}^{(+)} = \begin{pmatrix} 0 & 1 \\ 1 & -u_+ \end{pmatrix}, \qquad \tilde{g}_{ab}^{(-)} = \begin{pmatrix} 0 & -1 \\ -1 & -u_- \end{pmatrix}.$$

Confining our attention for the moment to the metrics $ds_\pm^2$ as given here in terms of the co-ordinates $u_\pm$, $v_\pm$ (and ignoring their source in (63)), we take the co-ordinates $v_\pm$ to be cyclic and allow $u_\pm$ to take all real values. It is then easy to see that in the respective co-ordinate systems the curves $u_\pm = 0$ are limit cycles for the curve $x = \text{const.}$ "parallel" to the synchronous τ-axis and orthogonal to the orbits of the group $G = SO(2)$ (see Figure 120).

Note that in this example the co-ordinate change $u_+ = u_-$, $v_+ = -v_-$ transforms $\tilde{g}_{ab}^{(+)}$ to $\tilde{g}_{ab}^{(-)}$; there is however no correspondingly simple change in the case of the Taub–Misner solution.

31.7. More General Models

In the situation of non-empty space ($\varepsilon \neq 0$) there are known several other, more complex, homogeneous models also possessing fictional singularities, beyond which they extend to regions where the orbits are no longer space-like (and where the behaviour of the solutions can be exceedingly complex). Among these models the most interesting are those whose (3-dimensional) groups G have Lie algebras of types I, V, VII or IX in the classification given in Part I, §24.5; the defining relations of these Lie algebras are as follows:

$\mathbb{R}^3$ (type I): $[X_\alpha, X_\beta] = 0$;

$\mathfrak{g}_5$ (type V): $[X_1, X_2] = X_2, \quad [X_2, X_3] = 0, \quad [X_3, X_1] = -X_3$;

$\mathfrak{g}_7$ (type VII): $[X_1, X_2] = aX_2 + X_3, \quad [X_2, X_3] = 0,$
$[X_3, X_1] = X_2 - aX_3$;

$\mathfrak{g}_9$ (type IX): $[X_1, X_2] = X_3, \quad [X_2, X_3] = X_1, \quad [X_3, X_1] = X_2$.

These models are of primary interest among the general homogeneous models because they include the isotropic Friedman models (see §31.5) as special cases, and also because the idea of the universe "isotropifying" as it expands can be given precise meaning for them. The simplest example of the latter kind of solution, affording in a particular case a "naive" solution to the "isotropification" problem, is the Heckmann–Schücking solution (having associated Lie algebra of type I):

$$ds^2 = dt^2 - \sum_{i=1}^{3} c_i t^{2p_i}(t + t_0)^{(4/3) - 2p_i}\, dx_i^2,$$
$$c_i > 0, \qquad p_1 + p_2 + p_3 = p_1^2 + p_2^2 + p_3^2 = 1, \tag{64}$$

where the c_i and p_i are constants. It is clear that as $t \to \infty$ this solution is asymptotic to the isotropic Friedman solution with orbits $\mathbb{R}^3$ (see §31.5(c)), and is asymptotic as $t \to 0$ to the Kasner solution (§31.6(a)).

For Friedman's isotropic models the Lie algebras of the full groups of isometries have the following form:

orbit S^3; $G = G^6_+$; $[X_i, X_j] = \varepsilon_{ijk} X_k, \quad [Y_i, Y_j] = \varepsilon_{ijk} Y_k, \quad [X_i, Y_j] = 0$;

orbit $\mathbb{R}^3$; $G = G^6_0$; $[X_i, X_j] = \varepsilon_{ijk} X_k, \quad [X_i, Y_j] = \varepsilon_{ijk} Y_k, \quad [Y_i, Y_j] = 0$;

orbit L^3; $G = G^6_-$; Lie algebra isomorphic to $sl(2, \mathbb{C})$.

In these Lie algebras one can find subalgebras of dimension 3 of types I, V, VII and IX which are the Lie algebras of Lie subgroups of dimension 3 acting transitively on each orbit (S^3, $\mathbb{R}^3$ or L^3); for instance, the Lie subalgebra generated by Y_1, Y_2, Y_3 in the second of the above three cases (i.e. with orbits $\mathbb{R}^3$) is a type I Lie algebra of this sort. Thus as noted above, Friedman's isotropic models are indeed particular cases of the general homogeneous models with associated Lie algebras of types I, V, VII and IX, and consequently these latter models are useful for investigating homogeneous perturbations of the isotropic models. (Note incidentally that, as represented by their sets of structural constants, the Lie algebras of types I and V are "limiting" for those of types VII and IX. (Thus for instance for each $a \neq 0$ the Lie algebra defined by $[X_1, X_2] = aX_3$, $[X_2, X_3] = aX_1$, $[X_3, X_1] = aX_2$, is easily seen to be of type IX; on letting a become zero however we obtain a type I Lie algebra.))

It turns out that fictional singularities are not a typical feature of models of types I, VII or IX (nor, in all likelihood, of those of type V), and can be made disappear by means of small perturbations. From these homogeneous models there arises a whole series of non-trivial asymptotic solutions (as $t \to 0$), the most complex of which is the so-called "oscillatory regime" discovered relatively recently; although this particular theme has been intensively developed over the past 10 years, we shall not consider here the most recent questions concerning "oscillatory regimes", nor the associated problem of reformulating the theory of Einstein's equations as a qualitative theory of dynamical systems.†

For the model of type IX (with group $G \simeq SU(2)$), if the matter is assumed to be at rest "on the average", so that we may take $u = (1, 0, 0, 0)$, then, as noted earlier, by choosing a suitable synchronous local co-ordinate system the metric may be assumed to be at all times diagonal (with $g_{00} \equiv 1$); thus we seek the metric in the form

$$g_{\alpha\beta}(t) = q_\alpha^2 \delta_{\alpha\beta}, \qquad \alpha, \beta = 1, 2, 3.$$

If we further suppose $p = k\varepsilon$, where k is a constant between 0 and 1, then in terms of a new time co-ordinate τ, satisfying $q^k d\tau = dt$ (where $q = q_1 q_2 q_3$), and appropriately defined p_α, $\alpha = 1, 2, 3$, (not related to the pressure p!) the Einstein equations take the Hamiltonian form $\dot{p}_\alpha = -\partial H/\partial p_\alpha$, $\dot{q}_\alpha = \partial H/\partial p_\alpha$,

$$H = \frac{1}{4(q_1 q_2 q_3)^{1-k}} (P_2(p_1 q_1, p_2 q_2, p_3 q_3) + P_2(q_1^2, q_2^2, q_3^2)), \tag{65}$$

where the p_i are defined by $p_i = (d/dt)(q_j q_l) = q^{-k}(d/d\tau)(q_j q_l)$, i, j, l all distinct, and $P_2(x, y, z)$ is the quadratic expression

$$P_2(x, y, z) = 2(xy + yz + zx) - (x^2 + y^2 + z^2).$$

(We invite the reader to verify this!)

Remark. The analogous solution corresponding to the Lie algebra of type I also reduces after diagonalization to a Hamiltonian system, with Hamiltonian of the form

$$H^{\text{I}} = \frac{1}{4(q_1 q_2 q_3)^{1-k}} P_2(p_1 q_1, p_2 q_2, p_3 q_3).$$

† Throughout the 1970's, the early stages in the evolution of homogeneous models of the universe were studied using the methods of the modern qualitative theory of multi-dimensional dynamical systems; these investigations have made possible a precise formulation and solution of the problem of the "typical initial states" of the metric in relation to the expansion process, and furnished us with an answer (within the framework of the theory of homogeneous models) to the question as to what they can be in fact. (See the discussion below, concluding §31.)

Returning to our type IX solution, we note that the usual conditions (conservation of energy-momentum, non-negativity of energy density) yield the constraints

$$\varepsilon(q_1 q_2 q_3)^{1+k} = H(p_\alpha, q_\alpha) \geq 0. \tag{66}$$

It is therefore natural to take as the "physical region" $S \subset \mathbb{R}^6(p_\alpha, q_\alpha)$ of phase space, that defined by the conditions

$$q_\alpha > 0, \qquad H(p_\alpha, q_\alpha) \geq 0. \tag{67}$$

For arbitrary k $(0 \leq k \leq 1)$ it follows easily from the Hamiltonian equation $\dot{q}_\alpha = \partial H/\partial p_\alpha$ and the definition (65) of H, that

$$\dot{q} = \tfrac{1}{2} q^k (p_1 q_1 + p_2 q_2 + p_3 q_3), \tag{68}$$

where q is (as above) the volume element: $q = \sqrt{-g} = q_1 q_2 q_3$. It can be shown using this (in particular), that if t_0 is such that $\dot{q}(t_0) < 0$ then $\dot{q}(t) < 0$ for all $t > t_0$, and also that $(d^2/dt^2)(q^{1/3}) < 0$.

Exercise

Establish the following equations:

$$\frac{d^2}{d\tau^2}(q^{1/3}) = -R_0^0 q^{1/3}/3; \qquad R_0^0 = T_0^0 - \tfrac{1}{2} T_\alpha^\alpha = \varepsilon + 3p/2 \quad (\text{assuming } u = (1, 0, 0, 0)).$$

Hence if we let the time vary in the direction of contraction (which happens to be the positive direction for this time co-ordinate) we shall inevitably reach a time t_1 at which $q(t_1) = 0$ (where the metric will thus be either singular or at least fictionally singular). We shall now examine this situation in detail under the additional condition of axial isotropy $q_2 = q_3$, $p_2 = p_3$ (i.e. on the phase surface defined by these equations). We first reverse the direction of flow of the time co-ordinate (by replacing it by $t_1 - t$) so that the problem becomes that of the behaviour of the trajectories of the above system as $t \downarrow 0$, where here t is the new time co-ordinate in terms of which now $\dot{q} > 0$ (and $q > 0$ as before). It is easily verified that our Hamiltonian system admits the "scaling" transformations $q_\alpha \to \lambda q_\alpha$, $p_\alpha \to \lambda p_\alpha$, $H \to \lambda^{3k-1} H$; by means of these transformations the Einstein equations, in the guise of the above Hamiltonian system (and under the assumption $q_2 = q_3$, $p_2 = p_3$) can be reduced to the 3-dimensional dynamical system

$$\begin{aligned} \frac{du}{d\tau} &= \dot{u} = -w^2 + 2v^2 - 2uv^2 + (2u - 1)H_2, \\ \dot{w} &= w(u - 1 + 2H_2 - 2v^2), \\ \dot{v} &= \tfrac{1}{2} v(-k - (1-k)(u-1)^2 - (1-k)w^2 - 4kv^2), \\ H_2 &= \frac{1-k}{4}(1 - (u-1)^2 - w^2 + 4v^2), \end{aligned} \tag{69}$$

where the co-ordinates u, w, v and the time co-ordinate τ are given by

$$u = \frac{p_1 q_1}{p_2 q_2 + p_3 q_3} = \frac{p_1 q_1}{2p_2 q_2}, \qquad w = \frac{q_1^2}{2p_2 q_2},$$
$$v = \frac{q_2}{q_1} w, \qquad \frac{d\tau}{dt} = -\frac{w}{q_1 v^2} \tag{70}$$

The following conditions are imposed on this system:

$$H_2 \geq 0, \qquad w < 0, \qquad v < 0; \tag{71}$$

the first of these is easily seen (via (66)) to be equivalent to non-negativity of the energy density, while the other two represent a choice of region where the metric has no singularity. That portion of the "boundary" of this region where $v = 0$ is an invariant manifold of dimension 2 with co-ordinates u, v, on which the system (69) obviously reduces to the following one:

$$\dot{u} = -w^2 + (2u - 1)\bar{H}_2, \qquad \dot{w} = w(u - 1 + 2\bar{H}_2),$$
$$\bar{H}_2 = \frac{1-k}{4}(1 - (u-1)^2 - w^2). \tag{72}$$

It is almost immediate that as a vector field this system has exactly four singular points Φ, C, N, T, given by

Φ (saddle): $u = \frac{1}{2}$, $w = 0$, $v = 0$;

C (saddle): $u = 2$, $w = 0$, $v = 0$;

N (focal point): $u = \dfrac{3+k}{5-k}$, $w = -\dfrac{1}{5-k}\sqrt{(1+3k)(1-k)}$, $v = 0$;

T (node): $u = w = v = 0$.

The behaviour of the trajectories of the dynamical system (72) with respect to these singular points is indicated in Figure 121.

Returning to the 3-dimensional system (69) we observe that since $\dot{v}/v < 0$, the trajectories of this system will approach the surface $v = 0$ as $t \to 0$, i.e. as time flows in the direction of contraction. As a typical such trajectory approaches the boundary $v = 0$, its behaviour will approximate that of a trajectory of the 2-dimensional system (72) (see Figure 121). It can be shown that the separatrices (or "arms") of those trajectories which approach in this

Figure 121

way the singular points of the three possible types Φ, N or T from within the physical region (71), represent solutions of Einstein's equations which are asymptotic (as $t \to 0$) to the following metrics:

$$(\Phi) \quad q_i \sim c_i t^{4/[3(1-k)]},$$

(which in the case $q_1 = q_2 = q_3$ yields what is essentially Friedman's solution with orbits $\mathbb{R}^3$; see §35.1(c));

$$\begin{aligned} &(\mathrm{N}) \quad q_1 \sim c_1 t^{(1-k)/(1+k)}, \qquad && q_2 = q_3 \sim c_2 t^{(3+k)/[2(1+k)]}; \\ &(\mathrm{T}) \quad q_1 \sim c_1 t^2, && q_2 = q_3 \sim c_2, \end{aligned} \tag{73}$$

(cf. the example of a 2-dimensional metric concluding §31.6).

It can be inferred from (73) that there are many solutions (namely, those corresponding to trajectories approaching a singular point of type T) which are fictionally singular when $t = 0$ in the weak sense that although they can be extended smoothly beyond $t = 0$, such continuation will not be twice differentiable. The energy density ε of a "type T" solution will also have a "weak" singularity when $t = 0$, since in view of (66) and (73) we have $\varepsilon \sim \text{const.} \times t^{-2(1+k)}$. We remark incidentally also that without the condition of axial isotropy "type T" solutions turn out to be no longer "typical" among models with associated Lie algebra of type IX.

We now turn briefly to the comparatively simple homogeneous models with corresponding Lie algebras of types I and V. As before, on the assumption that the matter does not move (on the average), the metric takes diagonal form (with $g_{00} \equiv 1$) in terms of a suitable synchronous frame:

$$g_{\alpha\beta} = q_\alpha^2(t)\delta_{\alpha\beta}, \qquad \alpha, \beta = 1, 2, 3.$$

In each of the two cases in question the Einstein equations, without the condition of axial isotropy, reduce to a 2-dimensional dynamical system.

Type I. In terms of co-ordinates

$$u = \frac{p_1 q_1}{p_2 q_2 + p_3 q_3}, \qquad v = \frac{p_2 q_2 - p_3 q_3}{p_2 q_2 + p_3 q_3},$$

(where p_1, p_2, p_3 are suitably defined in advance), the Einstein equations reduce to the following system (with time increasing in the direction of contraction):

$$\frac{du}{d\tau} = (2u - 1)H_2, \qquad \frac{dv}{d\tau} = 2vH_2, \qquad H_2 = \frac{1-k}{4}(1 - (u-1)^2 - v^2),$$

$$\frac{d\tau}{dt} = -\frac{q_1^6}{(p_1 q_1 + p_2 q_2 + p_3 q_3)(q_1 q_2 q_3)}.$$

The singular points of this system are clearly given by

$$\mathrm{S}^1\colon \quad (u-1)^2 + v^2 = 1,$$

$$\Phi\colon \quad u = \tfrac{1}{2}, \quad v = 0,$$

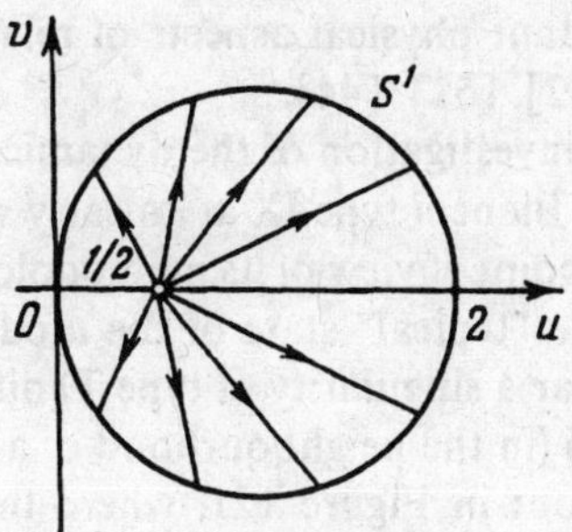

Figure 122

and the integral trajectories are as indicated in Figure 122.

Type V. In terms of co-ordinates

$$v = 2\frac{p_1 q_1 - p_2 q_2}{p_1 q_1 + p_2 q_2}, \qquad r = \frac{2q_1 q_2}{p_1 q_1 + p_2 q_2},$$

the Einstein equations reduce to the system

$$\frac{dv}{d\tau} = v(H_2 + 4r^2), \qquad \frac{dr}{d\tau} = r(-1 + H_2 + 4r^2),$$

$$H_2 = \frac{1-k}{4}(3 - v^2 - 12r^2), \qquad \frac{d\tau}{dt} = -\frac{p_1 q_1 + p_2 q_2}{2q_1 q_2 q_3},$$

which has singular points

$$i_\pm\colon\ r = 0,\ \ v = \pm\sqrt{3}; \qquad \Phi_0\colon\ r = v = 0; \qquad \Phi_1\colon\ r = \tfrac{1}{2},\ \ v = 0.$$

The integral trajectories are sketched in Figure 123.

Remark. A more general investigation of homogeneous models of types VII and IX leads to asymptotic regimes which already at an early stage of their expansion "isotropify" in a certain weak sense.

A more detailed exposition of these and other results from the theory of homogeneous cosmological models can be found in the book [36] (see also

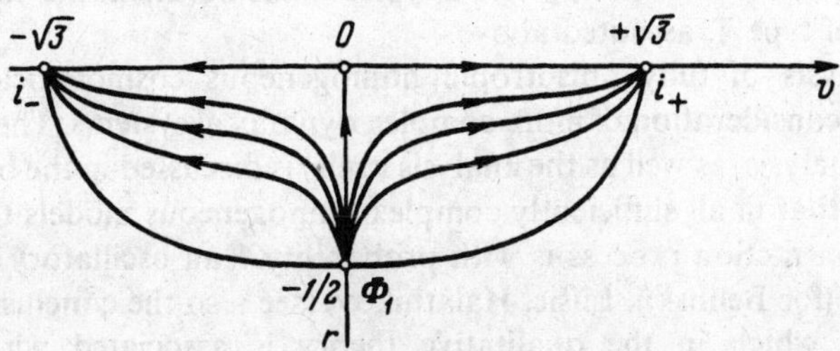

Figure 123

[37]). The various important physical aspects of relativistic cosmology are discussed in the books [42], [51], [44].

The above qualitative investigation of the dynamics of the homogeneous, axially isotropic model of Bianchi type IX at an early stage in its evolution, is of great interest from the point of view of its methodology. We may infer from that investigation that the "typical" state of the model during a process of contraction (occurring near a singularity of type T) differs from the "typical" state during an expansion (in the neighbourhood of a singularity of type N). (This is clearly brought out in Figure 121, where the arrows point in the direction of contraction.) The strict meaning here of the expression "typical state" (during contraction) is as follows: If we choose initial conditions randomly and solve Einstein's equations in the direction of contraction, then sufficiently near to the singularity we shall (in the model in question) with probability 1 find ourselves in a phase region of the type in the vicinity of T. The exact definition of a "typical state" during an expansion process is more complex; here it is particularly appropriate to use the three-dimensional phase manifold introduced above, co-ordinatized by u, w, v, where "infinitely early" states occur, namely on the piece of plane defined by $v = 0$, $\bar{H}_2 \geq 0$ (since $v \to 0$ as the spatial volume contracts to zero). This set of "infinitely early" states (which strictly speaking lies outside the "physical" region $v > 0$ (or $v < 0$), $H_2 > 0$, of the phase space) furnishes us with a natural approach to defining the "typical state" of the metric of our model of the universe during an expansion (and not during a contraction, contrary to one's expectations!). To see how this is, one imagines initial conditions prescribed randomly at a small distance $\varepsilon > 0$ from the boundary $v = 0$, i.e. at the points of the phase space where $|v| = \varepsilon$; then, solving the Einstein equations in the direction of expansion, one observes how the components of the metric vary. It may happen that after a small amount $t_0(\varepsilon)$ of time has elapsed (where $t_0(\varepsilon) \to 0$ as $\varepsilon \to 0$) the metric has become "concentrated" in a narrow region of phase space representing a neighbourhood of a "regime". It is such regimes that we define (under natural, not excessively wild assumptions on the distribution of initial values on $|v| = \varepsilon$) to be "typical early states" of an expansion process. Thus, referring to Figure 121 (with the arrows reversed) we see that in the axially isotropic, homogeneous model of type IX, a typical state during an expansion will occur in the vicinity of a singularity of type N, in contrast with the situation of a contraction, where a typical state occurs in the vicinity of a singularity of type T, as noted above.

The analysis of fully anisotropic homogeneous cosmological models requires the consideration of more complex dynamical systems. The outcome of such an analysis (as well as the analysis itself) is discussed in the book [36]. It turns out that in all sufficiently complex homogeneous models the typical state in a contraction process is with probability 1 an oscillatory regime of type "BLH" (for Belinskiĭ, Lifšic, Halatnikov) (see also the conclusion of the book [42]), which in the qualitative theory is associated with a very interesting, in a certain sense "strange", attractor of the Einsteinian dynamical system, lying on the boundary of the physical region of the phase space (suitably co-ordinatized), and to which all trajectories converge with contracting volume (i.e. as the singularity is approached).

We note that in GTR (assuming the value zero for the cosmological constant; see Part I, §39.3) it is known that a contraction process cannot continue isotropic, and also that any fluctuations inevitably lead to complex regimes of the type BLH, associated with analytically complicated singularities beyond which no continuation of solutions is possible. From this follows in particular the impossibility of investigating any "previous" stage of the universe, where a contraction may have preceded the present expansion.

It turns out that an expansion process (in an anisotropic homogeneous model) has completely different, more regular "typical early states" in its evolution than does a contraction. (The definition of these "typical states" is analogous to that given for the axially isotropic case of a type IX model considered above, though more complex.) These typical states are all asymptotic merely to "powers"; they include quasi-isotropic regimes of types Φ and of types N and T, among others (cf. (73)). In a few cases these "power" asymptotes are of a transitory character and successively replace one another in the early stages of the evolution of the model, depending rather feebly on the type of the homogeneous model. In sum it may be asserted that with large probability a regime of type Φ will be established at a quite early stage (in the exact sense indicated above), with the rate of expansion "almost" (i.e. as reflected by the dominant term of the asymptote) isotropic, although the components of the metric need not themselves be isotropic. A "real", "exact", isotropification of the universe (by which we mean a convergence at an early stage to Friedman's model with overwhelming probability) is not an inevitable consequence of classical GTR. Such are the present conclusions of the theory of homogeneous cosmological models.

§32. Some Examples of Global Solutions of the Yang–Mills Equations. Chiral Fields

32.1. General Remarks. Solutions of Monopole Type

Let $p\colon E \to M$ be a principal fibre bundle with Lie structure group G, and let U_α be an arbitrary chart of the base M with local co-ordinate system x; thus above U_α the bundle decomposes as a product: $p^{-1}(U_\alpha) \cong G \times U_\alpha$. A (general) *Yang–Mills field* $A_a(x)$ is then just (the local expression for) a G-connexion on the fibre bundle (cf. the definition in Part I, §42). Thus, recalling (from §25.1) the definition of a G-connexion (for a matrix Lie group G), $A_a(x)$ is for each $a = 1, \dots, n = \dim M$, a field on M (strictly speaking on an arbitrary cross-section of the bundle E) with values in the Lie algebra of G, which under the transition from one cross-section to another undergoes a "gauge transformation"

$$A_a(x) \to g(x)A_a(x)g^{-1}(x) - \frac{\partial g(x)}{\partial x^a} g^{-1}(x), \tag{1}$$

where the map $g(x)\colon M \to G$ is such that above each chart U_α the map $G \times U_\alpha \to G \times U_\alpha$ defined by $(h, x) \mapsto (hg(x), x)$, interchanges the two cross-sections. (This rule determines in particular how $A_a(x)$ transforms under the transition maps of the fibre bundle defined above the regions of overlap $U_\alpha \cap U_\beta$.)

We shall in fact assume (essentially as in Part I, §42) that $G = SU(2)$, $M = \mathbb{R}^n = U_\alpha$ (so that the bundle is trivial), and $n = 3$ or 4. We shall also suppose that the connexion "trivializes" as $|x| \to \infty$ (cf. the examples concluding §25.5), i.e. that

$$A_a(x) \sim -\frac{\partial g(x)}{\partial x^a} g^{-1}(x) \quad \text{as } |x| \to \infty, \tag{2}$$

(or equivalently that on some cross-section $A_a(x) \to 0$).

Apart from the field $A_a(x)$ we shall be considering another field $\psi(x)$ on M with values in a vector space V which comes with a representation of G by means of linear transformations: $G \to GL(V)$; we shall for simplicity take V to be (the vector-space structure of) the Lie algebra of G. Recalling that the *adjoint representation* of a Lie algebra by means of self-transformations is defined by

$$A \to \operatorname{ad} A\colon B \to [A, B],$$

we choose our *Lagrangian of the field ψ in the absence of any connexion* in the form (cf. Part I, §41.1)

$$L(\psi) = \tfrac{1}{2}\langle \partial\psi, \partial\psi \rangle - u(|\psi|^2), \tag{3}$$

where $|\psi(x)|^2 = \langle \psi(x), \psi(x) \rangle$ is defined at each $x \in M$ via the Killing form on the Lie algebra: $\langle A, B \rangle = -\operatorname{tr}(\operatorname{ad} A \operatorname{ad} B)$; and $\langle \partial\psi, \partial\psi \rangle$ is defined by

$$\langle \partial\psi, \partial\psi \rangle = g^{ab}\left\langle \frac{\partial\psi}{\partial x^a}, \frac{\partial\psi}{\partial x^b} \right\rangle, \tag{4}$$

where $g_{ab}(x)$ is a metric given on the base M and again $\langle \partial\psi/\partial x^a, \partial\psi/\partial x^b \rangle$ is defined by the Killing form; the function u (the "potential") is assumed to be non-negative with graph as indicated in Figure 124. On the other hand in the presence of a connexion $A_a(x)$, we replace the operator $\partial_a = \partial/\partial x^a$ as follows (cf. Part I, §41.1):

$$\partial_a \to \partial_a - \operatorname{ad} A_a(x) = \nabla_a,$$

and define the *full Lagrangian of the field ψ in the presence of the connexion A* by (cf. §§41.1, 42 of Part I)

$$L(\psi, A) = \tfrac{1}{2}\langle \nabla\psi, \nabla\psi \rangle - u(|\psi|^2) + \tfrac{1}{4}\operatorname{tr}(F_{ab}F^{ab}), \tag{5}$$

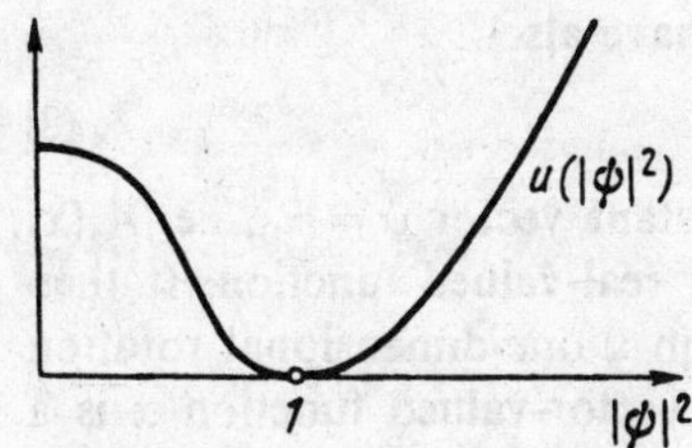

(a) Degenerate vacuum (the sphere S^2).

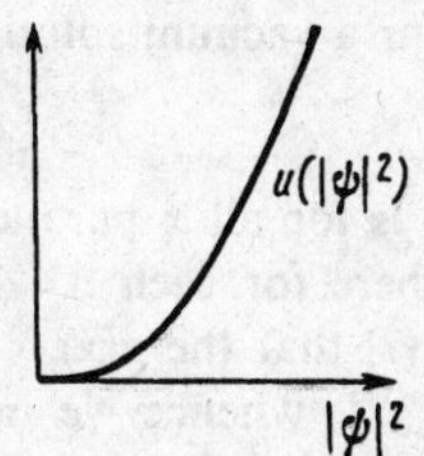

(b) Non-degenerate vacuum (a point).

Figure 124

where

$$F_{ab} = \frac{\partial A_b}{\partial x^a} - \frac{\partial A_a}{\partial x^b} + [A_a, A_b].$$

In order to formulate the global problems in which we are interested, we need the field ψ to be a cross-section of a vector bundle with fibre V and structure group G; the base, if not a region of $\mathbb{R}^n$, will be specified according to the context.

For the remainder of this subsection we shall assume $n = 3$, relegating the case $n = 4$ to the following subsection. Thus the structure group is $G = SU(2)$ (as always), the base is Euclidean 3-space $\mathbb{R}^3$, $\psi(x)$ is a vector field with three real components, and the Lie algebra operation is the ordinary vector or cross product. We are interested in the stationary problem for the quantity

$$S\{\psi, A\} = \int_{\mathbb{R}_3} d^3x\{\tfrac{1}{2}\langle\nabla\psi, \nabla\psi\rangle - u(|\psi|^2) + \tfrac{1}{4}\operatorname{tr}(F_{ab}F^{ab})\}, \tag{6}$$

the action corresponding to the Lagrangian $L(\psi, A)$.

32.1.1. Definition. A *vacuum solution* of the Yang–Mills equation $\delta S = 0$ is a pair (ψ, A) of fields satisfying:

(i) $F_{ab} = 0$;
(ii) $u(|\psi|^2) = \text{min.}$, $\psi = \psi_0 = \text{const.}$;
(iii) $\langle\nabla\psi, \nabla\psi\rangle = 0$.

Since the condition (i) implies that the connexion $A_a(x)$ is trivial (this is Exercise 1 of Part I, §41.2), it follows that for a vacuum solution we must have

$$A_a(x) = -\frac{\partial g(x)}{\partial x^a} g^{-1}(x), \qquad a = 1, 2, 3, \tag{7}$$

for some map $g\colon \mathbb{R}^3 \to G$. Conditions (ii) and (iii) together imply that $g^{ab}\langle[A_a(x), \psi], [A_b(x), \psi]\rangle \equiv 0$, where the scalar product is given by the Killing form, which in the present context defines the Euclidean metric (see Part I, §24.4), and where the metric g_{ab} is also Euclidean (by assumption); it follows that for a vacuum solution we must have also

$$[A_a(x), \psi] \equiv 0, \tag{8}$$

so that $A_a(x)$ is for all x parallel to the constant vector $\psi = \psi_0$, i.e. $A_a(x) = \alpha_a(x)\psi_0$, where for each $a = 1, 2, 3$, α_a is a real-valued function. It then follows from (7) that the $g(x)$, $x \in \mathbb{R}^3$, all lie in a one-dimensional rotation subgroup $\{e^{i\varphi(x)}\}$, whence we infer that the vector-valued function α is a gradient: $\alpha_a = \partial\varphi/\partial x^a$. Thus A_a can be made to vanish on M ($\cong \{0\} \times M$) simply by adding to it a gradient term. It follows that in the situation of a so-called "degenerate vacuum", where u is as indicated in Figure 124(a), the set of all vacuum solutions is essentially in one-to-one correspondence with the set of vectors $\psi_0 \in V \cong \mathbb{R}^3$ satisfying $|\psi_0|^2 = 1$, so that they form a "vacuum" sphere S^2.

In the variational problem $\delta S = 0$ for the functional (6), as it is usually posed, the variational class of pairs (ψ, A) is restricted by the requirement that as $|x| \to \infty$ an admissible pair (ψ, A) should approach a vacuum solution with $A \equiv 0$, $\psi = \psi_0$, where, however, the vector ψ_0 may depend on the manner in which $x \to \infty$. If we allow x to approach ∞ only along rays from the origin, so that $x = |x|v$ where v is a unit vector independent of x, then ψ_0 will depend only on the direction v of the ray, i.e. $\psi_0 = \psi_0(v)$, which in view of the condition $|\psi_0|^2 = 1$ (see above) defines a map

$$\psi_0: S^2 \to S^2,$$

where the value of ψ_0 at each point (or, equivalently, unit vector v) of S^2, namely $\psi_0(v)$, can be identified with a point of the "vacuum" manifold defined by $u(|\psi_0|^2) = \min$. Thus a given map ψ_0 can be regarded as specifying the boundary conditions at infinity on the admissible pairs (ψ, A), and its degree (see §13) affords an integral topological invariant of our variational problem in $\mathbb{R}^3$.

It can be verified directly that the Lagrangian (5) is invariant under gauge transformations, i.e. under transformations of the form

$$\begin{aligned} A_a(x) &\to g(x)A_a(x)g^{-1}(x) - \frac{\partial g}{\partial x^a} g^{-1}(x), \\ \psi &\to g\psi g^{-1} = T_g(\psi), \end{aligned} \tag{9}$$

where it is required that for each unit vector v, the map $g(x)$ (from $\mathbb{R}^3$ to G) have a limit as $|x| \to \infty$ along the ray defined by v, i.e. $x = |x|v$:

$$g(x) \to g_\infty(v): S^2 \to G. \tag{10}$$

(Note that g_∞ is homotopic to a constant since, as mentioned in example (iii) of §25.5, $\pi_2(G) = 0$ for all Lie groups G.) The latitude afforded by the invariance of the Lagrangian under the gauge transformations defined by (9) and (10), allows us to obtain the precise condition under which two sets of boundary conditions are equivalent.

32.1.2. Lemma. *If* $\psi_0^{(1)}: S^2 \to S^2$ *and* $\psi_0^{(2)}: S^2 \to S^2$ *are homotopic maps, then (and only then) there exists a map* $g_\infty: S^2 \to G$ *homotopic to a constant (or, equivalently, extendible to the whole of* $\mathbb{R}^3$*) such that*

$$\psi_0^{(2)}(v) = T_{g_\infty}(v)\psi_0^{(1)}(v) = g_\infty(v)\psi_0^{(1)}(v)g_\infty^{-1}(v), \tag{11}$$

for every unit vector v *(i.e. point of* S^2*).*

PROOF. The action of $G = SU(2)$ on its Lie algebra ($\simeq \mathbb{R}^3$ equipped with the cross product), defined for each $g \in G$ by $a \mapsto gag^{-1}$ (where gag^{-1} is the ordinary matrix product of the matrices g, a, g^{-1}), induces a (transitive) action of G on the set $S^2 \subset \mathbb{R}^3$ of unit vectors v; for the latter action we shall use the notation $v \mapsto gv$.

Consider the fibre bundle $\pi: G \times S^2 \to S^2 \times S^2$ defined by $\pi(g, v) = (gv, v)$, where gv is as just defined. Define maps $\Psi_0, \Psi_1: S^2 \to S^2 \times S^2$ by

$$\Psi_0(v) = (\psi_0^{(1)}(v), \psi_0^{(1)}(v)), \qquad \Psi_1(v) = (\psi_0^{(2)}(v), \psi_0^{(1)}(v)).$$

If $\psi_0^{(1)}$ and $\psi_0^{(2)}$ are homotopic, via a homotopy ψ_t say, then the family of maps Ψ_t defined by

$$\Psi_t(v) = (\psi_t(v), \psi_0^{(1)}(v))$$

constitute a homotopy between Ψ_0 and Ψ_1. Let $\hat{\Psi}_t: S^2 \to G \times S^2$ be a covering homotopy for Ψ_t (see §24.3), with the initial map $\hat{\Psi}_0$, covering Ψ_0, given by

$$\hat{\Psi}_0(v) = (1, \psi_0^{(1)}(v)), \qquad \pi\hat{\Psi}_0 = \Psi_0.$$

If we now define $g_\infty: S^2 \to G$ by

$$\hat{\Psi}_1(v) = (g_\infty(v), \psi_0^{(1)}(v)),$$

then it is clear from the choice of the initial map $\hat{\Psi}_0$ covering Ψ_0 that g_∞ is homotopic to a constant. That g_∞ satisfies (11) follows from

$$\begin{aligned}\pi(\hat{\Psi}_1(v)) &= \pi(g_\infty(v), \psi_0^{(1)}(v)) = (g_\infty(v)\psi_0^{(1)}(v), \psi_0^{(1)}(v)) \\ &= \Psi_1(v) = (\psi_0^{(2)}(v), \psi_0^{(1)}(v)).\end{aligned}$$

We leave the converse to the reader. □

This lemma tells us that a set of boundary conditions at infinity, as represented by a map $\psi_0: S^2 \to S^2$, may be replaced by any homotopic map, or, more precisely, that the boundary conditions correspond essentially to the homotopy classes

$$[\psi_0] \in \pi_2(S^2) \simeq \mathbb{Z},$$

which, in view of Theorem 13.3.1, are determined in turn by their degrees.

Examples. (a) If $\deg \psi_0 = 0$, then by the above lemma we can arrange, via a gauge transformation, that $\psi_0(v) = (0, 0, 1) \in S^2$ for all $v \in S^2$, or in other words that for every ψ in the corresponding variational class (of pairs (ψ, A)

with the given boundary conditions) we have $\psi \to (0, 0, 1)$ as $|x| \to \infty$. Thus in this case there is what they call a "symmetry loss" in the theory, in the sense that since the vector (0, 0, 1) must be preserved, the symmetries reduce to those in the subgroup $SO(2) \subset G$ of rotations of the plane spanned by the vectors (1, 0, 0) and (0, 1, 0) (the "small group of the vacuum"). With the aim in view of developing a theory of perturbations of the vacuum state, the following notation is usually introduced:

$$\psi = \psi_0 + \tilde{\psi}, \qquad f_a(x) = (A_a(x))^3, \qquad (B^1)_a = (A_a)^1, \qquad (B^2)_a = (A_a)^2,$$

in terms of which the Lagrangian is written as

$$\tilde{L}(f_a, B^1, B^2, \tilde{\psi}) = L(\psi, A).$$

We confine ourselves here to the remark that if the potential $u(\xi)$ (assumed, as hitherto, to be as in Figure 124(a)) satisfies $u_{\xi\xi}(1) = m^2 > 0$, then on setting $\tilde{\psi}^1 = \tilde{\psi}^2 = 0$ (see the exercise below), and expanding the function $u(\xi)$ in a power series in ξ about the point $|\psi_0|^2 = |(0, 0, 1)|^2 = 1$, we obtain

$$\tilde{L} = \tfrac{1}{2} \sum_a (\partial_a \tilde{\psi}^3)^2 - 2m^2(\tilde{\psi}^3)^2 + \tfrac{1}{2}(|B^1|^2 + |B^2|^2)$$
$$- \tfrac{1}{2}(|\text{curl } B^1|^2 + |\text{curl } B^2|^2 + |\text{curl } f|^2) + \cdots,$$

where the remainder involves only terms of degree ≥ 3 in the fields f, B and $\tilde{\psi} = (0, 0, \tilde{\psi}^3)$.

Exercise

Show that by means of a suitable gauge transformation it can be arranged that $\tilde{\psi}^1 = \tilde{\psi}^2 = 0$.

(b) If $\deg \psi_0 = 1$, then by Lemma 32.1.2 we may assume that $\psi_0: S^2 \to S^2$ is the identity map, so that the corresponding boundary conditions at infinity might be termed "spherically symmetric". Physicists have discovered an interesting spherically symmetric solution of the Yang–Mills equation $\delta S = 0$ (with S as in (6)), of the following form:

$$A_a^i = a(r)\varepsilon_{aij}x^j,$$
$$\psi^i = x^i \frac{u(r)}{r}, \qquad r = |x|, \tag{12}$$

where

$$u(r) \to u_\infty, \qquad a(r) \sim -\frac{1}{gr^2} \quad \text{as } r \to \infty,$$

$$u(r) \sim \text{const.} \times r, \qquad a(r) \to \text{const.} \quad \text{as } r \to 0.$$

Since the boundary conditions at infinity are given by the identity map

$$\psi_0: S^2 \to S^2, \qquad \psi_0(v) = v \in \mathbb{R}^3,$$

it follows (essentially) from Corollary 14.5.2 that any vector field $\psi(x)$ on $\mathbb{R}^3$ satisfying these boundary conditions must have a singular point in $\mathbb{R}^3$; thus in our present situation we are forced to the conclusion that every field $\psi(x)$ figuring in a solution (ψ, A) of the Yang–Mills equation, must vanish at some point x_0 of $\mathbb{R}^3$ (which we shall suppose in what follows to be unique).

There exists an interesting correspondence between solutions (ψ, A) of the Yang–Mills equation and vector-valued fields $f=(f_a)$ on $\mathbb{R}^3$ satisfying Maxwell's equations (i.e. behaving like the vector potential of an electromagnetic field) in empty space everywhere except where $\psi=0$ (i.e. away from x_0):

$$H_{ab}=\frac{\partial f_a}{\partial x^b}-\frac{\partial f_b}{\partial x^a}=\frac{1}{|\psi|}\psi^i F^i_{ab}-\frac{1}{|\psi|^3}\varepsilon_{ijk}\psi^i(\nabla_a\psi^j)(\nabla_b\psi^k). \tag{13}$$

This correspondence in fact associates each pair (ψ, A) (where ψ satisfies the above boundary conditions at infinity) with a fibre bundle over the region $\mathbb{R}^3\backslash\{x_0\}$ (where x_0 is the point, assumed unique, where ψ vanishes) with abelian structure group $SO(2)\cong S^1$, in such a way that solutions of the Yang–Mills equation become matched with solutions of Maxwell's equations for a stationary magnetic field $H=(H_{ab})=(\partial f_a/\partial x^b-\partial f_b/\partial x^a)$. It can be deduced from this correspondence that in the case of the solution (12), the surface integral of the corresponding stationary magnetic field $(H^i)=\operatorname{curl} f$ over a sphere of sufficiently large radius is 4π (in contrast with one of the geometric consequences of Maxwell's equations given in Part I, §26.3). In this way one is led via a non-singular solution of the Yang–Mills equation to the theoretical possibility of a "magnetic monopole" (cf. §25.5, Example (f)).

32.2. The Duality Equation

We now turn to the case $n=4$. Properly speaking the investigation of the Lagrangians arising in physics requires the solution of equations of Yang–Mills type ($\delta S=0$) (for fields A, ψ, where ψ may be a tensor or spinor field) in Minkowski space $\mathbb{R}^4_1$. However even in their "purest" form (i.e. in the absence of external fields ψ) the Yang–Mills equations in $\mathbb{R}^4_1$ are non-linear and rather complex (in contrast with the Maxwell equations); in fact no non-trivial real solutions of these equations in $\mathbb{R}^4_1$ are known. On the other hand in the physics literature there have appeared, on the grounds that they may prove to be physically relevant, several solutions in 4-dimensional Euclidean space $\mathbb{R}^4$ (some of which we shall describe below). Whether or not these solutions in $\mathbb{R}^4$ do find a use in physics, they are in any case, in view of their deep geometrical content, highly interesting from a purely mathematical point of view.

Thus we shall in this subsection consider the Yang–Mills functional (cf. Part I, §42)

$$S = \int_{\mathbb{R}^4} \operatorname{tr}(F_{ab}F^{ab})\, d^4x, \tag{14}$$

corresponding to the situation of a gauge field A alone, in the Euclidean space $\mathbb{R}^4$ with Euclidean co-ordinates x^0, x^1, x^2, x^3. (Here F_{ab} is defined in terms of the field A as before.) We shall impose the requirement that

$$A_a(x) \sim -\frac{\partial g(x)}{\partial x^a} g^{-1}(x) \quad \text{as } |x| \to \infty, \tag{15}$$

or equivalently that $F_{ab} \to 0$, and further, that in adjoining a point at infinity to $\mathbb{R}^4$ to obtain S^4 (with the north pole in the role of the point at infinity), the field $A(x)$ extends smoothly to S^4, i.e. once $A(x)$ is defined to be trivial at the north pole we have a gauge field $A(x)$ smooth on the whole of S^4. Thus now the fields $A(x)$ which we shall be considering, represent the local expressions of connexions on the trivial principal fibre bundle over S^4 with structure group $G = SU(2)$, from which the former bundle over $\mathbb{R}^4$ can be obtained by restricting to that part of the bundle over $\mathbb{R}^4 \subset S^4$.

The Lagrangian in (14) is as before invariant under gauge transformations of the form (9). In view of the asymptotic condition (15) the field $A(x)$ on S^4 will depend on the limit $g_\infty(v)$ of the map $g(x)$ in (15) as $|x| \to \infty$ (as before along half-lines $x = |x|v$, $|v| = 1$, emanating from the origin of $\mathbb{R}^4$), i.e. on the naturally associated map $g_\infty\colon S^3 \to G$. However unlike the case $n = 3$ where we had only one homotopy class of maps $S^2 \to G$, here, since $\pi_3(G) \simeq \mathbb{Z}$ (see §24.4), there are many homotopically distinct maps g_∞, and the degree of g_∞ affords a topological invariant of the connexion A on the bundle over S^4, from which it arose. It can be shown that the degree m of the map g_∞ coincides (essentially) with a characteristic class (see §25.5(64)):

$$m\{F\} = \frac{1}{8\pi^2} \int_{\mathbb{R}^4} \operatorname{tr} F_{ab} F_{cd} \varepsilon^{abcd}\, d^4x = \frac{1}{4\pi^2} \int_{\mathbb{R}^4} \operatorname{tr} F_{ab} (*F)^{ab}\, d^4x, \tag{16}$$

or in the notation of forms,

$$m\{F\} = \frac{1}{4\pi^2} \int_{\mathbb{R}^4} \operatorname{tr}(F \wedge *F).$$

(Recall from Part I, §19.3, that $*F$, the "dual" of $F = F_{ab}\, dx^a \wedge dx^b$, is given by $*F = \frac{1}{2}\varepsilon_{abcd} F^{ab}\, dx^c \wedge dx^d$.) Since for any non-zero matrix X in the Lie algebra of $SU(2)$ (see Part I, §24.2) we have $\operatorname{tr}(X^2) < 0$, and since the metric on $\mathbb{R}^4$ is Euclidean, it follows that the quantity

$$T = \frac{1}{2} \int_{\mathbb{R}^4} \operatorname{tr}[F_{ab} - (*F)_{ab}][F^{ab} - (*F)^{ab}]\, d^4x \tag{17}$$

is non-positive ($T \le 0$), and is zero precisely when the "duality equation" holds:

$$F_{ab} = (*F)_{ab}. \tag{18}$$

Now from (17) and (16) we have

$$T = \tfrac{1}{2}\int_{\mathbb{R}^4} \operatorname{tr}(F_{ab}F^{ab})\, d^4x + \tfrac{1}{2}\int_{\mathbb{R}^4} \operatorname{tr}(*F)_{ab}(*F)^{ab}\, d^4x$$
$$- \int_{\mathbb{R}^4} \operatorname{tr} F_{ab}(*F)^{ab}\, d^4x = S\{F\} - 4\pi^2 m\{F\} \le 0.$$

Since $m\{F\}$ is a characteristic class, and so has identically zero variational derivative (see Part I, §42 or §§25.4, 25.5), we infer from the equality $T = S\{F\} - 4\pi^2 m\{F\}$ that:

(i) the equations $\delta T = 0$ and $\delta S = 0$ are equivalent. Secondly, since $S\{F\} - 4\pi^2 m\{F\} \le 0$, with equality precisely if $F_{ab} = (*F)_{ab}$, it follows that

(ii) the duality equation $F_{ab} = (*F)_{ab}$ holds precisely if $S\{F\}$ is an absolute maximum value of the function S (among the class of all F with the given value of $m\{F\} = m$), and this absolute maximum value is equal to m.

Thus if corresponding to each integer m we can find at least one solution of the duality equation, we shall thereby have fully established that under any given set of boundary conditions at infinity the functional $S\{F\}$ does indeed attain the above upper bound. (Note that any solution of the duality equation will automatically be a solution of the corresponding Yang–Mills equation.) We now list particular solutions of the duality equation.

(a) For $m = 0$ we have the "trivial" solution, i.e. $F_{ab} = 0$, or, equivalently, A a trivial connexion.

(b) In the case $m = 1$ it is natural to seek a "spherically symmetric" solution in one or another of the forms

$$\pm A_a^i = \tfrac{1}{2}(\tilde{A}_a^{oi} \pm \tfrac{1}{2}\varepsilon_{ikl}\tilde{A}_a^{kl}), \qquad (\tilde{A}_a^{ij}) \in so(4). \tag{19}$$

It turns out that the following form of the $\tilde{A}_a^{ij}$ yields a solution (representing a so-called "instanton"):

$$\tilde{A}_a^{ij} = f(r)(x^i\delta_a^j - x^j\delta_a^i),$$
$$r = |x|, \qquad f(r) = \frac{1}{r^2 + \lambda^2}, \qquad \lambda = \text{const.} \tag{20}$$

Exercise

Show that for each $a = 0, 1, 2, 3$, the 4-vector $A_a(x) = (A_a^i(x))$ is determined by only 3 parameters (as it should be since $G = SU(2)$ has dimension 3).

(c) For each $m > 1$ the following solution is known:

$$A_a = -\frac{1}{\rho}\sum_{j=1}^{m}\frac{\lambda_j^2}{|x-x_j|^2}(\partial_a\omega_j)\omega_j^{-1}, \tag{21}$$

where

$$\rho = \sum_{j=1}^{m}\frac{\lambda_j^2}{|x-x_j|^2}, \qquad \omega_j = \frac{(x-x_j)^0 \times 1 + i(x-x_j)^k\sigma_k}{|x-x_j|},$$

the σ_k $(k = 1, 2, 3)$ being the Pauli matrices (see Part I, §§14.3, 40.2), and $(x-x_j)^0 \times 1$ denoting the identity 2×2 matrix multiplied by the first component of $x - x_j$. (Here the x_j are m particular points and the λ_j are m constants.)

A satisfactory form of the general solution (which will depend on $8m - 3$ parameters) has not yet been found. However there do exist several deep results, some of which we shall now briefly describe.

Observe first that since (as noted in Part I, §42) the Yang–Mills functional S, like the analogous Maxwellian functional, is conformally invariant (i.e. invariant under co-ordinate changes on $\mathbb{R}^4$ transforming the Euclidean metric into one proportional to it (with variable proportionality factor)), we may consider in place of $\mathbb{R}^4$ the whole of the 4-sphere S^4 with its usual (conformally Euclidean) metric, induced in the usual way from the Euclidean metric on $\mathbb{R}^5 \supset S^4$. (Note incidentally that the group of conformal transformations of S^4 is isomorphic to $O(5, 1)$; see Part I, §15.) Thus we may regard a solution of the duality equation (18) as defining a connexion on a principal fibre bundle over S^4 with structure group $SU(2)$.

We shall require in our further discussion a certain natural fibre bundle

$$p\colon \mathbb{C}P^3 \to S^4, \tag{22}$$

with fibre S^2, constructed as follows. Define an action of the group $SU(2)$ on $\mathbb{C}^4 = \mathbb{C}^2 \oplus \mathbb{C}^2$, by letting it act on each summand $\mathbb{C}^2$ in the usual way:

$$(z^1, z^2, w^1, w^2) \overset{g}{\mapsto} (g(z^1, z^2), g(w^1, w^2)), \qquad g \in SU(2). \tag{23}$$

From the very definition of $\mathbb{C}P^3$ it then follows that

$$\mathbb{C}P^3 = (\mathbb{C}^4 \setminus \{0\})/(\mathbb{R}^+ \times SO(2)), \tag{24}$$

where the right-hand side is the orbit space under the action (23) restricted to the subgroup $SO(2) \subset SU(2)$ (and the obvious action of the group $\mathbb{R}^+$ of positive reals). We then have further, essentially by Example (e) of §24.3, that

$$S^4 \cong (\mathbb{C}^4 \setminus \{0\})/(\mathbb{R}^+ \times SU(2)).$$

This and (24), together with the inclusion $SO(2) \subset SU(2)$, now yield the desired fibre bundle projection $p\colon \mathbb{C}P^3 \to S^4$, with fibres $p^{-1}(x) \cong \mathbb{C}P^1 \subset \mathbb{C}P^3$ situated in $\mathbb{C}P^3$ as projective lines.

We now return to the duality equation, each of whose solutions defines, as noted above, a connexion on a principal fibre bundle over S^4. By endowing the (naturally) associated vector bundle with the "associated" connexion in the manner described in §25.2, we may consider each such solution as corresponding to a connexion on a fibre bundle η over S^4 with fibre $\mathbb{C}^2$ and structure group $G = SU(2)$. We now "lift" the fibre bundle η in the natural way using the projection (22), to obtain a bundle $p^*(\eta)$ over the complex manifold $\mathbb{C}P^3$, endowed with the "lift" of the given connexion on η. (This lifted connexion will thus be trivial above each of the fibres $p^{-1}(x) \subset \mathbb{C}P^3$ of the original bundle η.) Now a connexion on any complex (not necessarily holomorphic) fibre bundle over a complex manifold (see §4.1), gives rise to a "quasi-complex structure" on the total space, meaning (in our present context) that in terms of the horizontal directions on the total space E of the bundle $p^*(\eta)$, determined by the connexion (see §25.1), there are defined certain "covariant" differential operators on functions on E. Let U be any chart of the base $\mathbb{C}P^3$ of $p^*(\eta)$, with complex co-ordinates z^1, z^2, z^3; $z^\alpha = x^\alpha + ix^{\alpha+3}$, which of course determine the differential operators $\partial/\partial z^1$, $\partial/\partial z^2$, $\partial/\partial z^3$; then by definition of a fibre bundle we have $E \supset q^{-1}(U) \cong \mathbb{C}^2 \times U$, where $q\colon E \to \mathbb{C}P^3$ is the projection map of the bundle $p^*(\eta)$. Denoting the co-ordinates of $\mathbb{C}^2$ by w^1, w^2, we then define the aforementioned "quasi-complex structure" on $p^*(\eta)$ to be given locally by the set of five operators

$$\frac{\partial}{\partial w^1}, \quad \frac{\partial}{\partial w^2}, \quad \frac{D}{Dz^1}, \quad \frac{D}{Dz^2}, \quad \frac{D}{Dz^3}, \tag{25}$$

where

$$\frac{D}{Dz^\alpha} = \frac{\partial}{\partial z^\alpha} + A_\alpha^{\mathbb{C}} = \frac{\partial}{\partial x^\alpha} - i\frac{\partial}{\partial x^{\alpha+3}} + A_\alpha - iA_{\alpha+3},$$

A_a being the given connexion on $p^*(\eta)$ (obtained originally as the lift of a connexion satisfying the duality equation).

Exercise

Show that the condition that the connexion on the fibre bundle η over S^4 satisfy the duality equation (18), is equivalent to the commutativity of the operators (25) on functions defined on the total space E of the bundle $p^*(\eta)$ over $\mathbb{C}P^3$.

It can be shown that if the operators (25) commute, then in fact E is a complex manifold with complex local co-ordinates z, w, and that the fibre bundle $p^*(\eta)$ over $\mathbb{C}P^3$ is then holomorphic. Thus in view of the above exercise, the problem of finding the solutions of the duality equation reduces to that of classifying the holomorphic fibre bundles over $\mathbb{C}P^3$, where the methods of algebraic geometry can be successfully applied.

32.3. Chiral Fields. The Dirichlet Integral

Among the non-linear fields which are of physical interest and involve topological considerations, there figure the so-called "chiral" fields. In its most general (local) form a chiral field is a map defined on $\mathbb{R}^k$ and taking its values in some (non-Euclidean) manifold M; globally, the most general chiral field is a cross-section of a fibre bundle with fibre M.

For those chiral fields of interest the manifold M is actually a homogeneous space of a Lie group:

$$M = G/H,$$

where the points of M are identified with the left cosets gH of a closed subgroup H, on which cosets G acts by left multiplication (see §5.1). (We shall assume throughout that G is compact, and moreover that it comes endowed with a two-sided invariant metric.) A chiral field of this sort is called *principal* if $H = \{1\}$, i.e. if M is a Lie group. Of particular importance are chiral fields of this type where $M = G/H$ is a symmetric space of the (compact) group G, with stationary group $H \subset G$ (see Definition 6.1.1). The simplest family of examples is given by

$$S^q \cong M = SO(q+1)/SO(q).$$

In this context if the points of the sphere S^q are represented as usual by the unit vectors n in $\mathbb{R}^{q+1}$, chiral fields taking their values in S^q are referred to as "n-fields".

Chiral fields arise in connexion with certain Lagrangians, of which we shall now consider (at length) the two most important kinds.

(i) Let $g(x) \in G$ (where G is a matrix Lie group) be a principal chiral field on $\mathbb{R}^k$, and set

$$A_a(x) = \frac{\partial g(x)}{\partial x^a} g^{-1}(x), \qquad a = 1, \ldots, k.$$

(For each $x \in \mathbb{R}^k$, $A_a(x)$ will then be an element of the Lie algebra of G.) The first type of functional we wish to consider is given by

$$S = \sum_{a=1}^{k} \int_{\mathbb{R}^k} \tfrac{1}{2} \langle A_a(x), A_a(x) \rangle \, dx^1 \wedge dx^2 \wedge \cdots \wedge dx^k, \tag{26}$$

where the scalar product is given by the Killing form on the Lie algebra of G (cf. §32.1(5)).

Thus in the case $G = SO(2) = \{e^{i\varphi} | 0 \leq \varphi < 2\pi\}$, the field A takes the form (essentially) of the gradient of a scalar-valued function $\varphi(x)$ (cf. §25.2(15) et seqq.):

$$\begin{gathered} g(x) = \exp\{i\varphi(x)\}, \\ A_a(x) = i \frac{\partial \varphi}{\partial x^a}, \end{gathered} \tag{27}$$

and the equation $\delta S = 0$ is here equivalent to Laplace's equation $\sum \partial^2 \varphi / \partial (x^a)^2 = 0$.

In the case $G = SU(2)$ the equation $\delta S = 0$ does not have so simple a form.

For each a we introduce scalar-valued functions $A_a^i(x)$, $i = 1, 2, 3$, defined by $A_a(x) = A_a^i(x)X_i$, where X_1, X_2, X_3 form a basis for the Lie algebra of $SU(2)$, satisfying

$$[X_1, X_2] = X_3, \qquad [X_2, X_3] = X_1, \qquad [X_3, X_1] = X_2.$$

The equation $\delta S = 0$ then reduces to the equation

$$\frac{\partial A_a^i}{\partial x^a} = 0, \qquad i = 1, 2, 3, \tag{28}$$

(verify this!) together with the equation

$$F_{ab} = \frac{\partial A_b}{\partial x^a} - \frac{\partial A_a}{\partial x^b} + [A_a, A_b] = 0, \tag{29}$$

which arises from the fact that the connexion defined by the fields A_a is trivial, and therefore has zero curvature.

On any non-abelian group G (of arbitrary dimension) one can define a standard two-sided invariant 3-form Ω by specifying its value on an ordered triple of tangent vectors at the identity of G (i.e. on a triple of elements of the Lie algebra of G) to be the "mixed" or "triple" product of those vectors:

$$\Omega(X, Y, Z) = \langle [X, Y], Z \rangle$$

where [,] denotes as usual the Lie commutator and $\langle\ ,\ \rangle$ the Killing form. The form Ω is closed (this follows directly from Cartan's formula (see Part I, Theorem 25.2.3), or alternatively from the two-sided invariance of Ω), but can be shown to be never cohomologically equivalent to zero (i.e. Ω is never exact: there is no 2-form $\hat{\Omega}$ such that $\Omega = d\hat{\Omega}$). (Note that in the case $G = SU(2)$ the form Ω clearly defines (to within a constant scalar factor) the volume element on $SU(2) \cong S^3$ and can be normalized so that $\int_{S^3} \Omega = 1$.)

For the remainder of our discussion of the Lagrangian in (26) we specialize to the case $k = 3$. Thus we now consider a principal chiral field $g(x)$ on $\mathbb{R}^3$, on which we impose the requirement that as $|x| \to \infty$, $g(x) \to g_\infty \in G$ in such a way that the field $g(x)$ extends to a smooth field on $S^3 = \mathbb{R}^3 \cup \{\infty\}$. With each such chiral field we can then associate the topological invariant afforded by the homotopy class $[g] \in \pi_3(G)$. (Note that in the case $G = SU(2)$, by Theorem 13.3.1 this invariant reduces to the degree of the map $g\colon S^3 \to SU(2) \cong S^3$, which in view of Theorem 14.1.1 and the above-mentioned property $\int_{S^3} \Omega = 1$, is given by the formula

$$\deg g = \int_{\mathbb{R}^3} g^*(\Omega), \tag{30}$$

$g^*(\Omega)$ being the pull-back of the form Ω to $S^3 = \mathbb{R}^3 \cup \{\infty\}$.)

The functional (26), in the present context sometimes called the "Dirichlet integral" (cf. Part I, §37.5, and below), becomes (in brief notation)

$$S\{g\} = \frac{1}{2} \int_{\mathbb{R}^3} \langle A, A \rangle \, d^3x, \tag{31}$$

(where as before $A_a(x) = (\partial g(x)/\partial x^a) g^{-1}(x))$, and the Euler–Lagrange equation $\delta S = 0$ is given by

$$\frac{\partial}{\partial x^a}\left(\frac{\partial g(x)}{\partial x^a} g^{-1}(x)\right) = 0, \tag{32}$$

or, equivalently (cf. (28) and (29)),

$$\frac{\partial A_a}{\partial x^a} = 0, \qquad \frac{\partial A_b}{\partial x^a} - \frac{\partial A_a}{\partial x^b} + [A_a, A_b] = 0. \tag{33}$$

For the Dirichlet functional (31) it turns out that (unlike the situation for n-fields on $\mathbb{R}^2$ to be considered below) there is no "topological" criterion ensuring the attainment of the absolute minimum for a given $[g]$.

We now turn to the "adjusted" chiral Lagrangian (of the "Skyrme model"), with functional

$$S_\alpha\{g\} = \frac{1}{2}\int_{\mathbb{R}^3} (\langle A, A\rangle + \alpha^2 \langle [A, A], [A, A]\rangle)\, d^3x, \tag{34}$$

where α is a non-zero real constant, and $[A, A]$ is regarded as a 2-form (with components $[A, A]_{ab} = [A_a, A_b]$) taking its values in the Lie algebra of G (so that in more detailed notation the integrand in (34) is actually $\sum_a \langle A_a, A_a\rangle + \alpha^2 \sum_{a,b} \langle [A, A]_{ab}, [A, A]_{ab}\rangle$.) The question of interest here is the following one: In the case $G = SU(2)$ how does one find a field $g(x)$ of prescribed degree $d = \deg[g\colon \mathbb{R}^2 \cup \{\infty\} \to G]$, minimizing the functional S_α (among all chiral fields of degree d)? In attempting to answer this one might consider (analogously to the preceding section) the functional

$$S_\alpha + T = \frac{1}{2}\sum_a \int_{\mathbb{R}^3} \langle A_a + \alpha\varepsilon^{abc}[A_b, A_c], A_a + \alpha\varepsilon^{abc}[A_b, A_c]\rangle\, d^3x,$$

which is obviously non-negative: $S_\alpha + T \geq 0$. From this and the formula (30) we obtain

$$S_\alpha + \sum_a \int_{\mathbb{R}^3} \alpha \langle A_a, \varepsilon^{abc}[A_b, A_c]\rangle\, d^3x = S_\alpha + \text{const.} \times d \geq 0. \tag{35}$$

Hence a chiral field $g(x)$ of degree d for which the lower bound of zero is attained, must satisfy the equations

$$A_a + \alpha\varepsilon^{abc}[A_b, A_c] = 0,$$

that is,

$$A_1 = -\alpha[A_2, A_3], \qquad A_2 = +\alpha[A_1, A_3], \qquad A_3 = -\alpha[A_1, A_2], \tag{36}$$

where

$$A_a = \frac{\partial g(x)}{\partial x^a} g^{-1}(x), \qquad \frac{\partial A_a}{\partial x^b} - \frac{\partial A_b}{\partial x^a} = [A_a, A_b]. \tag{37}$$

It follows that $A = \alpha \operatorname{curl} A$, whence $A_a(x) \equiv 0$. From this we conclude that if for a prescribed degree d an adjusted chiral Lagrangian of the form (34) actually attains its absolute minimum at some $g: S^3 \to SU(2)$ (of degree d) then that absolute minimum exceeds the lower bound given by (35). (This situation contrasts with those of the Yang–Mills fields on $\mathbb{R}^4$ considered in the preceding subsection, and the n-fields on $\mathbb{R}^2$ to be considered below.)

Finally we note also that, in further contrast with the Yang–Mills Lagrangian, the adjusted chiral Lagrangian in (34) is not conformally invariant, so that different solutions of the variational problem are obtained for this Lagrangian depending on whether it is considered over S^3 or $\mathbb{R}^3$; for instance over S^3 the equations (36) are satisfied by the identity map $g: S^3 \to S^3 \cong SU(2)$, so that the lower bound afforded by (35) *is* attained if we work over S^3 rather than $\mathbb{R}^3$ (verify this!).

(ii) The second type of Lagrangian that we wish to investigate involves *n-fields*, i.e. chiral fields $n(x)$ on $\mathbb{R}^k$ with values unit vectors $n(x) \in S^q \subset \mathbb{R}^{q+1}$ (see the beginning of this subsection). The functional in question (also called the "Dirichlet integral") is given by

$$S\{n(x)\} = \int_{\mathbb{R}^k} \left\langle \frac{\partial n}{\partial x^a}, \frac{\partial n}{\partial x^a} \right\rangle d^k x, \tag{38}$$

where $\langle\ ,\ \rangle$ denotes the Euclidean scalar product in $\mathbb{R}^{q+1}$ (and where the index a is summed over). The requirement is usually imposed on the n-fields $n(x)$ that as $|x| \to \infty$, $n(x) \to n_\infty$ in such a way that $n(x)$ extends smoothly to $S^k = \mathbb{R}^k \cup \{\infty\}$, thus yielding a map $n: S^k \to S^q$, whose degree will then afford a topological invariant of the n-field. In view of Theorem 14.1.1, when $k = q$ the degree of this map $n: S^q \to S^q$ is given by

$$d = \deg(n) = \int_{\mathbb{R}^q} n^*(\Omega), \tag{39}$$

where Ω is the volume element of $S^q \subset \mathbb{R}^{q+1}$ (normalized so that $\int_{S^q} \Omega = 1$).

For the remainder of this section we shall be concerned only with the case $q = k = 2$. As always our interest lies in the problem of finding the n-fields of prescribed degree d for which the above Dirichlet functional assumes its absolute minimum (among all n-fields of degree d). Let u^α ($\alpha = 1, 2$) be local co-ordinates on the sphere S^2, and x^a ($a = 1, 2$) be Euclidean co-ordinates on the plane $\mathbb{R}^2$; in terms of such co-ordinates the functional (38) has the form (cf. Part I, §37.5)

$$S\{n(x)\} = \int_{\mathbb{R}^2} g^{ab} \tilde{g}_{\alpha\beta} \frac{\partial u^\alpha}{\partial x^a} \frac{\partial u^\beta}{\partial x^b}\, dx^1 \wedge dx^2, \tag{40}$$

where $g_{ab} = \delta_{ab}$ is the Euclidean metric on $\mathbb{R}^2$, $\tilde{g}_{\alpha\beta}$ is the metric on S^2 ($\subset \mathbb{R}^3$) in the co-ordinates u^1, u^2, and the map $n(x)$ is written as

$$n(x) = (u^\alpha(x^1, x^2)), \qquad \alpha = 1, 2.$$

Remark. The formula (40) extends in the obvious way to give the Dirichlet integral $S\{n\}$ corresponding to any map $n: M \to N$ of manifolds M and N endowed respectively with metrics g_{ab} and $\tilde{g}_{\alpha\beta}$ in terms of local co-ordinate systems (x^a) and (u^α).

EXERCISE

Show (with reference to this remark) that if we take N to be a Lie group G, then the Dirichlet integral (40) takes the form (31) of the action of a principal chiral field (which explains the use of the term "Dirichlet integral" in both situations).

Returning to the case $M = \mathbb{R}^2$, $N = S^2$ of present interest, let u^1, u^2 now be conformally Euclidean co-ordinates on $S^2 \setminus \{\infty\}$, in terms of which the metric on $S^2 \setminus \{\infty\}$ has the form (derived in Part I, §9)

$$\tilde{g}_{\alpha\beta}\, du^\alpha\, du^\beta = \frac{4((du^1)^2 + (du^2)^2)}{(1 + (u^1)^2 + (u^2)^2)^2} = \frac{4\, dz\, d\bar{z}}{(1 + |z|^2)^2}, \tag{41}$$

where $z = u^1 + iu^2$, $w = x^1 + ix^2$. In terms of z and w the functional (40) becomes

$$S\{n\} = 4i \int_{\mathbb{R}^2} \frac{\left|\frac{\partial z}{\partial w}\right|^2 + \left|\frac{\partial z}{\partial \bar{w}}\right|^2}{(1 + |z|^2)^2}\, dw \wedge d\bar{w}, \tag{42}$$

(where the operators $\partial/\partial z$, $\partial/\partial w$ are defined as usual; see Part I, §12.1(6)), and the formula (39) for the degree of a map $n: S^2 \to S^2$, becomes

$$\deg(n) = \int_{\mathbb{R}^2} n^*(\Omega) = \frac{-1}{2\pi i} \int_{\mathbb{R}^2} n^*\left(\frac{dz \wedge d\bar{z}}{(1 + |z|^2)^2}\right) = \frac{1}{\pi} \int_{\mathbb{R}^2} \frac{u_x v_y - u_y v_x}{(1 + |z|^2)^2}\, dx\, dy, \tag{43}$$

where we have now set $u = u^1$, $v = u^2$, $x = x^1$, $y = x^2$. From (42) and (43) it follows immediately that

$$S - 2\pi \deg(n) = \int_{\mathbb{R}^2} \frac{(u_x - v_y)^2 + (u_y + v_x)^2}{(1 + |z|^2)^2}\, dx\, dy \geq 0, \tag{44}$$

whence we draw the following conclusions:

(1) For any n-field of degree d, we have $S - 2\pi d \geq 0$, where S is the functional (40);

(2) For an n-field $n(x)$ of degree $d \geq 0$ to be such that the functional S attains at $n(x)$ the lower bound $2\pi d$ given by (44), it is necessary and sufficient that

$$u_x = v_y, \qquad u_y = -v_x,$$

i.e. that u and v satisfy the Cauchy–Riemann equations.

Thus the absolute minimum of the functional S (on the variational class of all smooth n-fields of prescribed degree d) is attained for precisely those n-

fields $n: S^2 \to S^2$ which are holomorphic (and therefore of the form $z = P(w)/Q(w)$ where P and Q are polynomials). This important result, which was obtained first in a geometrical context, and then in physics, has application to the theory of ferromagnetism.

We shall now consider the same example from a different angle. The 2-sphere S^2 can of course be realized as a homogeneous space

$$S^2 \cong SO(3)/SO(2) \cong SU(2)/U(1) = G/H,$$

and moreover is clearly symmetric (see Definition 6.1.1). Hence by Lemma 6.2.1 the Lie algebra L of G decomposes as a direct sum $L = L_0 + L_1$ of subspaces (L_0 being the Lie algebra of the stationary subgroup $H \subset G$) with the following properties:

$$[L_0, L_0] = L_0, \qquad [L_0, L_1] \subset L_1, \qquad [L_1, L_1] \subset L_0,$$

and from these properties it follows that the subspaces L_0 and L_1 are orthogonal with respect to the Killing form on L. The theory of n-fields (on $\mathbb{R}^2$ with values in S^2) described above, can be alternatively expounded as follows. Consider the totality of fields $g(x)$ on $\mathbb{R}^2$ with values in $G = SU(2)$; we shall regard fields $g(x)$ and $e^{i\varphi(x)}g(x)$, differing by a factor $e^{i\varphi(x)}$ (which for each x is regarded as lying in $H = SO(2) \subset SO(3)$) as being *equivalent*. (Transformations of a field $g(x)$ of the form $g(x) \to e^{i\varphi(x)}g(x)$ are "gauge transformations" in the nomenclature first introduced in Part I, §41.1.) The corresponding equivalence classes may then be identified with the n-fields $n(x)$ with values in G/H. Consider the following functional of the field $g(x)$ (the "chiral functional"; cf. (38)):

$$S\{g(x)\} = \int_{\mathbb{R}^2} \langle A_a, A_a \rangle_{L_1} \, d^2x, \tag{45}$$

(with summation over a understood) where $A_a(x) = (\partial g(x)/\partial x^a)g^{-1}(x)$, and $\langle\, , \rangle_{L_1}$ denotes the scalar product on L determined by the Killing form on the subspace L_1 and zero on L_0. Direct calculation (bearing in mind that for each x, $e^{i\varphi(x)}$ represents an element of $SO(2) \subset SO(3)$, and $g(x) \in SO(3)$) shows that under a gauge transformation

$$g(x) \to e^{i\varphi(x)}g(x) = \tilde{g}(x),$$

we have

$$A \to \tilde{A} = e^{i\varphi(x)}Ae^{-i\varphi(x)} + i\nabla\varphi(x), \tag{46}$$

where the term $i\,\nabla\varphi(x)$ lies in L_0. From this it follows easily that the functional (45) is gauge-invariant:

$$S\{g(x)\} = S\{\tilde{g}(x)\},$$

so that it is well defined as a functional of the equivalence classes of fields $g(x)$, i.e. of the n-fields with values in $SO(3)/SO(2)$.

Since the fields A_a ($a = 1, 2$) take their values in the Lie algebra L of $SO(3)$, they each have three components, say $A_a = A_a^0 e_0 + A_a^1 e_1 + A_a^2 e_2$, where e_0, e_1, e_2 form a canonical basis for L:

$$[e_0, e_1] = e_2, \qquad [e_1, e_2] = e_0, \qquad [e_2, e_0] = e_1, \tag{47}$$

chosen so that e_0 generates L_0, e_1, e_2 span L_1, and e_1, e_2 are of length 1 with respect to the scalar product $\langle\ ,\ \rangle_{L_1}$; thus

$$\langle e_0, e_0\rangle_{L_1} = 0; \qquad \langle e_1, e_1\rangle_{L_1} = \langle e_2, e_2\rangle_{L_1} = 1; \qquad \langle e_i, e_j\rangle_{L_1} = 0, \qquad i \neq j.$$

Regarding A as defining a trivial connexion on the principal G-bundle over $\mathbb{R}^2$, we deduce (as earlier) that $F_{ab} = 0$ (see e.g. Part I, §41.2), i.e.

$$\frac{\partial A_b^\beta}{\partial x^a} - \frac{\partial A_a^\beta}{\partial x^b} + [A_a, A_b]^\beta = 0, \qquad a, b = 1, 2; \quad \beta = 0, 1, 2. \tag{48}$$

Setting

$$B_a = A_a^1 + iA_a^2, \qquad \bar{B}_a = A_a^1 - iA_a^2,$$
$$A_a^0 = if_a, \qquad f_{ab} = \frac{\partial A_a^0}{\partial x^b} - \frac{\partial A_b^0}{\partial x^a}, \qquad a, b = 1, 2, \tag{49}$$

we obtain from (48) by direct calculation (using (47)) that

$$f_{12} = \frac{i}{2}(B_1\bar{B}_2 - B_2\bar{B}_1), \tag{50}$$

and also

$$\frac{\partial B_1}{\partial x^2} - \frac{\partial B_2}{\partial x^1} = B_2 f_1 - B_1 f_2. \tag{51}$$

Defining a "covariant" differential operator $D = (D_1, D_2)$ in the usual way by (cf. §25.3 and Part I, §41.1)

$$D_a = \frac{\partial}{\partial x^a} + f_a,$$

(51) becomes

$$D_1 B_2 - D_2 B_1 = 0.$$

The Euler–Lagrange equations for the functional (45) turn out to have the form

$$\frac{\partial B_a}{\partial x^a} = 2f_a B_a, \qquad a = 1, 2. \tag{52}$$

Since the field $f_a = -iA_0^a$ is a gauge field (which, using (46), can be shown to have its transformation rule (under gauge transformations $g(x) \to e^{i\varphi(x)}g(x)$) of the purely gradient form $f_a \to f_a + \partial\varphi/\partial x^a$), and since $B = (B_a)$ can be shown (again via (46)) to transform according to the rule

$$B_a \to e^{i\varphi} B_a,$$

it follows that we have reduced our variational problem for n-fields to that of solving the equations (52) for the complex vector field (B_a) and the gauge field f_a, on whose stress tensor f_{ab} there is imposed the constraint

$$f_{12}=\frac{\partial f_1}{\partial x^2}-\frac{\partial f_2}{\partial x^1}=\frac{i}{2}(B_1\bar{B}_2-B_2\bar{B}_1). \tag{53}$$

It can be shown that for any n-field n we have

$$n^*(\Omega)=\text{const.}\times f_{12}\,dx^1\wedge dx^2, \tag{54}$$

where as before Ω is the $SO(3)$-invariant 2-form on S^2 given by the element of area on S^2, normalized so that $\int_{S^2}\Omega=1$. (Thus

$$\Omega=\text{const.}\times\frac{dz\wedge d\bar{z}}{(1+|z|^2)^2} \tag{55}$$

in terms of a suitable complex co-ordinate z on $S^2\backslash\{\infty\}$ (cf. (41).) Hence the homotopy invariant d, the degree of the n-field, is also given (to within a constant factor) by the integral of the stress tensor:

$$\begin{aligned} d&=\int_{\mathbb{R}^2}n^*(\Omega)=\text{const.}\times\int_{\mathbb{R}^2}f_{12}\,dx^1\wedge dx^2\\ &=\text{const.}\times\oint_{\Gamma\cup\Gamma_\infty}(f_1\,dx^1+f_2\,dx^2)=\lambda\int_{\mathbb{R}^2}(B_1\bar{B}_2-B_2\bar{B}_1)\,dx^1\wedge dx^2, \end{aligned} \tag{56}$$

where λ is a constant depending only on the degree d, and where to obtain the second equality we have used Stokes' formula (see Part I, §26.3), Γ_∞ denoting a circle in $\mathbb{R}^2$ of sufficiently large radius R and $\Gamma=\cup_i\Gamma_i$ consisting of appropriately oriented circles of small radius ε around the singular points x_i of the fields B or f. (Fields B, f with finitely many singular points in the "finite" part of S^2, i.e. in $\mathbb{R}^2$, must be admitted for consideration in order for the present formulation of the theory (of n-fields) to be valid, since in the present formalism the points x_i such that $n(x_i)=\infty$, will be singular for (B_a, f_a). A given map $n: S^2\to S^2$ need not be globally covered by a map $g: S^2\to S^3$, so that at one point (∞) of S^2 the corresponding cross-section (defining g) of the bundle $S^3\to S^2$ may be many-valued, i.e. $g(x)$ may be many-valued at the points x_i in $n^{-1}(\infty)$, and consequently B or f will fail to be defined at these x_i.) (Note also that the form $f_1\,dx^1+f_2\,dx^2$ appearing in (56) can be regarded as a connexion on the standard Hopf bundle $S^3\to S^2$ with group $G\cong S^1$, and projection given by the natural map $SU(2)\to SU(2)/U(1)$; see §24.3, Example (a).)

We conclude the present discussion by finding an explicit expression for the B_a in terms of a holomorphic field n. Let w be the complex co-ordinate on $\mathbb{R}^2$ given by $w=x^1+ix^2$; if the function $n=z(w)$ defined by a given n-field is holomorphic, then by (55) and Part I, Lemma 12.2.2,

$$n^*(\Omega)=\text{const.}\times\left|\frac{dz}{dw}\right|^2\frac{dw\wedge d\bar{w}}{(1+|z(w)|^2)^2}. \tag{57}$$

Since

$$S=\int_{\mathbb{R}^2}\langle A, A\rangle_{L_1}\,d^2x=\int_{\mathbb{R}^2}(B_1\bar{B}_1+B_2\bar{B}_2)\,dx^1\wedge dx^2, \tag{58}$$

it follows (via 56) that

$$\begin{aligned}S+\frac{i}{\lambda}\int_{\mathbb{R}^2}n^*(\Omega)&=\int_{\mathbb{R}^2}(B_1\bar{B}_1+B_2\bar{B}_2+i(B_1\bar{B}_2-B_2\bar{B}_1))\,dx^1\wedge dx^2\\&=\int_{\mathbb{R}^2}|B_1+iB_2|^2\,dx^1\wedge dx^2\geq 0.\end{aligned} \tag{59}$$

Since we are assuming $n=z(w)$ to be holomorphic, it follows (essentially from the result established earlier to the effect that the functional (58) attains the lower bound given by (59) precisely for the holomorphic n-fields of degree d) that

$$B_1+iB_2=0. \tag{60}$$

Equations (53), (54), (57) and (60) together give

$$\begin{aligned}\left|\frac{dz}{dw}\right|^2\frac{dw\wedge d\bar{w}}{(1+|z|^2)^2}&=\text{const.}\times(B_1\bar{B}_2-B_2\bar{B}_1)\,dw\wedge d\bar{w}\\&=\text{const.}\times B_1^2\,dw\wedge d\bar{w},\end{aligned}$$

whence finally

$$B_1=\frac{\text{const.}}{1+|z|^2}\frac{dz}{dw}. \tag{61}$$

We end this section with a proof of the following recent result: *The Lagrange–Euler equation $\delta S=0$ for the extremals of the functional S given by (40), with the metric g_{ab} taken to be the Minkowski metric (so that the n-fields are defined on $\mathbb{R}^2_1$) rather than the Euclidean one, is equivalent to the "sine–Gordon equation"*. (The latter equation was derived in Part I, §30.4, as a necessary and sufficient condition for a surface of constant negative Gaussian curvature to be embeddable in Euclidean space $\mathbb{R}^3$.)

To prove this we first change from the usual Minkowski co-ordinates x, y of $\mathbb{R}^2_1$, in terms of which the metric has the form $ds^2=dx^2-dy^2$, to the co-ordinates $\eta=x+y$, $\xi=x-y$, in terms of which the metric is given by $ds^2=d\eta\,d\xi$. If we write as before n^1, n^2, n^3 for the components of $n(x)\in\mathbb{R}^3\supset S^2$, where $(n^1)^2+(n^2)^2+(n^3)^2=1$, the functional (40) (with g_{ab} now given by the Minkowski metric) becomes

$$S\{n\}=\int_{\mathbb{R}^2}\left(\sum_{\alpha=1}^{3}n^\alpha_\xi n^\alpha_\eta\right)d\xi\,d\eta. \tag{62}$$

To obtain the explicit form of the Euler–Lagrange equation $\delta S=0$ for this functional (subject to the constraint $(n^1)^2+(n^2)^2+(n^3)^2=1$), let μ be an arbitrary real-valued function of η and ξ, to be determined subsequently (i.e. a "Lagrange multiplier"), and consider the functional

$$S_\mu\{n\}=\int_{\mathbb{R}^2}\left(\sum_{\alpha=1}^{3} n_\eta^\alpha n_\xi^\alpha+\mu\sum_{\alpha=1}^{3}(n^\alpha)^2\right)d\eta\, d\xi$$

$$=\int_{\mathbb{R}^2}\Lambda_\mu(n, n_\eta, n_\xi)\, d\eta\, d\xi.$$

The general Euler–Lagrange equations given in Part I, §37.1, become in the present context

$$\frac{\partial}{\partial\xi}\left(\frac{\partial\Lambda_\mu}{\partial n_\xi^\alpha}\right)+\frac{\partial}{\partial\eta}\left(\frac{\partial\Lambda_\mu}{\partial n_\eta^\alpha}\right)=\frac{\partial\Lambda}{\partial n^\alpha},\qquad \alpha=1,2,3,$$

(to be taken together with the constraint $(n^1)^2+(n^2)^2+(n^3)^2=1$), which simplify to

$$n_{\xi\eta}^\alpha=\mu n^\alpha,\qquad \alpha=1,2,3.$$

Bringing in the condition $\sum(n^\alpha)^2=1$ (or $\langle n, n\rangle=1$ where $\langle\ ,\ \rangle$ denotes the Euclidean scalar product on $\mathbb{R}^3$), we obtain from this

$$\mu=\langle n, n_{\xi\eta}\rangle,\qquad n_{\xi\eta}=\langle n_{\xi\eta}, n\rangle. \tag{63}$$

It follows that

$$\frac{1}{2}\frac{\partial}{\partial\eta}\langle n_\xi, n_\xi\rangle=\langle n_{\xi\eta}, n_\xi\rangle=\langle n_{\xi\eta}, n_\xi\rangle=\langle n_{\xi\eta}, n\rangle\langle n_\xi, n\rangle=0,$$

(invoking the orthogonality of n_ξ and n in $\mathbb{R}^3$), and similarly that $(\partial/\partial\xi)\langle n_\eta, n_\eta\rangle=0$. Hence the functions $|n_\xi|$ and $|n_\eta|$ are "integrals", in the sense that

$$\frac{\partial|n_\xi|}{\partial\eta}=0,\qquad \frac{\partial|n_\eta|}{\partial\xi}=0,$$

and therefore depend respectively on ξ and η alone:

$$|n_\xi|=f(\xi),\qquad |n_\eta|=g(\eta).$$

If we define the angle ω by

$$\cos\omega=\frac{\langle n_\xi, n_\eta\rangle}{f(\xi)g(\eta)},$$

then the second equation in (63) is equivalent to

$$\frac{\partial^2\omega}{\partial\xi\,\partial\eta}=fg\sin\omega,$$

which by means of a local change of co-ordinates of the form $\xi\to\hat{\xi}(\xi)$, $\eta\to\hat{\eta}(\eta)$, can be brought into the form of the "sine–Gordon" equation

$$\varphi_{\xi\bar{\eta}} = \sin \varphi. \tag{64}$$

(However the investigation of this equation presents difficulties significantly greater than those encountered above in finding the n-fields of prescribed degree for which the functional (40) attains its absolute minimum.)

§33. The Minimality of Complex Submanifolds

Recall from Part I, §27.2 that a complex manifold M is called *Kählerian* if it comes endowed with a Hermitian metric (see loc. cit.) $g_{ij}\,dz^i\,d\bar{z}^j$ with the property that the associated form

$$\omega = \frac{i}{2} g_{ij}\,dz^i \wedge d\bar{z}^j$$

is closed.

33.1. Theorem. *Let M be a Kählerian manifold of n complex dimensions, and let $X \subset M$ be an arbitrary complex k-dimensional submanifold. (By Theorem 4.1.2 the manifolds M and X automatically have orientations defined on them by virtue of their complex structure; we may assume that the complex co-ordinates on X are such that its orientation is induced from that on M.) Consider the variational class consisting of all those real $2k$-dimensional submanifolds Y of the $2n$-dimensional (realized) manifold M, which coincide with X outside a compact region of X, and with each of which it is possible to associate a "region of deformation" in the form of a real oriented $(2k+1)$-dimensional submanifold $Z \subset M$ with boundary $\partial Z = X \cup (-Y)$ where $-Y$ is the manifold Y with orientation opposite to that induced on it from M (see Figure 125). Then every such Y has volume $v(Y)$ at least as large as $v(X)$ (and if X is not compact this remains true with $v(X)$ and $v(Y)$ denoting instead the volumes of those regions of X and Y respectively, where they differ). Furthermore if $v(X) = v(Y)$, then Y also is a complex manifold.*

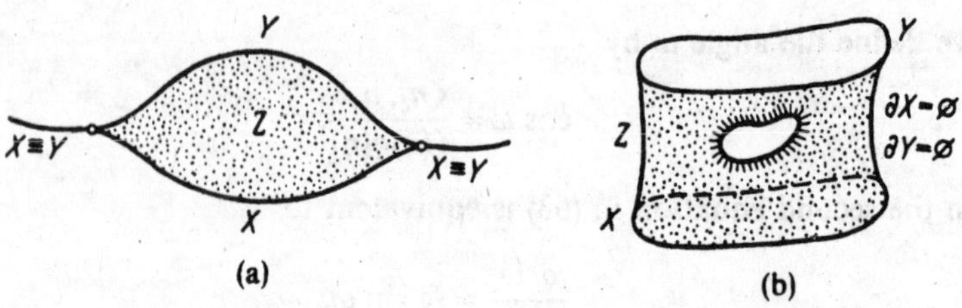

Figure 125

Thus for a Kählerian manifold the compact complex submanifolds are in this sense globally smallest among the submanifolds of the same real dimension (or in other words they are precisely the extremals for the "least-volume" problem of the higher-dimensional calculus of variations). In particular since the manifolds $\mathbb{C}P^n$ and $\mathbb{C}^n$ are Kählerian (once endowed with naturally defined Hermitian metrics), any complex submanifold of either of them will have least volume among all perturbations leaving it fixed outside a compact region. (Recall incidentally from §4.1 that all complex submanifolds of $\mathbb{C}^n$ of positive dimension one non-compact.)

For the proof of the above theorem we shall need the following two lemmas.

33.2. Lemma. *Given any skew-symmetric, bilinear form ω on the vector space $\mathbb{R}^{2n}$ (equipped with the Euclidean scalar product), there exists an orthonormal basis $\{e_1, \ldots, e_{2n}\}$ for $\mathbb{R}^{2n}$ in terms of which ω has the form*

$$\omega = \lambda_1 \omega_1 \wedge \omega_2 + \cdots + \lambda_n \omega_{2n-1} \wedge \omega_{2n}, \tag{1}$$

where $\lambda_1, \ldots, \lambda_n$ are non-negative real numbers, and $\{\omega_1, \ldots, \omega_{2n}\}$ is the dual basis to $\{e_1, \ldots, e_{2n}\}$.

PROOF. Let $\hat{e}_1, \ldots, \hat{e}_{2n}$ be any orthonormal basis for $\mathbb{R}^{2n}$ and form the matrix $A = (a_{ij})$ with entries $a_{ij} = \omega(\hat{e}_i, \hat{e}_j)$. Since the form ω is completely determined by its values on ordered pairs of basis vectors, it is determined by its associated matrix A (together with the basis $\{\hat{e}_1, \ldots, \hat{e}_{2n}\}$). Since A is skew-symmetric there is (by a standard result of linear algebra) an orthonormal basis $\{e_1, \ldots, e_{2n}\}$ in terms of which the linear transformation defined by A has matrix of the form

$$\begin{pmatrix} \begin{matrix} 0 & \lambda_1 \\ -\lambda_1 & 0 \end{matrix} & & 0 \\ & \ddots & \\ 0 & & \begin{matrix} 0 & \lambda_n \\ -\lambda_n & 0 \end{matrix} \end{pmatrix},$$

with $\lambda_1, \ldots, \lambda_n$ non-negative, and then relative to this basis the form ω will be as in (1). □

33.3. Lemma. *Let $\langle\ ,\ \rangle$ be a Hermitian scalar product on the space $\mathbb{C}^n = \mathbb{R}^{2n}$, and let ω be the skew-symmetric bilinear form on $\mathbb{R}^{2n}$ determined by this metric, i.e. given at each point of $\mathbb{R}^{2n}$ by the formula*

$$\omega(v_1, v_2) = \langle iv_1, v_2 \rangle,$$

for arbitrary vectors $v_1, v_2 \in \mathbb{R}^{2n}$. *Writing*

$$\sigma_k = \frac{1}{k!}\omega^k = \frac{1}{k!}\underbrace{\omega \wedge \omega \wedge \cdots \wedge \omega}_{k}, \qquad k \le n,$$

we shall then have

$$|\sigma_k(v_1, \ldots, v_{2k})| \le 1, \tag{2}$$

for every orthonormal $2k$*-frame* $(v_1, \ldots, v_{2k})$ *of vectors in* $\mathbb{R}^{2n}$, *with equality precisely when the subspace of* $\mathbb{R}^{2n}$ *spanned (over the reals) by* $v_1, \ldots, v_{2k}$, *is actually a complex subspace of* $\mathbb{C}^n$ *(which condition reduces to the preservation of that linear span under multiplication by the complex scalar i).*

PROOF. We first deal with the case $k = 1$. Thus let v_1, v_2 be a pair of orthogonal vectors in $\mathbb{R}^{2n}$ of length 1. Then

$$|\omega(v_1, v_2)| = |\langle iv_i, v_2\rangle| \le |iv_1|\cdot|v_2| = 1,$$

establishing the first claim. Here equality obtains precisely when $\pm v_2 = iv_1$, i.e. when the (2-dimensional) $\mathbb{R}$-linear span of v_1 and v_2 is a (one-dimensional) complex subspace.

Turning now to the general case, let $V \subset \mathbb{R}^{2n}$ be any subspace of dimension $2k$, let $\{v_1, \ldots, v_{2k}\}$ be an orthonormal basis for V, and denote by $\tilde{\omega}$ the restriction $\omega|_V$ of the form ω to V. By Lemma 33.2 we can find an orthonormal basis $\{e_1, \ldots, e_{2k}\}$ for V in terms of whose dual basis $\omega_1, \ldots, \omega_{2k}$, the form $\tilde{\omega}$ is given by

$$\tilde{\omega} = \lambda_1\omega_1 \wedge \omega_2 + \cdots + \lambda_k\omega_{2k-1} \wedge \omega_{2k},$$

where $\lambda_1, \ldots, \lambda_k$ are non-negative reals. Since then $\tilde{\omega}(e_{2p-1}, e_{2p}) = \lambda_p$ for $p = 1, \ldots, k$, we infer from the case $k = 1$ already treated above that $\lambda_p \le 1$ for all p, and moreover that $\lambda_p = 1$ precisely if $ie_{2p-1} = \pm e_{2p}$. Denoting the restriction of the form $\sigma_k = (1/k!)\omega^k$ to the subspace V by $\tilde{\sigma}_k$, we then have

$$|\tilde{\sigma}_k(e_1, \ldots, e_{2k})| = \left|\frac{1}{k!}\tilde{\omega}^k(e_1, \ldots, e_{2k})\right| = \lambda_1 \ldots \lambda_k \le 1,$$

with equality precisely if $ie_{2p-1} = \pm e_{2p}$, which clearly implies that V is a complex subspace of $\mathbb{C}^n$. (We leave the reverse implication to the reader.) □

PROOF OF THEOREM 33.1. Let φ be a skew-symmetric multilinear form of degree l on the vector space $\mathbb{R}^{2n}$, let V be any l-dimensional subspace of $\mathbb{R}^{2n}$, and let $(v_1, \ldots, v_l)$, $(\hat{v}_1, \ldots, \hat{v}_l)$ be a pair of orthonormal frames for V in the same orientation class. From the transformation rule for a skew-symmetric multilinear form of degree l on a vector space of dimension l under a linear transformation (namely multiplication by the determinant of the linear transformation), we infer immediately that $\varphi(v_1, \ldots, v_l) = \varphi(\hat{v}_1, \ldots, \hat{v}_l)$, so that the restriction of the form φ to V gives rise to a well-defined function of the

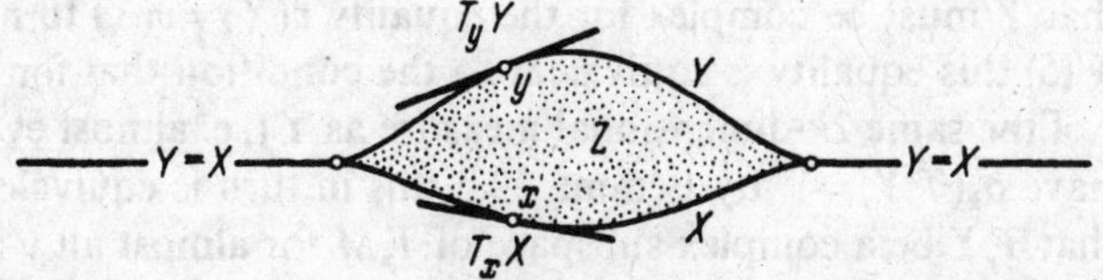

Figure 126

two orientation classes of orthonormal frames for V. If we now allow V to vary, we see that the l-form φ on $\mathbb{R}^{2n}$ determines in this way a function (also denoted by φ) defined on the Grassmannian manifold $\hat{G}_{2n,l}$ whose points are the oriented l-dimensional subspaces of $\mathbb{R}^{2n}$ (cf. §5.2). For each l-dimensional subspace $V \subset \mathbb{R}^{2n}$ we shall denote by $\hat{V}$ the subspace taken together with the orientation induced on it by a specified orientation of $\mathbb{R}^{2n}$. (Thus $\hat{V}$ is identifiable with a point of $\hat{G}_{2n,l}$.)

Now let X, a complex submanifold (of a given complex manifold M), Y, a permitted perturbation of X, and Z, having boundary $\partial Z = X \cup (-Y)$, all be as in the theorem (see Figure 126). As usual we denote by T_xX and T_yY the respective tangent spaces to the submanifolds X and Y at the points x and y. We denote by ω the closed 2-form $(i/2)g_{ij}\, dz^i \wedge d\bar{z}^j$ on M, afforded by a given Hermitian metric g_{ij}, and write $\sigma_k = (1/k!)\omega^k$. The closure of ω implies that of σ_k: $d\sigma_k = 0$ (see Part I, §25.2), whence by the general Stokes formula (Part I, §26.3)

$$0 = \int_Z d\sigma_k = \int_{\partial Z} \sigma^k = \int_{X \cup (-Y)} \sigma_k = \int_X \sigma_k - \int_Y \sigma_k,$$

yielding

$$\int_X \sigma_k = \int_Y \sigma_k. \tag{3}$$

Denoting by dx and dy the $2k$-dimensional volume elements (induced from the given metric g_{ij}) on the submanifolds X and Y, respectively, we have (see Part I, §26.1)

$$\int_X \sigma_k = \int_X \sigma_k(\hat{T}_xX)\, dx, \qquad \int_Y \sigma_k = \int_Y \sigma_k(\hat{T}_yY)\, dy. \tag{4}$$

The fact that the complex manifold M contains X as a complex submanifold is equivalent to the containment (for each $x \in X$) of T_xX as a complex subspace of $T_xM \simeq \mathbb{C}^n$; hence by Lemma 33.3 we have $\sigma_k(\hat{T}_xX) = 1$, while $\sigma_k(\hat{T}_yY) \le 1$ (since the submanifold Y is not necessarily complex). Consequently, in view of (3) and (4) we have

$$v(X) = \int_X dx = \int_X \sigma_k(\hat{T}_xX)\, dx = \int_Y \sigma_k(\hat{T}_yY)\, dy \le \int_Y dy = v(Y). \tag{5}$$

whence the first assertion of the theorem.

To see that Y must be complex for the equality $v(X)=v(Y)$ to hold, note first that by (5) this equality is equivalent to the condition that for all y in a subset of Y of the same $2k$-dimensional measure as Y (i.e. almost everywhere on Y), we have $\sigma_k(\hat{T}_y Y)=1$. By Lemma 33.3 this in turn is equivalent to the condition that $T_y Y$ be a complex subspace of $T_y M$ for almost all y in Y, and therefore, by continuity, for all $y \in Y$. Hence Y is complex and the proof of the theorem is complete. □

It is clear from this proof that the (implicit) assumption that the submanifold $X \subset M$ have no singularities is not crucial; the proof (in terms of subsets of "full measure") goes through for complex algebraic surfaces $X \subset M$ (i.e. for surfaces defined by systems of polynomial equations on M), even though such surfaces may possess singular points (for example, cones over smooth manifolds). In this situation the assumption in the above theorem that X and Y are "cobordant", needs to be replaced by the more general requirement that X be homologous to Y in the "homology group" $H_{2k}(M, \partial X)$ (see Part III for the definition of these groups), i.e. that X and Y define the same element of this group.

We conclude by mentioning some facts, offered without proof, concerning the Kählerian manifolds $\mathbb{C}P^n$ (obtained in the usual way as a quotient space of $\mathbb{C}^{n+1}\backslash\{0\}$; see §2.2). In $\mathbb{C}P^n$ the submanifold $\mathbb{C}P^k$ (defined in the natural way for each $k=1, \ldots, n$) represents a generator of the homology group $H_{2k}(\mathbb{C}P^n, \mathbb{Z}) \simeq \mathbb{Z}$, and is globally of smallest volume in its homology class. It can also be shown (though with greater difficulty) that if $Y \subset \mathbb{C}P^n$ is any (real) $2k$-dimensional submanifold representing the same generator of $H_{2k}(\mathbb{C}P^n, \mathbb{Z}) \simeq \mathbb{Z}$ and having the property that $v(Y)=v(\mathbb{C}P^k)$, then there is a transformation in $SU(n+1)$ which transforms Y into $\mathbb{C}P^k$. Consequently the submanifold $\mathbb{C}P^k \subset \mathbb{C}P^n$ is, to within an isometry of $\mathbb{C}P^n$, the unique solution of the higher-dimensional variational problem of finding, among the $2k$-dimensional submanifolds in a generating homology class for $H_{2k}(\mathbb{C}P^n, \mathbb{Z}) \simeq \mathbb{Z}$, one with absolute minimum volume. Furthermore, it turns out that if Y is a submanifold of dimension $2k$ representing a non-generator of $H_{2k}(\mathbb{C}P^n, \mathbb{Z}) \simeq \mathbb{Z}$ (i.e. corresponding to an integer $m \neq \pm 1$), then $v(Y) > v(\mathbb{C}P^k)$.

Bibliography

(i) Textbooks on Geometry and Topology

*1. Aleksandrov, A. D., 1948. *Intrinsic Geometry of a Convex Surface*. Gostehizdat: Moscow–Leningrad.
2. Bishop, R. L. and Crittenden, R. J., 1964. *Geometry of Manifolds*. Pure and Applied Math., Vol. 25. Academic Press: New York–London.
3. Chern, S. S., 1959. *Complex Manifolds*. Textos de Matemática, No. 5, Instituto de Física e Matemática, Universidade do Recife.
*4. Efimov, N. V., 1971. *Higher Geometry*. Nauka: Moscow.
*5. Finikov, S. P., 1952. *A Course in Differential Geometry*. Gostehizdat: Moscow.
6. Gromoll, D., Klingenburg, W., and Meyer, W., 1968. *Riemannsche Geometrie im Grossen*. Lecture Notes in Mathematics, No. 55. Springer-Verlag: Berlin–New York.
7. Helgason, S., 1962. *Differential Geometry and Symmetric Spaces*. Academic Press: New York.
8. Hilbert, D. and Cohn-Vossen, S., 1952. *Geometry and the Imagination*. Chelsea Publishing Co.: New York. (Translated from the German by P. Neményi.)
9. Lefschetz, S., 1942. *Algebraic Topology*. Am. Math. Soc. Colloquium Publications, Vol. 27.
10. Milnor, J. W., 1963. *Morse Theory*. Ann. of Math. Studies, No. 51. Princeton Univ. Press: Princeton, N.J.
11. Milnor, J. W., 1968. *Singular Points of Complex Hypersurfaces*. Ann. of Math. Studies, No. 61. Princeton Univ. Press: Princeton, N.J.
12. Nomizu, K., 1956. *Lie Groups and Differential Geometry*. Math. Soc. Japan.
*13. Norden, A. P., 1956. *The Theory of Surfaces*. Gostehizdat: Moscow.
*14. Novikov, S. P., Miščenko, A. S., Solov'ev, Ju. P., and Fomenko, A. T., 1978. *Problems in Geometry*. Moscow State University Press: Moscow.
15. Pogorelov, A. V., 1967. *Differential Geometry*. P. Noordhoff: Groningen. (Translated from the first Russian edition by L. F. Boron.)
16. Pogorelov, A. V., 1973. *Extrinsic Geometry of Convex Surfaces*. Translations of Math. Monographs, Vol. 35. A.M.S.: Providence, R.I.

*In Russian.

17. Pontrjagin, L. S., 1959. *Smooth Manifolds and Their Applications to Homotopy Theory*. Am. Math. Soc. Translations, Series 2, Vol. 11, pp. 1–114. A.M.S.: Providence, R.I.

18. Pontrjagin, L. S., 1966. *Topological Groups*. Gordon & Breach: New York–London–Paris. (Translation of the second Russian edition by Arlen Brown.)

*19. Raševskiĭ, P. K., 1956. *A Course in Differential Geometry*. Nauka: Moscow.

*20. Raševskiĭ, P. K., 1967. *Riemannian Geometry and Tensor Analysis*. Nauka: Moscow.

*21. Rohlin, V. A. and Fuks, D. B., 1977. *A Beginning Course in Topology. Chapters in Geometry*. Nauka: Moscow.

*22. Rozendorn, E. R., 1971. *Problems in Differential Geometry*. Nauka: Moscow.

23. Seifert, H. and Threlfall, W., 1980. *A Textbook of Topology*. Academic Press: New York. (Translated from the German by M. A. Goldman.)

24. Seifert, H. and Threlfall, W., 1932. *Variationsrechnung im Grossen*. Hamburger Math. Einzelschr., No. 24. Teubner: Leipzig. (Reprinted by Chelsea: New York, 1951.)

25. Serre, J.-P., 1964. *Lie Algebras and Lie Groups*. Lectures given at Harvard Univ. W. A. Benjamin: New York.

26. Springer, G., 1957. *Introduction to Riemann Surfaces*. Addison-Wesley: Reading, Mass.

27. Steenrod, N. E., 1951. *The Topology of Fibre Bundles*. Princeton Math. Series, Vol. 14. Princeton Univ. Press: Princeton, N.J.

28. Struik, D. J., 1950. *Lectures on Classical Differential Geometry*. Addison-Wesley: Reading, Mass.

(ii) Texts on Differential Equations and Classical Mechanics

29. Arnol'd, V. I., 1978. *Mathematical Methods of Classical Mechanics*. Graduate Texts in Math. Springer-Verlag: New York–Heidelberg–Berlin. (Translated from the Russian by K. Vogtmann and A. Weinstein.)

*30. Arnol'd, V. I., 1978. *Supplementary Chapters to the Theory of Ordinary Differential Equations*. Nauka: Moscow.

*31. Golubev, V. V., 1953. *Lectures on the Integration of the Equations of Motion of a Heavy Rigid Body About a Fixed Point*. Gostehizdat: Moscow.

32. Landau, L. D. and Lifšic, E. M., 1960. *Mechanics*. Course of Theoretical Physics, Vol. 1. Pergamon Press: Oxford–London–New York–Paris; Addison-Wesley: Reading, Mass. (Translated from the Russian by J. B. Sykes and J. S. Bell.)

*33. Pontrjagin, L. S., 1970. *Ordinary Differential Equations*. Nauka: Moscow.

34. Coddington, E. A. and Levinson, N., 1955. *Theory of Ordinary Differential Equations*. McGraw-Hill: New York–Toronto–London.

(iii) Supplementary Texts

35. Ahiezer, A. I. and Beresteckiĭ, V. B., 1965. *Quantum Electrodynamics*. Interscience Monographs and Texts in Physics and Astronomy, Vol. 11. Interscience Publishers (John Wiley and Sons): New York–London–Sydney. (Translated from the second Russian edition by G. M. Volkhoff.)

*36. Bogojavlenskiĭ, O. I., 1980. *Methods of the Qualitative Theory of Dynamical Systems in Astrophysics and the Dynamics of Gases*. Nauka: Moscow.

*37. Bogojavlenskiĭ, O. I., 1979. *Nonlinear Waves*. Nauka: Moscow.

38. Bogoljubov, N. N. and Širkov, D. B., 1959. *Introduction to the Theory of Quantum Fields*. Interscience: New York. (Translation.)

39. Coxeter, H. S. M. and Moser, W. O. J., 1972. *Generators and Relations for Discrete Groups*. Ergebnisse der Mathematik und ihrer Grenzgebiete, Bd. 14. Springer-Verlag: New York–Heidelberg–Berlin.

*40. Delone, B. N., Aleksandrov, A. D., and Padurov, N. N., 1934. *Mathematical Foundations of the Lattice Analysis of Crystals*. ONTI: Leningrad-Moscow.

41. Feynman, R. P., Leighton, R. B., and Sands, M., 1963. *The Feynman Lectures on Physics*. Addison-Wesly: Reading, Mass.

42. Landau, L. D. and Lifšic, E. M., 1971. *The Classical Theory of Fields*. Third revised English edition. Course of Theoretical Physics, Vol. 2. Addison-Wesley: Reading, Mass.; Pergamon Press: London. (Translated from the Russian by Morton Hamermesh.)

43. Landau, L. D. and Lifšic, E. M., 1960. *Electrodynamics of Continuous Media*. Course of Theoretical Physics, Vol. 8. Pergamon Press: Oxford–London–New York–Paris; Addison-Wesley: Reading, Mass. (Translated from the Russian by J. B. Sykes and J. S. Bell.)

44. Misner, C. W., Thorne, K. S., and Wheeler, J. A., 1973. *Gravitation*. W. H. Freeman: San Francisco.

45. Peierls, R. E., 1960. *Quantum Theory of Solids. Theoretical Physics in the Twentieth Century* (Pauli Memorial Volume), pp. 140–160, Interscience: New York.

*46. Sedov, L. I., 1976. *Mechanics of a Continuous Medium*. Nauka: Moscow.

*47. Slavnov, A. A. and Faddeev, L. D., 1978. *Introduction to the Quantum Theory of Gauge Fields*. Nauka: Moscow.

48. Wintner, A., 1941. *Analytical Foundations of Celestial Mechanics*. Princeton Univ. Press: Princeton, N.J.

*49. Zaharov, V. E., Manakov, S. V., Novikov, S. P., and Pitaevskiĭ, L. P., 1980. *The Theory of Solitons* (under the general editorship of S. P. Novikov). Nauka: Moscow.

*50. Zel'dovič, Ja. B. and Novikov, I. D., 1977. *Relativistic Astrophysics*. Nauka: Moscow.

*51. Zel'dovič, Ja. B. and Novikov, I. D., 1975. *The Structure and Evolution of the Universe*. Nauka: Moscow.

39. Coxeter, H. S. M. and Moser, W. O. J. 1972 *Generators and Relations for Discrete Groups.* Ergebnisse der Mathematik und ihrer Grenzgebiete, Bd 14. Springer-Verlag, New York-Heidelberg-Berlin.
*40. Delone, B. N., Alexandrov, A. D. and Padurov, N. N. 1934 *Mathematical Foundations of the Lattice Analysis of Crystals.* ONTI, Leningrad-Moscow.
41. Feynman, R. P., Leighton, R. B. and Sands, M. 1964 *The Feynman Lectures on Physics.* Addison-Wesley, Reading, Mass.
*42. Landau, L. D. and Lifshitz, E. M. 1976 *The Classical Theory of Fields.* Third revised English edition. Course of Theoretical Physics, Vol. 2. Addison-Wesley, Reading, Mass.; Pergamon Press, Oxford. (Translated from the Russian by Morton Hamermesh.)
*43. Landau, L. D. and Lifshitz, E. M. 1960 *Electrodynamics of Continuous Media.* Course of Theoretical Physics, Vol. 8. Pergamon Press, Oxford-London-New York-Paris; Addison-Wesley, Reading, Mass. (Translated from the Russian by J. B. Sykes and J. S. Bell.)
44. Misner, C. W., Thorne, K. S. and Wheeler, J. A. 1973 *Gravitation.* W. H. Freeman, San Francisco.
45. Peierls, R. E. 1960 *Quantum Theory of Solids. Theoretical Physics in the Twentieth Century* (Pauli Memorial Volume), pp. 140–160. Interscience, New York.
*46. Sedov, L. I. 1976 *Mechanics of a Continuous Medium.* Nauka, Moscow.
*47. Slavnov, A. A. and Faddeev, L. D. 1978 *Introduction to the Quantum Theory of Gauge Fields.* Nauka, Moscow.
48. Wintner, A. 1941 *Analytical Foundations of Celestial Mechanics.* Princeton Univ. Press, Princeton, N.J.
*49. Zakharov, V. E., Manakov, S. V., Novikov, S. P. and Pitaevskii, L. P. 1980 *The Theory of Solitons* (under the general editorship of S. P. Novikov). Nauka, Moscow.
*50. Zel'dovich, Ya. B. and Novikov, I. D. 1971 *Relativistic Astrophysics.* Nauka, Moscow.
*51. Zel'dovich, Ya. B. and Novikov, I. D. 1975 *The Structure and Evolution of the Universe.* Nauka, Moscow.

Index